INTRODUCTION
MINERAL EXPLORATION

INTRODUCTION TO MINERAL EXPLORATION

SECOND EDITION

Edited by Charles J. Moon, Michael K.G. Whateley
& Anthony M. Evans

With contributions from
William L. Barrett
Timothy Bell
Anthony M. Evans
John Milsom
Charles J. Moon
Barry C. Scott
Michael K.G. Whateley

Blackwell
Publishing

BLACKWELL PUBLISHING
350 Main Street, Malden, MA 02148-5020, USA
9600 Garsington Road, Oxford OX4 2DQ, UK
550 Swanston Street, Carlton, Victoria 3053, Australia

First edition published   995 by Blackwell Publishing Ltd
Second edition published 200

6   2010

Library of Congress Cataloging-in-Publication Data

Introduction to mineral exploration.–2nd ed. / edited by Charles J. Moon, Michael K.G. Whateley & Anthony M.
Evans; with contributions from William L. Barrett . . . [et al.].
        p. cm.
    Evans's appears first on the previous ed.
    Includes bibliographical references and index.
    ISBN 978-1-4051-1317-5 (pbk. : acid-free paper) 1. Prospecting. 2. Mining geology. I. Moon, Charles J.
    II. Whateley, M.K.G. III. Evans, Anthony M.

    TN270.I66 2006
    622'.1—dc22

                                                                                    2005014825

A catalogue record for this title is available from the British Library.

Set in 9/11pt Trump Mediaeval
by Graphicraft Limited, Hong Kong
Printed and bound in Singapore
by Fabulous Printers Pte Ltd

For further information on
Blackwell Publishing, visit our website:
www.blackwellpublishing.com

CONTENTS

CONTRIBUTORS

WILLIAM L. BARRETT *Tarmac Quarry Products Ltd, Millfields Road, Ettingshall, Wolverhampton WV4 6JP*. Previously Geological Manager, Estates and Environment Department of Tarmac (now Consultant) but he has enjoyed a varied career in applied geology. His first degree was taken at the University of Wales and he obtained his doctorate from Leeds University. Much of his early career was spent in base metal exploration and underground mining in East Africa and this culminated with a spell as Chief Geologist at Kilembe Mines. Latterly he has been concerned with all aspects of Tarmac's work in the field of industrial minerals.

TIMOTHY BELL *Bell Hanson, Foss Road, Bingham NG11 7EP*. Graduated from West London College with a first class degree (cum laude) in Geology and Geography before obtaining an MSc degree with distinction in Mineral Exploration and Mining Geology at Leicester University in 1989. He joined CP Holdings (opencast coal contractors) as geologist in charge of computer applications in the mining division before moving to SURPAC Software International where he was Technical Manager in their Nottingham office and is now running his own business.

ANTHONY M. EVANS *19 Hall Drive, Burton on the Wolds, Loughborough LE12 5AD*. Formerly Senior Lecturer in charge of the MSc Course in Mining Geology and Mineral Exploration and the BSc Course in Applied Geology at Leicester University. A graduate of Liverpool University in Physics and Geology he obtained a PhD at Queen's University, Ontario and worked in Canada for the Ontario Department of Mines spending much time investigating uranium and other mineralisation. In Europe he has carried out research on many aspects of mineralisation and industrial mineralogy and is the editor and author of a number of books and papers in these fields. For a number of years he was Vice-President of the International Society for Geology Applied to Mineral Deposits.

JOHN MILSOM *Department of Geological Sciences, University College, Gower St, London WC1E 6BT*. Senior Lecturer in Exploration Geophysics at University College. He graduated in Physics from Oxford University in 1961 and gained the Diploma of Imperial College in Applied Geophysics the following year. He joined the Australian Bureau of Mineral Resources as a geophysicist and worked in the airborne, engineering, and regional gravity sections. He obtained a PhD from Imperial College for his research on the geophysics of eastern Papua and joined the staff there in 1971. Then came a spell as Chief Geophysicist for Barringer Research in Australia followed by one as an independent consultant when his clients included the UNDP, the Dutch Overseas Aid Ministry, and BP Minerals. In 1981 he joined the staff at Queen Mary College to establish the Exploration Geophysics BSc Course which later transferred to University College. He is the author of a textbook on field geophysics.

CHARLES J. MOON *Department of Geology, University of Leicester, Leicester, LE1 7RH*. A graduate in Mining Geology from Imperial College, London in 1972, he returned some years later to take his PhD in Applied Geochemistry. He has had wide experience in the fields of mineral deposit assessment and exploration all over the globe. He has worked for Consolidated Gold Fields searching for tin deposits in Cornwall, the South African Geological Survey, Esso Minerals Africa Inc., with whom he was involved in the discovery of three uranium deposits. Besides his general expertise in mineral exploration he has developed the use of raster and vector based GIS systems and courses in these techniques, particularly in ArcGIS for exploration integration. His present position is that of Senior Lecturer in Mineral Exploration at the University of Leicester.

BARRY C. SCOTT *Knoyle House, Vicarage Way, Gerrards Cross, Bucks SL9 8AS.* His most recent position was Lecturer in Mining Geology and Mineral Exploration at Leicester University from which he retired in 1993. Previously he had occupied senior positions with a number of mining companies and academic institutions. These included Borax Consolidated Ltd, Noranda Mines Ltd, G. Wimpey plc, the Royal School of Mines, and Imperial College. Dr Scott obtained his BSc and PhD from Imperial College and his MSc from Queen's University, Ontario. He was a Member of Council of the Institution of Mining and Metallurgy for many years, was its President and Treasurer.

MICHAEL K.G. WHATELEY *Rio Tinto Technical Services, Principal Consultant-Manager.* He graduated in Geology and Geography from London University in 1968, gained an MSc from the University of the Witwatersrand in 1980 and a PhD from Leicester University in 1994 for his work on the computer modeling of the Leicestershire Coalfield. He worked in diamond exploration in Botswana and on gold mines near Johannesburg before joining Hunting Geology and Geophysics in South Africa. He then joined Utah International Inc. in charge of their Fuel Division in southern Africa. In 1982 he returned to the UK to work for Golder Associates before taking up a post at the University of Leicester as Lecturer in Mining Geology. Since 1996 he has been a consultant with Rio Tinto Technical Services based in Bristol.

PREFACE TO THE SECOND EDITION

Since the first edition was published much has changed in the mineral exploration business. It has been through one of the sharpest downturns since the 1930s and the industry has become truly global, with many mergers and acquisitions of mining and exploration companies. Indeed, conditions at the time at the time of writing appear to have moved back to boom driven by increasing demand.

We have tried to maintain a balance between principles and case histories, based on our experience of teaching at senior undergraduate and postgraduate levels. We have attempted to reflect the changing mineral exploration scene by rewriting the sections on mineral economics (Chapter 1) and the increased emphasis on public accountability of mineral exploration companies (Chapters 10 and 11), as well as updating the other chapters. One of the major areas of advance has been the widespread use of information technology in mineral exploration. We have added a chapter (Chapter 9) dealing with the handling of mineral exploration data. Although we do not advocate geologists letting their hammers get rusty, the correct compilation of data can help expedite fieldwork and improve understanding of deposits for feasibility studies.

The case histories are largely the same as the first edition although we have replaced a study on tin exploration with one on diamond exploration. This details one of the major exploration successes of the 1990s near Lac de Gras in arctic Canada (Chapter 17).

Another of the trends of the last few years has been the decline in the teaching of mining-related subjects at universities, particularly in Europe and North America. We hope that the contents of this book will help to educate the geologists needed to discover the mineral deposits required for future economic growth, whether or not they are studying at a university.

The two of us have taken over editorship from Tony Evans who has completely retired. We are both based in western Europe, although we manage to visit other continents when possible, and one of us was commuting to southern Africa when this edition was completed. Inevitably the examples we mention are those with which we are familiar and many readers' favorite areas will be missing. We trust, however, that the examples are representative.

We owe a debt of thanks to reviewers of the additions: Ivan Reynolds of Rio Tinto for comments on Chapter 2; Jeremy Gibbs of Rio Tinto Mining and Exploration Limited for his help with Chapter 6; Peter Fish, John Forkes, and Niall Weatherstone of Rio Tinto Technical Services for their helpful comments on Chapters 10 and 11; Paul Hayston of Rio Tinto Mining and Exploration Limited for comments on John Milsom's update of Chapter 7; and Andy Davy of Rio Tinto Mining and Exploration Limited and Gawen Jenkin of the University of Leicester for comments on Chapter 17. They are, of course, not responsible for any errors that remain our own. Lisa Barber is thanked for drawing some of the new figures.

Finally, we would like to thank our families for their forbearance and help. Without them the compilation of this second edition would not have been possible.

Charlie Moon and Mike Whateley
Leicester and Bristol

PREFACE TO THE FIRST EDITION

It is a matter of regret how little time is devoted to teaching mineral exploration in first degree university and college courses in the United Kingdom and in many other countries. As a result the few available textbooks tend to be expensive because of the limited market. By supplying our publisher with camera ready copy we have attempted to produce a cheaper book which will help to redress this neglect of one of the most important aspects of applied geology. We hope that lecturers will find here a structure on which to base a course syllabus and a book to recommend to their students.

In writing such a book it has to be assumed that the reader has a reasonable knowledge of the nature of mineral deposits. Nevertheless some basic facts are discussed in the early chapters.

The book is in two parts. In the first part we discuss the principles of mineral exploration and in the second, case histories of selected deposit types. In both of these we carry the discussions right through to the production stage. This is because if mining projects are to progress successfully from the exploration and evaluation phases to full scale production, then exploration geologists must appreciate the criteria used by mining and mineral processing engineers in deciding upon appropriate mining and processing methods, and also by the financial fraternity in assessing the economic viability of a proposed mining operation. We have tried to cover these and many other related subjects to give the reader an overall view of mineral exploration. As this is only an introductory textbook we have given some guidance with regard to further reading rather than leaving the student to choose at random from references given in the text.

This textbook arises largely from the courses in mineral exploration given in the Geology Department of Leicester University during the last three decades and has benefitted from the ideas of our past and present colleagues and to them, as well as to our students and our friends in industry, we express our thanks. They are too many to name but we would like in particular to thank Dave Rothery for his help with Chapter 6; Martin Hale for discussion of Chapter 8; Nick Laffoley, Don Moy and Brendon Monahan for suggestions which have improved Chapter 9; and John Rickus for a helping hand with Chapter 12. Steve Baele of US Borax gave us good advice and Sue Button and Clive Cartwright performed many miracles of draughting in double quick time. In addition we are grateful to the Director-General of MTA for permission to reproduce parts of the maps used to prepare Figs 12.2 and 12.3, and to the Director-General of TKI for permission to publish the data on Soma.

UNITS, ABBREVIATIONS, AND TERMINOLOGY

NOTE ON UNITS

With few exceptions the units used are all SI (Système International), which has been in common use by engineers and scientists since 1965. The principal exceptions are: (a) for commodity prices still quoted in old units, such as troy ounces for precious metals and the short ton (= 2000 lb); (b) when there is uncertainty about the exact unit used, e.g. tons in certain circumstances might be short or long (2240 lb); and (c) Chapter 16 where all the original information was collected in American units, but many metric equivalents are given.

SI prefixes and suffixes commonly used in this text are k = kilo-, 10^3; M = mega-, 10^6 (million); G = giga-, 10^9.

SOME ABBREVIATIONS USED IN THE TEXT

AAS	atomic absorption spectrometry
ASTM	American Society for Testing Materials
CCT	computer compatible tapes
CIF	carriage, insurance, and freight
CIPEC	Conseil Inter-governmental des Pays Exportateurs de Cuivre (Intergovernmental Council of Copper Exporting Countries)
CIS	Commonwealth of Independent States (includes many of the countries formerly in the USSR)
d.c.	direct current
DCF ROR	discounted cash flow rate of return
DTH	down-the-hole (logs)
DTM	digital terrain model
EEC	see EU
EIS	environmental impact statement
EM	electromagnetic
ERTS	Earth Resources Technology Satellite
EU	European Union sometimes still referred to as EEC
FOB	freight on board
FOV	field of view
GA	Golder Associates
GIS	Geographical Information Systems
GPS	global positioning satellites
GRD	ground resolution distance

HRV	high resolution visible
ICP-ES	inductively coupled plasma emission spectroscopy
IFOV	instantaneous field of view
IGRF	International Geomagnetic Reference Field
INPUT	induced pulse transient system
IP	induced polarization
IR	infrared
ITC	International Tin Council
MS	multispectral
MSS	multispectral scanners
MTA	Maden Tetkik ve Arama
NAA	neutron activation analysis
NAF	North Anatolian Fault
NPV	net present value
OECD	Organization for Economic Co-operation and Development
OPEC	Organization of Petroleum Exporting Countries
PFE	percent frequency effect
PGE	platinum group elements
PGM	platinum group metals
ppb	parts per billion
ppm	parts per million
REE	rare earth elements
RMR	rock mass rating
ROM	run-of-mine
RQD	rock quality designation
SEM	scanning electron microscopy
SG	specific gravity
SI units	Système International Units
SLAR	side-looking airborne radar
TDRS	Tracking and Data Relay System
TEM	transient electromagnetics
TKI	Turkiye Komur Isletmekeri Kurumu
TM	thematic mapper
t p.a.	tonnes per annum
t p.d.	tonnes per diem
USGS	United States Geological Survey
USSR	the geographical area that once made up the now disbanded Soviet Republics
VLF	very low frequency
VMS	volcanic-associated massive sulfide (deposits)
XRD	X-ray diffraction
XRF	X-ray fluorescence

NOTE ON TERMINOLOGY

Mineralisation: "Any single mineral or combination of minerals occurring in a mass, or deposit, of economic interest. The term is intended to cover all forms in which mineralisation might occur, whether by class of deposit, mode of occurrence, genesis, or composition" (Australasian JORC 2003). In this volume we have spelt the term mineralisation to be consistent with the JORC code although it is usually spelt mineralization in the USA. For further discussion see section 1.2.1.

PART I

PRINCIPLES

1

ORE, MINERAL ECONOMICS, AND MINERAL EXPLORATION

CHARLES J. MOON AND ANTHONY M. EVANS

1.1 INTRODUCTION

A large economic mineral deposit, e.g. 200 Mt underlying an area of 2 km², is minute in comparison with the Earth's crust and in most countries the easily found deposits cropping out at the surface have nearly all been found. The deposits for which we now search are largely concealed by weathered and leached outcrops, drift, soil, or some other cover, and sophisticated exploration methods are required to find them. The target material is referred to as a mineral deposit, unless we use a more specific term such as coal, gas, oil, or water. Mineral deposits contain mineral resources. What sort of mineral deposit should we seek? To answer this question it is necessary to have some understanding of mineral economics.

1.2 MINERAL ECONOMICS

1.2.1 Ore

Ore is a word used to prefix reserves or body but the term is often misused to refer to any or all in situ mineralisation. The Australasian Joint Ore Reserves Committee (2003) in its code, the JORC Code, leads into the description of ore reserves in the following way: "When the location, quantity, grade, geological characteristics and continuity of mineralisation are known, and there is a concentration or occurrence of the material of intrinsic economic interest in or on the Earth's crust in such form and quantity that there are reasonable prospects for eventual economic extraction, then this

deposit can be called a mineral resource." "Mineral Resources are subdivided, in order of increasing geological confidence, into Inferred, Indicated, and Measured categories." A more complete explanation is given in section 10.4.1.

The JORC Code then explains that "an ore reserve is the economically mineable part of a Measured or Indicated Mineral Resource. It includes diluting materials and allowances for losses that may occur when the material is mined. Appropriate assessments, which may include feasibility studies (see section 11.4), have been carried out, and include consideration of and modification by realistically assumed mining, metallurgical, economic, marketing, legal, environmental, social, and governmental factors. These assessments demonstrate at the time of reporting that extraction could reasonably be justified." "Ore Reserves are sub-divided in order of increasing confidence into Probable Ore Reserves and Proved Ore Reserves."

"The term 'economic' implies that extraction of the ore reserve has been established or analytically demonstrated to be viable and justifiable under reasonable investment assumptions. The term ore reserve need not necessarily signify that extraction facilities are in place or operative or that all governmental approvals have been received. It does signify that there are reasonable expectations of such approvals." An orebody will be the portion of a mineralized envelope within which ore reserves have been defined.

Ore minerals are those metallic minerals, e.g. galena, sphalerite, chalcopyrite, that form the economic portion of the mineral deposit.

These minerals occur in concentrations that range from parts per million (ppm) to low percentages of the overall mineral deposit.

"Industrial minerals have been defined as any rock, mineral or other naturally occurring substance of economic value, exclusive of metallic ores, mineral fuels and gemstones" (Noetstaller 1988). They are therefore minerals where either the mineral itself, e.g. asbestos, baryte, or the oxide, or some other compound derived from the mineral, has an industrial application (end use). They include rocks such as granite, sand, gravel, and limestone, that are used for constructional purposes (these are often referred to as aggregates or bulk materials, or dimension stone if used for ornamental cladding), as well as more valuable minerals with specific chemical or physical properties like fluorite, phosphate, kaolinite, and perlite. Industrial minerals are also frequently and confusingly called nonmetallics (e.g. Harben & Kuzvart 1997), although they can contain and be the source of metals, e.g. sodium derived from the industrial mineral halite. On the other hand, many deposits contain metals such as aluminum (bauxite), ilmenite, chromite, and manganese, which are also important raw materials for industrial mineral end uses.

The JORC definition covers metallic minerals, coal, *and* industrial minerals. This is the sense in which the terms mineral resource and ore reserve will normally be employed in this book, except that they will be extended to include the instances where the whole rock, e.g. granite, limestone, salt, is utilized and not just a part of it.

The term mineralisation is defined as "any single mineral or combination of minerals occurring in a mass, or deposit, of economic interest" (IMM Working Group 2001). This group used the term mineral reserve for the mineable part of the mineralisation (for further discussion see section 10.4). The Australian Joint Reserve Committee has a similar approach but prefers the term ore reserve.

Another useful discussion on the economic definition of ore is given by Lane (1988). This text explains the principles of cut-off grade optimization used in mining and processing. This is an essential part of extracting maximum value out of finite geological resources.

Gangue material is the unwanted material, minerals, or rock, with which ore minerals are usually intergrown. Mines commonly possess processing plants in which the run-of-mine (ROM) ore undergoes comminution before the ore minerals are separated from the gangue minerals by various processes. This provides a saleable product, e.g. ore concentrates and tailings which are made up of the gangue material.

1.2.2 The relative importance of metallic and industrial minerals

Metals always seem to be the focus of attention for various reasons, such as their use in warfare, rapid and cyclical changes in price, occasional occurrence in very rich deposits (e.g. gold bonanzas), with the result that the great importance of industrial minerals to our civilization is overlooked. As flints, stone axes, bricks, and pottery they were the first earth resources to be exploited by humans. Today industrial minerals permeate every segment of our society (McVey 1989). They occur as components in durable and nondurable consumer goods. The use of industrial minerals is obvious but often unappreciated, e.g. the construction of buildings, the manufacture of ceramic tables, and sanitary ware. The consumer is frequently unaware that industrial minerals play an essential role in numerous other goods, ranging from books to pharmaceuticals. In developed countries such as the UK and USA, but also on a world-wide basis (Tables 1.1 & 1.2), industrial mineral production is far more important than metal production from both the tonnage and financial viewpoints.

Graphs of world production of the traditionally important metals (Figs 1.1–1.3) show interesting trends. The world's appetite for the major metals appeared to be almost insatiable after World War II. Postwar production increased rapidly. However, in the mid seventies an abrupt slackening in demand occurred triggered by the coeval oil crisis. Growth for many metals, such as copper, zinc, and iron ore, resumed in the 1980s and 1990s. Lead, however, shows a different trend, with overall production increasing slightly. Mine production has declined with more than 50% of overall production now derived from recycling. Other factors affecting the change in growth

TABLE 1.1 World production of mineral resources in 1998 by quantity; in 000 tonnes (ore is given in metal equivalent, natural gas in million m³). (From Wellmer & Becker-Platen 2002.)

Commodity	000 tonnes
Diamonds	0.02
Platinum group metals	0.35
Gold	2.5
Electronic metals	3.2
Silver	16
Cobalt	24
Columbium	26
Tungsten	32
Uranium	35
Vanadium	45
Antimony	118
Molybdenum	135
Mica	205
Tin	209
Magnesium	401
Kyanite and related minerals	431
Zirconium	456
Graphite	648
Boron	773
Nickel	1100
Asbestos	1970
Diatomite	2060
Titanium	2770
Lead	3040
Chromium	4180
Fluorite	4810
Barite	5890
Zinc	7470
Talc and pyrophyllite	7870
Manganese	8790
Feldspar	8800
Bentonite	9610
Copper	12,200
Magnesite	20,100
Peat	25,300
Potash	26,900
Aluminum	27,400
Kaolin	36,600
Phosphate	44,000
Sulfur	55,600
Gypsum anhydrite	103,700
Rock salt	191,100
Industrial sand	300,000
Clay	500,000
Iron	562,000
Lignite	848,000
Natural gas	2,357,000
Crude oil	3,578,000
Coal	3,735,000
Aggregates	4,100,000
Sand and gravel	>15,000,000

TABLE 1.2 World minerals production in 1998 by value in million euro. (From Wellmer & Becker-Platen 2002.)

Commodity	000,000a
Kyanite and related minerals	64
Mica	69
Zirconium	130
Antimony	146
Tungsten	158
Graphite	194
Electronic metals	200
Columbium	237
Titanium	345
Bentonite	388
Talc and pyrophyllite	439
Diatomite	440
Lead	556
Fluorite	576
Asbestos	618
Peat	658
Gypsum anhydrite	689
Chromium	727
Uranium	862
Magnesium	864
Vanadium	882
Tin	918
Molybdenum	926
Cobalt	1040
Boron	1560
Silver	1590
Manganese	1610
Sulfur	1820
Nickel	1890
Zinc	2930
Rock salt	3030
Phosphate	3140
Potash	3510
Platinum group metals	3760
Kaolin	3950
Diamonds	3960
Industrial sand	4910
Barite	5890
Feldspar	8800
Clay	9500
Copper	9800
Iron	15,200
Magnesite	20,100
Gold	20,900
Aggregates	23,100
Lignite	30,300
Aluminium	39,400
Sand and gravel	97,000
Coal	122,000
Natural gas	161,000
Crude oil	412,000

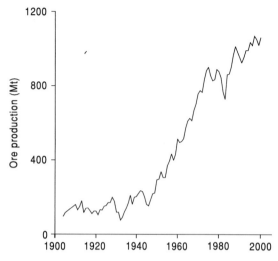

FIG. 1.1 World production of iron ore, 1900–2000. (Data from Kelly et al. 2001.)

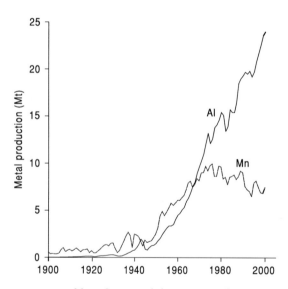

FIG. 1.2 World production of aluminum and manganese metal, 1900–2000. (Data from Kelly et al. 2001.)

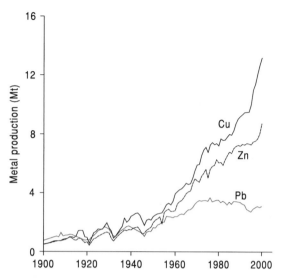

FIG. 1.3 World production of copper, lead, and zinc metal, 1900–2000. The lead data are for smelter production 1900–54 and for mine production 1955–2000. (Data from Kelly et al. 2001.)

increases in production of selected metals and industrial minerals provide a striking contrast and one that explains why for some years now some large metal mining companies have been moving into industrial mineral production. In 2002 Anglo American plc derived 8% of its operating profits and Rio Tinto plc in 2003 derived 11% of its adjusted earnings from industrial minerals. An important factor in the late 1990s and early 2000s was the increasing demand for metals from China, e.g. aluminum consumption grew at 14% a year between 1990 and 2001 (Humphreys 2003).

1.2.3 Commodity prices – the market mechanism

Most mineral trading takes place within the market economy. The prices of minerals or mineral products are governed by supply and demand. If consumers want more of a mineral product than is being supplied at the current metal and mineral prices price, this is indicated by their "bidding up" the price, thus increasing the profits of companies supplying that product. As a result, resources in the form of capital investment are attracted into the

rate include more economical use of metals and substitution by ceramics and plastics, as industrial minerals are much used as a filler in plastics. Production of plastics rose by a staggering 1529% between 1960 and 1985, and a significant fraction of the demand behind this is attributable to metal substitution. The

industry and supply expands. On the other hand if consumers do not want a particular product its price falls, producers make a loss, and resources leave the industry.

World markets

Modern transport leads to many commodities having a world market. A price change in one part of the world affects the price in the rest of the world. Such commodities include wheat, cotton, rubber, gold, silver, and base metals. These commodities have a wide demand, are capable of being transported, and the costs of transport are small compared with the value of the commodity. The market for diamonds is worldwide but that for bricks is local.

Over the last few centuries formal organized markets have developed. In these markets buying and selling takes place in a recognized building, business is governed by agreed rules and conventions and usually only members are allowed to engage in transactions. Base metals are traded on the London Metal Exchange (LME) and gold and silver on the London Bullion Market. Similar markets exist in many other countries, e.g. the New York Commodity Exchange (Comex). Because these markets are composed of specialist buyers and sellers and are in constant communication with each other, prices are sensitive to any change in worldwide supply and demand.

The prices of some metals on Comex and the LME are quoted daily by many newspapers and websites, whilst more comprehensive guides to current metal and mineral prices can be found in *Industrial Minerals*, the *Mining Journal*, and other technical journals. Short- and long-term contracts between buyer and seller may be based on these fluctuating prices. On the other hand, the parties concerned may agree on a contract price in advance of production, with clauses allowing for price changes because of factors such as inflation or currency exchange rate fluctuations. Contracts of this nature are common in the cases of iron, uranium, and industrial minerals. Whatever the form of sale is to be, the mineral economists of a mining company must try to forecast demand for, and hence the price of, a possible mine product well in advance of mine development, and such considerations will usually play a decisive role in formulating a company's mineral exploration strategy. A useful recent discussion of mineral markets can be found in Crowson (1998).

Forces determining prices

Demand and supply

Demand may change over a short period of time for a number of reasons. Where one commodity substitutes to a significant extent for another and the price of this latter falls then the substituting commodity becomes relatively expensive and less of it is bought. Copper and aluminum are affected to a degree in this way. A change in technology may increase the demand for a metal, e.g. the use of titanium in jet engines, or decrease it in the case of the development of thinner layers of tin on tinplate and substitution (Table 1.3). The expectation of future price changes or shortages will induce buyers to increase their orders to have more of a commodity in stock.

Supply refers to how much of a commodity will be offered for sale at a given price over a set period of time. This quantity depends on the price of the commodity and the conditions of supply. High prices stimulate supply and investment by suppliers to increase their output. A fall in prices has the opposite effect and some mines may be closed or put on a care-and-maintenance basis in the hope of better times in the future. Conditions of supply may change fairly quickly through: (i) changes due to abnormal circumstances such as natural disasters, war, other political events, fire, strikes at the mines of big suppliers; (ii) improved techniques in exploitation; (iii) discovery and exploitation of large new orebodies.

Government action

Governments can act to stabilize or change prices. Stabilization may be attempted by building up a stockpile, although the mere building up of a substantial stockpile increases demand and may push up the price! With a substantial stockpile in being, sales from the stockpile can be used to prevent prices rising significantly and purchases for the stockpile may be used to prevent or moderate price falls. As commodity markets are worldwide it is in most cases impossible for one country acting on its own to control prices. Groups of

Aluminum	28	*Cobalt*	35	*Copper*	16	
Diatomite	29.1	Feldspar	81.5	*Gold*	8.8	
Gypsum	37.6	*Iron ore*	12	*Lead*	−5.5	
Mica	18.9	*Molybdenum*	8.3	*Nickel*	16.8	
Phosphate	42.5	*PGM*	80.8	Potash	39.1	
Silver	13.9	Sulfur	19	Talc	44	
Tantalum	143	*Tin*	−9.8	Trona	44	
Zinc	26					

TABLE 1.3 Percentage increase in world production of some metals and industrial minerals 1973–88; metals are in italics. Recycled metal production is not included.

countries have attempted to exercise control over tin (ITC) and copper (CIPEC) in this way but with little success and, at times, signal failure (Crowson 2003).

Stockpiles may also be built up by governments for strategic reasons and this, as mentioned above, can push up prices markedly. Stockpiling policies of some leading industrialized nations are discussed by Morgan (1989).

An action that has increased consumption of platinum, palladium, and rhodium has been the adoption of regulations on the limitation of car exhaust fumes by the EU countries. The worldwide effort to diminish harmful exhaust emissions resulted in a record industrial purchase of 3.2 million ounces of platinum and 3.7 million ounces of palladium in 2003. Comparable actions by governments stimulated by environmental lobbies will no doubt occur in the coming years.

Recycling
Recycling is already having a significant effect on some product prices. Economic and particularly environmental considerations will lead to increased recycling of materials in the immediate future. Recycling will prolong resource life and reduce mining wastes and smelter effluents. Partial immunity from price rises, shortages of primary materials, or actions by cartels will follow. A direct economic and environmental bonus is that energy requirements for recycled materials are usually much lower than for treating ores, e.g. 80% less electricity is needed for recycled aluminum. In the USA the use of ferrous scrap as a percentage of total iron consumption rose from 35% to 42% over the period 1977–87 and aluminum from 26% to 37%; but both copper and aluminum were approximately 30% for the western world in 1999 (Crowson 2003). Of course the poten-

tial for recycling some materials is much greater than for others. Contrary to metals the potential for recycling industrial minerals is much lower. Aggregate recycling is currently being promoted within the European Union. Other commodities such as bromine, fluor-compounds, industrial diamonds, iodine and feldspar and silica in the form of glass are recyled but industrial mineral prices will be less affected by this factor (Noetstaller 1988).

Substitution and new technology
These two factors may both lead to a diminution in demand. We have already seen great changes such as the development of longer-lasting car batteries that use less lead, substitution of copper and plastic for lead water pipes, and a change to lead-free petrol; all factors that have contributed to a downturn in the demand for lead (Fig. 1.3). Decisions taken by OPEC in 1973 affected all metals (Figs 1.1–1.3). They led to huge increases in the prices of oil and other fuels, pushed demand towards materials having a low sensitivity to high energy costs, and favored the use of lighter and less expensive substitutes for metals (Cook 1987).

In the past, base metal producers have spent vast sums of money on exploration, mine development, and production, but have paid too little attention to the defence and development of markets for their products (Davies 1987, Anthony 1988). Producers of aluminum, plastics, and ceramics, on the other hand, have promoted research for new uses including substitution for metals. Examples include tank armour, now frequently made of multilayer composites (metal, ceramic, and fibers) and ceramic-based engine components, widely used in automobiles. It has been forecast that by 2030 90% of engines used in cars, aeroplanes, and power stations will be made from novel ceramics. A useful article on developments in ceramic technology is by Wheat (1987).

Metal and mineral prices

Metals

Metal prices are erratic and hard to predict (Figs 1.4–1.6). In the short run, prices fluctuate in response to unforeseen news affecting supply and demand, e.g. strikes at large mines or smelters, unexpected increases in warehouse stocks. This makes it difficult to determine regular behavior patterns for some metals. Over the intermediate term (several decades) the prices clearly respond to the rise and fall in world business activity, which is some help in attempts at forecasting price trends (Figs 1.4–1.6). The dramatic oil price rises caused by OPEC in 1973, besides setting off a severe recession, led to less developed countries building up huge debts to pay for the increased costs of energy. This led to a reduction in living standards and the purchasing of fewer durable goods. At the same time many metal-producing, developing countries such as Chile, Peru, Zambia, and the Democratic Republic of Congo increased production irrespective of metal prices to earn hard currencies for debt repayment. A further aggravation from the supply and price point of view has been

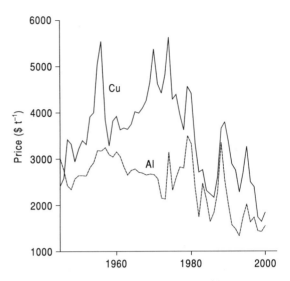

FIG. 1.5   Copper and aluminum metal prices, 1950–2000, New York in US$ 1998. (Data from Kelly et al. 2001.)

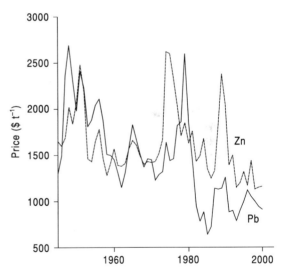

FIG. 1.6   Lead and zinc metal prices, 1950–2000, New York in US$ 1998. (Data from Kelly et al. 2001.)

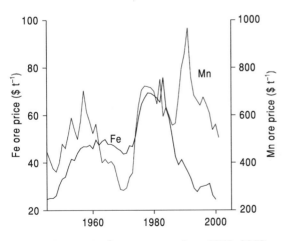

FIG. 1.4   Iron ore and manganese prices, 1950–2000, New York in US$ 1998. The prices have therefore been deflated or inflated so that they can be realistically compared. (Data from Kelly et al. 2001.)

the large number of significant mineral discoveries since the advent of modern exploration methods in the fifties, many of which are still undeveloped (Fig. 1.7). Others, that have been developed since 1980, such as the large disseminated copper deposits in Chile, are producing metal at low cost. Metal explorationists

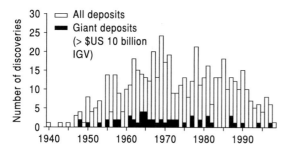

FIG. 1.7 Mineral exploration discovery rate. (From Blain 2000.)

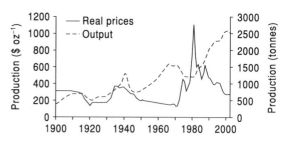

FIG. 1.8 Gold price (in US$ in real 1999 terms) and production. (Updated from Crowson 2003.)

have, to a considerable extent, become victims of their own success. It should be noted that the fall off in nongold discoveries from 1976 onwards is largely due to the difficulty explorationists now have in finding a viable deposit in an unfavorable economic climate.

During the 1990s the general outlook was not promising for many of the traditional metals, in particular manganese (Fig. 1.4), lead (Fig. 1.6), tin, and tungsten, although this seems to have changed at the time of writing (2004). Some of the reasons for this prognostication have been discussed above. It is the minor metals such as titanium, tantalum, and others that seemed likely to have a brighter future. For the cyclicity of price trends over a longer period (1880–1980) see Slade (1989) and Crowson (2003). Nevertheless a company may still decide to make one of the traditional metals its exploration target. In this case evaluate the potential for readily accessible, high grade, big tonnage orebodies, preferably in a politically stable, developed country – high quality deposits of good address as Morrissey (1986) has put it. This is a tall order but well exemplified by the discovery at Neves-Corvo in southern Portugal of a base metal deposit with 100 Mt grading 1.6% Cu, 1.4% Zn, 0.28% Pb, and 0.10% Sn in a well-explored terrane. The discovery of such deposits is still highly desirable.

Gold has had a different history since World War II. From 1934 to 1972 the price of gold remained at $US 35 per troy ounce. In 1971 President Nixon removed the fixed link between the dollar and gold and left market demand to determine the daily price. The following decade saw gold soar to a record price of $US 850

per ounce, a figure inconceivable at the beginning of the seventies; it then fell back during the late 1990s to a price lower in real terms than that of the 1930s (Fig. 1.8). (Using the US Consumer Price Index the equivalent of $35 in 1935 would have been $480 in 2004.) Citizens of many countries were again permitted to hold gold either as bars or coinage and many have invested in the metal. Unfortunately for those attempting to predict future price changes, demand for this metal is not determined so much by industrial demand but by fashion and sentiment, both notoriously variable and unpredictable factors. The main destinations of gold at the present day are carat jewellery and bars for investment purposes.

The rise in the price of gold after 1971 led to an increase in prospecting and the discovery of many large deposits (Fig. 1.7). This trend continued until 1993 and gold production increased from a low point in 1979 to a peak in 2001 (Fig. 1.8). Many diversified mining companies adopted a cautious approach and, like the major gold producers, are not opening new deposits without being sure that they could survive on a price of around $US 250 per ounce, whilst others are putting more emphasis in their exploration budgets on base metals.

Industrial minerals
Most industrial minerals can be traded internationally. Exceptions are the low value commodities such as sand, gravel, and crushed stone which have a low unit value and are mainly produced for local markets. However, minor deviations from this statement are beginning to appear, such as crushed granite being shipped from Scotland to the USA, sand from Western Australia to Japan, and filtration

sand and water from the UK to Saudi Arabia. Lower middle unit value minerals from cement to salt can be moved over intermediate to long distances provided they are shipped in bulk by low cost transport. Nearly all industrial minerals of higher unit value are internationally tradeable, even when shipped in small lots.

The cost to the consumer of minerals with a low unit value will increase greatly with increasing distance to the place of use. Consequently, low unit value commodities are normally of little or no value unless available close to a market. Exceptions to this rule may arise in special circumstances such as the south-eastern sector of England (including London) where demand for aggregates cannot now be met from local resources. Considerable additional supplies now have to be brought in by rail and road over distances in excess of 150 km. For high unit value minerals like industrial diamonds, sheet mica, and graphite, location is largely irrelevant.

Like metals, industrial minerals respond to changes in the intensity of business activities, but as a group, to nothing like the extent shown by metals, and their prices are generally much more stable. One reason for the greater stability of many industrial mineral prices is their use or partial use in consumer nondurables for which consumption remains comparatively stable during recessions, e.g. potash, phosphates, and sulfur for fertilizer production; diatomite, fluorspar, iodine, kaolin, limestone, salt, sulfur, talc, etc., used in chemicals, paint, paper, and rubber. The value of an industrial mineral depends largely on its end use and the amount of processing it has undergone. With more precise specifications of chemical purity, crystalline perfection, physical form, hardness, etc., the price goes up. For this reason many minerals have quite a price range, e.g. kaolin to be used as coating clay on paper is four times the price of kaolin for pottery manufacture.

Individual commodities show significant price variations related to supply and demand, e.g. potash over the last 40 years. When supplies were plentiful, such as after the construction of several large Canadian mines, prices were depressed, whereas when demand has outstripped supply, prices have shot up.

According to Noetstaller (1988) already discovered world reserves of most industrial minerals are adequate to meet the expected demand for the foreseeable future, and so no significant increases in real long-term prices are expected. Exceptions to this are likely to be sulfur, barite, talc, and pyrophyllite. Growth rates are expected to rise steadily, rates exceeding 4% p.a. are forecast for nine industrial minerals and 2–4% for 29 others. These figures may well prove to be conservative estimates. Contrary to metals, the recycling potential of industrial minerals, with some exceptions, is low, and competing substitutional materials are frequently less efficient (e.g. calcite for kaolinite as a cheaper paper filler) or more expensive.

An industrial mineral, particularly one with a low or middle unit value, should be selected as an exploration target only if there is an available or potential market for the product. The market should not be from the exploration area to keep transport costs at a sustainable level.

1.3 IMPORTANT FACTORS IN THE ECONOMIC RECOVERY OF MINERALS

1.3.1 Principal steps in the exploration and exploitation of mineral deposits

The steps in the life cycle of a mineral deposit may be briefly summarized as follows:

1 *Mineral exploration*: to discover a mineral deposit.

2 *Feasibility study*: to prove its commercial viability.

3 *Mine development*: establishment of the entire infrastructure.

4 *Mining*: extraction of ore from the ground.

5 *Mineral processing*: milling of the ore, separation of ore minerals from gangue material, separation of the ore minerals into concentrates, e.g. copper concentrate; separation and refinement of industrial mineral products.

6 *Smelting*: recovering metals from the mineral concentrates.

7 *Refining*: purifying the metal.

8 *Marketing*: shipping the product (or metal concentrate if not smelted and refined at the mine) to the buyer, e.g. custom smelter, manufacturer.

9 *Closure*: before a mine has reached the end of its life, there has to be a closure management plan in place that details and costs the proposed closure strategies. Significant expenditure could be incurred with clean up and remediation of mining and smelting sites, the costs of employee retrenchment, and social and community implications.

The exploration step can be subdivided as follows:

(i) *Study phase*: choice of potential target, study of demand, supply, commodity price trends, available markets, exploration cost, draw up budget.

(ii) *Reconnaissance phase*: will start with a literature search and progress to a review of available remote sensing and photogeological data leading to selection of favorable areas, initial field reconnaissance, and land acquisition, probably followed by airborne surveys, geological mapping and prospecting, geochemical and geophysical surveys, and limited drilling (see Chapter 4).

(iii) *Target testing*: detailed geological mapping and detailed geochemical and geophysical surveys, trenching and pitting, drilling (see Chapter 5). If successful this will lead to an order of magnitude study which will establish whether there could be a viable project that would justify the cost of progressing to a prefeasibility study.

(iv) *Pre-feasibility*: major sampling and test work programs, including mineralogical examination of the ore and pilot plant testing to ascertain the viability of the selected mineral processing option and likely recoverability (see Chapter 11). It evaluates the various options and possible combinations of technical and business issues.

(v) *Feasibility study*: drilling, assaying, mineralogical, and pilot plant test work will continue. The feasibility study confirms and maximizes the value of the preferred technical and business option identified in the prefeasibility study stage.

It is at the end of the order-of-magnitude study that the explorationists usually hand over to the mining geologists, mineral processors and geotechnical and mining engineers to implement steps 1 to 9. Typical time spans and costs might be: stage (i) 1–2 years, US$0.25M; (ii) 2 years, US$0.5–1.5M; (iii) and (iv) 2–3 years, US$2.5–50M; (v) 2 years, US$2.5–50M (excluding actual capital cost for mine con-

struction). Some of these stages will overlap, but this is unlikely to reduce the time involved and it can be expected that around 12 years will elapse between the start of the exploration program and the commencement of mine production. In a number of cases the lead-in time has been less, but this has usually been the result of the involvement of favorable factors or a deliberate search for deposits (particularly of gold) which would have short lead-in times. Further information on costs of mineral exploration can be found Tilton et al. (1988) and Crowson (2003).

1.4 STRUCTURE OF THE MINING INDUSTRY

The structure of the mining industry changed greatly in the 1990s and early 2000s with the decline in government-funded mineral exploration, particularly in centrally planned economies of central Europe and the former USSR, and the merging and globalization of many mining companies. The producing section of the mining industry was dominated, in 2002, by three companies mining a range of commodities (BHP Billiton, Rio Tinto, and Anglo American) and by Alcoa, an aluminum producer (Fig. 1.9). Other major companies concentrate on gold mining (Newmont Mining, Barrick Gold, and AngloGold Ashanti), platinum (e.g. Anglo American Platinum and Impala Platinum), and nickel production (Norilsk, Inco). One major copper-producing company, Corporacion Nacional del Cobre (Codelco), is not shown as it is still owned by the Chilean state. Other smaller mining companies produce at regional or national levels.

Junior companies are a major feature of the mineral exploration industry. They are based largely in Canada, where more than 1000 companies are active, in Australia, and to a lesser extent in the USA and Europe. Their strategies are varied but can be divided into two subgroups: one is exclusively involved in mineral exploration and aims to negotiate agreements with major companies on any deposits they discover and the other to retain at least a share of any discovery and to control the production of any discovery (MacDonald 2002). The dependence of these small companies on speculative activities has led to some taking extreme risks

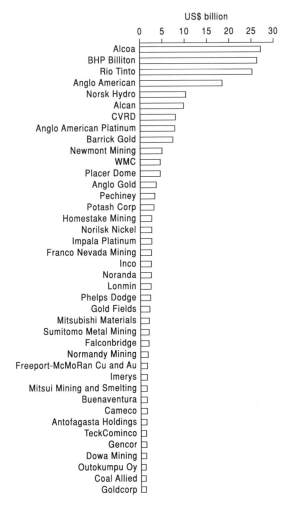

US$ billion

FIG. 1.9 Capitalization of major mining companies – a snapshot on 28 September 2001. Since then Homestake and Barrick Gold have merged as have Newmont, Franco Nevada, and Normandy. (From MMSD 2002.)

and giving the industry a bad name (e.g. Bre-X, see section 5.4).

1.5 SOME FACTORS GOVERNING THE CHOICE OF EXPLORATION AREAS

In Chapter 11, consideration is given in some detail to how mineralisation is converted into an ore reserve and developed into a mine. A brief introduction to some of these aspects is given here.

1.5.1 Location

Geographical factors may determine whether or not an orebody is economically viable. In a remote location there may be no electric power supply or water supply, roads, railways, houses, schools, hospitals, etc. All or some of these infrastructural elements will have to be built, the cost of transporting the mine product to its markets may be very high, and wages will have to be high to attract skilled workers.

1.5.2 Sustainable development

New mines bring prosperity to the areas in which they are established but they are bound to have an environmental and social impact. When production started at the Neves-Corvo copper mine in southern Portugal in 1989, it required a total labor force of about 1090. Generally, one mine job creates about three indirect jobs in the community, in service and construction industries, so the impact is clearly considerable. Such impacts have led to conflicts over land use and opposition to the exploitation of mineral deposits by environmentalists, particularly in the more populous of the developed countries. The resolution of such conflicts may involve the payment of compensation and planning for high closure costs, or even the abandonment of projects. A Select Committee (1982) stated "... whilst political risk has been cited as a barrier to investment in some countries, environmental risk is as much of a barrier, if not a greater in others." Opposition by environmentalists to exploration and mining was partially responsible for the abandonment of a major copper mining project in the Snowdonia National Park of Wales as early as 1973. Woodall (1992) has remarked that explorationists must not only prove their projects to be economically viable but they must also make them socially and therefore politically acceptable. A major attempt to understand the problem, and to suggest solutions, has been made by the Mining, Minerals and Sustainable Development project sponsored by most major mining companies (MMSD 2002). From this it is clear that the concerns of local inhabitants must be addressed from an early stage if mine development is to be successful (discussed further in section 4.3).

These new requirements have now led to a time gap, often of several years in developed countries, between the moment when a newly found deposit is proved to be economically viable, and the time when the complex regulatory environment has been dealt with and governmental approval for development obtained. During this time gap, and until production occurs, no return is being made on the substantial capital invested during the exploration phase (see section 11.2).

1.5.3 Taxation

Overzealous governments may demand so much tax that mining companies cannot make a reasonable profit. On the other hand, some governments have encouraged mineral development with taxation incentives such as a waiver on tax during the early years of a mining operation. This proved to be a great attraction to mining companies in the Irish Republic in the 1960s and brought considerable economic gains to that country.

Once an orebody is being exploited it has become a *wasting asset* and one day there will be no ore, no mine, and no further cash flow. The mine, as a company, will be wound up and its shares will have no value. In other words, all mines have a limited life and for this reason should not be taxed in the same manner as other commercial undertakings. When this is taken into account in the taxation structure of a country, it can be seen to be an important incentive to investment in mineral exploration and mining in that country.

1.5.4 Political factors

Political risk is a major consideration in the selection of a country in which to explore. In the 1970s and 1980s the major fear was nationalization with perhaps inadequate or even no compensation. Possible political turmoil, civil strife, and currency controls may all combine to increase greatly the financial risks of investing in certain countries. In the 1990s and 2000s perhaps more significant risks were long delays or lack of environmental permits to operate, corruption, and arbitrary changes in taxation. One of the most useful sources of information

on political risk in mining is the Fraser Institute in Vancouver (Fraser Institute 2003). It publishes an annual review of the investment attractiveness of many countries and regions based on a poll of mining company executives. The attractiveness is a combination of mineral potential and policy potential. Some countries, for example Chile, rank at the top of both indices, whereas others, such as Russia, have a very high mineral potential index but a very low policy potential index.

1.6   RATIONALE OF MINERAL EXPLORATION

1.6.1 Introduction

Most people in the West are environmentalists at heart whether engaged in the mineral extraction industry or some other employment. Unfortunately many are of the "nimby" (not in my backyard) variety. These and many other people fail to realize, or will not face up to the fact, that it is Society that creates the demand for minerals. The mining and quarrying companies are simply responding to *Society's* desire and demand for houses, washing machines, cars with roads on which to drive them, and so on.

Two stark facts that the majority of ordinary people and too few politicians understand are first that orebodies are wasting assets (section 1.5.3) and second that they are not evenly distributed throughout the Earth's crust. It is the depleting nature of their orebodies that plays a large part in leading mining companies into the field of exploration, although it must be pointed out that exploration *per se* is not the only way to extend the life of a mining company. New orebodies, or a share in them, can be acquired by financial arrangements with those who own them, or by making successful takeover bids. Many junior exploration companies are set up with the idea of allowing others to buy into their finds (farming out is the commercial term). The major mining companies also sell orebodies that they consider too small for them to operate.

The chances of success in exploration are tiny. Only generalizations can be made but the available statistics suggest that a success rating of less than a tenth of a percent is the norm and only in favorable circumstances will this rise above one percent. With such a high element of risk it might be wondered why any risk capital is forthcoming for mineral exploration. The answer is that successful mining can provide a much higher profitability than can be obtained from most other industrial ventures. Destroy this inducement and investment in mineral exploration will decline and a country's future mineral production will suffer. Increases in environmental constraints as well as an impression that the area has been thoroughly explored led, for example, to a flight of exploration companies from British Columbia to Latin America in the late 1990s.

1.6.2 Exploration productivity

Tilton et al. (1988) drew attention to this important economic measure of mineral exploration success and rightly pointed out that it is even more difficult to assess this factor than it is to determine trends in exploration expenditures. It is a measure that should be assessed on the global, national, and company scale.

For a company, success requires a reasonable financial return on its exploration investment. Exploration productivity can be determined by dividing the expected financial return by the exploration costs, *after* these have been adjusted to take account of inflation. For Society, on the global or national scale, the calculation is much more involved but has been attempted by a number of workers.

Data on exploration success is rare and scattered. The study of Blain (2000), originally based on a proprietary database, has attempted to analyse the mineral exploration success rate. Such an analysis is complicated by many discoveries only being recognized some years after initial drilling. The overall appearance of Fig. 1.7 is however of a peak in exploration success in the late 1960s and a distinct fall in the mid-1990s. Blain considers the discoveries as a series of waves, offset over time, in different commodities, uranium, nickel, copper, poly-metallic base metals, and gold (Fig. 1.10). Most of the discoveries in the 1980s and

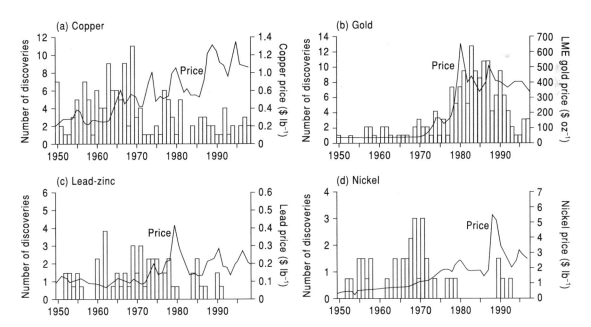

FIG. 1.10   Discovery rate by commodity: (a) copper; (b) gold; (c) lead-zinc; (d) nickel. The metal prices (in US$) are uncorrected for inflation, compare them with Figures 1.4, 1.5 and 1.8 in which prices have been corrected. (From Blain 2000.)

1990s have been of gold deposits but the rate of discovery does not appear to have been sustained from more recent data (BHP Billiton 2003).

The study by Mackenzie and Woodall (1988) of Australian and Canadian productivity is extremely penetrating and worthy of much more discussion than there is space for here. It should be emphasized that this analysis, and others in the literature, is concerned almost exclusively with metallic deposits, and comparable studies in the industrial mineral sector still wait to be made.

Mackenzie and Woodall studied *base metal* exploration only and compared the period 1955–78 for Australia with 1946–77 for Canada. They drew some striking conclusions. Australian exploration was found to have been uneconomic but Canadian financially very favorable. Although exploration expenditures in Canada were double those in Australia the resulting number of economic discoveries were eight times greater. Finding an economic deposit in Australia cost four times as much and took four times longer to discover and assess as one in Canada. By contrast, the deposits found in Australia were generally three times larger that those found in Canada. Comparable exploration expertise was used in the two countries. Australia is either endowed with fewer and larger deposits, or the different surface blankets (glacial versus weathered lateritic) render exploration, particularly geophysical, more difficult in Australia, especially when it comes to searching for small- and medium-sized deposits. The latter is the more probable

reason for the difference in productivity, which suggests that improved exploration methods in the future may lead to the discovery of many more small- and medium-sized deposits in Australia.

Mackenzie and Dogget (in Woodall 1992) have shown that the *average* cost of finding and proving up *economic* metallic deposits in Australia over the period 1955–86 was about $A51M or $A34M when discounted at the start of exploration. The average reward discounted in the same manner is $A35M. This is little better than a breakeven situation. Dissecting these data shows that the average exploration expenditure per deposit is gold $A17M, nickel $A19M, and base metals $A219M. By contrast, the expected deposit value at the start of exploration for gold is $A24M and nickel $A42M. This provided the reasonable rates of return of 21% and 19% on the capital invested. On average $53M more has been invested in finding and proving up an economic base metal deposit than has been realized from its subsequent exploitation. Clearly gold and nickel have on average been wealth creating whilst base metal exploration has not.

A study of uranium exploration by Crowson (2003) compared exploration expenditure for uranium with resources added over the period 1972–80 (Table 1.4). This suggests that, apart from the political requirement to have a domestic strategic supply, there was little justification for exploring in France. By contrast, the exploration success in Canada is probably understated, as there were major discoveries after 1980 (section 3.3).

TABLE 1.4   Exploration productivity for uranium in the 1970s. (Source: Crowson 2003.)

	Total expenditure 1972–1980 in US$ million 1982 terms	Resources April 1970 (000 t uranium)	Resources January 1981 (000 t uranium)	Spending per tonne of increased resources (US$)
Australia	300	34	602	530
Canada	625	585	1018	1440
France	340	73	121	6960
USA	2730	920	1702	3490
Brazil	210	2	200	1060
India	60	3	57	1150
Mexico	30	2	9	4000

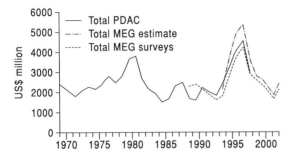

FIG. 1.11   Total global exploration spending in US$ 2001. Sources: MEG, Metals Economics Group; PDAC, Prospectors and Developers Association of Canada. (Updated from Crowson 2003.)

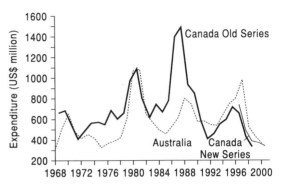

FIG. 1.12   Australian and Canadian exploration spending in $US 20001. The "New" and "Old" series indicate slightly different methods of collecting spending. (Compiled from Crowson 2003.)

1.6.3   Exploration expenditure

Some indication has been given in section 1.3.1 of the cost of individual mineral exploration programs. Here consideration is given to some statistics on mineral exploration expenditure worldwide (Fig. 1.11). Accurate statistics are hard to obtain, as it is often not clear whether overheads are included, and coal and industrial mineral exploration is generally excluded (Crowson 2003). However statistics released by the Metals Economics Group are generally comprehensive. What is clear is the strongly cyclic nature of spending with distinct peaks in 1980, 1988, and particularly in 1996. If Australian and Canadian data are examined (Fig. 1.12) they show similar peaks, but the 1996–97 Australian peak was higher than previous peaks and superimposed on a general increase from 1970 to 1996, whereas Canadian expenditure trended downwards from 1980.

In Fig. 1.13 the comparative total expenditures on exploration for metallic and industrial minerals (including coal) in Australia and Canada are shown, as are the sums spent on industrial minerals alone. Both countries are important producers of industrial minerals despite their considerably smaller expenditure on exploration for these commodities. The greater expenditure on exploration for metals seems to arise from a number of factors: firstly, it is in general cheaper to prospect for industrial minerals and the success rate is higher; secondly, metal exploration, particularly for gold, attracts more high risk investment.

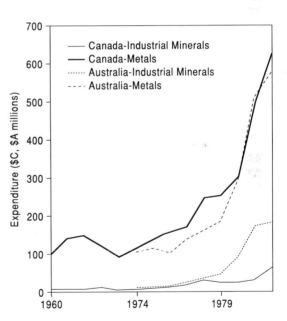

FIG. 1.13   Comparative total exploration expenditure for metals *and* industrial minerals (including coal) in Australia and Canada. The ordinate represents *current* Australian and Canadian dollars, i.e. no corrections have been made for inflationary effects. Also shown are the graphs for industrial mineral exploration in these countries. Canadian data are shown with solid lines and the Australian data with pecked lines. Australian expenditure is for companies only; Canadian expenditure includes government agencies. (Source Crowson 1988.)

Figures 1.12 and 1.13 show a general increase in exploration expenditure over recent decades, and Cook (1987) records that expenditure in the noncommunist world rose from about $400M in 1960 to over $900M in 1980 (constant 1982 US$). This increase is partially due to the use of more costly and sophisticated exploration methods, for example it has been estimated that of deposits found in Canada before 1950, 85% were found by conventional prospecting. The percentage then dropped as follows: 46% in 1951–55, 26% in 1956–65, 10% in 1966–75, and only 4% during 1971–75. A review by Sillitoe (1995, updated in 2000) of discoveries in the Circum-Pacific area showed that geological work, particularly intimate familiarity with the type of deposit being sought, was the key to success. Blain (2000) concurs on the importance of geological work and his analysis shows that geology took over from prospecting as the dominant factor in exploration success in approximately 1970.

1.6.4   Economic influences

Nowadays the optimum target (section 1.2.3, "Metal and mineral prices") in the metals sector must be a high quality deposit of good address, and if the location is not optimal then an acceptable exploration target will have to be of exceptional quality. This implies that for base metal exploration in a remote inland part of Australia, Africa, or South America a large mineral deposit must be sought with a hundred or so million tonnes of high grade resources or an even greater tonnage of near surface, lower grade material. Only high unit value industrial minerals could conceivably be explored for in such an environment. Anything less valuable would probably not be mineable under existing transportation, infrastructure, and production costs.

In existing mining districts smaller targets can be selected, especially close to working mines belonging to the same company, or a company might purchase any new finds. These brownfield finds can be particularly important to a company operating a mine that has only a few years of reserves left and, in such a case with an expensive milling plant and perhaps a smelter to supply, an expensive saturation search may be launched to safeguard the future of the operation. Expenditure may then be much higher for a relatively small target than could be justified for exploration in virgin territory.

1.7   FURTHER READING

An excellent summary of the definitions and usages of the terms mineral resources and ore reserves is to be found in the paper by Taylor (1989) and the more formal JORC (Australasian JORC 2003). The two books by Crowson – *Inside Mining* (1998) and *Astride Mining* (2003) – provide stimulating reading on mineral markets and more general aspects of the mineral industry, including mineral exploration. *Breaking New Ground* (MMSD 2002) provides a very detailed background to the mineral industry and how environmental and social issues might be addressed.

Readers may wish to follow developments in mineral exploration using websites such as *Infomine* (2004) and *Reflections* (2004), as well as material also available in printed form in the weekly *Mining Journal* and *Northern Miner*.

2

THE MINERALOGY OF ECONOMIC DEPOSITS

ANTHONY M. EVANS

2.1 INTRODUCTION

Ore minerals are the minerals of economic interest for which the explorationist is searching. They can be metallic or nonmetallic. Mineralogy is used to understand the relationships between the ore mineral and the uneconomic host rock for their eventual separation.

Economic mineral deposits consist of every gradation from bulk materials or aggregates, in which most of the rock or mineral is of commercial value, to deposits of precious metals (gold, silver, PGM) from which only a few ppm (or ppb in the case of diamond deposits) are separated and sold. The valuable mineral in one deposit may be a gangue mineral in another, e.g. quartz is valuable in silica sands, but is a gangue mineral in auriferous quartz veins. Thus the presentation of lists of ore and gangue minerals without any provisos, as given in some textbooks, can be very misleading to the beginner. This may lead to an erroneous approach to the examination of mineral deposits, i.e. what is recovered and what is discarded? An alternative question is, how can we process everything we are going to mine and market the products at a profit? There are few mineral operations where everything is mined gainfully. Fortunes can be made out of the waste left by previous mining and smelting operations, but not usually by the company that dumped it! A good example of a mine where everything is mined is at the King's Mountain Operation in the Tin–Spodumene Belt of North Carolina. It is in the world's most important lithium-producing area (Kunasz 1982). The spodumene occurs in micaceous granite–

pegmatites and in the mill the ore is processed to produce chemical grade spodumene and ceramic spodumene concentrates, mica and feldspar concentrates, and a quartz–feldspar mix marketed as sandspar. The amphibolite host rock is crushed, sized, and sold as road aggregate. Of course such comprehensive exploitation of all the material mined is not possible in isolated locations, but too often the potential of waste material is overlooked. A *comprehensive* mineralogical examination of a mineral deposit and its waste rocks may mean that additional valuable materials in the deposit are identified and the presence of deleterious substances detected. This may add value if the project is ever brought to the production stage and will help to avoid embarrassing undervaluation.

Ore minerals may be native metals (elements), of which gold and silver are examples, or compounds of metals with sulfur, arsenic, tellurium, etc., such as lead sulfide – the mineral galena (PbS) – or they may be carbonates, silicates, borates, phosphates. There are few common minerals that do not have an economic value in some mineralogical context or other. Some of the more important ore minerals are listed in Table 2.1 and those which are often classified as gangue minerals in Table 2.2. Ore minerals may be classed as primary (hypogene) or secondary (supergene). Hypogene minerals were deposited during the original period of rock formation or mineralisation. Supergene minerals were formed during a later period of mineralisation, usually associated with weathering and other near-surface processes, leading to precipitation of the secondary

TABLE 2.1   Some of the more important ores and ore minerals.

Commodity and principal ore minerals	Mineral formulae	% Metal	Primary	Supergene	Remarks	Major uses
Aggregates					Many different rocks and minerals used	Civil engineering, building
Antimony					Often a byproduct of base metal mining	Alloys
stibnite	Sb_2S_3	72	X			
tetrahedrite	$(Cu,Fe)_{12}Sb_4S_{13}$	36	X			
jamesonite	$Pb_4FeSb_6S_{14}$	33	X			
Arsenic					Normally a byproduct of metal mining	Herbicide, alloys, wood preservative
arsenopyrite	$FeAsS$	46	X			
enargite	Cu_3AsS_4	19	X			
löllingite	$FeAs_2$	73	X			
tennantite	$(Cu,Fe)_{12}As_4S_{13}$	13	X			
Asbestos					Dangerous materials, usage declining	Insulation
chrysotile	$(Mg,Fe)_3Si_2O_5(OH)_4$		X			
riebeckite	$Na_3(Fe,Mg)_4FeSi_8O_{22}(OH)_2$		X			
Baryte	$BaSO_4$	≈39	X		Industrial mineral uses increasing	Drilling muds, filler
Bauxite						Aluminum production
boehmite	$AlO(OH)$			X		
diaspore	$AlO(OH)$			X		
gibbsite	$Al(OH)_3$			X		
Bentonite and fuller's earth	Montmorillonite group minerals		X	X		Bleaching clays, drilling muds, binders
Beryllium					Bertrandite is now more important than beryl	Alloys, electronics
bertrandite	$Be_4Si_2O_7(OH)_2$		X			
beryl	$Be_3Al_2Si_6O_{18}$		X			
Bismuth					A byproduct of various metal mining operations	Pharmaceutical industry, alloys
bismuthinite	Bi_2S_3	81	X		Various unusual minerals	
Borates			X			Glass wool, borosilicate glasses, detergents, agriculture
Chromium chromite	$(Fe,Mg)(Cr,Al)_2O_4$	68(Cr)	X			Depends on Cr v. Al content. Steel industry, refractories, chemicals
Coal	C,O&H		X			Energy source
Cobalt						High temperature alloys, magnets, tool steels
carrollite cobaltiferous	$CuCo_2S_4$	56	X			
pyrite	$(Fe,Co)S_2$	Variable	X		Byproduct of some copper mines	
Common clay	See shale					

Mineral	Formula	%			Notes	Uses
Copper					Many other ores of copper	Copper production
bornite	Cu_5FeS_4	63		X		
chalcocite	Cu_2S	80		X		
chalcopyrite	$CuFeS_2$	34	X	X		
covellite	CuS	66	X	X		
Diamond	C			X	Considerable production of synthetic and industrial diamonds	Jewellery, cutting and grinding tools
Diatomite	SiO_2,nH_2O			X		Filters, filler, abrasive
Feldspar	$NaAlSi_3O_8$			X		Ceramics, glass making
	$KAlSi_3O_8$			X		
Fluorspar	CaF_2			X	Fluorspar is the raw ore, fluorite the pure mineral	Flux in steel making, fluorochemicals
Gold					Many other tellurides may occur in gold ores	Jewellery, hoarding, dentistry
native gold	Au	90–100		X		
calaverite	$AuTe_2$	39		X		
sylvanite	$(Au,Ag)Te_2$	variable		X		
Graphite	C			X		Steel making, refractories, foundries
Gypsum	$CaSO_4 \cdot 2H_2O$			X		Plasterboard, insulation
Iron						Iron production
hematite	Fe_2O_3	72	X	X		
magnetite	Fe_3O_4	70		X		
siderite	$FeCO_3$	48	X	X		
Kaolin					A constituent of many clays: ball clay, refractory clay, etc.	Paper manufacture, coating clay, filler, extender
kaolinite	$Al_4Si_4O_{10}(OH)_8$		X	X		
Limestone	$CaCO_3$		X	X	One of the world's most widely used materials	Constructional, agricultural, chemical and metallurgical industries
Lithium					Li brines are now important producers	Ceramics, glass, enamels, Li salts, Li chemicals, batteries
amblygonite	Li					
lepidolite	$K(LiAl)_3(Si,Al)_4O_{10}(F,OH)_2$			X		
petalite	$LiAlSi_4O_{10}$					
spodumene	$LiAlSi_2O_6$					
Lead					Galena may be argentiferous	Lead production
galena	PbS	86		X		
Magnesite	$MgCO_3$			X	Marketed mainly as magnesia	Refractories, animal feedstuffs, special cements

TABLE 2.1 (continued)

Commodity and principal ore minerals	Mineral formulae	% Metal	Primary	Supergene	Remarks	Major uses
Manganese					Most important ferro-alloy metal	Steel making. Over 1 Mt p.a. is used for industrial mineral purposes
pyrolusite	MnO_2	63	X	X		
psilomenane	$(BaMn)Mn_4O_8(OH)_2$	50	X	X		
braunite	Mn_7SiO_{12}	64	X	X		
manganite	$MnO.OH$	62	X	X		
Mercury						Mercury production
cinnabar	HgS	86	X			
Mica					Marketed as sheet or ground mica	Electrical insulator, furnace windows, wallpaper, paints, plasterboard
muscovite	$KAl_2[AlSi_3O_{10}](OH)_2$		X			
phlogopite	$KMg_3[AlSi_3O_{10}](F,OH)_2$			X		
Molybdenum					Mined as the principal metal or as a byproduct	Molybdenum production
molybdenite	MoS_2	60	X			
wulfenite	$PbMoO_4$	26	X			
Nepheline–syenite and sheet				X	Composed of nepheline, albite and microcline	Container glass, whitewares, glazes
Nickel						Nickel production
pentlandite	$(Fe,Ni)_9S_8$	28(max)	X			
garnierite	$(Ni,Mg)_3Si_2O_5(OH)_4$	<20		X		
Perlite	Rhyolitic composition		X		A volcanic glass that expands on heating	Insulation board, plaster, concrete
PGM					Several other PGM minerals may be present in ores	Catalysts, electrical industry, jewellery
nat. platinum	Pt	≈100	X			
sperrylite	$PtAs_2$	57	X			
braggite	$(Pt,Pd,Ni)S$	variable	X			
laurite	$(Ru,Ir,Os)S_2$	variable	X			
Phosphate rock						Fertilizer (90%), detergents, animal feedstuffs
apatite	$Ca_5[PO_4]_3(F,OH)$	P = 18	X	X		
Pyrophyllite	$Al_2Si_4O_{11}.H_2O$		X			Ladle linings in steel mills, ceramics, insecticides
Potash						About 95% goes into fertilizer manufacture, rest for soaps, glass, ceramics, etc.
sylvite	KCl		X			
carnallite	$KCl.MgCl_2.6H_2O$		X			
kainite	$4KCl.4MgSO_4.11H_2O$		X			
langbeinite	$K_2SO_4.2MgSO_4$		X			

Mineral	Formula				Remarks	Uses
REE					Many more REE in these minerals than shown in formulae	Catalysts, glass, ceramics, television tubes, permanent magnets, etc.
bastnäsite	(Ce,La)(CO₃)F → $(Ce,La)(CO_3)F$		X			
parisite	$(Ce,La)_2Ca(CO_3)_3F_2$		X			
monazite	$(Ce,La,Nd,Th)PO_4$		X			
Salt						Innumerable! Over half of all production used in the chemical industry, deicing agent, food preservative
halite	$NaCl$		X	X		
Shale and tiles, sewer, common clay	Clay and mica minerals, quartz			X	60–80% of all "clay" mined falls into this category	Bricks, pipes, cement, lightweight aggregate
Silica sand	SiO_2				Sand and gravel working taken worldwide represents one of the most important mining industries	Building, civil engineering, glass manufacture
Sillimanite minerals					Refractories account for 90% of production	
andalusite	Al_2SiO_5	100	X			
kyanite	Al_2SiO_5		X			
sillimanite	Al_2SiO_5		X			
Silver					Byproduct of many base metal mines particularly Pb-Zn tetrahedrite and many other minerals	Photography, electrical and electronic industries, sterling ware, jewellery, etc.
nat. silver	Ag		X	X		
acanthite	Ag_2S		87	X		
argentiferous tetrahedrite	$(Cu,Fe,Ag)_{12}Sb_4S_{13}$	75				
cerargyrite	$AgCl$		X			
and many other minerals						
Sodium carbonate					Bulk of soda ash is (trona) produced synthetically in Solvay plants	Production of soda ash, Na₂CO₃, for glass manufacture, sodium chemicals, paper production, etc.
trona	$Na_2CO_3 \cdot NaHCO_3 \cdot 2H_2O$					
Strontium minerals					Celestite is the only economically important Sr mineral. Used in industry as SrCO₃	Pyrotechnics, color TV tubes, permanent magnets, greases, soaps
celestite	$SrSO_4$	100	X	X		
Sulphur					Main source is crude oil. Native sulfur 30%, pyrite 16%	Sulfuric acid production, fertilizers, etc.
nat. sulphur	S		X	X		
pyrite	FeS_2		X			
Talc					Has very many important uses	Filler in paints, plastics, paper, rubber, porcelain, tiles, cosmetics
talc	$Mg_3Si_4O_{10}(OH)_2$	78	X			
Tin, cassiterite	SnO_2		X		Cassiterite is the main ore mineral	Solder, tin-plate, alloys, etc.
stannite	Cu_2FeSnS_4	27	X			

TABLE 2.1 (continued)

Commodity and principal ore minerals	Mineral formulae	% Metal	Primary	Supergene	Remarks	Major uses
Titania					Both an important metal and industrial material in form TiO_2	White pigment in paint, plastics, rubber, paper, etc.
ilmenite	$FeTiO_3$	TiO_2 = 45–65	X			
rutile	TiO_2	TiO_2 = 90–98		X		
Tungsten						Cutting materials, steels, electric light bulbs
scheelite	$CaWO_4$	64	X			
wolframite	$(Fe,Mn)WO_4$	61	X			
Uranium					Many other ore minerals	Nuclear-powered generators, special steels, military purposes
uraninite-pitchblende	UO_2	88	X			
uranophane	$Ca(UO_2)_2Si_2O_7.6H_2O$	56	X	X		
carnotite	$K_2(UO_2)_2(VO_4)_2.3H_2O$	55	X	X		
torbernite	$Cu(UO_2)_2(PO_4)_2.8H_2O$	51	X	X		
Vanadium					The main producer is now the RSA from Ti-V-magnetites in the Bushveld Complex	Vanadium steels, super alloys
carnotite	$K_2(UO_2)_2(VO_4)_2.3H_2O$	13	X	X		
tyuyamunite	$Ca(UO_2)_2(VO_4)_2.5-8H_2O$	15	X	X		
montroseite	$VO(OH)$	61	X			
roscoelite	vanadium mica		X			
magnetite	$(Fe,Ti,V)_2O_4$	V = 1.5–2.1	X			
Vermiculite	$(Mg,Fe,Al)_3(Al,Si)_4O_{10}(OH)_2.4H_2O$		X	X	Expands on heating	Acoustical and thermal insulator, carrier for fertilizer and agricultural chemicals
Wollastonite	$CaSiO_3$		X			Ceramics, filler, extender
Zeolites			X		Uses multiplying rapidly, much of the demand supplied by synthetic zeolites	Cements, paper filler, ion exchange resins, dietary supplement for animals, molecular "sieve", catalyst
Zinc						Alloy castings, galvanizing iron and steel, copper-based alloys, e.g. brass
hemimorphite	$Zn_4(OH)_2Si_2O_7.H_2O$	54		X		
smithsonite	$ZnCO_3$	52		X		
sphalerite	$(Zn,Fe)S$	up to 67	X			
Zirconium					Most zircon is converted to ZrO_2 for industrial use	Refractories, foundry sands, nuclear reactors
baddeleyite	ZrO_2	74	X			
zircon	$ZrSiO_4$	50	X			

TABLE 2.2   List of common gangue minerals.

Name	Composition	Primary	Supergene
Quartz	SiO_2	X	
Chert	SiO_2	X	
Limonite	$Fe_2O_3.nH_2O$		X
Calcite	$CaCO_3$	X	X
Dolomite	$CaMg(CO_3)_2$	X	X
Ankerite	$Ca(Mg,Fe)(CO_3)_2$	X	X
Baryte	$BaSO_4$	X	
Gypsum	$CaSO_4.2H_2O$	X	X
Feldspar	All types	X	
Fluorite	CaF_2	X	
Garnet	Andradite most common	X	
Chlorite	Several varieties	X	
Clay minerals	Various	X	X
Pyrite	FeS_2	X	
Marcasite	FeS_2	X	
Pyrrhotite	$Fe_{1-x}S$	X	
Arsenopyrite	$FeAsS$	X	

minerals from descending solutions. When secondary mineralisation is superposed on primary mineralisation the grade increases and this is termed supergene enrichment.

2.2   MINERALOGICAL INVESTIGATIONS

Before looking at some of the many methods that may be used, the economic importance of these investigations will be emphasized by discussing briefly the importance of mineralogical form and undesirable constituents.

Mineralogical form

The properties of a mineral govern the ease with which existing technology can extract and refine certain metals and this may affect the cut-off grade (see section 10.4.2). Thus nickel is far more readily recovered from sulfide than from silicate minerals and sulfide minerals can be extracted down to about 0.5%, whereas silicate minerals must assay about 1.5% to be economic.

Tin may occur in a variety of silicate minerals such as stanniferous andradite ($Ca_3Fe_2Si_3O_{12}$) and axinite (($Ca,Fe,Mn)Al_2BSi_4O_{15}OH$), from which it is not recoverable, as well as in its main ore mineral form, cassiterite (SnO_2).

Aluminum is of course abundant in many silicate rocks, but it must be usually in the form of hydrated aluminum oxides, the rock called bauxite, for economic recovery. The mineralogy of the ore mineral will also place limits on the maximum possible grade of the concentrate. For example, in a mineral deposit containing native copper it is theoretically possible to produce a concentrate containing 100% Cu but, if the ore mineral chalcopyrite ($CuFeS_2$) is the principal source of copper, then the best concentrate would only contain 34.5% Cu.

Undesirable substances

Deleterious elements may be associated with both ore and gangue minerals. For example, tennantite ($Cu_{12}As_4S_{13}$) in copper ores can introduce unwanted arsenic and sometimes mercury into copper concentrates. These, like phosphorus in iron concentrates and arsenic in nickel concentrates, will lead to custom smelters imposing financial penalties or refusing the shipment. The ways in which gangue minerals may lower the value of an ore are very varied. For example, an acid leach is normally employed to extract uranium from the crushed ore, but if the carbonate, calcite ($CaCO_3$), is present, there will be excessive acid consumption and the less effective alkali leach method may have to be used. Some primary tin deposits contain appreciable amounts of topaz which, because of its hardness, increases the abrasion of crushing and grinding equipment, thus raising the operating costs.

To summarize, the information that is required from a sample includes some, or all, of the following: (i) the grade of the economic minerals; (ii) the bulk chemical composition; (iii) the minerals present; (iv) the proportions of each of these and their chemical compositions; (v) their grain size; (vi) their textures and mineral locking patterns; (vii) any changes in these features from one part of an orebody to another.

2.2.1 Sampling

Mineralogical investigations will lose much of their value if they are not based on systematic and adequate sampling of all the material that might go through the processing plant, i.e. mineralized material *and* host rock. The basics of sound sampling procedures are discussed in Chapter 10. The material on which the mineralogist will have to work can vary from solid, coherent rock through rock fragments and chips with accompanying fines to loose sand. Where there is considerable variation in the size of particles in the sample it is advantageous to screen (sieve) the sample to obtain particles of roughly the same size, as these screened fractions are much easier to sample than the unsized material.

The mineralogist will normally subsample the primary samples obtained by geologists from the prospect to produce a secondary sample, and this in turn may be further reduced in bulk to provide the working sample using techniques discussed in Jones (1987) and recent technological innovations.

2.2.2 Mineral identification

Initial investigations should be made using the naked eye, the hand lens and a stereobinocular microscope to: (i) determine the ore types present and (ii) select representative specimens for thin and polished section preparation. At this stage uncommon minerals may be identified in the hand specimen by using the determinative charts in mineralogical textbooks such as Berry et al. (1983) or the more comprehensive method in Jones (1987).

The techniques of identifying minerals in thin section are taught to all geologists and in polished sections to most, and will not be described here. For polished section work the reader is referred to Craig and Vaughan (1994) and Ineson (1989), as well as the online manual of Ixer and Duller (1998). Modern optical microscopes have significantly increased resolution and oil immersion is not often used in commercial laboratories. Simple microscope and scanning electron microscope (SEM) methods are usually all that is required to effectively identify all the minerals in the samples. SEM and other methods requiring sophisticated equipment are discussed below.

X-ray diffraction

X-ray diffraction is used to identify clay mineral structure and properties, and for mineral analysis and mineral abundance measurements through spectroscopic sensing. Modern X-ray diffractometers can work well on solid specimens, compacted powder pellets representing whole rocks, or on a few grains on a smear mount. Multiple mounts can be automatically fed into the diffractometer.

The rock sample is normally powdered and packed into an aluminum holder. It is then placed in the diffractometer and bombarded with X-rays. The diffracted rays are collected by a detector and the information relayed to a computer where it is converted to d-values of specific intensities. This information can then be shown graphically in the form of a diffraction pattern or "diffractogram." The diffractograms from the unknown sample are then matched against a database of 70,000 recorded phases for mineral identification. The latest instruments allow for rapid recognition of the entire spectrum of the sample in minutes using a computer to match patterns and identify the minerals present.

Electron and ion probe microanalyzers

With this equipment a beam of high energy electrons is focused on to about 1–2 μm^2 of the surface of a polished section or a polished thin section. Some of the electrons are reflected and provide a photographic image of the surface. Other electrons penetrate to depths of 1–2 μm and excite the atoms of the mineral causing them to give off characteristic X-radiation which can be used to identify the elements present and measure their amounts. With this chemical information and knowledge of some optical properties reasonable inferences can be made concerning the identity of minute grains and inclusions. In addition important element ratios such as Fe:Ni in pentlandite ($(Fe,Ni)_9S_8$) and Sb:As in tetrahedrite ($(Cu,Fe,Ag,Zn)_{12}Sb_4S_{13}$) –tennantite ($(Cu, Ag,Fe,Zn)_{12}As_4S_{13}$), and small amounts of possible byproducts and their mineralogical location can be determined. Useful references are Goldstein et al. (1981), Reed (1993), and Zussman (1977).

The ion microprobe uses an ion rather an electron source and measures the mass of secondary ions from the sample rather than X-rays. This technique allows the measurement of much lower concentrations of heavy elements than the electron microprobe and has been widely used in the search for the location of gold and PGM in metallurgical testing. A summary of the technique can be found in Larocque and Cabri (1998).

Scanning electron microscopy (SEM)

SEM is of great value in the three-dimensional examination of surfaces at magnifications from ×20 to 100,000. Textures and porosity can be studied and, with an analytical facility, individual grains can be analyzed and identified *in situ*. Excellent microphotographs can be taken and, with a tilting specimen stage, stereographic pairs can be produced. This equipment is particularly valuable in the study of limestones, sandstones, shales, clays, and placer materials. The method is discussed in Goldstein et al. (1981) and Tucker (1988). Recent developments such as the QEM (Quantitative Evaluation of Minerals)*SEM allows the automatic quantification of mineral composition and size in a similar manner to point count and image analysis for optical microscopes, although the QEM*SEM is rapidly being overtaken by the Mineral Liberation Analyser (MLA) which has particular importance in applied mineralogy and metallurgical processing. Mineral Liberation Analyser data are

fundamental parameters used in the design and optimization of processing plants. Gu (2002) explained that the MLA system consists of a specially developed software package and a standard modern SEM fitted with an energy dispersive spectrum (EDS) analyzer. The on-line program of the MLA software package automatically controls the SEM, captures sample images, performs necessary image analysis (see Chapter 6), and acquires EDS X-ray spectra. Typically, 40–100 images (containing 4000–10,000 grains) are acquired for each sample block and a dozen blocks (of 30 mm diameter) are measured overnight. The MLA off-line processing program transforms the raw image into quality sample images, from which most important minerals can be differentiated using modern image analysis methods.

Differential thermal analysis

This method is principally used for clay and clay-like minerals which undergo dehydration and other changes on heating. Measurement of the differences in temperature between an unknown specimen and reference material during heating allow for the determination of the position and intensity of exothermic and endothermic reactions. Comparison with the behavior of known materials aids in the identification of extremely fine-grained particles that are difficult to identify by other methods (Hutchison 1974).

Autoradiography

Radioactive minerals emit alpha and beta particles which can be recorded on photographic film or emulsions in contact with the minerals, thus revealing their location in a rock or ore. In ores this technique often shows up the presence of ultra-fine-grained radioactive material whose presence might otherwise go unrecorded. The technique is simple and cheap and suitable for use on hand specimens and thin and polished sections (Robinson 1952, Zussman 1977).

Cathodoluminescence

Luminescence (fluorescence and phosphorescence) is common in the mineral kingdom.

In cathodoluminescence the exciting radiation is a beam of electrons and a helpful supplement to this technique is ultraviolet fluorescence microscopy. Both techniques are used on the microscopic scale to study transparent minerals. Minerals with closely similar optical properties or which are very fine-grained can be readily differentiated by their different luminescent colors, e.g. calcite *v.* dolomite, feldspar *v.* quartz, halite *v.* sylvite. Features not seen in thin sections using white light may appear, thin veins, fractures, authigenic overgrowths, growth zones in grains, etc. A good description of the apparatus required and the method itself is given in Tucker (1988).

2.2.3 Quantitative analysis

Grain size and shape

The *recovery* is the percentage of the *total* metal or industrial mineral contained in the ore that is recovered in the concentrate; a recovery of 90% means that 90% of the metal in the ore passes into the concentrate and 10% is lost in the tailings. It might be thought that if one were to grind ores to a sufficiently fine grain size then complete separation of mineral phases might occur to make 100% recovery possible. In the present state of technology this is not the case, as most mineral processing techniques fail in the ultra-fine size range. Small mineral grains and grains finely intergrown with other minerals are difficult or impossible to recover in the processing plant, and recovery may be poor. Recoveries from primary (bedrock) tin deposits are traditionally poor, ranging over 40–80% with an average around 65%, whereas recoveries from copper ores usually lie in the range 80–90%. Sometimes fine grain size and/or complex intergrowths may preclude a mining operation. The McArthur River deposit in the Northern Territory of Australia contains 200 Mt grading 10% zinc, 4% lead, 0.2% copper, and 45 ppm silver with high grade sections running up to 24% zinc and 12% lead. This enormous deposit of base metals remained unworked from its discovery in 1956 until 1995 because of the ultra-fine grain size and despite years of mineral processing research on the ore.

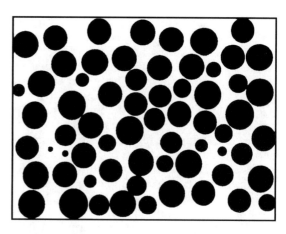

FIG. 2.1 A random section through a solid consisting of a framework of spheres of equal size.

Grain size measurement methods for loose materials, e.g. gravels and sands, placer deposits, or clays, vary, according to grain size, from calipers on the coarsest fragments, through sieving and techniques using settling velocities, to those dependent upon changes in electrical resistance as particles are passed through small electrolyte-filled orifices. These methods are described in Tucker (1988) and other books on sedimentary petrography.

Less direct methods have to be employed with solid specimens because a polished or thin section will only show random profiles through the grains (Fig. 2.1). Neither grain size, nor shape, nor sorting can be measured directly from a polished or thin section and when the actual grains are of different sizes microscopic measurements using micrometer oculars (Hutchison 1974) or other techniques invariably *overestimate* the proportion of small grains present. This bias can be removed by stereological methods *if* the grains are of simple regular shapes (cubes, spheres, parallelepipeds, etc.). Otherwise stereological transformation is not possible and great care must be taken in using grain size measurements taken from sectioned specimens.

An *indication* of grain shape can be obtained by measuring a large number of intercepts (lengths of randomly chosen grain diameters) and analyzing these, e.g. the presence of a substantial number of very small intercepts would indicate that the grains were angular and the opposite would show that the grains lacked edges and were smooth and convex (Jones 1987).

Modal analysis

Modal analysis produces an accurate representation of the distribution and volume percent of a given mineral in a thin or polished section. Two methods of analysis are normally used, namely:

1 *Area percentage.* The surface area of mineral grains of the same mineral are measured relative to the total surface area of the thin section, giving the areal proportions of each mineral type. Since volumes in this situation are directly proportional to areas, these are also the volume percentages.

2 *Point count.* Each mineral occurrence along a series of traverse line across a given thin section is counted. At least 2000 individual points must be counted for a statistically valid result.

The number of grains counted, the spacing between points, and successive traverse lines is dependent on the mean grain size of the sample.

Modal analysis can be used to compare rocks from different areas if there are only thin sections. No chemical analysis is required. The work can be achieved manually using a petrographic microscope. Modern optical microscopes and SEMs use image analysis (see Chapter 6) to count mineral grains and to calculate areal proportions automatically (Jones 1987, Sprigg 1987). Statistically representative numbers of points are achieved routinely using image analysis.

However, care must be taken with foliated, banded rocks which should only be sampled at right angles to the banding (Hutchison 1974). Experience shows that porphyritic rocks are difficult to count. Similarly, care must be taken to ensure that the total area of the sample is larger than the maximum diameter of the smallest grain size. Very coarse-grained rocks such as pegmatites can be measured with a grid drawn on transparent material and placed on outcrop, joint, or mine surfaces. A similar technique can be used to visually estimate the grade of a mineral deposit where the sampler is sure that no ore mineral will be missed, e.g. tungsten deposits where the scheelite is the only ore mineral and can be picked out using ultraviolet light.

The volume percentage of ore and/or deleterious minerals can be of crucial importance in mineral processing and a knowledge of whether a wanted metal is present in one or several minerals. If the latter is the case, their relative proportions may also be of great importance. Some examples of this are:

1 Gold ores – is all the gold present as native gold (free-milling gold) or is some in the form of tellurides or enclosed by sulfides (refractory gold)? Native gold is readily leached from milled ores by cyanide solutions, but refractory gold resists leaching and has to be roasted (after concentration) before cyaniding or leached under pressure, thereby increasing the cost of the treatment and of course decreasing the value of the ore.

2 Titanium ores in anorthosites will have significant amounts of titanium locked up in titaniferous magnetite, sphene, and augite from which it is not recoverable.

3 The skarn iron orebody at Marmoraton, Ontario assayed on average 50% Fe, but only 37.5% (in magnetite) was recoverable, the rest was locked in silicates. These and similar devaluing features are readily detected and quantified by microscopic investigations. Valuations based on assays alone may be grossly exaggerated.

If the chemical compositions of the minerals are known, dividing volume percentages by mineral densities (and converting to percent) provides the weight percentages of the minerals, and by using Table 2.1 (or by calculation) gives us an entirely independent way of obtaining an estimate of the grade. This may be of value as a check on chemical or X-ray fluorescence assays.

The mineral explorationist should train him or herself to make visual modal estimates in the field using hand specimens and natural exposures. By estimating the volume percentage of a metallic mineral such as chalcopyrite (often the only or principal copper mineral in a mineral deposit) and looking up the copper content in Table 2.1, a visual assay can be made. When the first laboratory assays become available for a prospect under investigation explorationists will have reference material with which they can compare their estimates and improve their accuracy for that particular mineral deposit type. The American Geological Institute Data Sheets (sheet 15.1) and various books (Spock 1953, Thorpe & Brown 1985, Tucker 1988, Barnes & Lisle 2003 and others) have comparison charts to help the field geologist in estimating percentage compositions in hand specimens.

2.2.4   Economic significance of textures

Mineral interlocking

Ores are crushed during milling to liberate the various minerals from each other (section 2.2.3) and for concentration a valuable mineral has to be reduced to less than its liberation size in order to separate it from its surrounding gangue. Crushing and grinding of rock is expensive and if the grain size of a mineral is below about 0.05 mm the cost may well be higher than the value of the liberated constituents. In addition there are lower limits to the degree of milling possible dictated by the separation processes to be employed because these are most effective over certain grain size ranges: e.g. magnetic separation, 0.02–2.5 mm; froth flotation, 0.01–0.3 mm; electrostatic separation, 0.12–1.4 mm.

In Fig. 2.2 a number of intergrowth patterns are illustrated. Further crushing of the granular textured grains in (a) will give good separation of ore (black) from gangue – this is an ideal texture from the processing point of view. In (b) further crushing of the tiny pyrite grain veined by chalcopyrite (black) is out of the question and this copper will be lost to the tailings. The chalcopyrite (black) occurring as spheroids in sphalerite grains (c) is too small to be liberated and will go as a copper loss into the zinc concentrate. The grain of pyrite coated with supergene chalcocite (black) in (d) will, during froth flotation, carry the pyrite as a diluting impurity into the copper concentrate. The grain is too small for separation of the two minerals by crushing. It must be noted that the market price for a metal does not apply fully or directly to concentrates. The purchase terms quoted by a custom smelter are usually based on a nominal concentrate grade and lower concentrate grades are penalized accord-

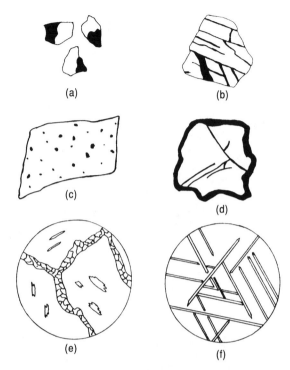

FIG. 2.2   (a)–(d) Grains from a mineral dressing plant. (a) Granular texture; black represents an ore mineral, unornamented represents gangue (×0.6). (b) A pyrite grain veined by chalcopyrite (black) (×177). (c) A sphalerite grain containing small rounded inclusions of chalcopyrite (black) (×133). (d) Pyrite grain coated with supergene chalcocite (black) (×233). (e) Grains of pyrrhotite with exsolved granular pentlandite in the interstices and flame exsolution bodies within the pyrrhotite (×57). (f) Exsolution blades of ilmenite in a magnetite grain (×163).

ing to the amount by which they fall below the contracted grade. Exsolution textures commonly devalue ores by locking up ore minerals and by introducing impurities. In (e) the tiny flame-shaped exsolution bodies of pentlandite (black) in the pyrrhotite grain will go with the pyrrhotite into the tailings and the ilmenite bodies (black) in magnetite (f) are likewise too small to be liberated by further grinding and will contaminate the magnetite concentrate. If this magnetite is from an ilmenite orebody then these interlocked ilmenite bodies will be a titanium loss.

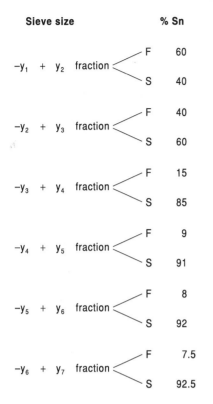

Sieve size	% Sn

FIG. 2.3 Mineral liberation size investigation of a possible tin ore. F, float; S, sink; y_1 to y_7, grain (sieve) sizes. For discussion see text.

Mineral liberation size

This is usually investigated as follows. Suppose we have a simple example such as a tin ore consisting of cassiterite ($\rho = 6.99$) and quartz ($\rho <$ of the minerals are separated, using heavy liquids such as ($\rho = 3.2$) sodium-polytunstate and lithium-tungstate. The float and sink fractions are analyzed for tin grade. Figure 2.3 shows a possible result, where little improvement in separation would be obtained by grinding beyond the y_5–y_6 fraction.

2.2.5 Deleterious substances

These have been discussed in under "Undesirable substances" in section 2.2 with reference to arsenic, mercury, phosphorus, calcite, and topaz in metallic ores. Among industrial min-

erals mention can be made of the presence of limonite coatings on quartz grains in sand required for glass making. Extra cost in exploitation will arise from the need for an acid leach, or other method, to remove the coatings. Coal fragments in gravel will render it valueless as gravel processing plants do not include equipment for eliminating the coal. Lastly it has become apparent recently that the alkaline-silica reaction, which produces "concrete cancer," is due in some cases to the presence of certain types of opaline silica in the aggregate.

2.2.6 Miscellaneous examples of the use of microscopy in ore evaluation and mineral processing

Chromite ores

Chromite is never pure $FeCr_2O_4$ and iron(III) may substitute for chromium, particularly along grain boundaries and fractures, producing "off color grains" which can be detected in polished sections by the experienced observer. The magnetite rims will introduce mineral processing difficulties, as will badly fractured chromite grains present in some podiform deposits; these may disintegrate rapidly into a very fine-grained powder on grinding.

Nickel sulfide ores

The normal opaque mineralogy is magnetite–pyrrhotite–pentlandite–chalcopyrite. The magnetite and pyrrhotite are separated magnetically and normally become waste. Nickel present in flame-like exsolution bodies (Fig. 2.2e) and in solid solution in the pyrrhotite will be lost. In an ore investigated by Stephens (1972) use of the electron probe microanalyzer revealed that 20% of the nickel would be lost in this way. Minute grains of PGM minerals are often included in the pyrrhotite, again leading to losses if this has not been noted. Also in such ores the chalcopyrite may contain large exsolved bodies of cubanite, a magnetic mineral, which could mean a substantial copper loss in the pyrrhotite concentrate. In such a case the ore must be roasted (at extra expense) to destroy the magnetism of the cubanite.

Tin ores

At mines where separate concentrates of cassiterite and copper–zinc–arsenic sulfides (often as a minor byproduct) are produced, cassiterite coated with stannite (Cu_2FeSnS_4) will pass into the sulfide concentrate.

Zinc loss

Lead–zinc ore from a new orebody was tested by being processed in a mill at a nearby mine. Despite being apparently identical to the ore at the mine, substantial zinc losses into the tailings occurred. It was discovered with the use of an electron probe microanalyzer that the siderite in the gangue carried 8–21% Zn in solid solution.

Mercury impurity

In 1976 Noranda Mines Ltd cut its copper–gold–silver concentrate purchases from Consolidated Rambler Mines Ltd, Newfoundland by nearly 50% because of "relatively high impurities" (Anon 1977). The major impurity was mercury. An electron probe investigation showed that this occurred in solid solution (1–2%) in the minor sphalerite in the ore. By depressing the zinc in the flotation circuit the mercury content of the concentrate was virtually eliminated.

Probably the first mine to recover mercury from copper concentrate was Rudnany in Czechoslovakia where it occurs in tetrahedrite. At the former Gortdrum Mine in Ireland the presence of cinnabar in the copper–silver ore, and of mercury in solid solution in the tennantite, remained unknown for several years and the smelting of concentrates from this mine at a custom smelter in Belgium presumably produced a marked mercury anomaly over a substantial part of western Europe. A mercury separation plant was then installed at Gortdrum and a record production of 1334 flasks was reached in 1973.

Sulfur in coal

Of the three sulfur types in coal (organic, sulfate, and sulfide) the sulfide is usually present as pyrite and/or marcasite. If it is coarse-grained then much can be removed during washing, but many coals, e.g. British ones, have such fine-grained pyrite that little can be done to reduce the sulfur content. Such coals are no longer easily marketable in this time of concern about acid rain.

2.3 FURTHER READING

General techniques in applied mineralogy are well discussed in Jones' *Applied Mineralogy – A Quantitative Approach* (1987). Hutchison's *Laboratory Handbook of Petrographic Techniques* (1974) is also an invaluable book, as is Zussman's (1977) *Physical Methods in Determinative Mineralogy*. Those working on polished sections should turn to *Ore Microscopy and Ore Petrography* by Craig and Vaughan (1994) and the online atlas of Ixer and Duller (1998). References to techniques not covered by these books are given in the text of this chapter.

3

MINERAL DEPOSIT GEOLOGY AND MODELS

ANTHONY M. EVANS AND CHARLES J. MOON

A detailed understanding of the geology of mineral deposits is required to explore effectively for them. As this is beyond the scope of this volume, only some key features are discussed here as well as how a more advanced appreciation of mineral deposit geology may be used in exploration programs. For further detailed consideration of mineral deposit geology the reader is referred to the companion volume in this series by Robb (2004) and references listed in section 3.4.

3.1 NATURE AND MORPHOLOGY OF OREBODIES

3.1.1 Size and shape of ore deposits

The size, shape, and nature of ore deposits affects the workable grade. Large, low grade deposits which occur at the surface can be worked by cheap open pit methods, whilst thin tabular vein deposits will necessitate more expensive underground methods of extraction. Open pitting, aided by the savings from bulk handling of large daily tonnages (say >30 kt), has led to a trend towards the large scale mining of low grade orebodies. As far as shape is concerned, orebodies of regular shape can generally be mined more cheaply than those of irregular shape, particularly when they include barren zones. For an open pit mine the shape and attitude of an orebody will also determine how much waste has to be removed during mining. The waste will often include not only overburden (waste rock above the orebody) but also waste rock around and in the orebody, which

has to be cut back to maintain a safe overall slope to the sides of the pit (see section 11.2.1). Before discussing the nature of ore bodies we must learn some of the terms used in describing them.

If an orebody viewed in plan is longer in one direction than the other we can designate this long dimension as its strike (Fig. 3.1). The inclination of the orebody perpendicular to the strike will be its dip and the longest dimension of the orebody its axis. The plunge of the axis is measured in the vertical plane ABC but its pitch or rake can be measured in any other plane, the usual choice being the plane containing the strike, although if the orebody is fault controlled then the pitch may be measured in the fault plane. The meanings of other terms are self-evident from the figure.

It is possible to classify orebodies in the same way as we divide up igneous intrusions according to whether they are discordant or concordant with the lithological banding (often bedding) in the enclosing rocks. Considering discordant orebodies first, this large class can be subdivided into those orebodies which have an approximately regular shape and those which are thoroughly irregular in their outlines.

Discordant orebodies

Regularly shaped bodies

Tabular orebodies. These bodies are extensive in two dimensions, but have a restricted development in their third dimension. In this class we have veins (sometimes called fissure-veins)

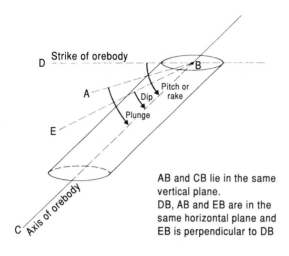

AB and CB lie in the same vertical plane.
DB, AB and EB are in the same horizontal plane and EB is perpendicular to DB

FIG. 3.1 Illustrations of terms used in the description of orebodies.

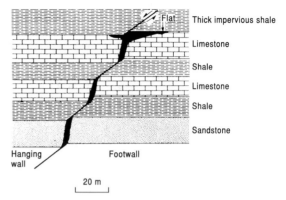

FIG. 3.2 Vein occupying a normal fault and exhibiting pinch-and-swell structure, giving rise to ribbon ore shoots. The development of a flat beneath impervious cover is shown also.

and lodes (Fig. 3.2). These are essentially the same and only the term vein is now normally used. Veins are often inclined, and in such cases, as with faults, we can speak of the hanging wall and the footwall. Veins frequently pinch and swell out as they are followed up or down a stratigraphical sequence (Fig. 3.2). This pinch-and-swell structure can create difficulties during both exploration and mining often because only the swells are workable. If these are imagined in a section at right angles to that in Fig. 3.2, it can be seen that they form ribbon ore shoots. Veins are usually developed in fracture systems and therefore show regularities in their orientation throughout the orefield in which they occur.

The infilling of veins may consist of one mineral but more usually it consists of an intergrowth of ore and gangue minerals. The boundaries of vein orebodies may be the vein walls or they may be assay boundaries within the veins.

Tubular orebodies. These bodies are relatively short in two dimensions but extensive in the third. When vertical or subvertical they are called pipes or chimneys, when horizontal or subhorizontal, "mantos." The Spanish word manto is inappropriate in this context for its literal translation is blanket; it is, however, firmly entrenched in the English geological literature. The word has been and is employed by some workers for flat-lying tabular bodies, but the perfectly acceptable word "flat" (Fig. 3.2) is available for these; therefore the reader must look carefully at the context when he or she encounters the term "manto." Mantos and pipes may branch and anastomose and pipes frequently act as feeders to mantos.

In eastern Australia, along a 2400 km belt from Queensland to New South Wales, there are hundreds of pipes in and close to granite intrusions. Most have quartz fillings and some are mineralized with bismuth, molybdenum, tungsten, and tin; an example is shown in Fig. 3.3. Pipes may be of various types and origins (Mitcham 1974). Infillings of mineralized breccia are particularly common, a good example being the copper-bearing breccia

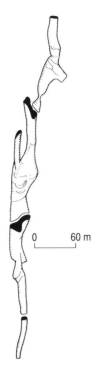

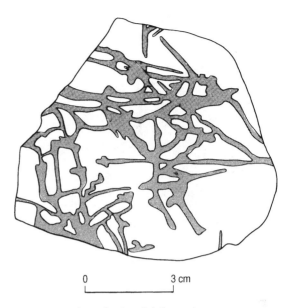

FIG. 3.4 Stockwork of molybdenite-bearing quartz veinlets in granite that has undergone phyllic alteration. Run of the mill ore, Climax, Colorado.

FIG. 3.3 Diagram of the Vulcan pipe, Herberton, Queensland. The average grade was 4.5% tin. (After Mason 1953.)

pipes of Messina in South Africa (Jacobsen & McCarthy 1976).

Irregularly shaped bodies

Disseminated deposits. In these deposits, ore minerals are peppered throughout the body of the host rock in the same way as accessory minerals are disseminated through an igneous rock; in fact, they often *are* accessory minerals. A good example is that of diamonds in kimberlites. In other deposits, the disseminations may be wholly or mainly along close-spaced veinlets cutting the host rock and forming an interlacing network called a stockwork (Fig. 3.4), or the economic minerals may be disseminated through the host rock along veinlets. Whatever the mode of occurrence, mineralisation of this type generally fades gradually outwards into subeconomic mineralisation and the boundaries of the orebody are assay limits. They are, therefore, often irregular in form and may cut across geological boundaries. The

overall shapes of some are cylindrical, others are caplike, whilst the mercury-bearing stockworks of Dubnik in Slovakia are sometimes pear-shaped.

Stockworks most commonly occur in porphyritic acid to intermediate plutonic igneous intrusions, but they may cut across the contact into the country rocks, and a few are wholly or mainly in the country rocks. Disseminated deposits produce most of the world's copper and molybdenum (porphyry coppers and disseminated molybdenums) and they are also of some importance in the production of tin, gold, silver (see Chapter 16), mercury, and uranium. Porphyry coppers form some of the world's monster orebodies. Grades are generally 0.4–1.5% Cu and tonnages 50–5000 Mt.

Irregular replacement deposits. Many ore deposits have been formed by the replacement of pre-existing rocks, particularly carbonate-rich sediments, e.g. magnesite deposits. These replacement processes often occurred at high temperatures, at contacts with medium-sized to large igneous intrusions. Such deposits have therefore been called contact metamorphic or pyrometasomatic; however, *skarn* is now

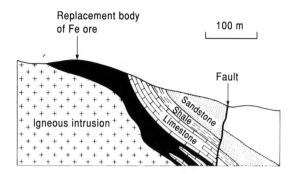

FIG. 3.5   Skarn deposit at Iron Springs, Utah. (After Gilluly et al. 1959.)

the preferred and more popular term. The orebodies are characterized by the development of calc-silicate minerals such as diopside, wollastonite, andradite garnet, and actinolite. These deposits are extremely irregular in shape (Fig. 3.5); tongues of ore may project along any available planar structure – bedding, joints, faults, etc., and the distribution within the contact aureole is often apparently capricious. Structural changes may cause abrupt termina-

tion of the orebodies. The principal materials produced from skarn deposits are iron, copper, tungsten, graphite, zinc, lead, molybdenum, tin, uranium, and talc.

Concordant orebodies

Sedimentary host rocks

Concordant orebodies in sediments are very important producers of many different metals, being particularly important for base metals and iron, and are of course concordant with the bedding. They may be an integral part of the stratigraphical sequence, as is the case with Phanerozoic ironstones – or they may be epigenetic infillings of pore spaces or replacement orebodies. Usually these orebodies show a considerable development in two dimensions, i.e. *parallel* to the bedding and a limited development *perpendicular* to it (Fig. 3.6), and for this reason such deposits are referred to as stratiform. This term must not be confused with strata-bound, which refers to any type or types of orebody, concordant or discordant, which are restricted to a particular part of the

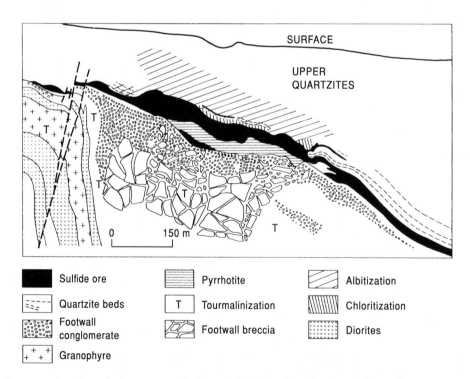

FIG. 3.6   Cross-section through the ore zone, Sullivan Mine, British Columbia. (After Sangster & Scott 1976.)

stratigraphical column. Thus the veins, pipes, and flats of the Southern Pennine orefield of England can be designated as strata-bound, as they are virtually restricted to the Carboniferous limestone of that region. A number of examples of concordant deposits which occur in different types of sedimentary rocks will be considered.

Limestone hosts. Limestones are very common host rocks for base metal sulfide deposits. In a dominantly carbonate sequence ore is often developed in a small number of preferred beds or at certain sedimentary interfaces. These are often zones in which the permeability has been increased by dolomitization or fracturing. When they form only a minor part of the stratigraphical succession, limestones, because of their solubility and reactivity, can become favorable horizons for mineralisation. For example the lead–zinc ores of Bingham, Utah, occur in limestones which make up 10% of a 2300 m succession mainly composed of quartzites.

Argillaceous hosts. Shales, mudstones, argillites, and slates are important host rocks for concordant orebodies which are often remarkably continuous and extensive. In Germany, the Kupferschiefer of the Upper Permian is a prime example. This is a copper-bearing shale a meter or so thick which, at Mansfeld, occurred in orebodies which had plan dimensions of 8, 16, 36 and 130 km². Mineralisation occurs at exactly the same horizon in Poland, where it is being worked extensively, and across the North Sea in north-eastern England, where it is subeconomic.

The world's largest, single lead–zinc orebody occurs at Sullivan, British Columbia. The host rocks are late Precambrian argillites. Above the main orebody (Fig. 3.6) there are a number of other mineralized horizons with concordant mineralisation. This deposit appears to be syngenetic and the lead, zinc and other metal sulfides form an integral part of the rocks in which they occur. The orebody occurs in a single, generally conformable zone 60–90 m thick and runs 6.6% Pb and 5.9% Zn. Other metals recovered are silver, tin, cadmium, antimony, bismuth, copper, and gold. This orebody originally contained at least 155 Mt of ore.

Other good examples of concordant deposits in argillaceous rocks, or slightly metamorphosed equivalents, are the lead–zinc deposits of Mount Isa, Queensland, many of the Zambian Copperbelt deposits, and the copper shales of the White Pine Mine, Michigan.

Arenaceous hosts. Not all the Zambian Copperbelt deposits occur in shales and metashales. Some bodies occur in altered feldspathic sandstones such as Mufulira, which consists of three extensive lenticular orebodies stacked one above the other and where the ore reserves in 1974 stood at 282 Mt assaying 3.47% Cu. The largest orebody has a strike length of 5.8 km and extends several kilometers down dip. Many other concordant sandstone-hosted orebodies occur around the world, such as those in desert sands (red bed coppers), which are very important in China where they make up nearly 21% of the stratiform copper reserves of that country (Chen 1988).

Many mechanical accumulations of high density minerals such as magnetite, ilmenite, rutile, and zircon occur in arenaceous hosts, usually taking the form of layers rich in heavy minerals in Pleistocene and Holocene sands. As the sands are usually unlithified, the deposits are easily worked and no costly crushing of the ore is required. These orebodies belong to the group called placer deposits. Beach sand placers supply much of the world's titanium, zirconium, thorium, cerium, and yttrium. They occur along present-day beaches or ancient beaches where longshore drift is well developed and frequent storms occur. Economic grades can be very low and sands running as little as 0.6% heavy minerals are worked along Australia's eastern coast.

Rudaceous hosts. Alluvial gravels and conglomerates also form important recent and ancient placer deposits. Alluvial gold deposits are often marked by "white runs" of vein quartz pebbles as in the White Channels of the Yukon, the White Bars of California, and the White Leads of Australia. Such deposits form one of the few types of economic placer deposits in fully lithified rocks, and indeed the majority of the world's gold is won from Precambrian deposits of this type in South Africa (see

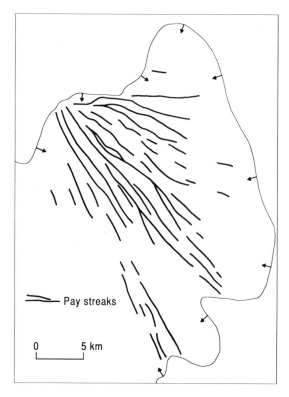

FIG. 3.7 Distribution of pay streaks (gold orebodies) in the Main Leader Reef in the East Rand Basin of the Witwatersrand Goldfield of South Africa. The arrows indicate the direction of dip at the outcrop or suboutcrop. For the location of this ore district see Fig. 14.2. (After Du Toit 1954.)

Chapter 14). Figure 3.7 shows the distribution of the gold orebodies in the East Rand Basin where the vein quartz pebble conglomerates occur in quartzites of the Witwatersrand Supergroup. Their fan-shaped distribution strongly suggests that they occupy distributary channels. Uranium is recovered as a byproduct of the working of the Witwatersrand goldfields. In the very similar Blind River area of Ontario uranium is the only metal produced.

Chemical sediments. Sedimentary iron and manganese formations and evaporites occur scattered through the stratigraphical column where they form very extensive beds conformable with the stratigraphy.

Igneous host rocks

Volcanic hosts. The most important deposit type in volcanic rocks is the volcanic-associated massive sulfide (see Chapter 15) or oxide type. The sulfide variety often consists of over 90% iron sulfide usually as pyrite. They are generally stratiform bodies, lenticular to sheetlike (Fig. 3.8), developed at the interfaces between volcanic units or at volcanic–sedimentary interfaces. With increasing magnetite content, these sulfide ores grade into massive oxide ores of magnetite and/or hematite such as Savage River in Tasmania, Fosdalen in Norway, and Kiruna in Sweden

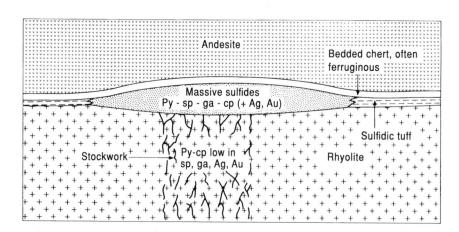

FIG. 3.8 Schematic section through an idealized volcanic-associated massive sulfide deposit showing the underlying feeder stockwork and typical mineralogy. Py, pyrite; sp, sphalerite; ga, galena; cp, chalcopyrite.

(Solomon 1976). They can be divided into three classes of deposit: (a) zinc–lead–copper, (b) zinc–copper, and (c) copper. Typical tonnages and copper grades are 0.5–60 Mt and 1–5%, but these are commonly polymetallic deposits often carrying other base metals and significant precious metal values which make them plum targets for exploration, e.g. Neves-Corvo (see section 1.2.3, "Metal and mineral prices").

The most important host rock is rhyolite and lead-bearing ores are only associated with this rock type. The copper class is usually, but not invariably, associated with mafic volcanics. Massive sulfide deposits commonly occur in groups and in any one area they are found at one or a restricted number of horizons within the succession (see section 15.2.5). These horizons may represent changes in composition of the volcanic rocks, a change from volcanism to sedimentation, or simply a pause in volcanism. There is a close association with volcaniclastic rocks and many orebodies overlie the explosive products of rhyolite domes. These ore deposits are usually underlain by a stockwork that may itself be ore grade and which appears to have been the feeder channel up which mineralizing fluids penetrated to form the overlying massive sulfide deposit. All these relationships are of great importance in the search for this orebody type.

Plutonic hosts. Many plutonic igneous intrusions possess rhythmic layering and this is particularly well developed in some basic intrusions. Usually the layering takes the form of alternating bands of mafic and felsic minerals, but sometimes minerals of economic interest such as chromite, magnetite, and ilmenite may form discrete mineable seams within such layered complexes. These seams are naturally stratiform and may extend over many kilometers, as is the case with the chromite seams in the Bushveld Complex of South Africa and the Great Dyke of Zimbabwe.

Another form of orthomagmatic deposit is the nickel–copper sulfide orebody formed by the sinking of an immiscible sulfide liquid to the bottom of a magma chamber containing ultrabasic or basic magma. These are known as liquation deposits and they may be formed in the bottom of lava flows as well as in plutonic intrusions. The sulfide usually accumulates in

hollows in the base of the igneous body and generally forms sheets or irregular lenses conformable with the overlying silicate rock. From the base upwards, massive sulfide gives way through disseminated sulfides in a silicate gangue to lightly mineralized and then barren rock (Fig. 3.9).

Metamorphic host rocks
Apart from some deposits of metamorphic origin such as the irregular replacement deposits already described and deposits generated in contact metamorphic aureoles – e.g. wollastonite, andalusite, garnet, graphite – metamorphic rocks are important for the metamorphosed equivalents of deposits that originated in sedimentary and igneous rocks and which have been discussed above.

Residual deposits
These are deposits formed by the removal of nonore material from protore (rock in which an initial but uneconomic concentration of minerals is present that may by further natural processes be upgraded to form ore). For example, the leaching of silica and alkalis from a nepheline–syenite may leave behind a surface capping of hydrous aluminum oxides (bauxite). Some residual bauxites occur at the present surface, others have been buried under younger sediments to which they form conformable basal beds. The weathering of feldspathic rocks (granites, arkoses) can produce important kaolin deposits which, in the Cornish granites of England, form funnel or trough-shaped bodies extending downwards from the surface for as much as 230 m.

Other examples of residual deposits include some laterites sufficiently high in iron to be worked and nickeliferous laterites formed by the weathering of peridotites.

3.2 WALL ROCK ALTERATION

Many ore deposits, particularly the epigenetic ones, may have beside or around them a zone or zones of wall rock alteration. This alteration of the host rock is marked by color, textural, mineralogical or chemical changes or any combination of these. The areal extent of the alteration can vary considerably, sometimes being

Surface

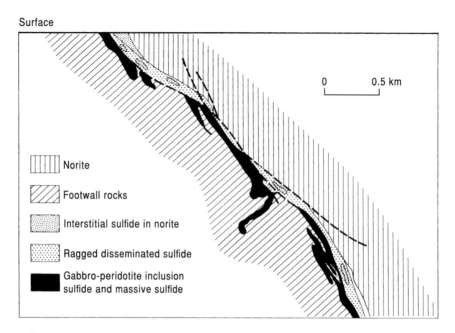

0   0.5 km

||||| Norite

//// Footwall rocks

Interstitial sulfide in norite

Ragged disseminated sulfide

Gabbro-peridotite inclusion
sulfide and massive sulfide

FIG. 3.9   Generalized section through the Creighton ore zone, Sudbury, Ontario, looking west. (After Souch et al. 1969.)

limited to a few centimeters on either side of a vein, at other times forming a thick halo around an orebody and then, since it widens the drilling target, it may be of considerable exploration value. Hoeve (1984) estimated that the drilling targets in the uranium field of the Athabasca Basin in Saskatchewan are enlarged by a factor of 10–20 times by the wall rock alteration. The *Atlas of Alteration* by Thompson and Thompson (1996) is a very well illustrated place to start your study.

3.3   GEOLOGICAL MODELS OF MINERAL DEPOSITS

One of the aims of the planning stage (see Chapter 4) is to identify areas for reconnaissance and to do this we must have some idea of how the materials sought relate to geological factors including geophysics and geochemistry. This is best achieved by setting up a model or models of the type of deposit sought. But what is a model? The term has been defined in various ways but a useful one is that of "functional idealization of a real world situation used to aid in the analysis of a problem." As such it is a

synthesis of available data and should include the most informative and reliable characteristics of a deposit type, identified on a variety of scales and including definition of the average and range of each characteristic (Adams 1985). It is therefore subject to uncertainty and change; each new discovery of an example of a deposit type should be added to the data base.

In mineral deposit models there are two main types which are often combined; the empirical model based on deposit descriptions and a genetic model which explains deposits in term of causative geological processes. The genetic model is necessarily more subjective but can be more powerful, as it can predict deposits not contained in the descriptive data base. Another type of model which is extremely useful for preliminary economic evaluations is a grade-tonnage model. This accumulates grade and tonnage data for known deposits and from this it is possible to estimate the size and grade of an average or large deposit and the cash flow if one were found. Examples of this type of modeling are given by Gorman (1994) for South American gold and copper deposits. Cost curves can also be calculated for differing deposit types. Examples for copper (Fig. 3.10) show the low cost of

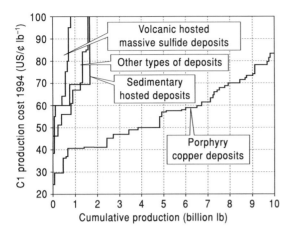

FIG. 3.10   Cost curves for copper production. The C1 cost is the cash operating cost. Note the predominance of porphyry copper production. (After Moore 1998.)

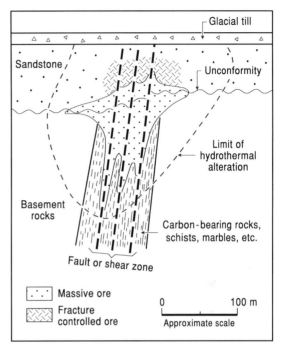

FIG. 3.11   Generalized diagram of an unconformity-associated uranium deposit. (After Clark et al. 1982.)

producing from porphyry copper deposits and the economies of scale from large production relative to sedimentary copper and volcanic-associated massive sulfide deposits.

To understand how geological models are constructed we will examine the deposit model for unconformity-related uranium deposits. These deposits are currently the main source of high grade uranium in the Western world and mainly occur in two basins; the Athabasca Basin of Northern Saskatchewan and the Alligator River area of the Northern Territory in Australia. Good accounts of most deposit models can be found in two publications by North American Geological Surveys, Cox and Singer (1986) and Eckstrand (1984), later revised as Eckstrand et al. (1995). Cox and Singer adopt a strictly descriptive approach, based on their classification of mineral deposits (see section 3.6), but also give grade-tonnage curves and geophysical signatures whereas Eckstrand is more succinct but gives brief genetic models. Some major deposit types are also dealt with in depth by Roberts and Sheahan (1988), Kirkham et al. (1993), and a special edition of the Australian Geological Survey Organisation journal (AGSO 1998). Internet versions of the USGS deposit models and a more extensive classification by the British Columbia Geological Survey are available at their web sites (BCGS 2004, USGS 2004).

Table 3.1 shows the elements that were used by Eckstrand and Cox and Singer in their unconformity-related deposit models. Both accounts agree that the deposits occur at or near the unconformable contact between regionally metamorphosed Archaean to Lower Proterozoic basement and Lower to Middle Proterozoic (1900–1200 Ma) continental clastic sediments, as shown in Fig. 3.11. The details of the controls on mineralisation are more subjective; the Canadian deposits occur both in the overlying sandstones and basement whereas the Australian deposits are almost exclusively in the basement. Most deposits in Canada are spatially associated very closely with graphitic schists whereas the Australian deposits are often in carbonates, although many are carbonaceous. Both models agree that the key alteration is chloritization together with sericitization, kaolinization, and hematitization along the intersection of faults and the unconformity. Generalization on the mineralogy of the deposits is difficult but the key mineral is pitchblende with lesser amounts of the

TABLE 3.1 Summary of elements used in the models for unconformity-related uranium deposits by Grauch and Moiser in Cox and Singer (1986) and Tremblay and Ruzicka in Eckstrand (1984).

Element	Cox & Singer	Eckstrand
Synonym: Commodities (Eckstrand)	Vein-like type U	U (Ni, Co, As, Se, Ag, Au, Mo)
Description	U mineralisation in fractures and breccia fills in metapelites, metapsammites, and quartz arenites below, across, and above an unconformity separating early and middle Proterozoic rocks	
Geological environment		
Rock types	Regionally metamorphosed carbonaceous pelites, psammites, and carbonates. Younger unmetamorphosed quartz arenites	Ores occur in clay sericite and chlorite masses at the unconformity and along intersecting faults and in clay-altered basement rocks and kaolinized cover rocks
Textures	Metamorphic foliation and later brecciation	
Age range	Early and middle Proterozoic affected by Proterozoic regional metamorphism	Most basement rocks are Aphebian, some may be Archaean. Helikian cover rocks. Ore: Helikian (1.28 ± 0.11 Ga)
Depositional environment: Associated rocks (Eckstrand)	Host rocks are shelf deposits and overlying continental sandstone	Basement: graphitic schist and gneiss, coarse-grained granitoids, calc silicate metasediments. Cover rocks: sandstone and shale
Tectonic setting (Cox & Singer) Geological setting (Eckstrand)	Intracratonic sedimentary basins on the flanks of Archaean domes. Tectonically stable since the Proterozoic	Relatively undeformed, intracratonic, Helikian sedimentary basin resting unconformably on intensely deformed Archaean and Aphebian basement. Deposits are associated with unconformity where intersected by faults. Palaeoregolith present in Saskatchewan
Associated deposit types	Gold- and nickel-rich deposits may occur	
Form of deposit		Flattened cigar shaped, high grade bodies oriented and distributed along the unconformity and faults, especially localized at unconformity–fault intersections. High grade bodies grade outwards into lower grade
Deposit description Mineralogy	Pitchblende + uraninite ± coffinite ± pyrite ± galena ± sphalerite ± arsenopyrite ± niccolite. Chlorite + quartz + calcite + dolomite + hematite + siderite + sericite. Late veins contain native gold, uranium, and tellurides. Latest quartz–calcite veins contain pyrite, chalcopyrite, and bituminous matter	Pitchblende coffinite, minor uranium oxides. Ni and Co arsenides and sulfides, native selenium and selenides, native gold and gold tellurides, galena, minor molybdenite, Cu and Fe sulfides, clay minerals, chlorite, quartz, graphite, carbonate

TABLE 3.1   (continued)

Element	Cox & Singer	Eckstrand
Texture and structure	Breccias, veins, and disseminations. Uranium minerals coarse and colloform. Latest veins have open space filling textures	
Alteration	Multistage chloritization dominant. Local sercitization, hematization, kaolinization, and dolomitization. Vein silicification in alteration envelope. Alteration enriched in Mg, F, REE, and various metals. Alkalis depleted	
Ore controls	Fracture porosity controlled ore distribution in metamorphics. Unconformity acted as disruption in fluid flow but not necessarily ore locus	1 At or near unconformity between Helikian sandstone and Archaean basement 2 Intersection of unconformity with reactivated basement faults 3 Associated with graphitic basement rocks and gray and/or multicolored shale and sandstone cover 4 Intense clay, chlorite, and sericite alteration of basement and cover rocks 5 Basement rocks with higher than average U content
Weathering	Various secondary U minerals	
Geochemical and geophysical Signature (Cox & Singer)	Increase in U, Mg, P, and locally Ni, Cu, Pb, Zn, Co, As; decrease in SiO_2. Locally Au with Ag, Te, Ni, Pd, Re, Mo, Hg, REE, and Rb. Anomalous radioactivity. Graphitic schists in some deposits are strong EM conductors	
Genetic model		Combinations of: 1 Preconcentration of U during deposition of Aphebian sediments and their anatexis 2 Concentration in lateritic regolith (Helikian) 3 Mobilization by heated oxidized solutions and precipitation in reducing environment at unconformity and fault locus 4 Additional cycles of mobilization and precipitation leading to redistribution of uranium
Examples	Rabbit Lake, Saskatchewan Cluff Lake Saskatchewan	Key Lake, Rabbit Lake, Cluff Lake, Canada. Jabiluka I and II, Ranger, N.T., Australia
Importance		Canada: 35% of current U production but 50% of reserves World: 15% of reserves
Typical grade and tonnage	see Fig. 3.12	Canada: small to 5 Mt of 0.3–3% U. Australia: maximum 200,000 t of contained U but grade lower than Canada

uranium silicate, coffinite. The more contentious issue is the associated minerals; in Canada high grade nickel arsenides are common, but by no means ubiquitous, whereas the mineralogy of the Australian deposits is simpler. However, gold is more common and may be sufficiently abundant to change the deposit model to unconformity-related uranium–gold. Selenides and tellurides are present in some deposits, although Cox and Singer omit the former but mention enrichment in palladium.

The genetic model for these deposits includes elements of the following:

1 Preconcentration of uranium and associated elements in basement sedimentary rocks.
2 Concentration during the weathering of the basement prior to the deposition of the overlying sediments.
3 Mobilization of uranium and the associated elements by oxidizing fluids and precipitation in a reducing environment at fault–unconformity intersections.
4 Additional cycles of oxidation and mobilization.

As an example of the differences between the descriptive and genetic models, the descriptive model would suggest exploration for graphitic conductors whereas the genetic model would broaden the possible depositional sites to any reducing environment. Descriptive and genetic models can be combined with the grade-tonnage curves of Cox and Singer (Fig. 3.12) to suggest the probable economic benefits. It should be noted that high tonnage does not necessarily mean low grade, and large deposits such as Cigar Lake (1.47 Mt of 11.9% U_3O_8) in Saskatchewan have grade and tonnage that rank in the top 10% of all deposits.

The history of exploration in the two areas clearly illustrates the uses and limitations of models. Unconformity vein deposits were first discovered as a result of exploration in known uraniferous areas of Northern Canada (Fig. 3.13) and Australia in the late 1960s. The initial Canadian discoveries were made by a French company which was exploring for uranium veins around the Beaverlodge deposit. These have no obvious relation to the mid Proterozoic sandstones and it was only the company's previous experience in Gabon and minor known occurrences near the unconformity which suggested that the sandstones might be miner-

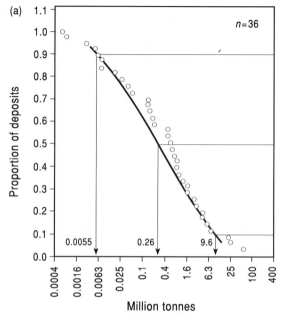

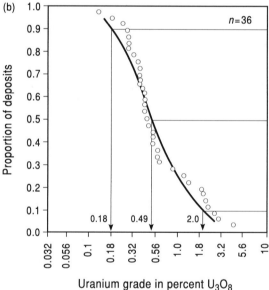

FIG. 3.12   (a,b) Grade–tonnage curves for unconformity associated deposits. (From Cox & Singer 1986.)

alized (Tona et al. 1985). On this premise the company (Amok) conducted an airborne radiometric survey which led them to boulders of mineralized sandstone and massive pitchblende close to the source deposit (Cluff Lake

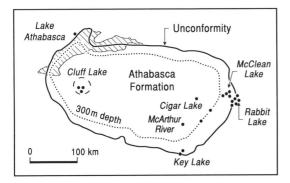

FIG. 3.13 Outline of the Athabasca Basin and distribution of the associated uranium deposits. (Modified after Clark et al. 1982.)

A). At roughly the same time a small Canadian company explored the eastern edge of the basin knowing that they had an aircraft available and that the sandstones might be prospective by comparison with the much younger Western US deposits (Reeves & Beck 1982). They also discovered glacial boulders which led them back to the deposit (Rabbit Lake). At roughly the same time (1969) in Australia airborne surveys in the East Alligator Valley detected major anomalies at Ranger, Koongarra, and Narbalek. The area selection was initially based on the similarity of the geology with known producing areas in South Alligator Valley, with acid volcanics as the postulated source rock, and by a re-interpretation of granites, that were previously thought to be intrusive, as basement gneiss domes (Dunn et al. 1990b). When drilling started it became clear that the mineralisation was closely associated with the unconformity. Thus the overlying sandstones, rather than being responsible for the erosion of the deposits or being a barren cover, became a target. In Canada the Key Lake and Midwest Lake deposits were discovered as a result of this change in model (Gatzweiler et al. 1981, Scott 1981). The exploration model has been further developed in Canada, to the point at which deposits with no surface expression (blind deposits), such as Cigar Lake and McArthur River, have been found by deep drilling of electromagnetic conductors caused by graphitic schists (McMullan et al. 1989). Similar developments could have been anticipated in Australia but for government control on uranium

exports, which inhibited uranium exploration. The importance of recognizing the elements associated with this type of deposit became obvious in the late 1980s with the discovery of the Coronation Hill gold–palladium deposit in the Northern Territory and similar prospects in Saskatchewan. Unfortunately for the company concerned Coronation Hill is on the edge of the Kakadu National Park and will not be mined in the foreseeable future.

The application of a particular deposit model will depend on the quality of the database and should not be regarded as a panacea for the exploration geologist. Some deposit types such as placer gold are easy to understand and have well-developed models, whereas others such as Besshi-style massive sulfides or the Olympic Dam models are not well developed and may be represented by a single deposit on which information is difficult to obtain.

The major pitfalls of using models are ably illustrated in cartoon form by Hodgson (1989). He identifies a number of problems which he likens to religious cults:

1 The cult of the fad or fashion. An obsession with being up to date and in possession of the newest model.
2 The cult of the panacea. The attitude that one model is the ultimate and will end all controversy.
3 The cult of the classicists. All new ideas are rejected as they have been generated in the hot house research environment.
4 The cult of the corporate iconoclasts. Only models generated within an organization are valid, all outside models are wrong.
5 The cult of the specialist. In which only one aspect of the model is tested and usually not in the field.

The onus is thus on the exploration geologist to ensure that a model is correctly applied and not to exclude deviations from the norm. Other examples of the application of models are given in Chapters 12–17.

3.4 FURTHER READING

Longer discussions of the subjects forming the main sections of this chapter can be found in Robb *Introduction to Ore-Forming Processes* (2004). Kesler's *Mineral Resources, Economics, and the Environment* (1994) provides a good introduction and a link with economics. Sawkins' *Metal Deposits in Relation to Plate Tectonics* (second edition 1990, Springer Verlag) gives an overview of the major structural controls on deposits.

3.5 SUMMARY OF CHAPTERS 1, 2, AND 3

The general nature of ore, industrial minerals, and orebodies are discussed in sections 1.1 and 1.2 and it is emphasized that these deposits must contain valuable constituents that can be economically recovered using suitable treatment. (Chapter 11 is largely devoted to a detailed coverage of this principle.) In the past too much interest in many mining circles has been placed on the more glamorous metallic deposits and the general importance of industrial minerals neglected (see section 1.2.2). The value of both is governed by demand and supply (see section 1.2.3) and many factors including government action, recycling, and substitution play a part in determining their market prices. In section 1.3 the principal steps involved in the exploration for, and development of, a mineral deposit are summarized and these are the subjects that are covered in more detail in the rest of this book. A sound knowledge of these is necessary in choosing exploration areas (see section 1.5.) leading to the development of a rationale of mineral exploration (see section 1.6).

Economic mineral deposits are extremely variable in their mineralogy and grade and the geologist must be able to identify, or have identified for him or her, *every mineral* in a possible orebody (see Chapter 2). This will help them to assess the full economic potential of any material to be mined from it and prevent them overlooking the presence of additional valuable constituents or deleterious substances that may render the deposit unworkable. There are many techniques now available for comprehensive mineralogical examination of mineral samples and these are discussed in section 2.2. However, such investigations must be quantitative as well as qualitative; grain size and shape, relative mineral amounts, and the manner of interlocking must be determined.

The nature and morphology of mineral deposits are very varied and only a restricted coverage of these subjects can be given in this book (see section 3.1). The tyro is therefore strongly recommended to acquire a broad knowledge of these subjects from extended reading so that when he or she detects signs of mineralisation they may soon develop a working hypothesis of the nature of the particular beast they have come upon.

Having recognized the tectonic setting of an exploration region, models of the deposits likely to be present, and particularly of those being sought, should be set up. Both empirical and genetic models are used, often in combination, to show how the materials sought relate to geological factors including geophysics and geochemistry. To understand how models are constructed and used, unconformity-related uranium deposits are discussed in some detail (see section 3.2.3).

3.6 APPENDIX: MINERAL DEPOSIT MODEL CLASSIFICATION OF COX AND SINGER (1986). ONLINE LINKS CAN BE FOUND AT USGS (2004).

Mafic and ultramafic intrusions

A. Tectonically stable area; stratiform complexes

Stratiform deposits
  Basal zone
    Stillwater Ni-Cu   1
  Intermediate zone
    Bushveld chromitite   2a
    Merensky Reef PGE   2b
  Upper zone
    Bushveld Fe-Ti-V   3

Pipe-like deposits
    Cu-Ni pipes   4a
    PGE pipes   4b

B. Tectonically unstable area

Intrusions same age as volcanic rocks
    Rift environment
        Duluth Cu-Ni-PGE   5a
        Noril'sk Cu-Ni-PGE   5b
    Greenstone belt in which lowermost rocks of sequence contain ultramafic rocks
        Komatiitic Ni-Cu   6a
        Dunitic Ni-Cu   6b
Intrusions emplaced during orogenesis
    Synorogenic in volcanic terrane
        Synorogenic-synvolcanic Ni-Cu   7a
    Synorogenic intrusions in nonvolcanic terrane
        Anorthosite-Ti   7b
    Ophiolite
        Podiform chromite   8a
        Major podiform chromite   8b
        (Lateritic Ni)   (38a)
        (Placer Au-PGE)   (39a)
    Serpentine
        Limassol Forest Co-Ni   8c
        Serpentine-hosted asbestos   8d
        (Silica-carbonate Hg)   (27c)
        (Low-sulfide Au-quartz vein)   (36a)
    Cross-cutting intrusions (concentrically zoned)
        Alaskan PGE   9
        (Placer PGE-Au)   (39b)

C. Alkaline intrusions in stable areas

    Carbonatite   10
    Alkaline complexes   11
    Diamond pipes   12

Felsic intrusions

D. Mainly phanerocrystalline textures

Pegmatitic
    Be-Li pegmatites   13a
    Sn-Nb-Ta pegmatites   13b
Granitic intrusions
    Wallrocks are calcareous
        W skarn   14a
        Sn skarn   14b
        Replacement Sn   14c
    Other wallrocks
        W veins   15a
        Sn veins   15b

Sn greisen	15c
(Low-sulfide Au-quartz vein)	(36a)
(Homestake Au)	(36c)
Anorthosite intrusions	
(Anorthosite Ti)	(7b)

E. Porphyroaphanitic intrusions present

High-silica granites and rhyolites	
Climax Mo	16
(Fluorspar deposits)	(26b)
Other felsic and mafic rocks including alkalic	
Porphyry Cu	17
Wallrocks are calcareous	
Deposits near contact	
Porphyry Cu, skarn-related	18a
Cu skarn	18b
Zn-Pb skarn	18c
Fe skarn	18d
Carbonate-hosted asbestos	18e
Deposits far from contact	
Polymetallic replacement	19a
Replacement Mn	19b
(Carbonate-hosted Au)	(26a)
Wallrocks are coeval volcanic rocks	
In granitic rocks in felsic volcanics	
Porphyry Sn	20a
Sn-polymetallic veins	20b
In calcalkalic or alkalic rocks	
Porphyry Cu-Au	20c
(Epithermal Mn)	(25g)
Wallrocks are older igneous and sedimentary rocks	
Deposits within intrusions	
Porphyry Cu-Mo	21a
Porphyry Mo, low-F	21b
Porphyry W	21c
Deposits within wallrocks	
Volcanic hosted Cu-As-Sb	22a
Au-Ag-Te veins	22b
Polymetallic veins	22c
(Epithermal quartz-alunite Au)	(25e)
(Low-sulfide Au-quartz vein)	(36a)

Extrusive rocks

F. Mafic extrusive rocks

Continental or rifted craton	
Basaltic Cu	23
(Sediment-hosted Cu)	(30b)
Marine, including ophiolite-related	
Cyprus massive sulfide	24a

Besshi massive sulfide	24b
Volcanogenic Mn	24c
Blackbird Co-Cu	24d
(Komatiitic Ni-Cu)	(6a)

G. Felsic-mafic extrusive rocks

Subaerial
  Deposits mainly within volcanic rocks

Hot-spring Au-Ag	25a
Creede epithermal vein	25b
Comstock epithermal vein	25c
Sado epithermal vein	25d
Epithermal quartz-alunite Au	25e
Volcanogenic U	25f
Epithermal Mn	25g
Rhyolite-hosted Sn	25h
Volcanic-hosted magnetite	25i
(Sn polymetallic veins)	(20b)

  Deposits in older calcareous rocks

Carbonate-hosted Au-Ag	26a
Fluorspar deposits	26b

  Deposits in older elastic sedimentary rocks

Hot-spring Hg	27a
Almaden Hg	27b
Silica-carbonate Hg	27c
Simple Sb	27d

Marine

Kuroko massive sulfide	28a
Algoma Fe	28b
(Volcanogenic Mn)	(24c)
(Volcanogenic U)	(25f)
(Low-sulfide Au-quartz vein)	(36a)
(Homestake Au)	(36b)
(Volcanogenic U)	(25f)

Sedimentary rocks

H. Clastic sedimentary rocks

Conglomerate and sedimentary breccia

Quartz pebble conglomerate Au-U	29a
Olympic Dam Cu-U-Au	29b
(Sandstone U)	(30c)
(Basaltic Cu)	(23)

Sandstone

Sandstone-hosted Pb-Zn	30a
Sediment-hosted Cu	30b
Sandstone U	30c
(Basaltic Cu)	(23)
(Kipushi Cu-Pb-Zn)	(32c)
(Unconformity U-Au)	(37a)

Shale-siltstone
   Sedimentary exhalative Zn-Pb   31a
   Bedded barite   31b
   Emerald veins   31c
   (Basaltic Cu)   (23)
   (Carbonate-hosted Au-Ag)   (26a)
   (Sediment-hosted Cu)   (30b)

I. Carbonate rocks

No associated igneous rocks
   Southeast Missouri Pb-Zn   32a
   Appalachian Zn   32b
   Kipushi Cu-Pb-Zn   32c
   (Replacement Sn)   (14c)
   (Sedimentary exhalative Zn-Pb)   (31a)
   (Karst bauxite)   (38c)
Igneous heat sources present
   (Polymetallic replacement)   (19a)
   (Replacement Mn)   (19b)
   (Carbonate-hosted Au-Ag)   (26a)
   (Fluorspar deposits)   (26b)

J. Chemical sediments

Oceanic
   Mn nodules   33a
   Mn crusts   33b
Shelf
   Superior Fe   34a
   Sedimentary Mn   34b
   Phosphate, upwelling type   34c
   Phosphate, warm-current type   34d
Restricted basin
   Marine evaporate   35a
   Playa evaporate   35b
   (Sedimentary exhalative Zn-Pb)   (31a)
   (Sedimentary Mn)   (34b)

REGIONALLY METAMORPHOSED ROCKS

K. Derived mainly from eugeosynclinal rocks

   Low-sulfide Au-quartz vein   36a
   Homestake Au   36b
   (Serpentine-hosted asbestos)   (8d)
   (Gold on flat faults)   (37b)

L. Derived mainly from pelitic and other sedimentary rocks

   Unconformity U-Au   37a
   Gold on flat faults   37b

Surficial and unconformity-related

M. Residual

Lateritic Ni	38a
Bauxite, laterite type	38b
Bauxite, karst type	38c
(Unconformity U-Au)	(37a)

N. Depositional

Placer Au-PGE	39a
Placer PGE-Au	39b
Shoreline placer Ti	39c
Diamond placers	39d
Stream placer Sn	39e
(Quartz pebble conglomerate Au-U)	(29a)

4

RECONNAISSANCE EXPLORATION

CHARLES J. MOON AND MICHAEL K.G. WHATELEY

Exploration can be divided into a number of interlinked and sequential stages which involve increasing expenditure and decreasing risk (Fig. 4.1). The terminology used to describe these stages is highly varied. The widely accepted terms used for the early stages of exploration are planning and reconnaissance phases. These phases cover the stages leading to the selection of an area for detailed ground work; this is usually the point at which land is acquired. The planning stage covers the selection of commodity, type of deposit, exploration methods, and the setting up of an exploration organization. The process of selecting drill targets within

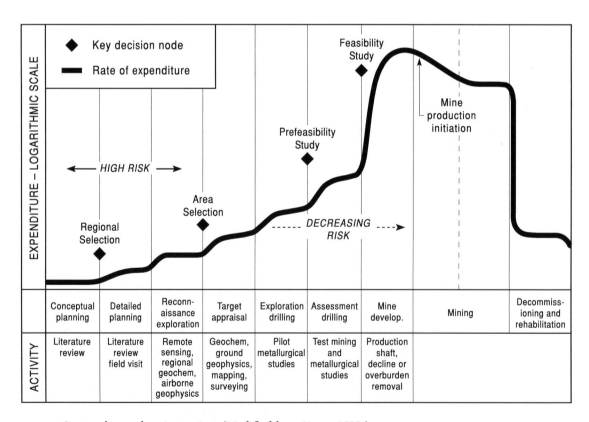

FIG. 4.1   Stages of an exploration project. (Modified from Eimon 1988.)

license blocks we term target selection and that of drilling, target testing (discussed in Chapter 5). The deposit is then at the stage of pre-development followed by a feasibility study (Chapter 11). Before we consider how exploration is planned we should discuss who explores.

The exploration players

Private sector
Most mineral exploration in developed countries is conducted by companies with a substantial capital base generated either from existing mineral production or from investors on stock markets. The size of the company can range from major multinational mining companies, such as Rio Tinto plc or Anglo American Corporation with operations on several continents, to small venture capital companies, usually known as a junior company, with one or two geologists. Exploration by individual prospectors has been an important factor in countries with large unexplored areas and liberal land tenure laws, e.g. Canada, Australia and Brazil. Here you can still meet the grizzled prospector or garimpeiro. Although they usually lack the sophisticated training of the corporate geologist, this can be compensated for by a keen eye and the willingness to expend a little boot leather.

State organizations
In more centrally directed economies, most exploration is carried out by state run companies, geological surveys, and often in the case of developing countries, international aid organizations. The role of a geological survey usually includes some provision of information on mineral exploration to government and the private sector. Usually this takes the form of reconnaissance work. By contrast, the Soviet Ministry of Geology had until 1991 exclusive prospecting rights in the former USSR, although exploration was carried out by a wide variety of organizations at the Union (federal) and republic levels. In China exploration is undertaken by a number of state and provincial groups, including the army.

These two groups essentially explore in similar ways although state enterprises are more constrained by political considerations and need not necessarily make a profit. The remainder of the chapter will be devoted to private sector organizations although much may be applicable to state organizations.

4.1 EXPLORATION PLANNING

Mineral exploration is a long-term commitment and there must be careful planning of a company's long-term objectives. This should take particular regard of the company's resources and the changing environment in which it operates (Riddler 1989). The key factors are:

1 Location of demand for products. This will depend on the areas of growth in demand. For metals the most obvious areas are the industrializing countries of the Pacific Rim, and who are resource deficient. China has assumed particular importance since the late 1990s for a wide range of commodities.

2 Metal prices. Price cycles should be estimated as far as possible and supply and demand forecast (see section 1.2.3).

3 Host country factors. The choice of country for operation is important in an industry which has seen substantial nationalization, such as copper in The Democratic Republic of Congo and Zambia and coal in the UK. The standing of foreign investment, the degree of control permitted, percentage of profits remittable to the home country, and most of all government stability and attitude are important. Other factors are availability of land, security of tenure, and supply of services and skilled labor (see section 1.4).

4 The structure of the mining industry. Barriers to the entry of new producers are competition from existing producers and the need for capital to achieve economies of scale.

These factors combined to encourage exploration companies in the 1980s to concentrate on precious metals in Australia, Canada, and the USA (see section 1.2.3 "Metal and mineral prices"). In the 1990s the net was more widely spread with access possible to the countries of the former USSR and exploration undertaken in countries with high political risk. The late 1990s saw a concentration of exploration activity in countries with proactive mineral policies in South America and selected parts of Africa and Asia. There was a revival in interest in base and ferrous metals in the early 2000s due to the initially lower gold price and boom in demand from China.

After the initial corporate planning, usually by senior executives, an exploration strategy must be chosen, a budget allocated, and desirable deposit type(s) defined.

The choice of exploration strategy varies considerably between companies and depends on the objects of the company and its willingness to take risks. For new entrants into a country the choice is between exploration by acquisition of existing prospects or grass roots (i.e. from scratch) exploration. Acquisition requires the larger outlay of capital but carries lower risk and has, potentially, a shorter lead time to production. Acquisitions of potential small producers are particularly attractive to the smaller company with limited cash flow from existing production. Potential large producers interest larger companies which have the capital necessary to finance a large project. This has been particularly marked in the early years of the twenty-first century in the consolidation of the gold industry into a few large producers that have devoured the medium sized producers. Larger companies tend to explore both by acquisition and by grass roots methods and often find that exploration presence in an area will bring offers of properties ("submittals"). Existing producers have the additional choice of exploring in the immediate vicinity of their mines, where it is likely that they will have substantial advantages in cost saving by using existing facilities. Most of the following section refers to grass roots exploration, although evaluation of potential acquisitions could run in parallel.

4.1.1 Organization

The key to exploration organization is to have the best available staff and adequate finance in order to create confidence throughout the organization (Woodall 1984, Sillitoe 1995, 2000). A number of factors that characterize a successful exploration team have been recognized by Snow and Mackenzie (1981), Regan (1971), and discussed in detail by White (1997).
1 High quality staff and orientation towards people. Successful organizations tend to provide more in-house training.

2 Sound basis of operations. The organization works within corporate guidelines towards objectives.
3 Creative and productive atmosphere. The group encourages independent creative and innovative thinking in an environment free from bureaucratic disruption.
4 High standard of performance, integrity, and ethics.
5 Entrepreneurial acumen. Innovation is fostered in a high risk, high reward environment.
6 Morale and team spirit. High morale, enthusiasm, and a "can do" attitude.
7 The quality of communication is high. The "top brass" are aware of the ideas of geologists.
8 Pre-development group. Successful organizations are more likely to have a specialist group responsible for the transition of a deposit from exploration to development.

All these points make it clear that the management must consist of flexible individuals with considerable experience of exploration.

If these are the optimum characteristics of an exploration group, is there an optimum size and what structure should it have? Studies such as those of Holmes (1977) show that the most effective size is in the range seven to ten geologists; larger organizations tend to become too formalized and bureaucratic, leading to inefficiency, whereas small groups lack the budgets and the manpower to mount a successful program. For the large mining group which wishes to remain competitive while spending a large budget, the solution is to divide its explorationists into semi-autonomous groups.

Exploration groups can be organized on the basis of geographical location or of deposit type. The advantage of having deposit specialists is that in-depth expertise is accumulated; however, the more usual arrangement is to organize by location with geologists in each region forming specialist subgroups. At the reconnaissance stage most work will be carried out in offices located in the national head office or state office, but as exploration focuses in more detail smaller district offices can be opened. A typical arrangement for a large company is shown in Fig. 4.2. In a company with existing production there may be close liaison with mines and engineering divisions.

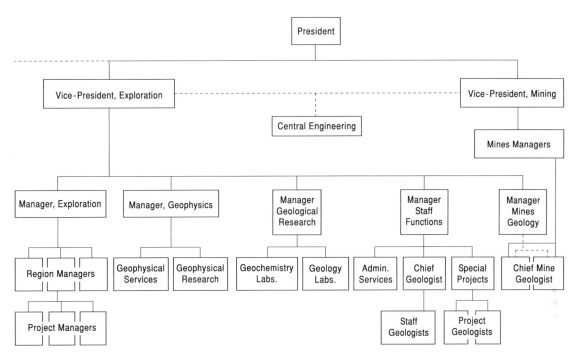

FIG. 4.2 Organization of exploration in a large mining group with producing mines.

4.1.2 Budgets

Exploration costs are considered in two ways: (i) as an expenditure within an organization and (ii) within the context of a project. It is usually the exploration manager that considers the former, but it is as a geologist on a specific exploration project that one becomes involved in the latter.

Corporate exploration expenditure

Finance for corporate exploration is derived from two main sources, revenue from existing production and by selling shares on the stock market. In the first case the company sets aside a percentage of its before-tax profits for exploration. The decision as to the percentage set aside is based upon how much the company wishes to keep as capital, for their running costs, as dividend for their shareholders, and for taxes. Exploration costs may range from 1% to 20% of the annual corporate cash flow but for large diversified companies this is an average 2.5% of sales and 6% for gold companies (Crowson 2003). This may be anything between US$0.5

million to $100 million per annum depending upon the size of the company and the size of the deposit for which they are searching (Table 4.1). For producing companies most exploration can be written off against tax. The smaller company is at a disadvantage in that money must be raised from shareholders. This is easy in times of buoyant share prices. For example, about $C2 billion were raised in the late 1980s for gold exploration in Canada, largely on the Vancouver Stock Exchange. This was aided by flow-through schemes which enabled share purchasers to offset this against tax liabilities. In a similar way, but without tax breaks, about US$100 million was raised on the Irish Stock Exchange from 1983 to 1989 (Gardiner 1989). At the peak of the 1996 boom $1.1 billion was raised in one year on the Vancouver Stock Exchange. This was through initial public offerings of companies and private placements to individuals and investment funds (Hefferman 1998). Although major mining groups should be able to maintain consistent budgets and avoid business cyclicity, there seems little evidence of this. In a study of major groups Eggert (1988) demonstrated that

TABLE 4.1 Corporate Exploration Spending in 2003 (Source: Corporate websites and *Mining Journal*, March 2004.)

Company	US$ million
De Beers	140
Rio Tinto	140
Barrick Gold	110
Newmont	86
Companhia Vale do Rio Doce	81
BHP Billiton	68
Anglogold	63
Anglo American (including platinum)	61
Placer Dome	60
Phelps Dodge (including research)	50
Noranda-Falconbridge	35
Teck-Cominco	30
Inco	27
Gold Fields	23
WMC Resources	20
Newcrest	19
Total spending	~2400

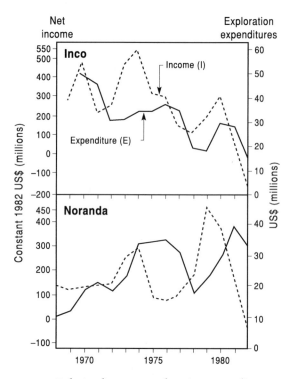

FIG. 4.3 Relation between exploration expenditure and income for two Canadian mining companies. (After Eggert 1988.)

spending was linked to income, and therefore metal prices, but lagged about 18 months behind the changes in income (Fig. 4.3).

The overall budget is then subdivided. For example, a multinational corporation may have several regional exploration centers, each of which may have anything from one or two persons to a fully equipped office with up to 50 people employed. The budget for the latter must include the salaries, equipment, office rentals, and vehicular leases before a single geologist sets foot in the field. Each project would be given a percentage of the regional office's budget, e.g. exploration for coal 20%, base metals 20%, uranium 15%, gold 25%, and industrial minerals 20%. Particular exploration projects then have to compete for funds. If a particular project is successful then it will attract additional spending. This distribution of funds is carried out on a regular basis and is usually coupled with a technical review of exploration projects in which the geologist in charge of a project has to account for money spent and put forward a bid for further funding.

Careful control of funds is essential and usually involves the nomination of budget holders and authorization levels. For example, the

exploration manager may be able to authorize expenditure of $100,000 but a project geologist only $2000, so all drilling accounts will be sent to head office, whereas the geologist will deal with vehicle hire and field expenses. In this case written authorization from the exploration manager would also be required before drilling starts. Summary accounts will be kept in head office under a qualified accountant or bookkeeper and will be subject to regular audit by external accountants. For most companies global expenditure on exploration will be included in annual accounts and will be written off against the year's income.

Project basis

In order to contain costs and provide a basis for future budgeting, costs will be calculated for each exploration project: firstly to monitor spending accurately, and secondly so that the expenses can be written off against production, if exploration is successful, or if the project is sold to another company. Budgets are normally

TABLE 4.2   Example of a budget worksheet used to estimate the annual budget requirement for a project.

Project Name:
Project No:

Month	Jan	Feb	Mar	Apr	May	Jun	Jul	Aug	Sep	Oct	Nov	Dec	Total
Salaries													
Wages													
Drilling													
Transport													
Office Lease													
Administration													
Analytical Costs													
Field Expenses													
Options – New													
Options – Renewals													
Surveys													
Other Contracts													
MONTH TOTALS													

calculated on a yearly basis within which projects can be allocated funds on a monthly basis. At the desk study stage the main costs are salaries of staff and these can be calculated from a nominal staff cost per month, the average of the salaries is usually multiplied by two to cover pensions, housing costs, and secretarial support. Other support costs must be included such as vehicles and helicopter transport in remote areas. Direct costs of exploration are easier to estimate and include geochemistry, remote sensing, and geophysical costs. These can be easily obtained by asking for quotations from contractors. Perhaps a more contentious matter is the allocation of overhead costs to each project; these cover management time and will include the time of head office staff who review projects for presentation to board level management. Often these can be frighteningly large, e.g. the costs of keeping a senior geologist in a metropolitan center will be much larger than in a small town. Table 4.2 shows a worksheet for the calculation of budget costs and Table 4.3 gives an example of a completed budget and typical costs for reconnaissance programs.

4.2 DESK STUDIES AND RECONNAISSANCE

4.2.1 Desk studies

Once the exploration organization is in place, initial finance budgeted, and target deposit type selected, then desk studies can start and areas can be selected for reconnaissance – but on what basis?

The first stage in a totally new program is to acquire information about the areas selected. Besides background information on geology, data on the occurrence of currently producing mines and prospects and their economic status are essential. This information will normally be based largely on published material but could

TABLE 4.3   Examples of budget forecasting for a low cost project and a project requiring drilling. Land costs are excluded.

Project	Cost in US$
Low cost project cost	
Time: 6 days @ US$200 per day	1200
Hotels and meals for 2 nights	350
Transport (vehicle hire)	200
Assays	200
Office staff time	100
TOTAL	2050
Medium cost project	
20 boreholes @ 200 m	
each = 4000 m	
4000 m at US$50 per meter	200,000
Overhead costs	200,000
TOTAL	400,000

also include open-file material from geological surveys and departments of mines, data from colleagues and from consultants with particular expertise in the area concerned.

Background geological information is available for most areas in the world although its scale and quality vary considerably. In some parts of Europe geological maps are published at 1:50,000 and manuscript field sheets at 1:10,000 are available. In less populated parts of the world, such as Canada or Australia, the base cover is 1:250,000 with more detailed areas at 1:100,000. For the mineral exploration geologist the published geological data will only be a beginning and he or she will interpret the geology using the geological features defined in the deposit model. This is best achieved by starting with a synoptic view of the geology from satellite imagery. Unless the area is extensively vegetated this is likely to be from Landsat or SPOT imagery (see Chapter 6). Landsat has the added advantage of highlighting areas of hydrothermal alteration within arid areas if processed correctly. In areas of dense vegetation side-looking radar can be used. For example, most of Amazonian Brazil has been mapped in this way. For smaller areas air photography provides better resolution often at considerably less cost, although air photographs are still regarded as top secret in some

less developed countries, in spite of the availability of satellite information. Besides surface mapping familiar to most geologists, a number of other sources of information on regional geology are widely used in exploration.

An invaluable addition to surface regional geology is the use of regional geophysics (see Chapter 7). Airborne magnetics, radiometrics, and regional gravity data are available for much of the developed world and help in refining geological interpretation and, particularly, in mapping deep structures. For example, a belt of the Superior Province in northern Manitoba which hosts a number of major nickel deposits, including the large deposit at Thompson, can be clearly followed under Palaeozoic cover to the south (Fig. 4.4). Regional seismic data are helpful but are usually only available if oil companies donate them to the public domain. Specially commissioned seismic surveys have greatly helped in deciphering the subsurface geology of the Witwatersrand Basin (see section 14.5.4). Subsurface interpretations of geophysical information can be checked by linking them with information from any available deep drill hole logs.

Regional geochemical surveys (see section 8.4) also provide much information in areas of poor outcrop and have defined major lithological provinces covered by boulder clay in Finland.

The sources of information for mineral occurrence localities are similar to those for regional geology. Geological surveys usually have the most comprehensive data base within a country and much of this is normally published (e.g. the summary of UK mineral potential of Colman 2000). Many surveys have collated all the mineral occurrences within their country and the results are available as maps, reports, or even on computerized databases. Two useful types of maps are mineral occurrence maps and metallogenic maps, many of which are now available in digital format. The former type merely shows the location of the occurrences whereas the latter attempts to show the form of the deposit and associated elements overlain on background geology using GIS (see section 9.2). Overlays of the mineral prospects with geology will provide clues to regional controls on mineralisation. At this point some economic input is required as it is often

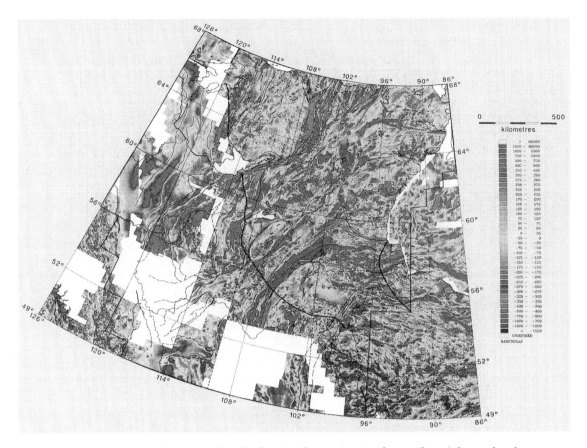

FIG. 4.4   Aeromagnetic map of western Canada showing the continuity of anomalies underneath sedimentary cover. The edge of the sedimentary cover is the bold dashed line striking NW–SE . The belt containing the Thompson nickel deposit runs NNE from 102°E, 52°N. Hudson Bay forms the northeast corner of the map. (Reproduced with permission from Ross et al. 1991.)

the case that significant prospects have different controls from weakly mineralized occurrences. The sort of economic information that is useful are the grades and tonnages mined, recovery methods, and the reasons for stopping mining. This can be obtained from journals or from company reports.

An example of a desk study is the recognition of target areas for epithermal gold deposits in western Turkey. Here mineral exploration increased significantly after 1985 due to the reform of Turkish mining laws and the ability of non-Turkish mining companies to obtain a majority shareholding in any discovery. Western Turkey is of interest as its geological setting is similar to eastern Nevada (see section 16.4), with extension and graben formation in the

Miocene accompanied by extensive volcanism and much current hot spring activity. In addition a cursory glance at the Metallogenic Map of Europe shows that a number of gold occurrences, as well as other elements, such as arsenic, antimony, and mercury which are often associated with epithermal activity, are present (Fig. 4.5). The mineral map of Turkey (Erseçen 1989) provides further information on reserves at several of these localities. The compiler must test this information against the epithermal model in which gold is likely to be associated with graben bounding faults or volcanics. Most of the gold occurrences shown in western Turkey are within high grade metamorphic rocks and represent metamorphosed quartz veins of little interest. However the gold

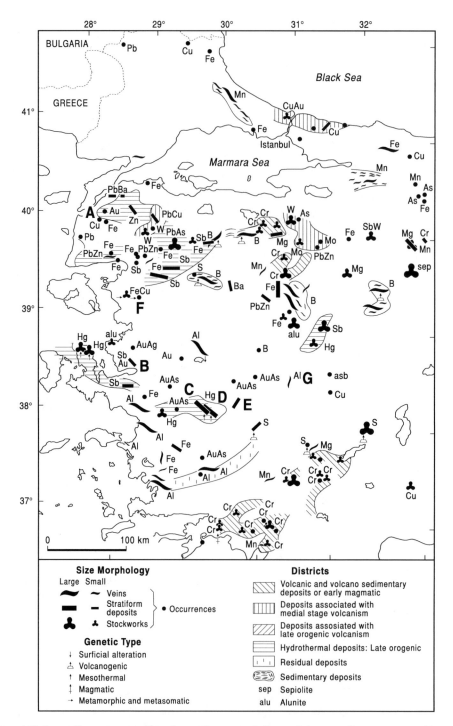

FIG. 4.5 Simplified metallogenic map of Turkey. The original sheet of the Metallogenic Map of Europe shows different deposit types and commodities in different colors as well as underlying geology. (From Lafitte 1970, Sheet 8.)

occurrences within volcanics to the south of Çanakkale and to the northwest of Izmir (points A and B, Figs 4.5, 4.6) remain of interest, as do the mercury and antimony occurrences along the graben bounding faults (points C, D, E). The area south of Çanakkale was mined by a British company until the outbreak of the 1914–18 World War and has lain nearly dormant since then. The first company to appreciate the potential of this area has defined a resource of 9 Mt at 1.25 g t^{-1} Au within one of the areas of brecciated volcanics and a number of other targets have been identified in the vicinity. The area near Izmir has been mined since Roman times but the gold mineralisation is largely confined to a vein structure. Recent exploration by the Turkish geological survey (MTA) who hold the license has defined a further resource of 1.4 Mt at 1.3 g t^{-1} Au. The mercury and antimony deposits are largely the concessions of the state mining group, Etibank, but at least one exploration company was able to agree a joint venture and found significant, although subeconomic, gold grades at the Emirli Antimony Mine (point C, Figs 4.5, 4.6). The most significant gold discovery in the area, Ovacik (point F), is not shown on either map, although a number of veins in the area had, at least in part, been mined by ancient workers. The Ovacik deposit was discovered during a regional reconnaissance of favorable volcanic areas using stream sediment geochemistry and prospecting. Larson (1989) provides a further overview of the potential of the area including much on logistics. Since the Ovacik discovery exploration has been much reduced as the company (originally Eurogold, a joint venture beteeen Australian and Euopean interests) was unable to get permission to mine even though it had constructed all the facilities. The mine was, however, brought into production in 2001 after local opposition had been overcome. The limited exploration in west Turkey since 1990 has also identified a major porphyry sytem at Kisladag (point G) which contains a reserve of 5M ounces of gold at 1.1 g t^{-1} Au (Eldorado Gold 2003).

4.2.2 Reconnaissance

The aim of reconnaissance is to evaluate areas of interest highlighted in the desk study rapidly and to generate other, previously unknown, targets preferably without taking out licenses.

In areas of reasonable outcrop the first stage will be to check the geology by driving along roads or from the air if ground access is impossible. A considerable number of disseminated gold prospects have been found by helicopter follow-up of Landsat data processed to highlight argillic alteration and siliceous caps in arid areas, particularly in Chile. Airborne geology can be extremely effective in solving major questions quickly, but it is expensive, US$500 per hour, and is best undertaken by experienced geologists with specific problems to solve. In any case a preliminary visit to the field should be made as soon as practicable to check the accessibility of the area, examine on the ground some of the data emerging from the desk study, and check on competitor activity.

Reconnaissance techniques can be divided into those which enable a geologist to locate mineral prospects directly and those which provide background information to reduce the search area.

Airborne geophysics and stream or lake sediment geochemistry are the principal tools for directly detecting mineralisation. Perhaps the most successful of these has been the use of airborne electromagnetic techniques in the search for massive sulfide deposits within the glaciated areas of the Canadian Shield. This has led to the discovery of a number of deposits, such as Kidd Creek (see section 15.3.2), by drilling the airborne anomalies after very little ground work. However, each method has drawbacks. In the case of airborne electromagnetic techniques it is the inability to distinguish base-metal-rich sulfides from pyrite and graphite and interferences from conductive overburden. Airborne radiometrics have been very successful in finding uranium deposits in nonglaciated areas and led to the discovery of deposits such as Ranger 1 and Yeelirrie in Australia (Dunn et al. 1990a,b). In glaciated areas radiometric surveys have found a number of boulder trains which have led indirectly to deposits. In areas of residual overburden and active weathering stream sediment sampling has directly located a number of deposits, such as Bougainville in Papua New Guinea, but it is more likely to highlight areas for ground follow-up and licensing.

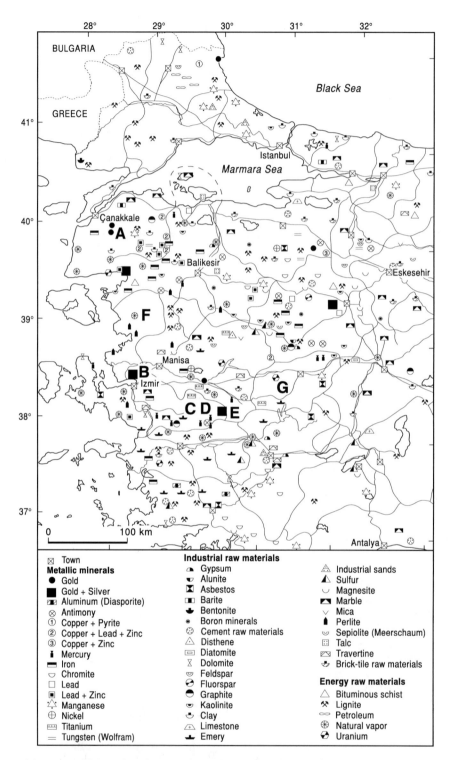

FIG. 4.6   Part of the Mineral Map of Turkey. (Redrawn from Erseçen 1989.)

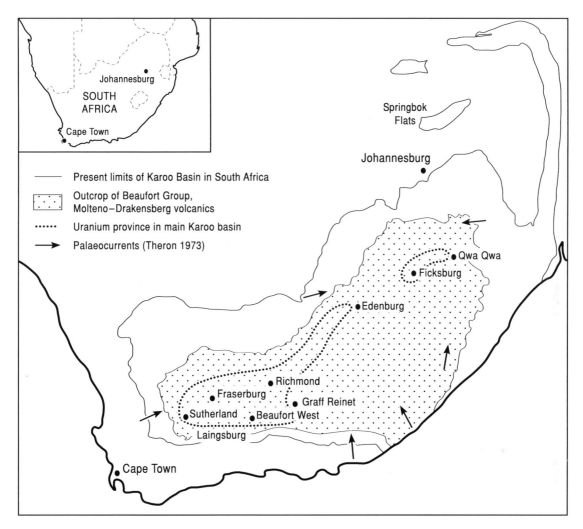

FIG. 4.7   The Karoo Basin in South Africa. (From Moon & Whateley 1989.)

An example of an exploration strategy can be seen in the search for sandstone-hosted, uranium deposits in the Karoo Basin of South Africa (Fig. 4.7) during the 1970s and early 1980s. The initial reconnaissance was conducted by Union Carbide in the late 1960s. They became interested in the basin because of the similarity of the Karoo sediments to those of the Uravan area in Colorado, where the company had uranium mines, and because of the discovery of uranium anomalies in Karoo sediments in Botswana and Zimbabwe. As little of the basin, which is about 600,000 km^2, had been mapped, the company chose a carborne radiometric survey as its principal technique. In this technique a scintillometer is carried in a vehicle and gamma radiation for the area surrounding the road is measured. Although a very limited area around the road is measured it has the advantages that the driver can note the geology through which he is passing, can stop immediately if an anomaly is detected, and it is cheap. In this case a driver was chosen with experience in Uravan geology and he was able to use that experience to compare the geology traversed with favorable rocks in the western USA. After a good deal of traversing a significant anomaly (Fig. 4.8) within sandstones was

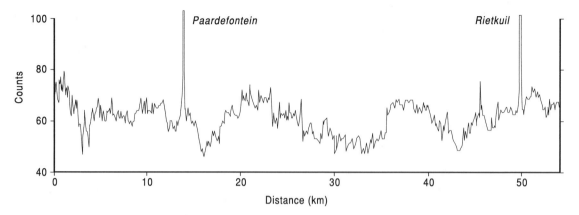

FIG. 4.8 Trace of a carborne radiometric survey similar to that which discovered the first major radiometric anomaly in the Karoo Basin. (After Moon 1973.)

discovered to the west of the small town of Beaufort West (Fig. 4.7). Once the initial discovery was made the higher cost reconnaissance technique of fixed wing airborne radiometrics was used as this provided much fuller coverage of the area, about 30% at 30 m flying height. This immediately indicated to Union Carbide a number of areas for follow up work and drilling.

It became rapidly apparent that the deposits are similar to those of the Uravan area. In both the uranium is in the fluviatile sandstones in an alternating sequence of sandstone, siltstone, and mudstone; in the South African case, in the Upper Permian Beaufort Group of the Karoo Supergroup (Turner 1985). These sediments were deposited by braided to meandering streams which resulted in interfingering and overlapping of sandstone–mudstone transitions and made stratigraphical correlation difficult. Significant mineralisation is confined to thicker multistorey sandstones, usually >14 m thick, and is epigenetic (Gableman & Conel 1985). The source of the uranium is unknown, possibly it was leached from intrabasinal volcanic detritus but certainly mobilized as uranyl carbonate complexes by oxygen-rich, alkaline ground waters which then migrated into the porous sandstone (Turner 1985, Sabins 1987). The migrating ground water then encountered reducing conditions in the presence of organic material, bacterially generated hydrogen sulfide, and pyrite. This change from oxic to anoxic conditions caused the uranium to precipitate as minerals (coffinite and uraninite)

that coated sand grains and filled pore spaces. Diagenesis rendered the deposits impermeable to further remobilization. The outcrop characteristics of the deposits are either calcareous, manganese-rich sandstone, or bleached alteration stained with limonite and hematite, that locally impart yellow, red, and brown coloration. Typical sizes of deposits, which occur as a number of pods, are 500 by 100 m. Average grades are about 0.1% U_3O_8 and thicknesses vary from 0.1 to 6 m but are typically 1 m.

From the airborne radiometrics and follow-up drilling Union Carbide defined one area of potential economic interest and inevitably word of the discovery leaked out. In addition the price of uranium rose following the 1973 Arab–Israeli war leading to interest on the part of a large number of other companies. The reconnaissance approach of competitor companies was however markedly different in that they knew that the basin hosted significant mineralisation and that Union Carbide had important areas under option to purchase. In general these companies selected at least one area around the known Union Carbide discovery and a number of areas elsewhere within the basin. These were chosen on the basis of comparison with deposits in the western USA where the deposits are preferentially hosted at specific horizons and within thicker fluvial or lacustrine sandstones. Mosaics of the, then (1973) newly available, Landsat MSS images, as well as government-flown air photography, were used to derive a regional stratigraphy and to highlight thicker sandstone packages.

FIG. 4.9 View from a helicopter traversing in the central Karoo Basin, northwest of Beaufort West.

Landsat imagery offered a hitherto unavailable synoptic view and the ability to use the spectral characteristics of the different bands to discriminate alteration and rock types. On false color images (see sections 6.2.3 & 6.2.6) sandstone and mudstone rocks could be differentiated, but it was not possible to distinguish the altered sandstone from sandstone that had a natural desert varnish of limonite. To enhance the appearance of the altered rocks, color ratio images were produced by combining the ratios of bands 4/7, 6/4, and 7/4 in blue, green, and red respectively. The altered sandstones were identified by light yellow patches, which were used by field geologists as targets for ground follow-up surveys. This was less successful than flying radiometric surveys over areas of favorable sandstones. In more rugged areas helicopters were used to trace the sandstones and check for radiometric anomalies (Fig. 4.9).

Although airborne radiometric surveys were extremely successful in areas of good outcrop, they were ineffective in locating deposits concealed under inclined sedimentary rock cover. The most successful exploration methods in the search for subsurface mineralisation were detailed surface mapping coupled with hand-held surface radiometric surveys and down the hole logging (see section 7.13) of all available boreholes drilled for farm water supplies (Moon & Whateley 1989).

4.2.3 Land acquisition

The term land is deliberately used as the actual legal requirements for exploration and mining varies from country to country. What the explorer needs to acquire is the right (preferably exclusive) to explore and to mine a deposit if the exploration is successful. Normally a company will obtain the right to explore the property for a particular period of time and the option to convert this into a right to mine, if desired, in return for an annual payment and in some cases the agreement to expend a minimum amount on exploration and to report all results. Unfortunately most legal systems are extremely complicated and the explorer may not be able to obtain the exact right that he or she requires, for example gold or energy minerals may be excluded and the right to the surface of the land (surface rights) may be separate from the right to mine (mineral rights). Two end-member legal systems can be distinguished as far as mineral rights are concerned: the first in which all mineral rights are owned by the state and the explorer can mine with no regard to the current occupier of the surface rights, and the second in which all mineral and surface rights are privately owned. The first normally results from governmental decree or revolution whereas the second is typical of many former British colonies.

Private model

In the private situation, for example in Britain, the company's lawyer will negotiate with a private mineral rights owner and obtain an exploration or option agreement under which the company will be able to explore for a minimum period, normally 3 years, and then be able to renew the option; or to buy the mineral rights for a fixed sum, normally in excess of the free market value. In exchange, the mineral rights owner will receive a fixed annual sum option payment and compensation for any damage to the surface if he or she owns the surface rights. If the surface rights are separately owned then an agreement must be made with that owner. In the case of Britain the rights to gold are owned by the Crown and must be covered in a further separate agreement. There is no legal obligation to report the results to government although summary drillhole results must be reported to the British Geological Survey. Most physical exploration of any significance and drilling of greater than 28 days duration requires consent from the local planning authority.

State model

In the case of state ownership the state will normally own the mineral rights and be able to grant access to the surface. In this case application for an exploration license will usually be made to the department of mines. The size of the exploration area may be fixed. For example the law in Western Australia limits a 2-year prospecting license to 200 ha although a 5-year exploration license can cover between 10 and 200 km². Annual work commitments on the former are $A40 ha^{-1} and $A300 km^{-2} on the latter. Mining leases are granted for a renewable period of 21 years with a maximum area of 1000 ha. Normally in areas with state ownership a full report of exploration results must be filed to the mines department every year and at the termination of the lease. Such reports often provide information for future exploration as well as data for government decision making.

New mineral laws

An example of a new mineral law designed to encourage exploration is the 1985 law of Turkey. Under this exploration licenses are granted for 30 months; these can be converted into pre-operation licenses for 3 years and then into an operating or mining license. The cost of an exploration license in 1992 was 4000 Turkish Lira (US$0.10) ha^{-1}, which is refundable when the license is relinquished. The rights of small miners are protected by a right of denunciation under which any Turkish citizen who can demonstrate a previous discovery in the area can claim 3% of the gross profit. The whole operation is policed by a unit in the geological survey, which uses a computerized system to monitor license areas.

Problem countries

In some countries the rule of law is less secure. For example, following the collapse of the Soviet Union it was not clear who was responsible for, or owned, mineral rights in the newly created countries or in Russia. Deposits had been explored by state-financed organizations, which were left without funds and effectively privatized. Although these organizations were keen to sell rights to the deposits it was by no means certain that national governments recognized their title. Even when title could be agreed, some governments tore up agreements, without compensation, when they realised the value of the deposits. An example of this confusion was the Grib diamond pipe in northern Russia, which was discovered by a junior north American company in 1996 and worth more than $5 billion in situ (see section 17.1.6). The junior was in a 40% joint venture with a Russian expedition which held the license. Subsequent to the discovery the junior company was unable to get the license transferred to a new joint venture company between them and the Russian expedition as had been previously understood. This was probably partly because the assets of the expedition had in the mean time been taken over by a major Russian oil company and a Russian entrepreneur. A few joint ventures have been successful under these conditions, notably those in which governments have a significant stake. A company 67% owned by the government of Kyrgyzstan and 33% by Cameco Corporation of Canada (now Centerra Gold) commissioned, and operates, the large Kumtor gold mine in central Asia.

Legal advice is a virtual necessity in land acquisition and most exploration groups have a lawyer or land person on the staff or on a retainer. If legal tenure is insecure then all exploration effort may be wasted and all revenue from the property go elsewhere. The lawyer will also be responsible for checking the laws that need to be observed, e.g. information from drillholes may need to be reported to government. Corporate administration should be checked so that there is no insider dealing or any activities that could be considered unethical. A famous recent case was that in which Lac Minerals lost control of part of the large and profitable Hemlo deposit in Ontario. It was alleged that Lac Minerals had bought claims from the widow of a prospector after receiving confidential information from another company, Corona Corp., and in spite of a verbal agreement not to stake that ground. Besides losing control of the deposit, Lac was faced with a large legal bill.

A more recent example was the dispute between Western Mining Corporation and Savage Exploration Pty Ltd over control of the Ernest Henry deposit in northern Queensland (Western Mining 1993). Leases in the area, including a tenement named ML2671, had originally been pegged in 1974 by Savage Exploration for iron ore on the basis of the results of a government aeromagnetic survey. In 1989 Western Mining decided to explore the area for base metals and obtained information in the public domain including the government airborne geophysics. When Western Mining selected areas for further work they found that much of the land was held by Hunter Resources Ltd and therefore entered into a joint venture with Hunter but operated by Western Mining.

One of the areas targeted for further work by the joint venture was in the general area of ML2671 as described in the Mines Department files. A baseline was pegged in July 1990 at 100 m intervals and crossed the described position of ML2671 although no pegs were seen on the ground to indicate the position of the lease. A magnetic survey was undertaken including readings within the described area of ML2671 and an initial TEM survey in August 1990. Savage Exploration was approached in March 1991 and agreement made over the terms of an option by which the Western Mining–Hunter joint venture could acquire a number of leases, including ML2671. In May 1991 further TEM surveys were made and a formal option signed in October 1991. In late October 1991 the first hole was drilled and this intersected strong mineralisation.

After further drilling the discovery of the Ernest Henry deposit was announced and Savage Exploration were advised in June 1992 that the joint venture wished to exercise its option and buy the lease of ML2671. Savage Exploration then commenced court proceedings alleging Western Mining had misrepresented the situation when agreeing an option on ML2671 and that Western Mining had trespassed on ML2671 during the ground surveys. In the ensuing court case it transpired that the pegs for ML2671 were not in the place indicated on the Mines Dept files and were 850 m north of the stated location and the lease was rotated several degrees anticlockwise from its plotted position. The case was settled out of court with Western Mining agreeing to give up any claim to ML2671 and paying Hunter $A17 million and certain legal costs.

4.3 SUSTAINABLE DEVELOPMENT

One of the major changes in exploration in the last 10 years has been the difficulty in obtaining a permit to develop a new mine. As discussed in section 1.5.2 much research has been undertaken on the socioeconomic aspects of opening and operating a new mine (see section 11.2.7). The major components of sustainable development in addition to the technical and economic components are: (i) environmental; (ii) social; (iii) governmental. Most major companies issue annual reports on their progress in regard to these three aspects of their operations and these are available on their websites. These form a good place to delve more deeply into specific examples and policies in different countries.

Environmental aspects

Obtaining an environmental permit to operate a mine has become a vital part of the feasibility process but also needs to be addressed during the exploration phases. Obtaining a permit involves the preparation of an environmental impact statement (EIS) describing the problems that mining will cause and the rehabilitation program that will be followed once mining is complete (Hinde 1993). Such an impact statement requires that the condition of the environment in the potential mining area before development began is recorded (a baseline survey). Thus it is essential to collect data during the exploration stage for use in these EISs. Initial data might include surface descriptions and photographs, and geochemical analyses indicating background levels of metals and acidity as well as water levels and flows. It is of course essential to minimize damage during exploration and to set a high standard for environmental management during any exploitation. Trenches and pits should be filled and any damage by tracked vehicles should be minimized and if possible made good. It may be that a more expensive method of access, e.g. helicopter-supported drilling, may be necessary to minimize impact during exploration. Checklists for various exploration activities and discussion of best practices are available at the E3 website (E3 2004). Data collection and baseline surveys become more intense as a prospect becomes more advanced and the prospect of an EIS looms.

One example of a major environmental problem in mining that has had impact in exploration has been the use of cyanide in gold extraction. This has been controversial particularly in Europe where cyanide extraction has been prohibited in the Czech Republic since 2000, causing the abandonment of gold exploration including the advanced prospect at Kaperske Hory. In other areas, the use of cyanidation will be restricted, particularly after two spills, in 1998 in Kyrgyzstan from a delivery truck and in 2000 in Romania from a tailings dam. Strong objections to cyanidation were raised at Ovacik in Turkey and this was one of the reasons for the long delay in commissioning that mine after development (section 4.2.1).

Although mining has had a poor record in environmental impact, modern mines are capable of being designed to minimize these impacts. The very limited impact of the Ekati and Diavik mines (see section 17.2) and modern zinc mines in Ireland are good examples of what can, and should be, achieved.

Social aspects

Another aspect of environmental studies is public relations, particularly keeping the local population informed of progress and obtaining their active approval for any development project. The past few years are littered with examples of projects that are technically excellent but have failed to obtain permission for development, and others that have been significantly delayed causing them to become economically unviable. It is at the exploration stage that the local population form impressions of the nature of the exploration group and whether they wish to be involved in its activities. Establishing good relation and communications with the local community is the first step in gaining their backing for future mining – "a social license to operate" (MMSD 2002). A good summary of the problems is freely available from the PDAC website (E3 2004). Initial concerns of the local community are the transient nature of exploration and the lack of knowledge of the local population to the techniques used and their scale. In addition, the local population may have economic expectations and the arrival of exploration from another part of, or from outside the, country may cause cultural stress. It is always advisable to obtain local advice. For example, in Australia, sacred Aboriginal sites will be known to the local population but not to most geologists. Their unintentional desecration has caused the development of intense opposition to further exploration.

The initial contact with the local population should be carefully planned and, if possible, be enabled by an intermediary, such as a local official trusted by both parties and probably after consulting someone with well-developed skills in dealing with local government and community leaders in the area. The process of exploration and possible outcomes should be carefully explained so that unrealistic expectations are not raised. Local labor and purchasing should be used wherever possible and training should be provided. Major mining companies are now aware of the problems and provide training for their field geologists, emphasizing the importance of contacts with local communities in the exploration process.

Governmental aspects

Relations with governments can be problematic particularly with major projects that will generate a large part of a developing country's export earnings. If large amounts of money are involved in countries where public servants are poorly paid there is a tendency for significant amounts to be appropriated, either in the form of taxes that do not reach the government treasury or bribes. Governmental relations need to be handled carefully at a senior level and care should be taken that they are not in conflict with relations at a community level (see sections 11.2.6 & 11.2.8). There are a number of recent (2004) examples, such as the Esquel gold deposit in Argentina, where the national government is keen on the development but local or provincial bodies are opposed.

4.3.1 Health and safety

A key aspect of a company's reputation, both as an employer and with the local community, is the health and safety of its staff. Although problems at the exploration stage are less severe than during mining, serious injuries and deaths have occurred. Assessments of hazards should be made so that high risk activities can be recognized and mitigated. First aid training should also be provided. At least one major mining group has linked staff pay to safety record and claim that this is the only way to effectively improve their safety record. All contractors should behave in a similar way to company staff and safety record should be a significant factor in choosing contractors. Probably the major source of serious accidents in the authors' experiences is road transport, especially in remote areas. Staff should be provided with training in driving on poor road surfaces.

4.4 SUMMARY

Mineral exploration is conducted by both the private and public sectors in both market and most centrally planned economies. Groups from both sectors must be clear about what commodity and type of deposit they are seeking before setting up an exploration organization. Mineral exploration is a long-term commitment and there must be careful planning of the participant's long-term objectives to ensure viable budgeting. With the objective decided upon an exploration organization must be set up and its success will be enhanced by developing all the factors listed in section 4.1.1.

Budgets must be carefully evaluated and not be figures drawn out of the blue. An underfunded project will in most cases be a failure. Careful control of funds is essential and it provides a basis for future budgeting.

With the organization in place and the target deposit type selected, desk studies can start in earnest and areas for reconnaissance be chosen. Relevant information must be acquired, assessed, and selected using published works and open-file material from government institutions (section 4.2.1). The aim of reconnaissance is to evaluate rapidly areas highlighted in the desk study and to identify targets for follow-up work and drilling. If this is successful then the exclusive rights to explore and to mine any deposits found in the target area must be acquired (section 4.2.3).

4.5 FURTHER READING

Literature on the design and execution of exploration programs is sparse. Peters (1987) provides a good introduction. *Management of Mineral Exploration* by White (1997) is well worth reading with a wealth of practical experience. Papers edited by Huchinson and Grauch (1991) discuss the evolution of genetic concepts as well as giving some case histories of major discoveries and are updated in the volume of Goldfarb and Nielsen (2002). The two volumes of Sillitoe (1995, 2000) provide succinct case histories of discoveries in the Pacific Rim area.

An enthusiastic view from industry, tempered by economic reality, is provided by Woodall (1984, 1992). Much data on the economics of exploration can be found in the volume edited by Tilton et al. (1988), updated in Crowson (2003).

5

FROM PROSPECT TO PREFEASIBILITY

CHARLES J. MOON AND MICHAEL K.G. WHATELEY

Once land has been acquired the geologist must direct his or her efforts to proving whether or not a mineral prospect is worthy of commercial evaluation. Proving a discovery to be of sufficient size and quality inevitably involves a subsurface investigation and the geologist usually faces the task of generating a target for drilling. In exceptional cases, such as very shallow mineralisation, a resource may be proved by digging pits or trenches or, in mountainous, areas by driving adits into the mountainside.

Whatever the method used the key requirement is to explore the area at the lowest cost without missing significant targets. Fulfilling this requirement is not easy and the mineral deposit models, such as those discussed in Chapter 3, must be modified to include economic considerations. There should be a clear idea of the size of the deposit sought, the maximum depth of interest, and whether underground mining is acceptable. Finding a drilling target normally involves the commissioning of a number of different surveys, such as a geophysical survey, to locate the target and indicate its probable subsurface extension. The role of the geologist and the exploration management is to decide which are necessary and to integrate the surveys to maximum effect.

5.1 FINDING A DRILLING TARGET

5.1.1 Organization and budgeting

Once land has been acquired, an organization and budget must be set up to explore it and bring the exploration to a successful conclu-sion. The scale and speed of exploration will depend on the land acquisition agreements (see section 4.2.3) and the overall budget. If a purchase decision is required in 3 years, then the budget and organization must be geared to this. Usually the exploration of a particular piece of land is given project status and allocated a separate budget under the responsibility of a geologist, normally called the project geologist. This geologist is then allocated a support team and he or she proceeds to plan the various surveys, usually in collaboration with in-house experts, and reports his or her recommendations to the exploration management. Typically, reporting of progress is carried out in monthly reports by all staff and in 6-monthly reviews of progress with senior exploration management. These reviews are often also oral presentations ("show and tell") sessions and linked to budget proposals for the next financial period.

Reconnaissance projects are normally directed from an existing exploration office, but once a project has been established serious consideration should be given (in inhabited areas) to setting up an office nearer the project location. Initially this may be an abandoned farmhouse or caravan but for large projects it will be a formal office in the nearest town with good communications and supplies. A small office of this type is becoming much easier to organize following the improvement of communication and computing facilities during the last decade. Exploration data can be transferred by modem, fax, and satellite communications to even the remotest location. If the project grows into one with a major drilling commitment then it is probable that married staff will work more effectively if their families are moved to this town. In general, the town should be less than 1 hour's commuting time so that a visit to a drilling rig is not a chore. In the remotest areas staff will be housed in field camps and will commute by air to their home base. If possible exploration staff should not be expected to stay for long periods in field camps as this is bad for morale and efficiency declines.

Budgeting for the project is more detailed than at the reconnaissance stage and will take account of the more expensive aspects of exploration, notably drilling. A typical budget sheet is shown in Table 4.2. Usually the major expenditure will be on labor and on the various surveys. If the area is remote then transport costs, particularly helicopter charter, can become significant. Labor costs are normally calculated on the basis of man months allocated and should include an overhead component to cover office rental, secretarial and drafting support. Commonly overheads equal salary costs. For geophysical, and sometimes geochemical surveys, contractors are often hired as their costs can be accurately estimated and companies are not then faced with the possibility of having to generate work for staff. Estimates are normally made on the basis of cost per line-kilometer or sample. In remote areas careful consideration should be given to contractor availability. For example in northern Canada the field seasons are often very short and a number of companies will be competing for contractors. For other types of survey and for drilling, estimates can be obtained from reputable operators.

5.1.2 Topographical surveys

The accurate location of exploration surveys relative to each other is of crucial importance and requires that the explorationist knows the basics of topographical surveying. The effort put into the survey varies with the success and importance of the project. At an early stage in a remote area a rough survey with the accuracy of a few meters will be adequate. This can be achieved using aerial photographs and handheld Global Positioning Satellite (GPS) receivers (Ritchie et al. 1977, Sabins 1987) where only a few survey points are needed. For a major

drilling program surveying to a few millimeters will be required for accurate borehole location.

The usual practice when starting work on a prospect is to define a local grid for the prospect using GPS to relate the local grid to the national or international grid systems (see section 9.1.7). Some convenient point such as a wind pump or large rock is normally taken as the origin. The orientation of the grid will be parallel to the regional strike if steeply dipping mineralisation is suspected (from geophysics) but otherwise should be N–S or E–W.

Surveying methods

Geologists generally use simple and cheap techniques in contrast to those used by professional surveyors. Surveying has recently been transformed by the advent of GPS, based on a network of satellites installed by the US Department of Defense. These enable a fix of approximately 5 m accuracy to be made anywhere on earth, with an inexpensive (currently approximately $US200) receiver, where three satellites can be viewed. The major problems are in forested areas.

One of the simplest and most widely used techniques, before the advent of GPS, is the tape and compass survey. This type of survey starts from a fixed point with directions measured with a prismatic or Brunton compass and distances measured with a tape or chain. Closed traverses, i.e. traverses returning to the initial point, are often used to minimize the errors in this method. Errors of distance and orientation may be distributed through the traverse (section 5.1.4). Grids are often laid out using this technique with baselines measured along a compass bearing. Distances are best measured with a chain which is less vulnerable to wear and more accurate than a tape. Longer baselines in flat country can be measured using a bicycle wheel with a cyclometer attached. Lightweight hip-mounted chains are commonly used in remote areas. Straight grid lines are usually best laid out on flat ground by back and forward sighting along lines of pegs or sticks. Sturdy wood or metal pegs should be used, the grid locations marked with metallic tags and flagged with colored tape. The tape should be animal proof; goats have a particular fondness for colored tape.

Leveling is the most accurate method of obtaining height differences between stations and is used for example to obtain the elevation of each station when undertaking gravity surveys. This method measures the height difference between a pair of stations using a surveyor's level. Leveling is required in an underground mine to determine the minimum slope required to drain an adit or drive and in a surface mine the maximum gradient up which load, haul, dump (LHD) trucks can climb when fully laden (usually <10%).

More exact ground surveys use a theodolite as a substitute for a compass. It is simple enough for geologists to learn to use. Numerical triangulation is carried out using angles measured by theodolite to calculate x and y coordinates. By also recording the vertical angles, the heights of points can be computed (Ritchie et al. 1977). Professional surveyors are readily available in most parts of the world and they should be contracted for more exact surveys. They will use a theodolite and electronic distance measuring (EDM) equipment, often combined in one instrument, and can produce an immediate printout of the grid location of points measured.

5.1.3   Geological mapping

One of the key elements during the exploration of a prospect is the preparation of a geological map. Its quality and scale will vary with the importance of the program and the finance available. Initial investigation of a prospect may only require sketch mapping on an aerial photograph, whereas detailed investigations prior to drilling may necessitate mapping every exposure. Mapping at the prospect scale is generally undertaken at 1:10,000 to 1:2500. For detailed, accurate mapping a telescopic alidade and plane table or differential GPS may be used. The principle of the alidade is the same as for a theodolite, except that the vertical and horizontal distances between each point are calculated in the field. The base of the alidade is used to plot the position of the next point on the waterproof drafting film covering the plane table.

The process of geological mapping of mineral prospects is similar to that of general geological mapping, but is more focused, and is well described in the book by Majoribanks (1997). Although the regional environment of the prospect is important, particular attention will fall on known mineralisation or any discovered during the survey. The geological relationship of the mineralisation must be assessed in detail, in particular whether it has any of the features of the geological model sought. For example, if the target is a volcanic-associated massive sulfide deposit then any sulfides should be carefully mapped to determine if they are concordant or cross cutting. If the sulfides are cross cutting it should be established whether the sulfides are in a stockwork or a vein. Particular attention should be paid to mapping hydrothermal alteration, which is described in detail by Pirajno (1992).

Detailed guidance on geological mapping is beyond the scope of this book but can be found in a series of handbooks published by the Geological Society of London (Fry 1991, McClay 1991, Thorpe & Brown 1993, Tucker 2003), particularly in the summary volume of Barnes and Lisle (2003).

One of the key elements of mapping is its final presentation. Conventionally this was in the form of a map drafted by Indian ink pen on to transparent film. From film, multiple copies either on film or on paper can be made using the dyeline process and the map can be overlaid on other maps of the same scale, allowing easy comparison of features and selection of targets.

The conventional pen and paper approach has been superseded by computerized drafting that allows the storage of information in digital form. Computer packages, such as AUTOCAD, a computer-aided drafting package, are widely used in industry (see section 9.2). Maps and plans can be produced to scale and different features of the overall data set, held on different layers in the computer, can be selected for viewing on the computer screen or printed as a hard copy. The data can also be transmitted to more sophisticated Geographical Information Systems (GIS), for example ArcGIS or MapInfo, that allow the inquisition of data (see section 9.2).

5.1.4   Mapping and sampling of old mines

Many prospects contain or are based around old mines. They may become attractive as exploration targets because of rising commodity prices, cheaper mining and processing costs, the development of new technology which may improve recovery, or the development of a new geological model which could lead to undiscovered mineralisation. The presence of a mining district indicates mineral potential, which must reduce the exploration risk. However, there will be a premium to pay, as the property will probably already be under option to, or owned by, a rival company.

The type of examination warranted by an old mine will depend on its antiquity, size, and known history. In Europe and west Asia old mines may be over 2000 years in age and be the result of Roman or earlier activity. In this case there are few, if any, records and the target commodity can only be guessed at. In such cases small areas of disturbed ground, indicating the presence of old prospecting pits or trenches, can best be found from aerial photography. Field checking and grab sampling will confirm the presence and indicate the possible type of mineralisation.

Nineteenth or twentieth century mines are likely to be larger and have more extensive records. Available records should be obtained but should be treated with some caution, as many reports are unreliable and plans likely to be incomplete. An aim of this type of investigation is to check any records carefully by using systematic underground sampling above the water level, as old mines are frequently flooded. Evaluation of extensive underground workings requires considerable planning and will be more expensive than surface exploration because equipment and labor for development and securing old underground workings are costly. A key consideration is safety and access to the old workings must be made safe before any sampling program is established. It may be necessary to undertake trenching and pitting in areas adjacent to the old workings, and eventually drilling may be used to examine the deeper parts of the inaccessible mineralisation. Guidelines on safe working practices in old workings can be found in Peters (1987) and Berkman (2001), and in the UK in publications of the National Association of Mining History Organizations (NAMHO 1985) and the Institution of Geologists (now the Geological Society of London) (IG 1989).

Before a sampling program can be put into operation, a map of the old workings will be required, if none is available from archives. The exploration geologist is often the first person at the site and it is up to him or her to produce a plan and section of the old workings. This would be done using a tape and compass survey (Ritchie et al. 1977, Reedman 1979, Peters 1987, Majoribanks 1997) (Table 5.1, Figs 5.1 & 5.2). Once the layout of the old workings is known, the mapping and complementary sampling program can begin. The survey pegs established during the surveying will be used to locate the sample points and guide mapping. With the tape held between the pegs, the sample points are marked on the drive or crosscut walls and the distance from one peg to the sample point recorded in the field note book along with the sample number. The same number is written on a sample ticket and included with the sample in the sample bag. The samples are normally collected at regular intervals from channels cut normal to the dip of the mineralized rock (Fig. 5.3). The sample interval varies depending upon the type of mineralisation. A vein gold deposit may well be sampled at 1 m intervals along every drive, while a copper deposit may only be sampled every 5 or 10 m.

If old records are available and reliable, then their data should be evaluated in conjunction with new sample data acquired during the remapping and resampling exercises. Geological controls on mineralisation should be established using isopach and structure contour maps as discussed in section 5.2.2. It may be necessary to apply a cut-off value below which the mineralisation is not considered mineable. In Table 5.2 two cut-off parameters (Lane 1988) are used, in one a direct cut-off and in the second a weighted average value is used.

In the first case the upper sample cut-off is taken where the individual grade falls below 1.5%. Some assays within this sample are also below 1.5%, but they are surrounded by higher values, which, when averaged out locally, have a mean greater than 1.5%. The samples between the two lines are then averaged using the sample thickness as the weighting function, giving an average of 2.26% Zn over 2.10 m. In the second case, the samples are averaged from the base upwards using thickness as the

TABLE 5.1   An example of balancing a closed traverse. (Modified from Reedman 1979.)

Line	Bearing Ø	Quadrant	Distance (D) m	±sinØ	±cosØ	ΔE ±D.sinØ	ΔN ±D.cosØ	Unadjusted coords E	N	Final coords E	N	
AB	56	1	82	+0.8290	+0.5592	+67.9811	+45.8538	10,000	10,000	10,000	10,000	A
BC	331	4	74	−0.4848	+0.8746	−35.8759	+64.7219	10,067.98	10,045.76	10,069.76	10,045.53	B
CD	282	4	91	−0.9781	+0.2079	−89.0114	+18.9200	10,032.10	10,110.58	10,035.49	10,109.97	C
DE	172	2	100	+0.1392	−0.9903	+13.9173	−99.0268	9943.09	10,129.50	9948.46	10,128.53	D
EA	130	2	45	+0.7660	−0.6428	+34.4720	−28.9254	9957.01	10,030.47	9964.54	10,029.11	E
								9991.48	10,001.54	10,000.00	10,000.00	A'
TOTAL			392					−8.52	+1.54			

Corrections for Eastings

Correction for B	82/392	× 8.52 =	1.78
Correction for C	74/392	× 8.52 + B = 1.61 + 1.78 =	3.39
Correction for D	91/392	× 8.52 + C = 1.98 + 3.39 =	5.37
Correction for E	100/392	× 8.52 + D = 2.17 + 5.37 =	7.54
Correction for A	45/392	× 8.52 + E = 0.98 + 7.54 =	8.52

Quadrants

N

4		1
sin −ve		sin +ve
cos +ve		cos +ve
sin −ve		sin +ve
cos −ve		cos −ve
3		2

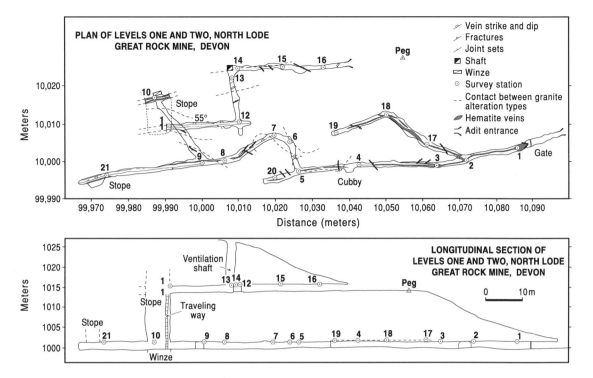

FIG. 5.1   Typical underground survey of a small vein mine using tape and compass.

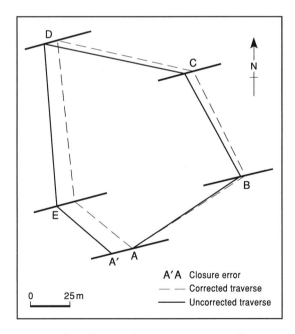

FIG. 5.2   Plot of a typical compass traverse. The errors in closure of the traverse can be corrected using the calculations shown in Table 5.1. (Modified from Reedman 1979.)

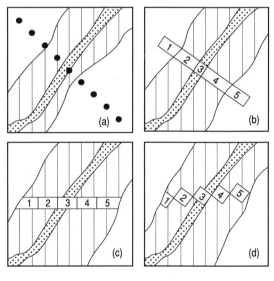

FIG. 5.3   Sampling strategies across a mineralized vein. (a) Grab samples across the vein. (b) Channel samples normal to the vein. (c) Channel samples at the same height above the floor. (d) Channel samples at the same height but with individual samples oriented normal to the strike of the vein. In all cases the vein is shown in vertical section.

Assay width	Zn%	Average grade at a 1.5% cut-off	Grade using a 1.5% weighted average cut-off	
0.30	0.01			TABLE 5.2 Example of the use of a grade cut-off to establish the mining width of a Pb-Zn vein deposit.
0.30	0.05			
0.30	0.13		———	
0.20	0.17			
0.20	0.25			
0.20	0.64			
0.20	0.92			
0.20	1.10	———		
0.20	2.30			
0.20	1.20			
0.20	2.10	2.26%	1.59%	
0.20	5.30	over	over	
0.23	1.40	2.10 m	3.40 m	
0.15	2.00			
0.20	1.90			
0.20	1.80			
0.20	1.30			
0.20	2.60			
0.12	3.50	———	———	
4.00 m	1.35% (overall average)			

weighting function. The weighted average is calculated until the average falls above the cut-off, which in this case makes 1.59% Zn over 3.40 m. The inclusion of just one more sample would bring the average down below 1.5%. The overall weighted average is 1.35% Zn over 4.00 m.

5.1.5 Laboratory program

Once the samples have been collected, bagged, and labeled, they must be sent to the laboratory for analysis. Not only must the elements of interest be specified, but the type of analytical procedure should be discussed with the laboratory. Cost should not be the overriding factor when choosing a laboratory. Accuracy, precision, and an efficient procedure are also needed. An efficient procedure within one's own office is also required. Sample numbers and the analyses requested should be noted down rigorously on a sample control form (Box 5.1).

An alternative practice is to state clearly the instructions for test work on the sample sheet (Box 5.2). In coal analytical work there are several ways in which the proximate analyses (moisture, volatile, ash, and fixed carbon contents) can be reported; e.g. on an as received, moisture free or dry, ash free basis (Stach 1982, Speight 1983, Ward 1984, Thomas 1992).

A similar form may well be utilized for sand and gravel or crushed rock analyses where the geologist requires special test work on his or her samples, such as size analysis, aggregate crushing value (ACV), or polished stone value (PSV) to name but a few.

In today's computerized era, results are often returned to the company either on a floppy disk or direct to the company's computer from the laboratory's computer via a modem and a telephone link. Care must be taken when entering the results thus obtained into the company's database, that the columns of data in the laboratory's results correspond with the columns in your own data base (discussed in detail in section 9.1). Gold values of several percent and copper values in the ppb range should alert even the most unsuspecting operator to an error!

BOX 5.1   An example of a sample control form used to ensure correct analyses are requested.

Submitted by: Submitted to:
Sample control file: Date submitted:
Returned to: Job File:

Job no. /Area: Date returned:
Personnel file:

Channel or borehole no.	Sample no.	From (m)	To (m)	Thickness (m)	Description, analyses and elements requested

BOX 5.2   An example of a form used to give instructions to the laboratory for detailed coal test work.

To:	Date:
	Order no:
Please analyse the following samples as per the attached instructions	

SAMPLE IDENTIFICATION	DETERMINATION REQUIRED
	On individual samples marked B or M undertake the following: 1. Specific gravity of sample 2. Crush to 4.75 mm and screen out the <0.5 mm fraction 3. Cut out a representative sample of the raw coal and undertake: 4. Proximate analysis 5. Ultimate analysis (H, N, C, & O) on a moisture free basis 6. Analyze for inorganic and organic forms of S and Cl 7. Calorific value on a moisture free basis 8. Free silica % 9. Combined silica %
Submitted by:	Signature:

Element selection

Before samples are submitted to the laboratory, discussions between the project manager, chief geologist, and the field staff should take place to ensure that all the elements that may be associated with the mineral deposit in question are included on the analytical request sheet and that analysis includes possible pathfinder elements. Typical elemental associations are discussed in detail in Chapter 8.

Analytical techniques

There are a wide variety of analytical techniques available to the exploration geologist. The method selected depends upon the element which is being analysed and upon the amount expected. Amongst the instumental methods available are atomic absorption spectrometry (AAS), X-ray fluorescence (XRF), X-ray diffraction (XRD), neutron activation analysis (NAA), and inductively coupled plasma emission or mass spectrometry (ICP-ES/MS). These analytical methods and their application to different mineral deposit types are discussed in Chapter 8. AAS is a relatively inexpensive method of analysis and some exploration camps now have an instrument in the field; this ensures rapid analysis of the samples for immediate follow-up. The other methods involve purchasing expensive equipment and this is usually left to specialist commercial laboratories.

Detailed identification of individual minerals is usually undertaken using a scanning electron microscope (SEM) or an electron microprobe, and these and other techniques are discussed in section 2.2.2.

5.1.6 Prospecting

One of the key exploration activities is the location of surface mineralisation and any old workings. Although this often results from the follow-up of geochemical and geophysical anomalies or is part of routine geological mapping, it can also be the province of less formally trained persons. These prospectors compensate for their lack of formal training with a detailed knowledge of the countryside and an acute eye – an "eye for ore." This eye for ore is the result of experience of the recognition of weathered outcrops and the use of a number of simple field tests.

Many deposits which crop out have been recognized because they have a very different appearance from the surrounding rocks and form distinct hills or depressions. A classic example of this is the Ertsberg copper deposit in Irian Jaya, Indonesia. It was recognized because its green-stained top stood out through the surrounding jungle. Its presence was first noted by two oil exploration geologists on a mountaineering holiday in 1936 and reported in a Dutch university geological journal (Dozy et al. 1939). A literature search by geologists working for a Freeport Sulfur–East Borneo company joint venture found the report and investigated the discovery resulting in one of the largest gold deposits outside South Africa. The Carajas iron deposits in the Brazilian Amazon were also recognized because they protruded through rain forest (Machamer et al. 1991). On a smaller scale silicification is characteristic of many hydrothermal deposits, resulting in slight topographical ridges. It is said that the silicification of disseminated gold deposits is so characteristic that one deposit in Nevada was discovered by the crunchy sound of a geologist's boots walking across an altered zone after dark. Karst-hosted deposits in limestone terrain often form distinct depressions, as do many kimberlite pipes, which often occur under lakes in glaciated areas (see section 17.2).

Besides forming topographical features most outcropping mineral deposits have a characteristic color anomaly at the surface. The most common of these is the development of a red, yellow or black color over iron-rich rocks, particularly those containing sulfides. These altered iron-rich rocks are known generically as ironstones, and iron oxides overlying metallic sulfide deposits as gossans or iron hats (an example is shown in Fig. 5.4). These relic ironstones can be found in most areas of the world, with the exception of alpine mountains and polar regions, and result from the instability of iron sulfides, particularly pyrite. Weathering releases SO_4^{2-} ions, leaving relic red iron oxides (hematite) or yellow-brown oxyhydroxides (limonite) that are easily recognized in the field. Other metallic sulfides weather to form even more distinctively colored oxides

FIG. 5.4 Gossan (above adit) overlying sulfides exposed in the pit wall of the San Miguel deposit, Rio Tinto, southern Spain.

alternation of color and texture giving rise to the term "live," in contrast to ironstones of nonsulfide origin that show little variation and are known as "dead" ironstones. Ironstones also preserve chemical characteristics of their parent rock, although these can be considerably modified due to the leaching of mobile elements.

The recognition of weathered sulfides overlying base metal or gold deposits and the prediction of subsurface grade is therefore of extreme importance. This particularly applies in areas of laterite development. These areas, such as much of Western Australia, have been stable for long periods of geological time, all rocks are deeply weathered, and the percentage of ironstones that overlie nonbase metal sulfides (known rather loosely as false gossans) is large. In Western Australia the main techniques for investigating these are (Butt & Zeegers 1992):

1 visual description of weathered rocks;
2 examination for relic textures;
3 chemical analysis.

Visual recognition requires experience and a good knowledge of primary rock textures. Visual recognition can also be supplemented by chemical analysis for trace and major elements. Hallberg (1984) showed that Zr-TiO_2 plots are extremely useful in confirming rock types. Both elements are immobile and easy and cheap to determine at the levels required. They may be supplemented by Cr determinations in ultrabasic areas, although Cr is more mobile than the other two elements.

Relic textures can sometimes be observed in hand specimen but this is usually supplemented by microscopic examination using either a binocular microscope or a petrographic microscope and an impregnated thin section. In a classic study of relic textures, Blanchard (1968) described the textures resulting from the weathering of sulfides. Major types are shown in Fig. 5.5. Although boxwork textures are diagnostic they are, unfortunately, not present in every gossan overlying base metal sulfides and textural examination must be supplemented by chemical analysis.

The choice of material to be sampled *and* the interpretation need to be carefully scrutinized. Andrew (1978) recommends taking 20 samples as representative of each gossan and using a

or secondary minerals. For example, copper sulfides oxidize to secondary minerals that have distinctive green or blue colors such as malachite and azurite. Metals such as lead and zinc normally form white secondary minerals in carbonate areas that are not easily distinguished on color grounds from the host carbonates. A fuller list is given in Table 5.3. Besides providing ions to form secondary minerals, sulfides often leave recognizable traces of their presence in the form of the spaces that they occupied. These spaces are relic textures, often known as boxworks from their distinctive shapes, and are frequently infilled with limonite and goethite. Ironstones overlying sulfides have a varied appearance with much

TABLE 5.3 Colours of minerals in outcrop.

Mineral or metal	Outcrop color	Mineral/compound in outcrop
Iron sulfides	Yellows, browns, chestnuts, reds	Goethite, hematite, limonite, sulfates
Manganese	Blacks	Mn oxides, wad
Antimony	White	Antimony bloom
Arsenic	Greenish, greens, yellowish	Iron arsenate
Bismuth	Light yellow	Bismuth ochres
Cadmium	Light yellow	Cadmium sulfide
Cobalt	Black, pink, sometimes violet	Oxides, erythrite
Copper	Greens, blues	Carbonates, silicates, sulfates, oxides, native Cu
Lead	White, yellow	Cerussite, anglesite, pyromorphite
Mercury	Red	Cinnabar
Molybdenum	Bright yellow	Molydenum oxides, iron molybdate
Nickel	Green	Annabergite, garnierite
Silver	Waxy green, yellow	Chlorides, native Ag
Uranium	Bright green, yellow	Torbernite, autunite
Vanadium	Green, yellow	Vanadates
Zinc	White	Smithsonite

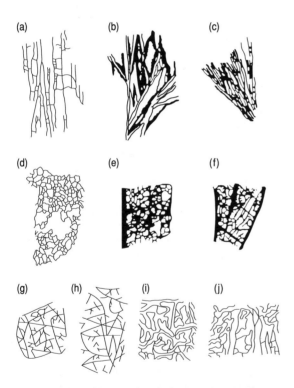

FIG. 5.5 Typical boxwork structures. Primary ore minerals were: (a–c) galena (a – cleavage, b – mesh, c – radiate); (d,e) sphalerite (d – sponge structure, e – cellular boxwork); (f) chalcopyrite; (g,h) bornite (triangular cellular structure); (i,j) -tetrahedrite (contour boxwork). Approximately 4× magnification. (After Blanchard & Boswell 1934.)

multielement analysis, either XRF or ICP–ES, following a total attack to dissolve silica. The interpretation needs to be treated with care. A large amount of money was wasted in Western Australia at the height of the nickel boom in the early 1970s because ironstones were evaluated for potential on the basis of their nickel content. While most ironstones with very high nickel contents do overlie nickel sulfide deposits, a number have scavenged the relatively mobile nickel from circulating ground waters or overlie silicate sources of nickel and a number of deposits have a weak nickel expression at surface. More careful research demonstrated that it is better to consider the ratio of nickel to more immobile elements, such as copper, or even better immobile iridium that is present at the sub-ppm level in the deposits. These elements can be combined into a discriminant index (Travis et al. 1976, Moeskops 1977). Similar considerations also apply to base metal deposits; barium and lead have been shown to be useful immobile tracers in these (summarized in Butt & Zeegers 1992).

Field tests

Although a large number of field tests have been proposed in the literature only a few of the simplest are in routine use. At present only two geophysical instruments can be routinely carried by man, a scintillometer to detect gamma

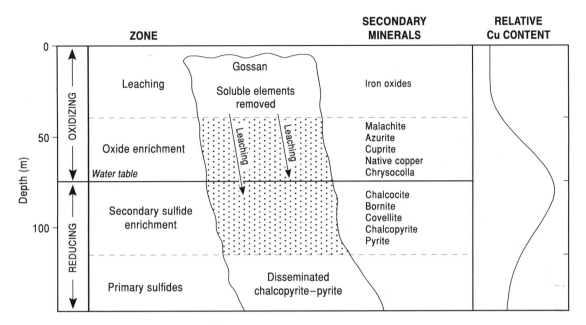

FIG. 5.6 Diagram showing the gossan and copper minerals formed by weathering of a copper sulfide vein. (After Levinson 1980.)

radiation and a magnetic susceptibility meter to determine rock type. The application of the former is discussed in section 7.5. Ultraviolet lamps to detect fluorescence of minerals at night are commonly used, particularly for scheelite detection, although a number of other minerals glow (Chaussier & Morer 1987). In more detailed prospecting darkness can be created using a tarpaulin.

A number of simple field chemical tests aid in the recognition of metal enrichments in the field, usually by staining. Particularly useful tests aid the recognition of secondary lead and zinc minerals in carbonate areas. Lead minerals can be identified in outcrops by reaction with potassium iodide following acidification with hydrochloric acid. Lead minerals form a bright yellow lead iodide. A bright red precipitate results when zinc reacts with potassium ferrocyanide in oxalic acid with diethylaniline. Gray copper minerals can be detected with an acidified mixture of ammonium pyrophosphate and molybdate and nickel sulfides with dimethyl glyoxine. Full details are given in Chaussier and Morer (1987).

Supergene enrichment

As metallic ions are leached at the surface they are moved in solution and deposited at changes in environmental conditions, usually at the water table, where active oxidation and generation of electrical charge takes place. This leaching and deposition has very important economic implications that are particularly significant in the case of copper deposits as illustrated in Fig. 5.6. Copper is leached from the surface, leaving relic oxidized minerals concentrated at the water table in the form of secondary sulfides. A simple Eh–pH diagram enables the prediction of the occurrence of the different copper minerals. From the surface these are malachite, cuprite, covellite, chalcocite, bornite, and chalcopyrite (the primary phase). The changes in mineralogy have important implications for the recovery of the minerals and the overall grade of the rock. Oxide minerals cannot be recovered by the flotation process used for sulfides and their recovery needs a separate circuit. The overall grade of the surface rock is much lower than the primary ore and sampling of the surface outcrops does not reflect that of the primary ore. Near the water table the copper grade increases, as the result of the conversion of chalcopyrite (35% Cu) to chalcocite (80% Cu). This enrichment, known as supergene enrichment, often provides high grade zones in disseminated copper deposits, such as those of the porphyry type. These high grade zones provide much extra revenue and may even be the basis of mines with low primary grades. In contrast to the behavior of copper, gold is less mobile and is usually concentrated in oxide zones. If these oxide zones represent a significant amount of leaching over a long period then the gold grades may be become economic. An example of this is the porphyry deposit of Ok Tedi discussed in Box 11.4. These near surface, high grade zones are especially important as they provide high revenue during the early years of a mine and the opportunity to repay loans at an early stage. These gold-rich gossans can also be mechanically transported for considerable distances, e.g. the Rio Tinto Mine in southern Spain.

Float mapping

The skill of tracing mineralized boulders or rock fragments is extremely valuable in areas of poor exposure or in mountainous areas. In mountainous areas the rock fragments have moved downslope under gravity and the lithology hosting the mineralisation can be matched with a probable source in a nearby cliff and a climb attempted. Float mapping and sampling is often combined with stream sediment sampling and a number of successful surveys have been reported from Papua New Guinea (Lindley 1987). In lowland areas mineralized boulders are often disturbed during cultivation and may be moved to nearby walls. In this case it is often difficult to establish a source for the boulders and a soil survey and a subsequent trench may be necessary. Burrowing animals may also be of help. Moles or rabbits, and termites in tropical areas, often bring small fragments to the surface.

In glaciated areas boulders may be moved up to tens of kilometers and distinct boulder trains can be mapped. Such trains have been followed back to deposits, notably in central

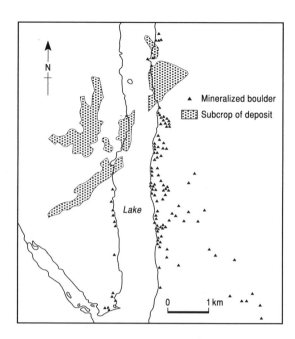

FIG. 5.7   Train of mineralized boulders from the Laisvall lead deposit, Sweden.

Canada, Ireland, and Scandinavia (Fig. 5.7). In Finland the technique proved so successful that the government offered monetary rewards for finding mineralized boulders. Dogs are also trained to sniff out the sulfide boulders as their sense of smell is more acute than that of the exploration geologist!

5.1.7   Physical exploration: pitting and trenching

In areas of poor to moderate outcrop a trench (Fig. 5.8) or pit is invaluable in confirming the bedrock source of an anomaly, be it geological, geochemical, or geophysical. The geology of a trench or pit wall should be described and illustrated in detail (Fig. 5.9). For further details see section 9.2. Trenches and pits also provide large samples for more accurate grade estimates as well as for undertaking pilot processing plant test work to determine likely recoveries.

Some operators in remote areas, particularly in central Canada, strip relatively large areas of the overburden to enable systematic mapping of bedrock. However, this would now (probably) be regarded as environmentally unfriendly.

FIG. 5.8 Typical exploration trench, in this case on the South African Highveld.

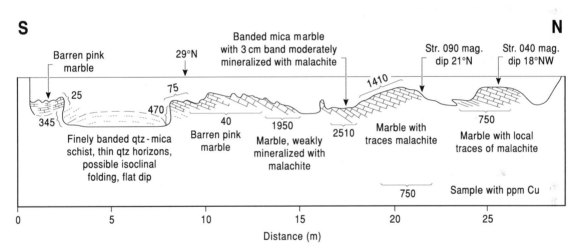

FIG. 5.9 Sketch of a trench testing a copper anomaly in central Zambia. (From Reedman 1979.)

An alternative to disturbing the environment by trenching is to use a hand-held drill for shallow drilling. This type of drill is lightweight and can be transported by two people. It produces a small core, usually around 25–30 mm in diameter. Penetration is usually limited, but varies from around 5 m to as much as 45 m depending on the rig, rock type, and skill of the operators!

5.1.8 Merging the data

The key skill in generating a drilling target is integrating the information from the various surveys. Not all the information obtained will be useful, indeed some may be misleading, and distinction should be clearly made between measured and interpreted data. Geophysical and geochemical surveys can indicate the surface or subsurface geology of the potential host rocks or more directly the presence of mineralisation as discussed in Chapters 7 and 8. Traditionally the results of these surveys have been combined by overlaying colored transparent copies of the data on a topographical or geological paper base. It is then possible to determine the interrelation between the various data sets. More recent developments have allowed the use of computerized methods (GIS) which allow rapid integration and interrogation of the databases and are discussed in detail in section 9.2. A straightforward example of the overlaying of different data types is shown in Fig. 5.10 taken from the work of Ennex plc in Northern Ireland as detailed in Clifford et al. (1990, 1992). The area was chosen for investigation because of the presence of favorable geology, placer gold in gravels and anomalous arsenic geochemistry discovered during government surveys. A licence was applied for in 1980. Ennex geologists investigated the anomalous areas and by careful panning and float mapping identified the source of the gold as thin quartz veins within schist. Although the veins identified assayed about $1 \, g \, t^{-1}$ Au and were not of economic grade, they were an indication that the area was mineralized and prospecting was continued over a wider area. In late 1983 economically significant gold in bedrock was found and subsequently three other gold-bearing quartz veins were found where a small stream has eroded to bedrock. More than 1300 mineralized boulders were mapped over a strike of 2700 m. Channel samples of the first identified vein gave assays of $14 \, g \, t^{-1}$ Au over 3.8 m with selected samples assaying over $150 \, g \, t^{-1}$ Au.

Detailed follow-up concentrated on building up a systematic picture of the area where there is no outcrop. This focused on boulder mapping and sampling, deep overburden geochemistry and geophysical mapping using very low frequency electromagnetic resistivity (VLF–EM/R) surveys. Deep overburden geochemistry was undertaken using a small core overburden drill to extract 250 g samples on a $50 \times 20 \, m$ grid. This spacing was decided on following an orientation study over a trenched vein. Geochemical sampling defined 33 anomalies (>100 ppb Au) (Fig. 5.10a). A variety of geophysical methods were tested including induced polarization and VLF–EM/R. No method located the quartz veins but the VLF–EM/R method did define previously unidentified gold bearing shears (Fig. 5.10b) and the subcrop of graphitic pelites. The nature of the geophysical and geochemical anomalies was defined by a 2850 m trenching program (Fig. 5.10c). Besides increasing the number of gold-bearing structures to 16, this trenching was also instru-

mental in demonstrating the mineralogy of the mineralized structures. The quartz veins contain significant sulfides, mainly pyrite but with lesser chalcopyrite, galena, sphalerite, native copper, and tetrahedrite–tennantite. By contrast, the shears are richer in arsenopyrite with chloritization often observed. Thus geochemistry defined the approximate location of the quartz veins and auriferous shears, geophysics the location of the shears, and trenching the subcrop of the quartz veins.

Once the gold-bearing structures had been defined their depth potential was tested by diamond drilling (Fig. 5.10c). The diamond drilling (6490 m in 63 holes) took place during 1985–86 and defined a resource of 900,000 t at $9.6 \, g \, t^{-1}$. Subsequent underground exploration confirmed the grades and tonnages indicated by drilling but the deposit is undeveloped because of a ban on the regular use of explosives in Northern Ireland by the security authorities.

5.2   WHEN TO DRILL AND WHEN TO STOP

One of the hardest decisions in exploration is to decide when to start drilling, and an even harder one is when to stop.

The pressure to drill will be evident when the program has identified surface mineralisation. Management will naturally be keen to test this and gain an idea of subsurface mineralisation as soon as possible. However, this pressure should be resisted until there is a reasonable idea of the overall surface geology and the inferences that can be made from this knowledge concerning mineralisation in depth.

5.2.1   Setting up a drilling program

The geologist in charge of a drilling program is faced with a number of problems, both logistical and geological. There must be a decision on the type of drilling required, the drillhole spacing (see section 10.4.4), the timing of drilling, and the contractor to be used. The logistics of drilling should be considered carefully as the drill will need drill crews, consumables, and spare parts; this will require helicopter support in remote areas and vehicle access in more populated areas. Many drills require vehicle access and access roads must be made and pads

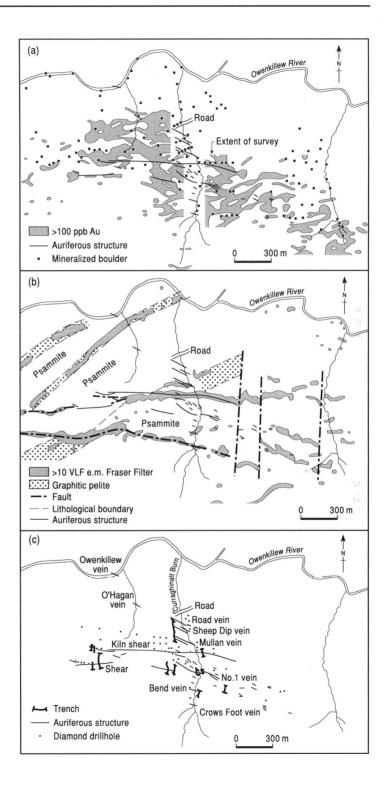

FIG. 5.10 Drilling based on a combination of (a) geochemistry, (b) geophysics, and (c) trenching and prospecting, Sperrin Mountains, Northern Ireland. (After Clifford et al. 1992.)

for drilling constructed so that the drill rigs can be set on an almost horizontal surface.

The pattern of drilling used is dependent on the assumed attitude and thickness of the drilling target. This depends on the available information which may, of course, be inaccurate. Drilling often causes reconsideration of geological ideas and prejudices. Vertical boreholes are the easiest and cheapest to drill and widely used for mineralisation with a shallow dip or for disseminated deposits. However, inclined holes are usually preferred for targets with steep dips. The aim will be to cut the mineralisation at 90 degrees with the initial hole, cutting immediately below the zone of oxidation (weathered zone) (Fig. 5.11).

Drilling is used to define the outlines of any deposit and also the continuity of mineralisation for purposes of resource estimation. The initial pattern of drilling will depend on surface access, which may be very limited in moun-

tainous areas. In areas without access problems, typical drill hole patterns are square with a regular pattern or with rows of holes offset from adjacent holes. The first hole normally aimed at the down dip projection of surface anomalies or the interpreted centre of subsurface geophysics. Most programs are planned on the basis of a few test holes per target with a review of results while drilling. The spacing between holes will be based on anticipated target size, previous company experience with deposits of a similar type, and any information on previous competitor drilling in the district (Whateley 1992). The subsequent drilling location and orientation of the second and third holes will depend on the success of the first hole. Success will prompt step outs from the first hole whereas a barren and geologically uninteresting first hole will suggest that another target should be tested.

Once a deposit has been at least partly defined then the continuity of mineralisation must be assessed. The spacing between holes depends on the type of mineralisation and its anticipated continuity. In an extreme case, e.g. some vein deposits, boreholes are mainly of use in indicating structure and not much use in defining grade, which can only be accurately determined by underground sampling (see Chapter 10). Typical borehole spacing for a vein deposit is 25–50 m and for stratiform deposits anything from 100 m to several hundreds of meters. Examples of drill spacing and orientation for a variety of deposits is shown in Fig. 5.12 (pp. 87–89).

5.2.2 Monitoring drilling programs

Monitoring the geology and mineralisation intersected during a drilling program is vital in controlling costs. In the initial phases of drilling this may involve the geologist staying beside the rig if it is making rapid progress, e.g. when using percussion drilling, and logging material as it comes out of the drill hole. In the case of diamond drilling twice daily visits to examine core, make initial logs, and decide on the location of the next drill holes are usually sufficient, although longer visits will be needed when cutting potentially mineralized zones or nearing the scheduled end of the hole. Often the geologist will be required by contractors to

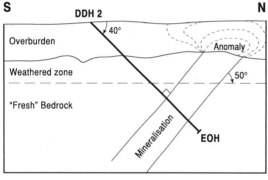

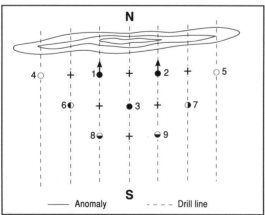

FIG. 5.11   Idealized initial drill grid. EOH, end of hole. (After Annels 1991.)

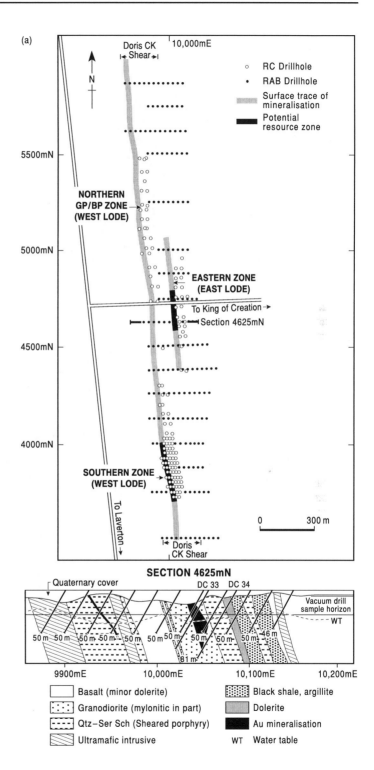

FIG. 5.12 (a) Typical combination of shallow and deeper drilling over the Doris Creek vein gold deposit in Western Australia. The shallow (rotary air blast, RAB) drilling defines the subcrop of the vein whereas the deeper (reverse circulation, RC) drilling defines grades. (After Jeffery et al. 1991.)

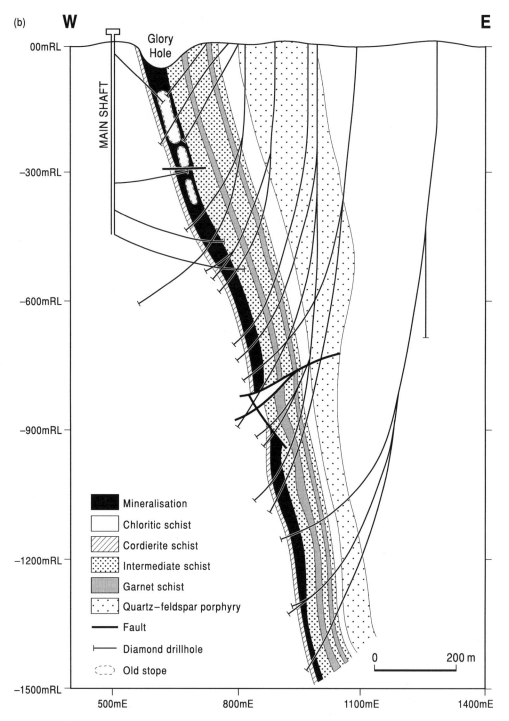

FIG. 5.12 (b) Typical drilling of a previously mined deep vein gold deposit, Big Bell, Western Australia. Note the deflections on the deeper holes, the abandoned holes, and combination of drilling both from surface and underground. (After Handley & Carrey 1990.)

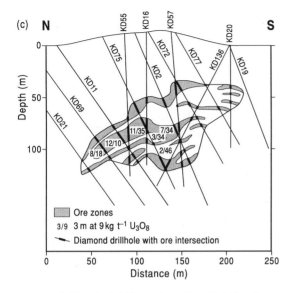

FIG. 5.12   (c) Typical drill section of an ill-defined mass, Kintyre uranium deposit, Western Australia. (After Jackson & Andrew 1990.)

sign for progress or the use of casing. Besides detailed core logs, down hole geophysical logging (see section 7.14) is often used and arrangements should be made with contractors for timely logging as holes can become rapidly blocked and cleaning of holes is an expensive undertaking.

Data on mineralisation, the lithologies, and structures hosting it should be recorded and plotted on to a graphic log as soon as the information becomes available. Initially, strip drill logs (Fig. 5.13) can be used and sections incorporating the known surface geology. Assay information will usually be delayed for a few days as the cores will need to be split or cut and sampled before analysis (see Chapter 10). However estimates of the location and importance of mineralisation can be plotted alongside the lithological log (Fig. 5.15e). As further drilling proceeds the structural and stratigraphical controls on mineralisation should become

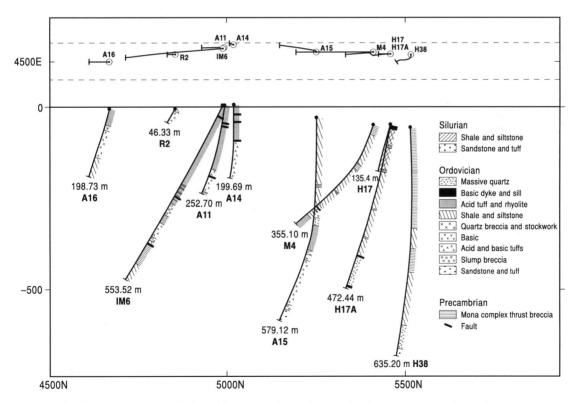

FIG. 5.13   Geological strip logs and plot of drill holes, Parys Mountain zinc prospect, Anglesey, Wales. (Data courtesy Anglesey Mining, plot by L. Agnew.)

clearer. Ambiguities concerned with the interpretation of drilling are common and often cannot be resolved until there has been underground development. A typical example is shown in Fig. 5.14 in which three different interpretations are possible from the information available.

Once a stratigraphy has been established and several boreholes are available, more sophisticated plotting techniques can be used. Typical methods used to plot drillhole information are structure contour plans, isopach, grade (quality), thickness, and grade multiplied by thickness, known as the grade–thickness product or accumulation maps. Grade and grade–thickness product maps are extremely useful in helping to decide on the areal location of oreshoots and of helping direct drilling towards these shoots (Fig. 5.15) (pp. 91, 92).

One of the key issues in any drilling program is continuity of mineralisation. This determines the spacing of drill holes and the accuracy of any resource estimation (as discussed in Chapter 10). In most exploration programs the continuity can be guessed at by comparison with deposits of a similar type in the same district. However, it is usual to drill holes to test continuity once a reasonable sized body has been defined. Typical tests of continuity are to drill holes immediately adjacent to others and to test a small part of the drilled area with a closer spacing of new holes.

5.2.3 Deciding when to stop

Usually the hardest decision when directing a drilling program is to decide when to stop. The main situations are:

1 No mineralisation has been encountered.

2 Mineralisation has been intersected, but it is not of economic grade or width.

3 Drill intercepts have some mineralisation of economic grade but there is limited continuity of grade or rough estimates show that the size is too small to be of interest.

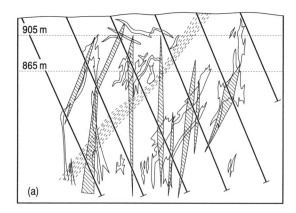

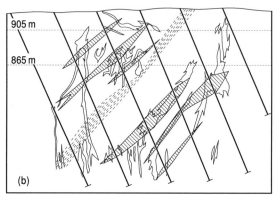

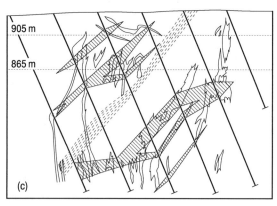

☐ Actual orebody sections

▨ Sections obtained by interpolation

▨ Zone of dislocation

⌐ Borehole

FIG. 5.14 (opposite) (a–c) Three different interpretations of drill intersections in an irregular deposit compared with outlines defined by mining. (After Kreiter 1968.)

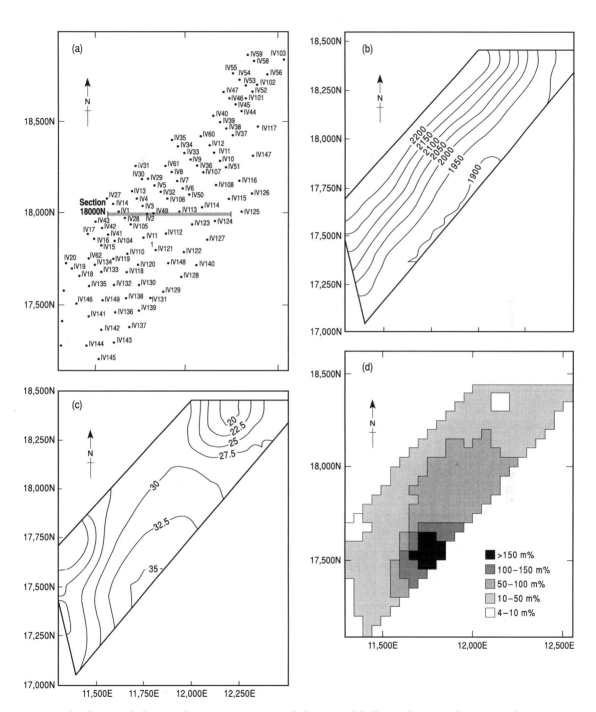

FIG. 5.15 (a–d) Typical plots used in interpretation and planning of drilling. The example is a stratiform copper deposit of the Zambian type: (a) drillhole plan and numbers; (b) structure contour of the top of the mineralized interval (meters above sea level); (c) isopach of the mineralized interval (m); (d) grade thickness product (accumulation) of the mineralized interval (m%).

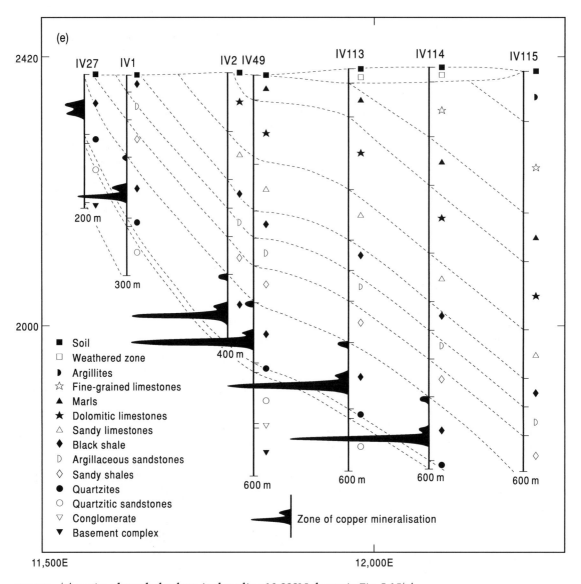

FIG. 5.15   (e) section through the deposit along line 18,000N shown in Fig. 5.15(a).

4 A body of potentially economic grade and size has been established.

5 The budget is exhausted.

The decision in the first case is easy, although the cause of the surface or subsurface anomalies that determined the siting of the drill hole should be established. The second case is more difficult and the possibility of higher grade mineralisation should have been eliminated. The presence of some interesting intersections in a prospect suggests that the processes forming economic grade mineralisation were operative but not at a sufficient scale to produce a large deposit, at least not in the area drilled. The encouragement of high grade mineralisation will often result in several phases of drilling to test all hypotheses and surrounding areas to check whether the same processes may have taken place on a larger scale. This type of prospect is often recycled (i.e. sold) and investigated by several different companies before being abandoned or a deposit discovered. Often it is only some factor such as the impending deadline on a mineral rights option that will result in a decision to exercise the option or drop this type of prospect.

The definition of a potentially economic deposit is an altogether more agreeable problem, and the exact ending of the exploration phase will depend on corporate policy. If a small company is having to raise money for the evaluation phase, it will probably leave the deposit partly open (i.e. certain boundaries undetermined), as a large potential will encourage investors. Larger companies normally require a better idea of the size, grade, and continuity of a deposit before handing over the deposit to their pre-development group.

It is usual for a successful program to extend over several years, resulting in high costs. Exhausting a budget is common; however overspending may result in a sharp reprimand from accountants and the painful search for new employment!

5.3 RECYCLING PROSPECTS

The potential of many prospects is not resolved by the first exploration program and they are recycled to another company or entrepreneur. This recycling has many advantages for the explorer. If for some reason the prospect does not meet their requirements, they are often able to recover some of the investment in the prospect with the sale of the option or the data to a new investor. Often it is not an outright sale but one in which the seller retains some interest. This is termed a joint venture or farm out. Typical agreements are ones in which the buyer agrees to spend a fixed amount of money over a fixed period of time in order to earn a pre-agreed equity stake in the prospect. The period over which the funding takes place is known as the earn-in phase. Normally the seller retains an equity stake in the project and the right to regain complete control if the buyer does not meet its contractual obligations. The type of equity retained by the seller varies; it may include a duty to fund its share of development costs or it may have no further obligations – a free or carried interest. The seller may as an alternative choose merely to have a royalty if the project is eventually brought to production.

Joint venturing is normally carried out by circulating prospective partners with brief details of the project, but this may exclude location and other sensitive details. Seriously interested parties will then be provided with full details and the possibility of visiting the prospect to gain a more accurate impression before committing any money.

Buying or re-examining a prospect that a previous party has discarded is tempting, but the buyer must be satisfied that they can be more successful than the other explorer. This success may result from different exploration models or improved techniques.

Mulberry Prospect

A good example of this is the exploration of the area on the north side of the St Austell Granite in southwestern England, here termed Mulberry after the most productive old mine (Dines 1956). It was examined by three different companies from 1963 to 1982, as reported in British Geological Survey (BGS) open file reports. The Mulberry area is of obvious interest to companies as it is one of the few in Cornwall that supported open pit mining of disseminated tin deposits in the nineteenth century and might therefore be more efficiently mined using advances in mining technology. The deposits worked (Fig. 5.16) were the Mulberry open pit with a N–S strike and a series of open pits on the Wheal Prosper structure which runs E–W. All the open pits were based on a sheeted vein complex or stockwork of thin veins containing cassiterite, tourmaline, and quartz. In addition there were a number of small underground workings based on E–W copper- and N–S iron-bearing veins. All the deposits are within the metamorphic aureole of the St Austell Granite, hosted in slate and another metamorphic rock known locally as calc-flinta. This consists of garnet, diopside, actinolite, quartz, and calcite and was probably originally an impure tuff or limestone.

The initial examination in 1964 resulted from a regional appraisal in which Consolidated Gold Fields (CGF) sought bulk mining targets that were amenable to improved mining and processing techniques. CGF immediately identified Mulberry as being of interest and confirmed this by sampling the old open

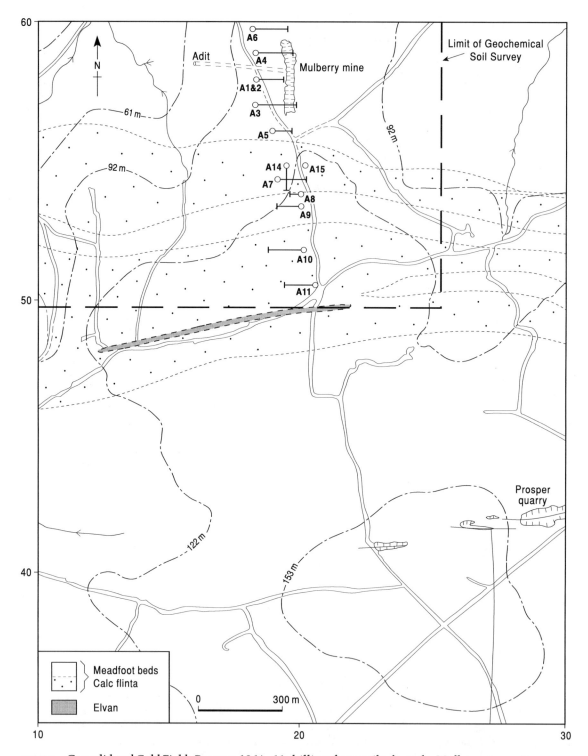

FIG. 5.16   Consolidated Gold Fields Program 1964–66: drilling along strike from the Mulberry open pit.

pit and the underground accesses. In addition the company investigated the potential of the area around the open pit and particularly its strike extensions by taking soil samples for geochemical analysis along E–W lines (Hosking 1971). These suggested that the deposit continued to the south into the calc flintas. CGF then undertook a drilling program to test the potential of the deposit at depth (Fig. 5.16). Drilling over the period 1964–66 confirmed that the mineralisation continued at roughly the width of the open pit but that the cassiterite was contained in thin erratic stringers. Drilling to the north of the open pit showed a sharp cut-off whereas drilling to the south showed that tin values continued from the slate host into a calc flinta host where mineralisation became much more diffuse. Although the grades intersected were similar to those in the open pit, the size and grade were not considered economic, especially when compared with Wheal Jane (eventually a producing mine) which CGF was also investigating, and the lease was terminated.

The next phase of exploration was from 1971 to 1972 when the area was investigated by Noranda–Kerr Ltd, a subsidiary of the Canadian mining group Noranda Inc. This exploration built on the work of CGF and investigated the possibility that the diffuse mineralisation within the calc silicates might be stratabound and more extensive than previously suspected. Noranda conducted another soil survey (Fig. 5.17) that showed strong Sn and Cu anomalies over the calc silicates. Four drillholes were then sited to test these anomalies after pits were dug to confirm mineralisation in bedrock. The drillholes intersected the calc silicates but the structure was unclear and not as predicted from the surface. In addition although extensive (>20 m thick) mineralisation was intersected it was weak (mainly 0.2–0.5% Sn) with few high grade (>1% Sn) intersections. The program therefore confirmed the exploration model but was terminated.

As part of an effort to encourage mineral exploration in the UK, the British Geological Survey investigated the southern part of the area (Bennett et al. 1981). Their surveys also defined the southern geochemical anomaly to the SW of Wheal Prosper, as well as an area of alluvial tin in the west of the area. The extension of the Mulberry trend was not as clearly

detected, probably because of the use of a wider sampling spacing. BGS also drilled two holes (Fig. 5.19) to check the subsurface extension of the western end of the Wheal Prosper system as well as a copper anomaly adjacent to the old pit. Limited assays gave similar results (up to 0.3% Sn) to surface samples.

The third commercial program was a more exhaustive attempt from 1979 to 1982 by Central Mining and Finance Ltd (CMF), a subsidiary of Charter Consolidated Ltd, themselves an associate company of the Anglo American group. Charter Consolidated had substantial interests in tin and wolfram mining, including an interest in the South Crofty Mine and an interest in Tehidy Minerals, the main mineral rights owners in the area. CMF's approach included a re-evaluation of the Mulberry pit and a search for extensions, as well as a search for extensions of the Wheal Prosper system and an attempt to identify any buried granite cupolas with their associated mineralisation. Their program contained four phases. In the first phase the Mulberry pit was resampled and regional soil geochemistry and gravity surveys were undertaken. The gravity survey identified a perturbation in the regional gravity field along the Mulberry trend (Fig. 5.18). Regional shallow soil geochemistry samples were taken on 100 m centers and analysed for Sn, Cu, As, Mo, and W. They showed an ill-defined southern continuation of the Mulberry system in Cu data and a large Cu and As anomaly to the southwest of the Wheal Prosper system. Resampling of the Mulberry pit led to a re-estimation of the resource to 1.1 Mt at 0.44% Sn using a 0.2% Sn cut off. This would form part of the resource for any mine.

The Phase 2 program in 1980 consisted of a detailed geochemical soil survey follow-up of anomalous areas identified in phase 1, as well as the planning of two holes to test the geophysical anomaly. The first hole (CMF1, Fig. 5.19) was a 540 m deep diamond-cored borehole that failed to find any indication of granite or associated contact metamorphism, although it did cut vein mineralisation of 0.51% Sn over 3.1 m at 206.5 m depth. Because of the lack of granite the planned second hole was cancelled.

The results of the soil geochemistry survey encouraged CMF to progress to phase 3 with a deep overburden sampling program to the

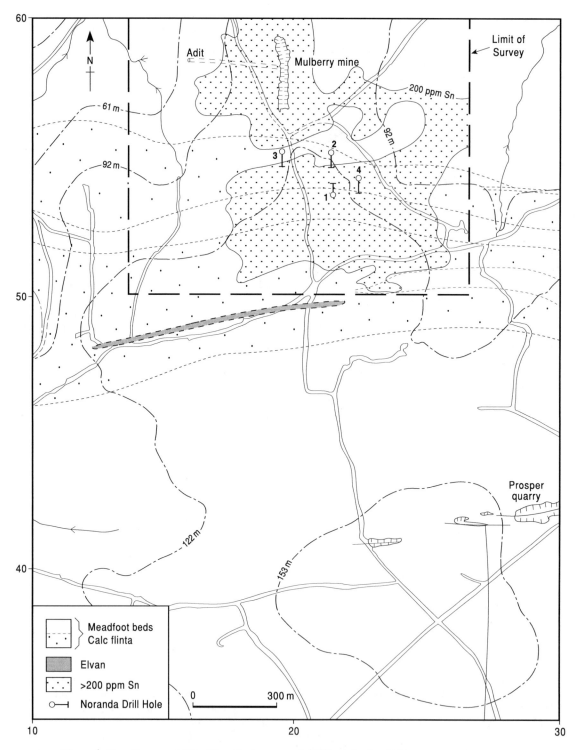

FIG. 5.17   Noranda Kerr Program 1971–72: soil surveys and drilling of calc silicates.

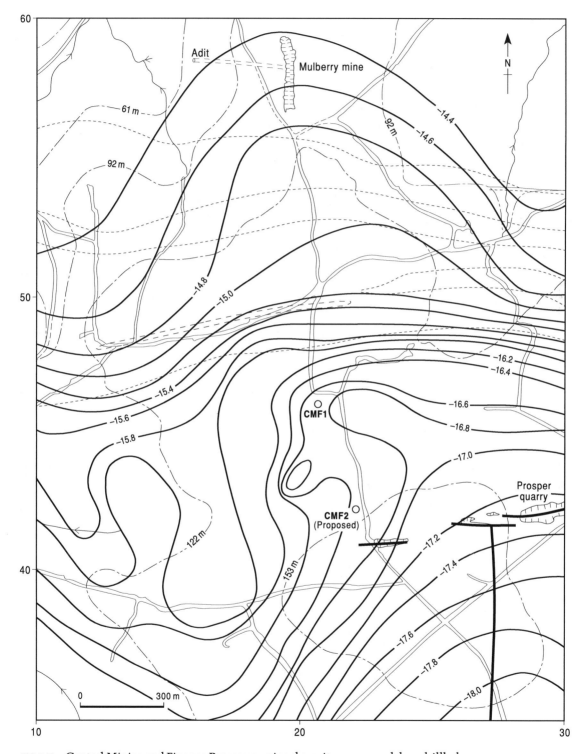

FIG. 5.18 Central Mining and Finance Program: regional gravity survey and deep drillholes.

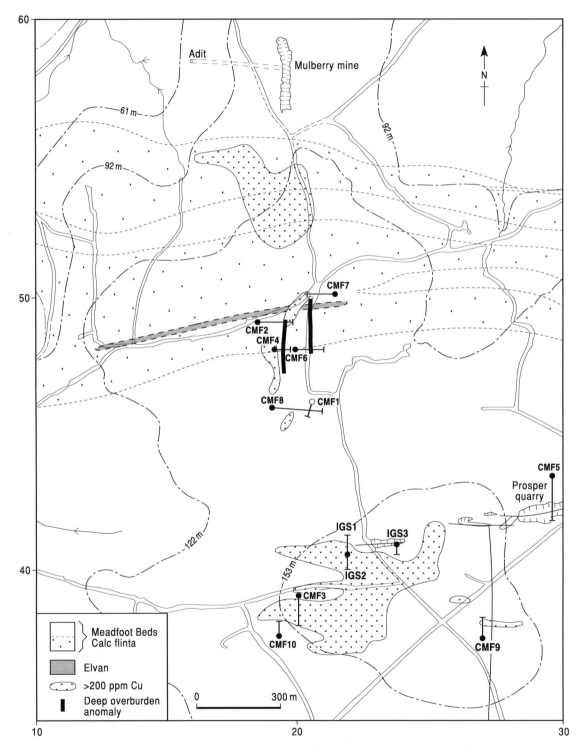

FIG. 5.19   Central Mining and Finance Program 1980–82: drilling to test deep overburden anomalies. The British Geological Survey drillholes to test the Wheal Prosper system (IGS 1–3) are also shown.

south of the Mulberry trend. This defined two parallel mineralized zones with encouraging assays in bedrock. Their trends were tested with five holes (CMF 2,4,6–8, Fig. 5.19), which demonstrated that the mineralized zones are steeply dipping veins. Although some higher grade intersections were made the overall tenor did not appear to be viable, especially when the mineralogy was carefully examined. Laboratory mineral processing trials of the higher grade intersections failed to recover sufficient cassiterite to account for the total tin content determined. It therefore seemed likely that some tin was contained in silicates, probably partly in the tin garnet, malayaite. This tin is not recoverable during conventional mineral processing and considerably devalued these intersections. Two holes were also drilled to test the subsurface extension of the Wheal Prosper pit (CMF 5) and the major copper anomaly (CMF 3). Hole CMF 5 confirmed that surface grades extended to depth but the grade was too low to be of further interest. By contrast, CMF 3 was more encouraging with several intersections, including 1.5 m at 0.5% Sn at 95 m.

The encouragement of Phase 3 in the south of the area led to a Phase 4 (1982) program which further examined the large copper anomaly at the intersection of the Wheal Prosper and Mulberry trends. Two holes were drilled (CMF9 and CMF10) and although one of these cut high grade mineralisation (0.85 m of 7.0% Sn at 15.9 m) the program was terminated.

5.3.1 Valuation of prospects

The valuation of mineral properties at the exploration stage is not straightforward. If a resource is defined then the methods discussed in Chapter 11 can be used. If however the property is an early stage of exploration then valuation is more subjective.

Very little has been published formally on valuation of early stage exploration properties, although a useful starting point is the short course volume published by the Prospectors and Developers Association of Canada (PDAC 1998). In a review of Australian practice, Lawrence (1998) cites a number of methods:

1 Comparable transaction method. In this method the value of similar transactions is compared.

2 Multiple of exploration spending. The value of the prospect is some multiple (usually between 0.5 and 3) of the amount that has been spent on the property.
3 Joint venture comparison. A value is assigned based on the value of a joint venture of the property or a similar one when the joint venture does not involve a company associated with the holding company.
4 Replacement value. In this method the value is a multiple of the cost of acquiring and maintaining the property, e.g. taxes and pegging costs, corrected for inflation.
5 Geoscience rating method. The value is based on a series of geological parameters such as alteration, width, and grade of any mineralized intersections and proximity to deposits using weightings. The basis for one method of weightings is discussed in Kilburn (1990).

In a practical example Ward and Lawrence (1998) compared these approaches and preferred a comparable transaction approach. They attempted to value a block of ground prospective for gold in northern Ireland based on a proprietary database of European and global transactions. This gave valuations between $US2.1M and $US15.2M. However the most satisfactory estimate was based on the valuation of a similar property in Scotland at $US222 ha^{-1}. When this was applied to northern Ireland it gave a value of $US9.2M. They also cited examples from Alaska and Newfoundland, which showed an inverse relation between property size and value (Fig. 5.20) but

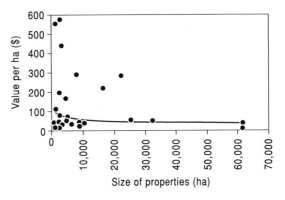

FIG. 5.20 Relation between property size and value for 26 grassroots properties, Labrador in 1995–97. (After Ward & Lawrence 1998.)

a strong positive correlation of value with decreasing distance from major deposits, in the latter case, the Voisey's Bay nickel deposit.

The reader should note the impact of exploration fads. When the Bre-X frenzy (discussed in detail below) was at its height the number of transactions of gold properties in Indonesia increased in number and value by a factor of 20 (Ward & Lawrence 1998). Even large companies can be sucked into paying a premium for properties, particularly when the property has a large resource. A good example is the Voisey's Bay nickel deposit which was bought in 1996 by Inco for $C4.3 billion from junior company Diamond Fields Resources after a bidding war with Inco's great rival Falconbridge (McNish 1999). In 2002 Inco was forced to take a $C1.55 billion write off of the value of the project which should eventually come into production in 2006.

5.4   THE BRE-X MINERALS SAGA

The events surrounding the rise and fall of Bre-X Minerals Ltd provide a good example of some the pitfalls surrounding mineral exploration and the need to audit data and information at every stage. Good accounts can be found in popular books by Goold and Willis (1997) and Wells (1999), as well as a more technical summary in Hefferman (1998).

The company was founded in 1989 by David Walsh, a stock trader in Calgary, Alberta, to explore for diamonds in the wake of the Diamet discovery (Chapter 17.2). In 1993, after being personally bankrupted, he spent some of his last $C10,000 on a visit to Jakarta, Indonesia, to meet an exploration geologist, John Felderhof, whom he knew to be active in gold exploration. Felderhof, who was a member of the team that had discovered the Ok Tedi deposit for Kennecott Copper in 1968, promoted a theory that gold in the Indonesian province of Kalimantan was related to volcanic diatremes. This was plausible as the area included significant gold deposits at Mount Mouro and Kelian. Felderhof and a Filipino geologist, Michael de Guzman, were hired by a Scottish-based company, Waverley Mining Finance, to investigate some properties that they had bought from Australian companies.

Amongst these was a prospect called Busang that had been investigated by drilling in 1988 and 1989, following its discovery by regional reconnaissance. The prospect was considered to report erratic and spotty gold. Following Felderhof's recommendation, Bre-X managed to raise $C80,100 to buy an option on the central 15,060 ha claim at Busang. The first two holes in the central block were barren but the third hole was reported to contain assays of up to 6.6 g t^{-1} Au.

In order to fund further exploration, Bre-X sold shares through a Toronto-based broker and by May 1994 had raised $C5.4M. Spectacular drill intersections were then announced and a resource of 10.3 Mt at 2.9 g t^{-1} Au for the Central Zone in the beginning of 1995. However, these announcements were eclipsed by the discovery of the Southeast zone and two sensational intersections in October 1995 of 301 m grading 4.4 g t^{-1} Au and 137 m grading 5.7 g t^{-1}. These results and the announcement by Walsh and Felderhof that the resources were of the order of 30–40 million ounces of gold drove the share price to $C286 per share and a company value of over $C6 billion (Fig. 5.21). Bre-X then hired a respected Canadian-based firm of consultants to calculate the resource and they confirmed Bre-X's estimate based on the core logs and assays provided by Bre-X of 47M oz of gold and an in situ value of $C16 billion.

It was at this stage that Indonesian politicians became involved, both in the confirmation of Bre-X's exploration licence for the Southeast zone and the business of finding a major mining company partner to develop the deposit. After a number of offers from two major Canadian companies, Bre-X decided after pressure from the Indonesian government over the licence renewal to enter a joint venture. The partners were to be the Indonesian government (10%), companies of Mohammed Hasan (an Indonesian friend of President Suharto) with 20%, and Freeport McMoRan, an American company that runs the Grasberg mine in Indonesia with 15%. As part of the deal Freeport was to invest $US400 million and immediately began a program to confirm the validity of the resource data, a process known as due diligence. Freeport drilled seven holes at Busang in an attempt to confirm Bre-X's drilling results, both by drilling four holes, collared

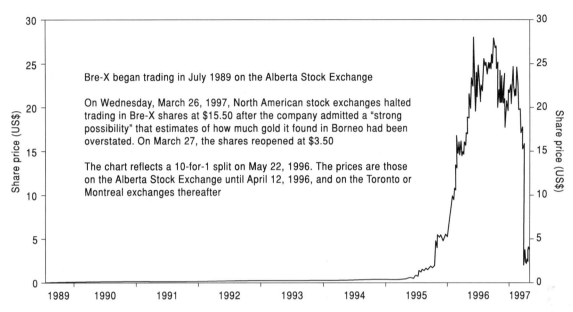

FIG. 5.21   Share price of Bre-X Minerals. (From Goold & Willis 1998. Source: stock prices from Datastream.)

2 m from high-grade Bre-X holes, two holes perpendicular to Bre-X holes, and one to infill between Bre-X holes. Samples were sent to four laboratories, the laboratory used by Bre-X, two in the Indonesian capital Jakarta, and one in New Orleans. In addition, the gold from the cores drilled by Freeport and from samples provided by Bre-X were also selected for SEM examination. Unusually Bre-X had assayed the whole core and not kept a portion as a record. The assay results from all laboratories showed low and erratic gold concentrations in the Freeport cores causing consternation to the Freeport geologists and management. The gold found in the Freeport cores was very fine and unlike the coarse rounded gold grains found in the Bre-X samples. As the due diligence program was bound by conditions of confidentiality, Freeport reported the results to Bre-X and waited for an explanation. Before Bre-X was able to investigate this claim, its site geologist, Michael de Guzman, fell or jumped from a helicopter on his way back to the Busang site from the Prospectors and Developers Association Conference in Toronto where Felderhof had been named Prospector of the Year.

On hearing the news from Freeport, Bre-X executives in Toronto immediately hired a major Canadian consultancy, Strathcona

Mineral Services, to undertake a technical audit of Busang. Strathcona mobilized a team led one of its partners, Graham Farquharson, to go to Jakarta and Busang. After arrival in Jakarta the Strathcona team reviewed the Freeport data and advised Bre-X to make an announcement of the problem on the Toronto stock exchange, on which Bre-X was now listed. The announcement said "there appears to be a strong possibility that the potential gold resources on the Busang project in East Kalimantan, Indonesia have been overstated because of invalid samples and assaying of those samples," This announcement was coupled with a 1-day suspension of trading of Bre-X at the end of which the share price dropped from $C15.80 (there had been a 1:10 split) to $C2.50, a loss of value in the company of approximately $C3 billion (Fig. 5.21). The Strathcona investigation which included drilling and assaying in Perth, Australia of 1470 m of core took a further month. At the end of this, Freeeport's results were confirmed. As the laboratory used by Bre-X was cleared, the only explanation was that gold was added to the samples after crushing and before analysis, a process generally known as salting. Further investigations suggested that this was probably carried out by at least one of the site geologists, including Michael de Guzman. The gold used appeared to have been alluvial and had been bought from local small scale miners. In this case, the salting had been difficult to detect as the gold had been added to sample interval that correlated with alteration and it appeared the fraudsters had used the core logs to determine which samples to salt. The only person to face justice was Felderhof who allegedly sold $C83 million in shares before the collapse and went to live in the Cayman Islands, a country with no extradition treaty with Canada. He was (2004) on trial for insider trading, alleging that he sold substantial shares knowing that the Bre-X ownership over Busang was in doubt. The major questions to be addressed in the wake of the investigation were how such a fraud could be detected and how further frauds of this nature could be prevented. There were a number of irregular procedures at Busang that should have been detected:

1 The whole core was assayed and no part was kept for a record.
2 The gold grains reported were more consistent with an alluvial derivation than from a bedrock source.
3 Pre-feasibility testing indicated a very high recovery by gravity methods – very unusual for epithermal gold deposits.
4 The reported occurrence of near surface sulfide in a deeply weathered environment.
5 The lack of alluvial gold in the stream draining the Southeast zone.
6 The wide spaced drilling on the Southeast zone (250 by 50 m) that was much larger than normal.

The result of the Bre-X fraud has been much stricter enforcement of disclosure of ore reserve calculations methods and the development of formal codes such as the JORC code (see section 10.4). The Bre-X affair also marked the end of the exploration boom of the late 1990s and many investors refused to invest in junior mining stocks, although some made similar mistakes in the subsequent dotcom boom.

5.5   SUMMARY

Following the identification of areas with possible mineralisation and the acquisition of land, the next step is to generate targets for drilling or other physical examination such as pitting and trenching or, in mountainous places, by driving adits into the mountainside. Budgeting and the organization of suitable staff are again involved.

If the area is remote then suitable topographical maps may not be available and the ground must be surveyed to provide a base map for geological, geochemical, and geophysical work. Some of the geological investigations may include the mapping and sampling of old mines and the dispatch of samples for assaying. Any of this work may locate areas of mineralisation which may be recognized by topographical effects, coloring, or wall rock alteration. Different ore minerals may give rise to characteristic color stains and relic textures in weathered rocks and gossans and below these supergene enrichment may have occurred. All these investigations together with pitting and trenching may indicate some promising drilling targets.

Geologists should never rush into drilling which is a most expensive undertaking! Very careful planning is necessary (section 5.2.1) and drilling programs require meticulous monitoring. Once the drilling is in full swing deciding when to stop may be very difficult and various possibilities are reviewed in section 5.2.3.

If an exploration program does not resolve the potential of some mineralized ground then the operating company may sell the option, with or without the data obtained on it, to a new investor. This is often nowadays termed recycling and it enables the original operator to recover some of its investment. Such recycling is exemplified by the Mulberry Prospect in Cornwall (section 5.3). The buying and selling of prospects can be fraught with problems, particularly of valuation (section 5.3.1) and also the possibility of fraud (section 5.4).

5.6   FURTHER READING

Basic field techniques are covered in Reedman (1979), Peters (1987), Chaussier and Morer (1987), and Majoribanks (1997). Reviews of geochemical, geophysical and remote sensing techniques, covered in Chapters 6–8, as well as some case histories of their integration, can be found in the volume edited by Gubins (1997).

Case histories of mineral exploration are scattered through the literature but the volume of Glasson and Rattigan (1990) compiled much Australian experience and those of Sillitoe (1995, 2000) that of the Pacific Rim area. The volume of Goldfarb and Neilsen (2002) provides an update of exploration in the 1990s and early 2000s. Evaluation techniques from exploration to production are well covered by Annels (1991).

6

REMOTE SENSING

MICHAEL K.G. WHATELEY

6.1 INTRODUCTION

Geologists have been using aerial photography to help their exploration efforts for decades. Since the advent of satellite imagery with the launch of the first earth resources satellite (Landsat 1) in 1972, exploration geologists are increasingly involved in interpreting digital images (computerized data) of the terrain. Recent technological advances now provide high resolution multispectral satellite and airborne digital data. More recently, geologists involved in research and commercial exploration have been seeking out the more elusive potential mineral deposits, e.g. those hidden by vegetation or by Quaternary cover. Usually geochemical, geophysical and other map data are available. It is now possible to express these map data as digital images, allowing the geologist to manipulate and combine them using digital image processing software, such as Erdas Imagine, ERMapper TNT MIPS, and geographical information systems (GIS). These latter techniques are discussed in detail in section 9.2.

As image interpretation and photogeology are commonly used in exploration programs today, it is the intention in this chapter to describe a typical satellite system and explain how the digital images can be processed, interpreted, and used in an exploration program to select targets. More detailed photogeological studies using aerial photographs or high resolution satellite images are then carried out on the target areas.

6.1.1 Data collection

Remote sensing is the collection of information about an object or area without being in physical contact with it. Data gathering systems used in remote sensing are:
1 photographs obtained from manned space flights or airborne cameras, and
2 electronic scanners or sensors such as multispectral scanners in satellites or aeroplanes and TV cameras, all of which record data digitally.

Most people are familiar with the weather satellite data shown on national and regional television. For most geologists and other earth scientists, multispectral imagery is synonymous with NASA's Landsat series. It is images from these satellites that are most readily available to exploration geologists and they are discussed below. The use of imagery and digital data from multispectral sources such as NASA's flagship satellite, Terra, or commercial satellites such as QuickBird and Ikonos, are becoming more widespread. Remote sensing data gathering systems are divided into two fundamental types, i.e. those with passive or active sensors.

Passive sensors

These sensors gather data using available reflected or transmitted parts of the electromagnetic (EM) spectrum, i.e. they rely on solar illumination of the ground or natural thermal radiation for their source of energy respectively. Some examples are:

1 Landsat Multispectral Scanner (MSS);

2 Landsat Thematic Mapper (TM) utilizes additional wavelengths, and has superior spectral and spatial resolution compared with MSS images;

3 Advanced Spaceborne Thermal Emission and Reflection Radiometer (ASTER) on NASA's satellite Terra;

4 SPOT, a French commercial satellite with stereoscopic capabilities;

5 high resolution commercial satellites, such as Ikonos (Infoterra 2004), the first of the next generation of high spatial resolution satellites, and QuickBird (Ball Aerospace & Technologies Corp 2002);

6 Space Shuttle;

7 airborne scanning systems that have even greater resolution and can look at more and narrower wavebands.

Different satellites collect different data with various degrees of resolution (Fig. 6.1) depending on their application. For geological applications resolution of the order of meters of Landsat is required in contrast to the tens of kilometers of NOAA (National Oceanic and Atmospheric Administration) weather satellites.

Active sensors

These sensors use their own source of energy. They emit energy and measure the intensity of energy reflected by a target. Some examples are Radar (microwave) and Lasers (European Space Agency 2004).

6.2   THE LANDSAT SYSTEM

The first Landsat satellite, originally called Earth Resources Technology Satellite (ERTS), was launched in July 1972. Seven satellites in the series have been launched (Table 6.1). Landsat has been superseded by the Terra satellite, hosting the ASTER sensors (Abrams & Hook 2002).

Each satellite is solar powered and has a data collecting system which transmits data to the home station (Fig. 6.2). On Landsat 1 to 3 data were recorded on tape when out of range of the home station for later transmission. Tracking and data relay system satellites (TDRSS) were operational from March 1984 (Fig. 6.2). Landsat 7 has only a direct downlink capability with recorders.

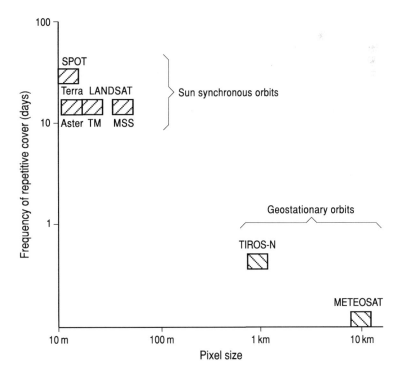

FIG. 6.1 Different resolution and frequencies of some satellite imagery.

TABLE 6.1   Characteristics of several remote sensing systems.

	Landsat MSS	Thematic mapper (TM)	Enhanced TM	ASTER	SPOT/HRV
Launch	1 July 1972–Jan 1978 2 Jan 1975–July 1983 3 Mar 1978–Sept 1983 4 July 1982–July 1987 5 Mar 1984–present 6 Oct 1993–Oct 1993	4 July 1982–June 2001 5 Mar 1984–present	6 Oct 1993–Oct 1993 7 April 1999–present	1 Dec 1999–present	1 Jan 1986–Dec 1993 2 Jan 1990–Dec 1997 3 Sept 1993–1998 4 1997–2001 5 1998–Present
Altitude (km)	918 (1–3)	705	705	705	822
Repeat coverage (days)	18	16	16	16	26
Spectral bands (µm)	4* 0.50–0.60 5 0.60–0.70 6 0.70–0.80 7 0.80–1.10	1 0.45–0.53 2 0.52–0.60 3 0.63–0.69 4 0.76–0.90 5 1.55–1.75 6 10.40–12.5 7 2.08–2.35	1 0.45–0.551 2 0.525–0.605 3 0.630–0.690 4 0.750–0.900 5 1.550–1.750 6 10.40–12.5 7 2.09–2.35 P 0.52–0.90	1 0.52–0.69 2 0.63–0.69 3 0.78–0.86 5 1.60–1.70 6 2.145–2.185 7 2.185–2.225 8 2.295–2.365 9 2.360–2.430 10 8.125–8.475 11 8.475–8.825 12 8.925–9.275 13 10.25–10.95 14 10.95–11.65	1 0.50–0.59 2 0.61–0.68 3 0.79–0.89 4 1.58–1.75 (SPOT 4) P 0.51–0.73
IFOV (m)	79×79 (1–3) 82×82 (4–5)	30×30 (bands 1–5, 7) 120×120 (band 6)	15×15 (P) 30×30 (bands 1–5, 7) 60×60 (band 6)	15×15 (bands 1–3) 30×30 (bands 4–9) 90×90 (bands 10–14)	20×20 (bands 1–3) 10×10 (P)
Scene dimensions (km)	185×185	185×185	183×175	60×60	60×60
Pixels per scene (×10⁶)	28	231	231	21	27 (bands 1–3)

* The MSS bands on Landsats 1–3 were numbered 4, 5, 6, and 7 because of a three-band RBV sensor on Landsats 1 and 2.
P, panchromatic mode.

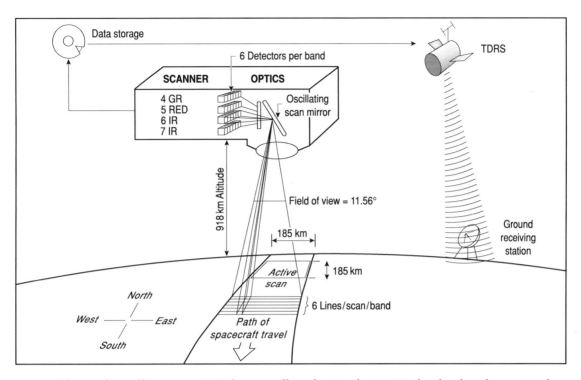

FIG. 6.2   Schema of a satellite system. MSS data are collected on Landsats 1–3 in four bands and transmitted via the TDRSS (relay satellite) to the ground receiving station. (Modified after Lo 1986, Sabins 1997.)

The Landsat series was designed initially to provide multispectral imagery for the study of renewable and nonrenewable resources, particularly land use resources using MSS and Return Beam Vidicon (RBV) cameras. RBV images are not commonly used today. Geologists immediately recognized the geological potential of the Landsat images and in July 1982 when Landsat 4 was launched it housed, in addition to MSS, a Thematic Mapper which scans in seven bands, two of which (5 and 7) were chosen specifically for their geological applicability (Table 6.1 & Fig. 6.6). Landsats 5 and 7 also have TM. Landsat 6 was lost at launch. The additional features of the ASTER sensors are their high spatial and radiometric resolution over similar parts of the spectrum as TM, broad spectral coverage (visible- through thermal-infrared), and stereo capability on a single path.

Not all wavelengths are available for remote sensing (Fig. 6.3). Even on an apparently clear day ultraviolet (UV) light is substantially absorbed by ozone in the upper atmosphere. UV, violet, and blue light are particularly affected by scattering within the atmosphere and are of little use in geology. Blue light is only usefully captured using airborne systems. The blanks in the spectrum are due mainly to absorption by water vapor, CO_2 and O_3 (Fig. 6.3). Only wavelengths greater than 3 μm, e.g. microwave wavelengths (or radar), can penetrate cloud (Fig. 6.3), so radar imagery is particularly useful in areas of considerable cloud cover (e.g. equatorial regions).

6.2.1   Characteristics of digital images

MSS data are recorded by a set of six detectors for each spectral band (Table 6.1). Six scan lines are simultaneously generated for each of the four spectral bands (Fig. 6.3). Data are recorded in a series of sweeps and recorded only on the eastbound sweep (Fig. 6.2). TM data are recorded on both east- and westbound sweeps and each TM band uses an array of 16 detectors (band 6 uses only four detectors). There is continuous data collection and the data are

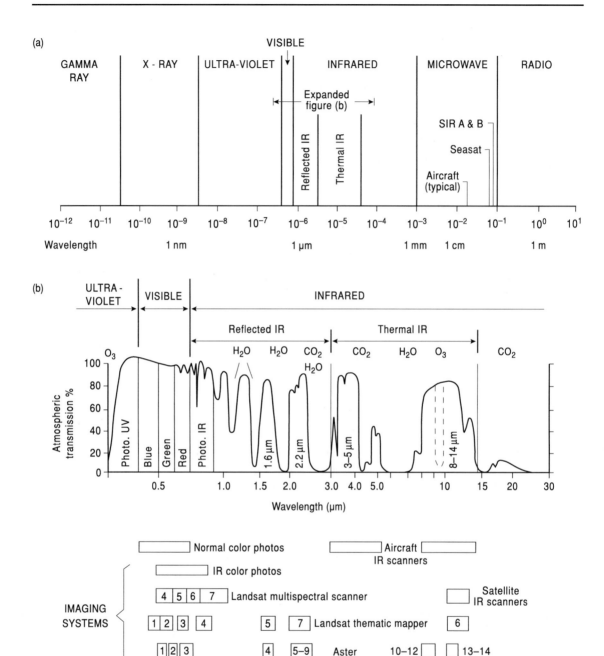

FIG. 6.3 (a) The electromagnetic spectrum. (b) Expanded portion of the spectrum showing the majority of the data-gathering wavelengths. Additional data are acquired in the microwave part of the spectrum for use by aircraft and to measure the surface roughness of either wave action or ice accumulation. Certain parts of the spectrum are wholly or partly absorbed by atmospheric gases. Atmospheric windows where transmission occurs are shown and the imaging systems that use these wavelengths are indicated. (Modified after Sabins 1997.)

transmitted to earth for storage on Computer Compatible Tapes (CCTs). These tapes are then processed on a computer to produce images. The width of scene dimensions on the surface under the satellite (its swath) is 185 km. The data are (for convenience) divided into sets that equal 185 km along its path.

The satellites have a polar, sun synchronous orbit, and the scanners only record on the southbound path (Fig. 6.2) because it is night-time on the northbound path. The paths of Landsats 1 to 3 shifted west by 160 km at the equator each day so that every 18 days the paths repeat resulting in repetitive, worldwide, MSS coverage. Landsats 4, 5 and 7 and the Terra satellite have a slightly different orbit result-ing in a revisit frequency of 16 days. Images are collected at the same local time on each pass – generally between 9.30 and 10.30 a.m. to ensure similar illumination conditions on adjacent tracks. Successive paths overlap by 34% at 40°N and 14% at the equator. Landsat does not give stereoscopic coverage, although it can be added digitally. ASTER does give stereoscopic coverage.

6.2.2 Pixel parameters

Digital images consist of discrete picture ele-ments, or pixels. Associated with each pixel is a number that is the average radiance, or bright-ness, of that very small area within the scene. Satellite scanners look through the atmosphere at the earth's surface, and the sensor therefore measures the reflected radiation from the sur-face and radiation scattered by the atmosphere. The pixel value represents both. Fortunately, the level of scattered radiation is nearly con-stant, so changes in pixel value are essenti-ally caused by changes in the radiance of the surface.

The Instantaneous Field of View (IFOV) is the distance between consecutive measure-ments of pixel radiance (79 m × 79 m in MSS), which is commonly the same as the pixel size (Table 6.1). However, MSS scan lines overlap, hence the pixel interval is less than the IFOV (56 m × 79 m). The TM and ASTER detectors have a greatly improved spatial resolution, giving an IFOV of 30 m × 30 m.

The radiance for each pixel is quantified into discrete gray levels and a finite number of bits are used to represent these data (Table 6.2). Table 6.2 shows how decimal numbers in the range 0–255 can be coded using eight indi-vidual bits (0 or 1) of data. Each is grouped as an 8 bit "byte" of information, with each bit used to indicate ascending powers of 2 from 2^0 (=1) to 2^7 (=128). This scheme enables us to code just 256 (0–255) values. On the display terminal, 256 different brightness levels are used, cor-responding to different shades of gray ranging

TABLE 6.2   The principle of binary coding using 8 bit binary digits (bits).

	Place values									
Binary	2^7	2^6	2^5	2^4	2^3	2^2	2^1	2^0		
Decimal	128	64	32	16	8	4	2	1		
	0	0	0	0	0	0	0	0	=	0
	0	0	0	0	0	0	0	1	=	1
	0	0	0	0	0	0	1	0	=	2
	0	0	0	0	0	1	0	0	=	4
	0	0	0	0	1	0	0	0	=	8
	0	0	0	1	0	0	0	0	=	16
	0	0	1	0	0	0	0	0	=	32
	0	1	0	0	0	0	0	0	=	64
	1	0	0	0	0	0	0	0	=	128
	1	0	0	0	0	1	0	1	=	128 + 4 + 1
									=	133
	1	1	1	1	1	1	1	1	=	128 + 64 + 32 + 16 + 8 + 4 + 2 + 1
									=	255

TABLE 6.3   False colors assigned to Landsat MSS bands.

MSS band	Wavelength (μm)	Natural color recorded	False color assigned
4	0.5–0.6	Green	Blue
5	0.6–0.7	Red	Green
6	0.7–0.8	IR	Red
7	0.8–1.1	IR	Red

from black (0) to white (255), giving us a gray scale. Commonly 6 bits per pixel (64 gray levels) are used for MSS and 8 bits per pixel (256 gray levels) for TM and ASTER. A pixel can therefore be located in the image by an x and y co-ordinate, while the z value defines the gray-scale value between 0 and 255. Pixel size determines spatial resolution whilst the radiance quantization affects radiometric resolution (Schowengerdt 1983).

6.2.3   Image parameters

An image is built up of a series of rows and columns of pixels. Rows of pixels multiplied by the number of columns in a typical MSS image (3240 × 2340 respectively) gives approximately 7M pixels per image. The improved resolution and larger number of channels scanned by the TM results in nine times as many pixels per scene, hence the need for computers to analyze the data. Remote sensing images are commonly multispectral, e.g. the same scene is imaged simultaneously in several spectral bands. Landsat MSS records four spectral bands (Table 6.3), TM records seven, and ASTER records 14 spectral bands. Visible blue light is not recorded by Landsat MSS or ASTER, but it is recorded in TM band 1. By combining three spectral bands and assigning a color to each band (so that the gray scale becomes a red, green, or blue scale instead), a multispectral false color composite image is produced.

The image intensity level histogram is a useful indicator of image quality (Fig. 6.4). The histogram describes the statistical distribution of intensity levels or gray levels in an image in terms of the number of pixels (or percentage of the total number of pixels) having each gray level. Figure 6.4 shows the general characteristics of histograms for a variety of images.

A photographic print of a geometrically corrected image (Fig. 6.5) is a parallelogram. The earth rotates during the time it takes the satellite to scan a 185 km length of its swath, resulting in the skewed image (Fig. 6.11f).

6.2.4   Availability of Landsat data

1 Landsat CCTs or CDs of MSS, TM, or ASTER imagery are available for computer processing (Box 6.1).
2 Black and white, single band prints in a standard 23 cm × 23 cm format at a scale of 1:1000,000 are available from the above sources.
3 Color composite prints in a similar format are also available. It is possible to enlarge Landsat MSS images up to 1:100,000, before the picture quality degrades severely.

6.2.5   Mosaics

Each satellite image has a uniform scale and relative lack of distortion at the edges. Black and white or color prints of each scene can be cut and spliced into mosaics. With the repetitive cover most of the world has cloud-free Landsat imagery available. The biggest difficulty is in matching the gray tone during printing, and the images from different seasons. Scenes can be merged digitally to alleviate gray tone differences.

6.2.6   Digital image processing

Data from the satellites are collected at the ground station on magnetic tape, i.e. in a digital form. Data are loaded into image processing computer systems from CCTs or CDs. The main functions of these systems in a geological

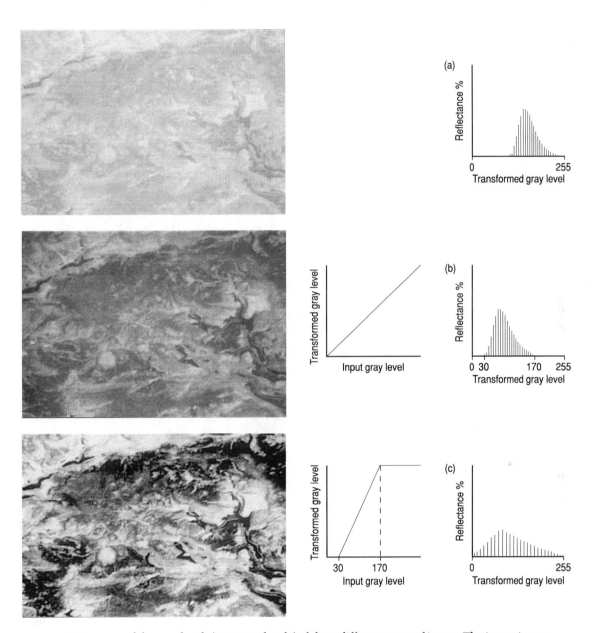

FIG. 6.4 Histograms of the gray levels (intensity levels) of three different types of image. The image is part of Landsat 5, TM band 3, taken on 26 April 1984 of the High Peak District of northern England. The area is approximately 15 × 15 km. The dark areas are water bodies; those in the the upper left corner are the Longdale reservoirs while those in the right center and lower right are the Derwent Valley reservoirs. The two E–W trending bright areas in the lower right of the images are the lowland valleys of Edale and Castleton. (From Mather 1987.) (a) A bright, low intensity image histogram with low spectral resolution. (b) A dark, low intensity, low spectral resolution, image histogram. (c) A histogram with high spectral resolution but low reflectance values. Increased spectral resolution has been achieved with a simple histogram transformation (also known as a histogram stretch). (Modified after Schowengerdt 1983.)

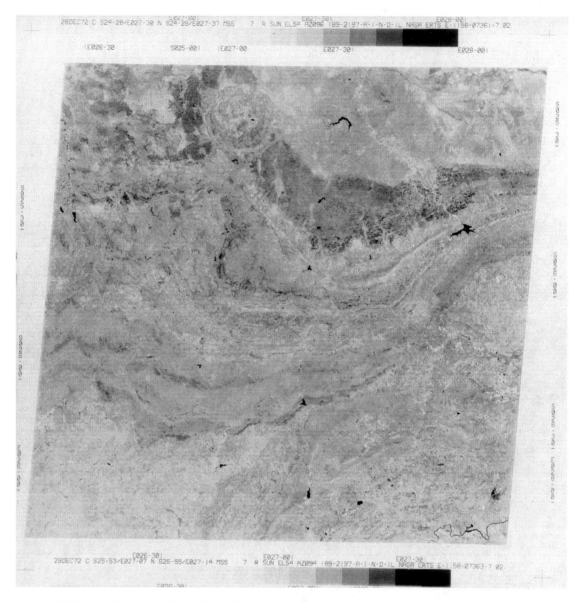

FIG. 6.5   A black and white Landsat 1 (ERTS), band 7 photograph of part of the Bushveld Igneous Complex in South Africa. This summer image (wet season) illustrates the strong geological control over vegetation and variations in soil moisture. The circular Pilanesberg Alkali Complex is seen in the north with the layered basic and ultra-basic intrusion to the southeast of it (dark tone). The banding in the Transvaal Supergroup dolomites, in the south center (medium gray tone) was unsuspected until viewed on Landsat photographs.

BOX 6.1   Some sources of remotely sensed data.

- US Geological Survey 1981. Worldwide directory of national earth science agencies and related international organizations. Geological Survey Circular 834, USGS, 507 National Center, Reston, Virginia, 22092, USA. http://www.usgs.gov and http://glovis.usgs.gov/
- SPOT-Image, 18 avenue Eduoard Belin, F31055, Toulouse Cedex, France. http://www.spot.com
- Meteorological satellite imagery.
US Department of Commerce, NOAA/NESDIS/NSDC, Satellite Data Services Division (E/CCGI), World Weather Building, Room 10, Washington DC, 20233, USA http://db.aoml.noaa.gov/dbweb/
- Landsat (pre-September 1985).
US Geological Survey, EROS Data Center (EDC), Mundt Federal Building, Sioux Falls, South Dakota, 57198, USA
- Landsat (post September 1985).
Earth Observation Satellite Company (EOSAT), 1901 North Moore Street, Arlington, Virginia, 22209, USA. http://edcimswww.cr.usgs.gov/pub/imswelcome/
- Landsat imagery is also available from local and national agencies and free from the GLCF. http://glcf.umiacs.umd.edu/intro/landsat7satellite.shtml
- Infoterra, Europa House, The Crescent, Southwood, Farnborough, Hampshire, GU14 0NL, UK. http://www.infoterra-global.com

---

exploration program are: image restoration, image enhancement, and data extraction (Sabins 1997). Drury (2001) explains the methods of digital image processing in more detail and shows how they can be applied to geological remote sensing.

Image restoration

Image restoration is the process of correcting inherent defects in the image caused during data collection. Some of the routines used to correct these defects are:

1 replacing lost data, i.e. dropped scan lines or bad pixels;
2 filtering out atmospheric noise;
3 geometrical corrections.

The last named correct the data for cartographic projection, which is particularly important if the imagery is to be integrated with geophysical, topographical, or other map-based data.

Image enhancement

Image enhancement transforms the original data to improve the information content. Some of the routines used to enhance the images are as follows:

1 *Contrast enhancement.* An image histogram has already been described in section 6.2.3 (Fig. 6.4). A histogram of a typical untransformed image has low contrast (Fig. 6.4a) and in this case the input gray level is equivalent to the transformed gray level (Schowengerdt 1983, Drury 2001). A simple linear transformation, commonly called a contrast stretch, is routinely used to increase the contrast of a displayed image by expanding the original gray level range to fill the dynamic range of the display device (Fig. 6.4b).

2 *Spatial filtering* is a technique used to enhance naturally occurring straight feature such as fractures, faults, joints, etc. Drury (2001) and Sabins (1997) described this in more detail.

3 *Density slicing* converts the continuous gray tone range into a series of density intervals (slices), each corresponding to a specific digital range. Each slice may be given a separate color or line printer symbol. Density slicing has been successfully used in mapping bathymetric variations in shallow water and in mapping temperature variations in the cooling water of thermal power stations (Sabins 1997).

4 *False color composite images* of three bands, e.g. MSS bands 4, 5, and 7, increase the amount of information available for interpretation.

Information extraction

Routines employed for information extraction use the speed and decision-making capability of the computers to classify pixels on some predetermined gray level. This is carried out interactively on computers by band ratioing, multispectral classification, and principal component analysis. These can all be used to enhance specific geological features.

1 *Ratios* are prepared by dividing the gray level of a pixel in one band by that in another band. Ratios are important in helping to recognize ferruginous and limonitic cappings (gossans) (Fig. 6.6). Rocks and soils rich in iron oxides and hydroxides absorb wavelengths less than 0.55 μm and this is responsible for their strong red coloration (Drury 2001). These iron oxides are often mixed with other minerals which mask this coloration. A ratio of MSS band 4 over band 5 will enhance the small contribu-

tion of iron minerals (Fig. 6.6). Similarly a ratio of MSS band 6 over band 7 will discriminate areas of limonite alteration. Unfortunately limonite also occurs in unaltered sedimentary and volcanic rocks. However, by ratioing TM bands, mineralogical spectral characteristics related to alteration can be detected. The Landsat TM scanner has two critical, additional bands (TM bands 5 and 7), which are important in being used to identify hydrothermally altered rocks. Absorption in the "clay band" causes low reflectance at 2.2 μm (TM band 7), but altered rocks have a high reflectance at 1.6 μm (TM band 5). A ratio of bands 5/7 will result in enhancement of altered rocks (Fig. 6.6). ASTER has even finer spectral bands in the "clay band" which gives enhanced alteration discrimination (Table 6.1).

2 *Multispectral classification.* Using this routine a symbol or color is given to a pixel or small group of pixels which have a high

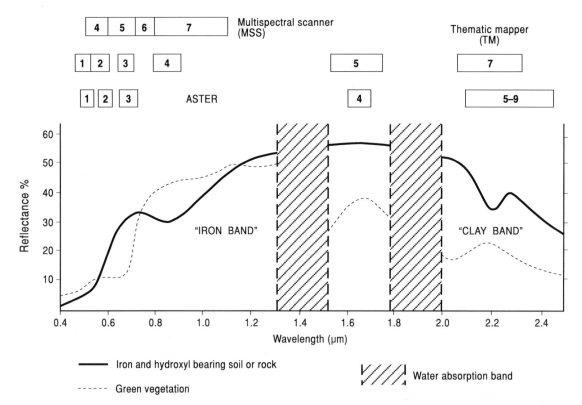

FIG. 6.6   A reflectance profile of the visible and IR parts of the EM spectrum, showing the changing reflectance profile of soil associated with "typical" hydrothermal alteration compared with that of green vegetation. (Modified after Settle et al. 1984.)

probability of representing the same kind of surface material, e.g. alteration. This decision is best made by computers. Multispectral classification is particularly useful for a large area or a region covered by dense vegetation, as differences in the vegetation reflect the underlying geology.

3 *Principal component analysis* is used to enhance or distinguish lithological differences, because spectral differences between rock types may be more apparent in principal component images than in single bands. The reflectances in different bands in MSS, TM, or ASTER images have a high degree of correlation. Principal component analysis is a commonly used method to improve the spread of data by redistributing them about another set of axes. This effectively exaggerates differences in the data.

6.2.7 Interpretation

Two approaches are used to extract geological information from satellite imagery.

1 *Spectral approach.* Spectral properties are used to separate units in image data based on spectral reflectance. This is done interactively on computers using multispectral data in areas with or without dense vegetation.

2 *Photogeological approach.* Known weathering and erosional characteristics (topographical expression) are used, on black and white or color prints, to imply the presence of geological structure or lithology. Photogeological elements include topography, erosion, tone, texture, drainage pattern, vegetation, and land use. These elements are discussed in section 6.4.

6.2.8 Applications of Landsat satellite imagery

The use of satellite imagery is now a standard technique in mineral exploration (Nash et al. 1980, Goetz & Rowan 1981, Peters 1983, Drury 2001). It has also been used in structural investigations (Drury 1986) and in hydrogeology (Deutsch et al. 1981). In mineral exploration Landsat imagery has been used to provide basic geological maps, to detect hydrothermal alteration associated with mineral deposits, and to produce maps of regional and local fracture patterns, which may have controlled mineralisation or hydrocarbon accumulations.

Structural maps

The synoptic view afforded by MSS and TM imagery is ideal for regional structural analyses, especially if the scene chosen is illuminated with a low sun angle (e.g. autumn or spring) which emphasizes topographical features. Manual lineament analysis on overlays of photographic prints is then carried out (usually after a spatial filtering technique has been employed) and lineaments are digitized. The original apparently random pattern can be quantified by computer processing, thus providing an objective method for evaluating lineaments. A rose diagram is prepared to depict the strike–frequency distribution. At the Helvetia porphyry copper test site in Arizona, the three trends identified by these methods correspond to deformation events in the Precambrian, the Palaeozoic, and the Laramide Orogeny (Abrams et al. 1984).

The most common structures seen on Landsat images are faults, fractures, lineaments of uncertain nature (see Fig. 9.10), and circular features. In conjunction with the statistical approach above, deductions regarding stress patterns in an area can be made. These structural studies may yield clues to the location of concealed mineral deposits. Linear zones, such as the 38th Parallel Lineament of the USA (Heyl 1972), consist of weak crustal areas that have provided favorable sites for upwelling intrusives and mineralizing fluids. These areas contain faults or fractures associated with deep-seated basement structures and the intersections of these features with other faults can provide favorable loci for mineralisation.

Lithology and alteration maps

Landsat imagery has made significant contributions to the advancement of geological mapping both in known and unmapped areas of the world. Many third world countries now have geological maps which were previously too expensive to produce by conventional field mapping. Today regional geological mapping often starts with Landsat imagery, followed by rapid field reconnaissance for verification where possible. The maps thus produced are used to select exploration target zones on the basis of favorable geology and structure. Detailed follow-up photogeology and field work then takes place.

Ferruginous residual deposits (gossans) which overlie mineralized ground can often be identified from color anomalies on enhanced false color ratio composites, whereas these may not be detected on standard imagery, e.g. at the Silver Bell porphyry copper site (Abrams et al. 1984) and the Cuprite Mining District in Nevada (Abrams & Hook 2002).

6.2.9   Advantages of Landsat imagery

1 This imagery provides a synoptic view of large areas of the Earth's crust (185 km × 185 km), revealing structures previously unrecognized because of their great extent.
2 The evolution of rapidly developing dynamic geological phenomena can be examined through the use of successive images of the same area produced at 16- to 26-day intervals, e.g. delta growth or glaciation.
3 Some geological features are only intermittently visible, e.g. under certain conditions of climate or vegetation cover. Landsat offers "revisit capability" and multitemporal coverage (Viljoen et al. 1975).
4 Digital images can be displayed in color. This is useful in mapping rock types and alteration products, either direct from rocks or from vegetation changes (Abrams & Hook 2002).
5 Landsat is valuable in providing a tool for rapid mapping of regional and local fracture systems. These systems may have controlled ore deposit location.
6 It is very cost effective, with an outlay of only a few pence km^{-2} for map production. The Global Land Cover Facility of University of Maryland (GLCF 2004) offers the largest free source of Landsat data.
7 Computer processing enables discrimination and detection of specified rocks or areas.

6.3   OTHER IMAGING SYSTEMS

6.3.1   SPOT

The first French satellite, SPOT, was launched in February 1986, followed by others with improved technologies (Table 6.1). They are more sophisticated than Landsat, having a high resolution visible (HRV) imaging system, with a 20 m (Spot 5, 10 m) ground resolution in MS mode and a 10 m (Spot 5, 5 m) resolution in panchromatic mode (Table 6.1). SPOT also has an off-nadir viewing capability which means that an area at 45°N can be imaged 11 times in the 26-day orbital cycle. Repeated off-nadir viewing introduces parallax from which stereoscopic image pairs can be produced. The commercial nature of SPOT means that images are slightly more expensive, but the increased resolution and stereoscopic capability mean that the additional cost may be warranted in an exploration program, although SPOT does not have as good a spectral range as TM.

6.3.2   Aster

NASA's satellite, Terra, was launched in December 1999. Terra is the first of a series of multi-instrument spacecraft that are part of NASA's Earth Observing System (EOS). The program comprises a science component and a data system supporting a coordinated series of polar-orbiting and low inclination satellites for long-term global observations of the land surface, biosphere, solid Earth, atmosphere, and oceans. It houses the Advanced Spaceborne Thermal Emission and Reflection Radiometer (ASTER) sensors. ASTER covers a wide spectral region with 14 bands from the visible to the thermal infrared parts of the spectrum with high spatial, spectral, and radiometric resolution (Abrams & Hook 2002). An additional backward-looking near-infrared band provides stereo coverage. The spatial resolution varies with wavelength (Table 6.1).

6.3.3   High resolution satellite systems

The Ikonos-2 satellite was launched in September 1999 and has been delivering commercial data since early 2000 (Infoterra 2004). Ikonos is the first of the next generation of high spatial resolution satellites. Ikonos data records four channels of multispectral data (0.45–0.53, 0.52–0.61, 0.64–0.72, and 0.77–0.88 μm) with a resolution of 4 m and one panchromatic channel with 1 m resolution (0.45–0.90 μm). Ikonos radiometric resolution is far greater than the Landsat scenes, with data collected as

11 bits per pixel (2048 gray tones). Specialist image processing software is required to process these images. Ikonos instruments have both cross- and along-track viewing capabilities, which allows frequent revisit capability, e.g. 3 days at 1 m resolution.

The high resolution imagery satellite QuickBird was launched in October 2001 (DigitalGlobe 2004). It provides a panchromatic channel with 0.6 m resolution (0.4–0.9 μm) and four channels of multispectral, stereoscopic data (0.45–0.52, 0.52–0.60, 0.63–0.69, and 0.76–0.90 μm) with a resolution of 2.44 m. QuickBird also collects 11 bits per pixel. The satellite operates in a 450 km 98 degrees sun-synchronous orbit, with each orbit taking 93.4 minutes. Each scene covers 16.5 × 16.5 km.

6.3.4 Hyperspectral airborne systems

Hyperspectral airborne systems acquire spectral coverage in the 0.4–2.4 μm range. These sensors are capable of acquiring any band combination ranging from an optimal number between 10 and 70 multispectral bands to a full hyperspectral data set of 286 bands (Spectrum Mapping 2003). Using a band configuration tool the user can define a band file for the desired number of bands and individual bandwidth for each band. This band file is then selected in-flight using flight operations software.

6.4 PHOTOGEOLOGY

Photogeology is the name given to the use of aerial photographs for geological studies. To get the best out of photographs geologists must plan the photogeological work in the office and in the field. A typical scheme is:

1 annotation of aerial photographs;
2 compilation of photogeology on to topographical base maps;
3 field checking;
4 re-annotation;
5 re-compilation for production of a final photogeological map.

The common types of aerial photos are: panchromatic black and white photographs (B&W), B&W taken on infrared (IR) sensitive film, color photographs, and color IR.

These films make use of different parts of the spectrum, e.g. visible light (0.4–0.7 μm) and photographic near infrared (0.7–0.9 μm). Panchromatic film produces a print of gray tones between black and white in the visible part of the spectrum. These are by far the most common and cheapest form of aerial photography. Color film produces a print of the visible part of the spectrum but in full natural color. It is expensive but very useful in certain terrains. By contrast, color infrared film records the green, red, and near infrared (to about 0.9 μm) parts of the spectrum. The dyes developed in each of these layers are yellow, magenta, and cyan. The result is a "false color" film in which blue areas in the image result from objects reflecting primarily green energy, green areas in the image result from objects reflecting primarily red energy, and red areas in the image result from objects reflecting primarily in the photographic near infrared portion of the spectrum. Vegetation reflects IR particularly well (Fig. 6.6) and so IR is used extensively where differences in vegetation may help in exploration. Aerial photographs are generally classified as either oblique or vertical.

Oblique photographs

Oblique photographs can be either high angle oblique photographs which include a horizon or low angle oblique photographs which do not include a horizon. Oblique photographs are useful for obtaining a permanent record of cliffs and similar features which are difficult to access, and these photographs can be studied at leisure in the office. Similarly, studies can be made of quarry faces to detect structural problems, to plan potential dam sites, etc.

Vertical photographs

Vertical photographs are those taken by a camera pointing vertically downward. A typical aerial photograph is shown in Fig. 6.7. The principal point is the point on the photograph that lies on the optical axis of the camera. It is found on a photograph by joining the fiducial marks. Normally a title strip includes bubble balance, flight number, photograph number, date and time of the exposure, sun elevation, flight height, and camera focal length.

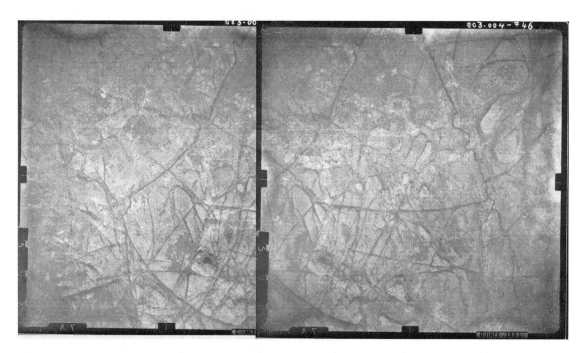

FIG. 6.7   A typical black and white, vertical, aerial photograph stereo pair from Guinea. If you view the pair in stereo you should see the stack in the center. The black marginal rectangles locate the fiducial marks.

Photographs are taken along flight lines (Fig. 6.8). Successive exposures are taken so that adjacent photographs overlap by about 60%. This is essential for stereoscopic coverage. Adjacent pairs of overlapping photographs are called stereopairs. To photograph a large block of ground a number of parallel flight lines are flown. These must overlap laterally to ensure that no area is left unphotographed. This area of sidelap is usually about 30%. Less sidelap is needed in flat areas than in mountainous terrain. Flight lines are recorded on a topographical map and drawn as a flight plan. Each principal point is shown and labeled on the flight plan. These are used for ordering photographs, from local and national Government agencies. At the start of a photogeological exercise, a photographic overview of the area is achieved by taking every alternate photograph from successive flight lines and roughly joining them to make a print laydown. The photographs are not cut, but simply trimmed to overlap and then laid down.

A slightly more sophisticated approach is to make an uncontrolled mosaic. For this, the best fit approach is used and photographs are cut and matched but of course they are then unusable for other purposes. The advantage is that apparent topographical mismatches and tonal changes are reduced to a minimum and one achieves a continuous photographic coverage. Unfortunately the scale varies across the mosaic and there are occasional areas of loss, duplication, or breaks in topographical detail.

These imperfections can be corrected by making a controlled mosaic using scale-corrected, rectified, matched (tonal) prints. However this can be very expensive. When photographs are scale corrected, matched, and produced as a series with contours on them, they are known as orthophotographs.

6.4.1   Scale of aerial photographs

The amount of detail in an aerial photograph (resolution) is dependent upon the scale of the photograph. The simplest way to determine a photograph's scale is to compare the distance between any two points on the ground and on the photographs.

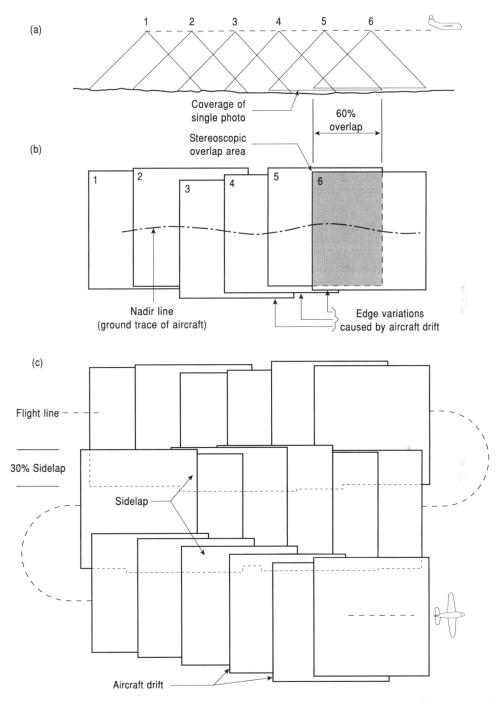

FIG. 6.8   Photographic coverage. (a) Showing the overlap of photographs during exposure. (b) A print laydown of the photographs from a single strip. (c) A print laydown of photographs from adjacent flight lines illustrating sidelap and drift. (Modified after Lillesand et al. 2004.)

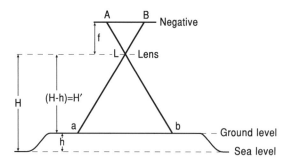

FIG. 6.9   The factors used to calculate the scale of aerial photographs. (Modified after Allum 1966.)

In vertical photographs taken over flat terrain, scale (S) is a function of the focal length (f) of the camera and the flying height above the ground (H′) of the aircraft (Fig. 6.9).

$$\text{Scale}(S) = \frac{f}{H'}$$

H′ is obtained by subtracting the terrain elevation (h) from the height of the aircraft above a datum (H), usually sea level which is the value given by the aircraft's altimeter. The important principle to understand is that photograph scale is a function of terrain elevation. A plane flies at a constant (or nearly constant) height. When a plane flies over varying terrain elevation, such as in mountainous areas, then the scale will vary rapidly across the photographs.

6.4.2   Parallax

All points on a topographical map are shown in their true relative horizontal positions, but points on a photograph taken over terrain of varying height are displaced from their actual relative position. This apparent displacement is known as parallax. Objects at a higher elevation lie closer to the camera and appear larger than similar objects at a lower elevation. The tops of objects are always displaced relative to their bases (Fig. 6.10). This distortion on aerial photographs is known as relief displacement and results in objects appearing to lean away radially from the principal point of a photograph (Fig. 6.7).

The effect of relief displacement is illustrated in Fig. 6.10 where the radial displace-

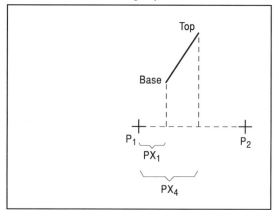

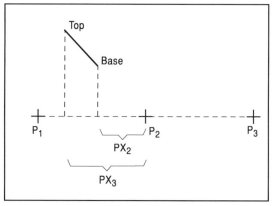

FIG. 6.10   Distortion on aerial photographs, where objects appear to lean away radially from the principal point (P$_1$ on photograph 1 and P$_2$ on photograph 2) of the photograph, is known as relief displacement. The principal points are joined to construct the baseline of the photograph and a perpendicular line is dropped from the base of the object to the baseline on each photograph. The distance from the principal point on photograph 1 to this intersection point is measured (PX$_1$). The same exercise is carried out on photograph 2 (PX$_2$). The parallax of the base of the object is given by the sum of PX$_3$ and PX$_4$. It is obvious that the parallax of the base is less than at the top. The top has greater parallax. (Modified after Allum 1966.)

ment differs on two adjacent photographs. It can be seen that the parallax of the base at the object is less than the parallax of the top of the object. Thus, a difference in elevation, in addition to producing radial displacement, also produces a difference in parallax. It is this difference in parallax that gives a three-dimensional effect when stereopairs are viewed stereoscopically. Allum (1966) gives a good description of relief and parallax.

6.4.3 Photographic resolution

The resolution of a photographic film is influenced by the following factors:
1 *Scale*, which has already been discussed.
2 *Resolving power of the film*. A 25 ISO film (a slow film) has a large number of silver halide grains per unit area and needs a long exposure time to obtain an image. A 400 ISO film (a fast film) has relatively fewer grains per unit area but requires less exposure time. The 25 ISO film has better resolution, while the 400 ISO film tends to be grainy.
3 *Resolving power of the camera lens*. Compare the results from a high quality Nikon lens with those of a cheap camera lens.
4 *Uncompensated camera motion* during exposure.
5 *Atmospheric conditions*.
6 *Conditions of film processing*.
  The effect of scale and resolution is expressed as ground resolution distance (GRD). Typically GRDs for panchromatic film vary from centimeters for low flown photography to about a meter for high flown photography.

6.4.4 Problems with aerial photographs and flying

1 *Drift* – edges of the photograph are parallel to the flight line but the plane drifts off course (Fig. 6.8c).
2 *Scale* – not uniform because of parallax, terrain elevation, aircraft elevation (Fig. 6.11a), and the factors described below.
3 *Tilt* – front to back or side to side, causing distortions of the photograph (Fig. 6.11b,c).
4 *Yaw* (or crab) – the plane turns into wind to keep to the flight line, resulting in photographs whose edges are not parallel to the flight line (Fig. 6.11d).

5 *Platform velocity* – changing speed of the aircraft during film exposure (Fig. 6.11e). This applies particularly to satellite imagery.
6 *Earth rotation* – see section 6.2.3 (Fig. 6.11f).

6.4.5 Photointerpretation equipment

The two main pieces of photointerpretation equipment that exploration geologists use are: (i) field stereoscopes capable of being taken into the field for quick field checking, and (ii) mirror stereoscopes which are used mainly in the office and can view full 23 cm × 23 cm photographs without overlap. These pieces of equipment and their use are fully described by Moseley (1981) and Drury (2001).
  Extra equipment which can be utilized in an exploration program includes a color additive viewer and an electronic image analyzer:
1 A color additive viewer color codes and superimposes three multispectral photographs to generate a false color composite. Multispectral photographs need three or four cameras taking simultaneous images in narrow spectral bands, e.g. 0.4–0.5, 0.5–0.6, 0.6–0.7 μm. Color is far easier to interpret because the human eye can differentiate more colors than gray tones.
2 The electronic image analyzer consists of a closed circuit TV scanner which scans a black and white image and produces a video digital image. The tape is then processed by computer and the results are shown on a TV monitor. This brings us back to the digital image processing described in section 6.2.6.
  High resolution scanners mean that aerial photographs can now be converted into digital images. To some extent these can be treated in a similar way to the satellite digital imagery. Low altitude aerial photography can also be acquired in a high resolution digital form, with numerous spectral bands. These high resolution images can be combined with lower resolution satellite digital imagery to preserve the best features in both. For example, a high spatial resolution aerial photography can be combined with high spectral resolution satellite imagery (Department of Land Information 2003). This produces a product that can distinguish roads, cultural and urban features derived from the aerial photograph, and also provide information on vegetation cover and geology derived from the satellite imagery.

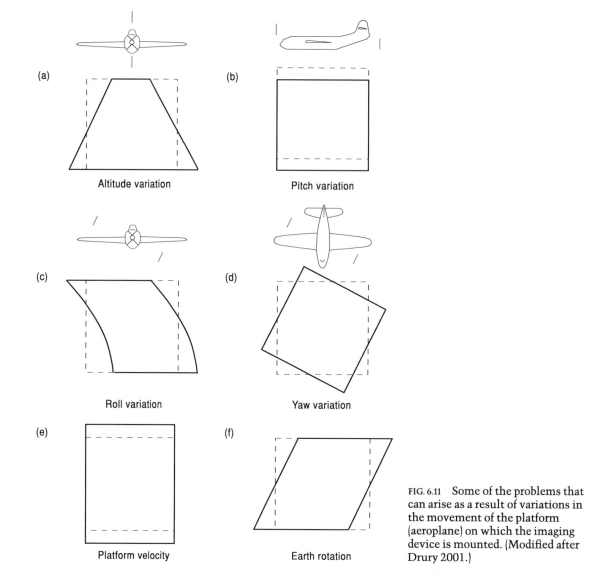

(a) Altitude variation

(b) Pitch variation

(c) Roll variation

(d) Yaw variation

(e) Platform velocity

(f) Earth rotation

FIG. 6.11 Some of the problems that can arise as a result of variations in the movement of the platform (aeroplane) on which the imaging device is mounted. (Modified after Drury 2001.)

6.4.6 Elements of aerial photograph interpretation

Aerial photograph interpretation is based on a systematic observation and evaluation of key elements, and involves the identification from pairs of stereophotographs of relief, tonal and textural variations, drainage patterns and texture, erosion forms, vegetation, and land use (Lillesand et al. 2004). Drury (2001) includes additional characteristics of the surface such as shape (of familiar geological features), context, and scale of a feature.

Topography (relief)

Each rock type generally has its own characteristic topographical form. There is often a distinct topographical change between two rock types, e.g. sandstone and shale. However, it is not an absolute quantity because a dolerite dyke may form a positive feature in certain climatic conditions but a negative feature elsewhere.

Vertical photographs with their 60% overlap result in an exaggeration of relief by three or four times. Consequently slopes and bedding dips appear steeper than they actually are. It is necessary to judge dips and group them into estimated categories, e.g. 0–5°, 6–10°, 11–25°, 26–45°, 46–85°, vertical. The vertical exaggeration works to the advantage of the photogeologist, as it may lead to the identification of subtle changes in slope in otherwise rounded and subdued topographical features.

Tone

Tone refers to the brightness at any point on a panchromatic photograph. Tone is affected by many factors such as: nature of the rock (sandstone is light, but shale is dark), light conditions at the time of photographing (cloud, haze, sun angle), film, filters, and film processing.

These effects mean that we are interpreting relative tone values. In general terms basic extrusive and intrusive igneous rocks (lava, dolerite) have a darker tone, while bedded sandstone, limestone, quartz schists, quartzite, and acid igneous rocks are generally lighter. Mudstone, shale, and slate have intermediate tones.

Subtle differences in rock colors are more readily detected using color photographs, but these are more expensive. Subtle differences in soil moisture and vegetation vigor can be more readily detected using color IR, but even these change with the time of the year.

Texture

There is a large variation in apparent texture of the ground surface as seen on aerial photographs. Texture is often relative and subjective, but some examples are limestone areas which may be mottled or speckled, whilst shale is generally smooth, sandstone is blocky, and granite is rounded.

Drainage pattern

This indicates the bedrock type which in turn influences soil characteristics and site drainage conditions. The six most common drainage patterns are: dendritic, rectangular, trellis, radial, centripetal, deranged (Fig. 6.12). Dendritic drainage (Fig. 6.12a) occurs on relatively homogenous material such as flat-lying sedimentary rock and granite. Rectangular drainage

(Fig. 6.12b) is a dendritic pattern modified by structure, such as a well-jointed, flat-lying, massive sandstone. Trellis drainage (Fig. 6.12c) has one dominant direction with subsidiary drainage at right angles. It occurs in areas of folded sedimentary rocks. Radial drainage (Fig. 6.12d) radiates outwards from a central area, typical of domes and volcanoes. Centripetal drainage (Fig. 6.12e) is the reverse of radial drainage where drainage is directed inwards. It occurs in limestone sinkholes, glacial kettle holes, volcanic craters, and interior basins (e.g. Lake Eyre, Australia and Lake Chad). Deranged drainage (Fig. 6.12f) consists of disordered, short, aimlessly wandering streams typical of ablation till areas.

These are all destructional drainage patterns. There are numerous constructional landforms such as alluvial fans, deltas, glacial outwash plains, and other superficial deposits. These are only indirectly of value in exploration studies (except for placer and sand and gravel exploration) and are described in detail by Siegal and Gillespie (1980) and Drury (2001).

Drainage texture

Drainage texture is described as either coarse or fine (Fig. 6.12h). The coarse texture develops on well-drained soil and rock with little surface run off, e.g. limestone, chalk. Fine texture develops where soils and rocks have poor internal drainage and high run off, e.g. lava and shale.

Erosion

Erosion is a direct extension of the description of drainage above, but gullies, etc., often follow lines of weakness and thus exaggerate features such as joints, fractures, and faults.

Vegetation and land use

The distribution of natural and cultivated vegetation often indicates differences in rock type, e.g. sandstone and shale may be cultivated, while dolerite is left as rough pasture. On the other hand forests may well obscure differences so great care must be taken to draw meaningful conclusions when annotating areas which have a dense vegetation cover. If an area

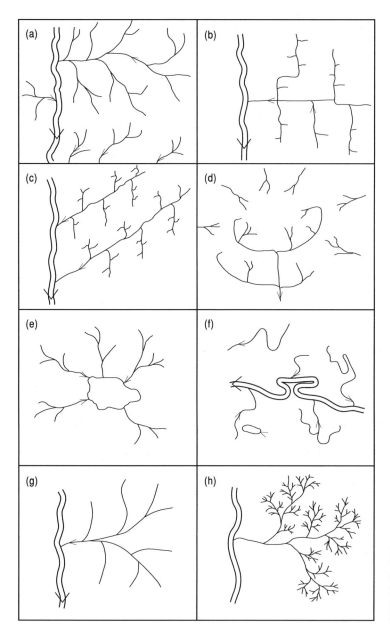

FIG. 6.12   Typical drainage patterns which can be used to interpret the underlying geology (a–f), see text for details. Drainage texture can be coarse (g) or fine (h). (Modified after Lillesand et al. 2004.)

is heavily forested, such as tropical jungles, then the use of conventional photography becomes limited and we have to look at alternative imagery such as reflected infrared (IR) and side-looking airborne radar (SLAR) (Sabins 1997, Drury 2001).

Scanners with narrow apertures viewing the reflectance from vegetation (thematic mappers) are now being used (Goetz et al. 1983). One way in which all imagery can be used is by taking pictures of the same area at different times of the year (multitemporal photography). This may emphasize certain features at a given time of the year. Lines of vegetation, e.g. bushes, trees, etc., are a good indicator of fractures, faults, veins, and joints. The joints are more porous and have more available ground water, producing lines of more vigorous vegetation.

Lineament

This is a word used in photogeology to describe any line on an aerial photograph that is structurally controlled by joints, fractures, faults, mineral veins, lithological horizons, rock boundaries, etc. These have already been discussed above, e.g. streams, gullies, lines of vegetation, etc. It is important to differentiate these from fence lines, roads, rails, etc.

6.4.7 Interpretation

By using the elements of aerial photograph interpretation described above we identify landforms, man-made lines versus photogeological lineaments, vegetation, rock outcrop boundaries, rock types, and rock structures (folds, faults, fractures). Once identified, the geological units and structure are transferred on to a topographical base map.

Most near-surface mineral deposits in accessible regions have been discovered. Emphasis is now on remote regions or deep-seated deposits. The advantages of using imagery in remote, inaccessible regions are obvious. Much information about potential areas with deep-seated mineralisation can be provided by interpretation of surface features, e.g. possible deep-seated faults and/or fractures which may have been pathways for rising mineralizing solutions.

Aerial photographs are normally taken between mid-morning and mid-afternoon when the sun is high and shadows have a minimal effect. Low sun angle photographs contain shadow areas which, in areas of low relief, can reveal subtle relief and textural patterns not normally visible in high sun elevation photographs. There are disadvantages in using low sun angles, especially in high relief areas, where loss of detail in shadow and rapid light change conditions occur very early and very late in the day.

Most exploration studies involve multi-image interpretation. Often satellite images are examined first (at scales from 1:1M, up to 1:250,000) for the regional view. If high altitude photography is available (1:120,000 to 1:60,000) then this may also be examined. Target areas are selected from the above studies and, in conjunction with literature reviews, it is then possible to select areas for detailed photogeological studies.

Aerial photographs are mainly used by various ordnance surveys for the production of topographic maps using photogrammetry. Photogrammetry is described in detail by Allum (1966). Some additional uses of aerial photography are regional geological mapping (1:36,000–1:70,000), detailed geological mapping (1:5000–1:20,000), civil engineering for road sites, dam sites, etc., open pit management for monthly calculations of volumes extracted during mining, soil mapping, land use and land cover mapping, agricultural applications, forestry applications, water resources applications (e.g. irrigation, pollution, power, drinking water, and flood damage), manufacturing and recreation, urban and regional planning, wetland mapping, wildlife ecology, archaeology, and environmental impact assessment.

6.5 SUMMARY

In remote sensing we are concerned with the collection of information about an area without being in contact with it, and this can be achieved using satellites or aeroplanes carrying electronic scanners or sensors or photographic and TV cameras.

Some satellite systems producing data useful for mineral exploration include NASA's Landsat and Terra (ASTER) and the French SPOT. Each has its advantages (section 6.3). The first Landsat satellite was launched in 1972 and SPOT 1 in 1986. Satellites are solar powered and transmit data to their home station in digital form which enables the geologist to manipulate, combine, and compare this data with geological, geochemical, and geophysical data which has itself been expressed as digital images. Satellite imagery is used in structural investigations, in hydrogeology, to provide basic geological maps, to detect hydrothermal alteration, and to produce maps of regional and local fracture patterns that may have controlled mineralisation or hydrocarbon accumulation.

Aerial photography has been used for much longer than satellite imagery and is important to the geologist for the production of topographical maps (photogrammetry) as well as in the making of geological maps (photogeology). In areas of good exposure aerial photographs yield much valuable geological information and even in areas of only 5% outcrop the amount of information they provide is often invaluable to the explorationist; some of this is discussed in section 6.4.6.

6.6   FURTHER READING

Good textbooks which cover most aspects of remote sensing in geology are Siegal and Gillespie (1980) and Drury (2001). The classic book about photogeology is Allum (1966), which despite its age is still worth reading. Detailed information on all aspects of remote sensing is covered by Lillesand et al. (2004). Recent developments in remote sensing appear monthly in the *International Journal of Remote Sensing* published by Taylor & Francis. A good introduction to digital image processing is given by Niblak (1986), and techniques for image processing and classification are explained well by Schowengerdt (1983).

7

GEOPHYSICAL METHODS

JOHN MILSOM

7.1 INTRODUCTION

Exploration geophysicists use measurements of physical quantities made at or above the ground surface or, more rarely, in boreholes to draw conclusions about concealed geology. Lines may have to be surveyed and cleared, heavy equipment may have to be brought on site, and detectors and cables may have to be positioned, so geophysical work on the ground is normally rather slow. Airborne geophysical surveys, on the other hand, provide the quickest, and often the most cost-effective, ways of obtaining geological information about large areas. In some cases, as at Elura in central New South Wales (Emerson 1980), airborne indications have been so clear and definitive that ground follow-up work was limited to defining drill sites, but this is unusual. More often, extensive ground geological, geochemical, and geophysical surveys are required to prioritize airborne anomalies.

For a geophysical technique to be useful in mineral exploration, there must be contrasts in the physical properties of the rocks concerned that are related, directly or indirectly, to the presence of economically significant minerals. Geophysical anomalies, defined as differences from a constant or slowly varying background, may then be recorded. Anomalies may take many different forms and need not necessarily be centered over their sources (Fig. 7.1). Ideally, they will be produced by the actual economic minerals, but even the existence of a strong physical contrast between ore minerals and the surrounding rocks does not guarantee a significant anomaly. The effect of gold, which is both dense and electrically conductive, is negligible in deposits suitable for large-scale mining because of the very low concentrations (see Corbett 1990). Diamonds (the other present-day "high profile" targets) are also present in deposits in very low concentration and have, moreover, no outstanding physical properties (see Chapter 17). In these and similar cases, geophysicists must rely on detecting associated minerals or, as in the use of seismic reflection to locate offshore placers, and magnetics and electromagnetics to locate kimberlites (see Macnae 1995), on defining favorable environments.

Geophysical methods can be classed as either *passive* (involving measurements of naturally existing fields) or *active*, if the response of the ground to some stimulus is observed. Passive methods include measurements of magnetic and gravity fields, naturally occurring alpha and gamma radiation, and natural electrical fields (static *SP* and magnetotellurics). All other electrical and electromagnetic techniques, seismic methods, and some downhole methods that use artificial radioactive sources are *active.*

In the discussion that follows, the general principles of airborne surveys are considered before describing the individual methods in detail. The final sections are concerned with practicalities, and particularly with the role of nonspecialist geologists in geophysical work.

7.2 AIRBORNE SURVEYS

Magnetic, electromagnetic, gamma-ray, and more recently gravity, measurements do not

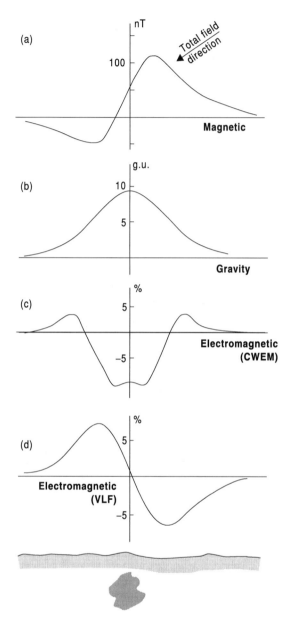

FIG. 7.1   Single body geophysical anomalies.
A massive sulfide orebody containing accessory
magnetite or pyrrhotite might produce the
magnetic anomaly (a), the gravity anomaly (b),
the Slingram electromagnetic anomaly (c), and
the VLF anomaly (d; see also Fig. 7.15). Values
shown on the vertical axes are typical, but the
ranges of possible amplitudes are very large for
some of the methods.

require physical contact with the ground and
can therefore be made from aircraft. Inevitably,
there is some loss of sensitivity, since detectors
are further from sources, but this may even
be useful in filtering out local effects from
manmade objects. The main virtue of airborne
work is, however, the speed with which large
areas can be covered. Surveys may be flown
either at a constant altitude or (more commonly in mineral exploration) at a (nominally)
constant height above the ground. Aircraft
are often fitted with multiple sensors and
most installations include a magnetometer
(Fig. 7.2).

Airborne surveys require good navigational
control, both at the time of survey and later,
when flight paths have to be plotted (recovered). Traditionally, the pilot was guided by a
navigator equipped with maps or photo-mosaics
showing the planned line locations. Course
changes were avoided unless absolutely necessary and in many cases the navigator's main job
was to ensure that each line was at least begun
in the right place. Although infills were (and
are) required if lines diverged too much, a line
that was slightly out of position was preferred
to one that continually changed direction and
was therefore difficult to plot accurately.

Low level navigation is not easy, since even
the best landmarks may be visible for only a
few seconds when flying a few hundred meters
above the ground, and navigators' opinions of
where they had been would have been very inadequate bases for geophysical maps. Tracking
cameras were therefore used to record images,
either continuously or as overlapping frames,
on 35 mm film. Because of the generally small
terrain clearance, very wide-angle ("fish-eye")
lenses were used to give broad, although dis-
torted, fields of view. Recovery was done directly from the negatives and even the most
experienced plotters were likely to misidentify
one or two points in every hundred, so rigorous
checking was needed. Fiducial numbers and
markers printed on the film provided the essential cross-references between geophysical data,
generally recorded on magnetic tape, and flight
paths. Mistakes could be made at every stage in
processing, and errors were distressingly common in aeromagnetic maps produced, and published, when these methods were being used.

FIG. 7.2 Airborne geophysical installation. The aircraft has been modified for combined magnetic and TEM surveys. The boom projecting from above the cockpit is the forward mounting of the electromagnetic transmitter loop, which extends from there to the wingtips and then to the tail assembly. The magnetometer sensor is housed in the fibreglass "stinger" at the rear, which places it as far as possible from magnetic sources in the aircraft.

Visual navigation in savannah areas was difficult enough, since one clump of trees looks much like another when seen sideways on, but plotting from vertical photographs in such areas was often easy because tree patterns are quite distinctive when viewed from above. Visual navigation and flight-path recovery were both very difficult over jungles or deserts and were, of course, impossible over water. If there were no existing radio navigational aids, temporary beacons might be set up, but since several were required and each had to be manned and supplied, survey costs were considerably increased. Self-contained Doppler or inertial systems, which could be carried entirely within the aircraft, produced significant improvements during the period from about 1975 to 1990, but airborne work has now been completely transformed by the use of global positioning satellites (GPS). GPS systems provide unprecedented accuracy in monitoring aircraft position and have largely eliminated the need for tracking cameras, although these are still sometimes carried for back-up and verification. Velocities can be estimated with great accuracy, making airborne gravimetry, which requires velocity corrections, usable for the first time in mineral exploration.

Survey lines must not only be flown in the right places but at the right heights. Ground clearance is usually measured by radar altimeter and is often displayed in the cockpits by limit lights on a head-up display. The extent to which the nominal clearance can be maintained depends on the terrain. Areas too rugged for fixed-wing aircraft may be flown by helicopter, but even then there may be lines that can only be flown downhill or are too dangerous to be flown at all. Survey contracts define allowable deviations from specified heights and line separations, and lines normally have to be reflown if these limits are exceeded for more than a specified distance, but contracts also contain clauses emphasizing the paramount importance of safety. In a detailed survey the radar altimeter data can be combined with GPS estimates of elevation to produce a high resolution digital terrain model (DTM) which may prove to be one of the most useful of the survey products (see section 9.2).

The requirement for ever smaller ground clearances and tighter grids is making demands on pilots that are increasingly difficult to meet. In the future some airborne surveys, and especially aeromagnetic surveys, may be flown by pilotless drones.

7.3   MAGNETIC SURVEYS

Magnetic surveys are the quickest, and often the cheapest, form of geophysics that can provide useful exploration information. A few minerals, of which magnetite is by far the most common, produce easily detectable anomalies in the Earth's magnetic field because the rocks containing them become magnetized. The magnetization is either temporary (induced) and in the same direction as the Earth's field, or permanent (remanent) and fixed in direction with respect to the rock, regardless of folding or rotation.

Since magnetite is a very minor constituent of sediments, a magnetic map generally records the distribution of magnetic material in the underlying crystalline basement. Even sediments that do contain magnetite have little effect on airborne sensors, partly because the fields from the randomly oriented magnetite grains typical of sediments tend to cancel out and partly because the fields due to thin, flat-lying sources decrease rapidly with height. Because the small pieces of iron scrap that are ubiquitous in populated areas strongly affect ground magnetometers, and also because ground coverage is slow, most magnetic work for mineral exploration is done from the air. Line separations have decreased steadily over the years and may now be as little as 100 m. Ground-clearance may also be less than 100 m.

Magnetic surveys are not only amongst the most useful types of airborne geophysics but are also, because of the low weight and simplicity of the equipment, the cheapest. The standard instrument is now the high sensitivity cesium vapor magnetometer. Proton magnetometers are still occasionally used but the cesium magnetometer is not only a hundred times more sensitive but provides readings every tenth of a second instead of every second or half second. Both cesium and proton instruments are (within limits) self-orienting and can be mounted either on the aircraft or in a towed bird. Sensors on aircraft are usually housed in specially constructed nonmagnetic booms (*stingers*) placed as far from the main aircraft sources of magnetic field as possible (Fig. 7.2). Aircraft fields vary slightly with heading and must be compensated by systems of coils and permanent magnets. The magnitude of any residual heading error must be monitored on a regular basis.

The Earth's magnetic field is approximately that of a dipole located at the Earth's center and inclined at about 10 degrees to the spin axis. Distortions covering areas hundreds of kilometers across can be regarded as due to a small number of subsidiary dipoles located at the core–mantle boundary. The practical unit of magnetic field for survey work is the nanotesla (nT, sometimes also known as the *gamma*). At the magnetic poles the field is about 60,000 nT and vertical, whereas at the equator it is about 30,000 nT and horizontal (Fig. 7.3). Slow variations in both magnitude and direction, which must be taken into account when comparing surveys made more than a few months apart, are described, together with large-scale variations with latitude and longitude, by a complicated experimentally determined formula, the International Geomagnetic Reference Field (*IGRF*). This is generally a reasonable approximation to the regional field in well-surveyed areas where the control on its formulation is good, but may be unsatisfactory in remote areas.

As well as long-term variations in field strength, there are cyclical daily (*diurnal*) changes which are normally of the order of 20–60 nT in amplitude. Variations are low at night but the field begins to increase at about dawn, reaches a peak at about 10 a.m., declines rapidly to a minimum at about 4 p.m., then rises more slowly to the overnight value. Although this pattern is repeated from day to day, the changes are not predictable in detail and must be determined by actual measurement. Diurnal effects are especially important in airborne surveys, which now typically use measuring precisions of the order of 0.1 nT and contour

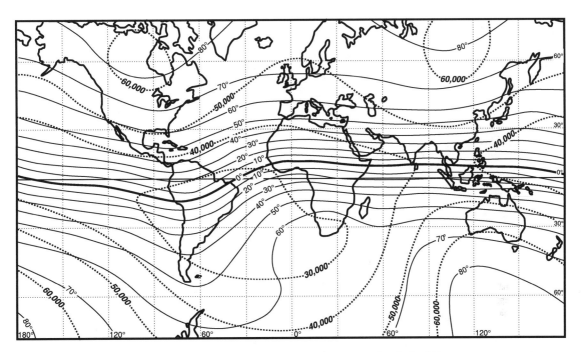

FIG. 7.3 The Earth's magnetic field. Solid lines are contours of constant dip, dashed lines are contours of constant total field strength (in nT).

intervals of 5 nT or less. During such surveys, fixed ground magnetometers provide digital records of readings taken every few minutes. Corrections are made either by using these records directly or by adjustments based on systems of tie lines flown at right angles to the main lines of the survey, or (preferably) by a combination of the two. Tie lines can be misleading if they cross flight lines in regions of steep magnetic gradients, and especially if the terrain is rugged and the aircraft was at different heights on the two passes over an intersection point.

During periods of sunspot activity, streams of high energy particles strike the upper atmosphere and enormously increase the ionospheric currents, causing changes of sometimes hundreds of nanotesla over times as short as 20 minutes. Corrections cannot be satisfactorily made for these "magnetic storms" (which are not related to meteorological storms and may occur when the weather is fine and clear) and work must be delayed until they are over.

Magnetized bodies produce very different anomalies at different magnetic latitudes (Fig. 7.4), the regional changes in dip being far more important from an interpretation point of view than the regional changes in absolute magnitude. Because of the bipolar nature of magnetic sources, all magnetic anomalies have both positive and negative parts and the net total magnetic flux involved is always zero. This is true even of vertically magnetized sources at a magnetic pole, but the negative flux is then widely dispersed and not obvious on contour maps. Magnetic maps obtained near the poles are therefore more easily interpreted than those obtained elsewhere, and reduction-to-pole (RTP) processing has been developed to convert low and middle latitude maps to their high latitude equivalents. If, however (as is usually the case with strongly magnetized bodies), there are significant permanent magnetizations in directions different from the present Earth field, RTP processing will distort and displace anomalies. It is far better that

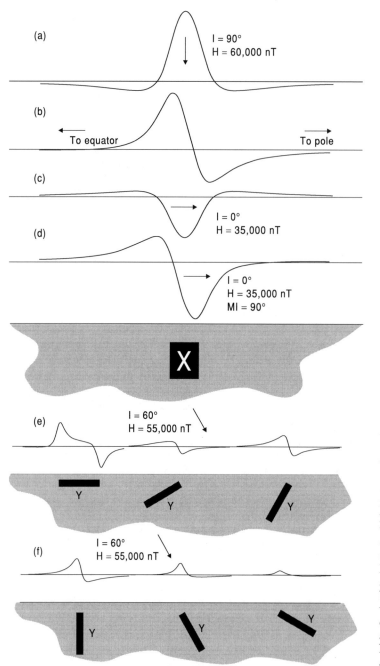

FIG. 7.4   Variations in shape of magnetic anomalies with latitude and orientation. (a–c) The variation with magnetic inclination of the anomaly produced by the body X when magnetized by induction only. (d) The anomaly at the equator if, in addition to the induced magnetization, there is a permanent magnetization in the vertical direction. This situation is common with vertical intrusions (dikes and volcanic pipes) in low latitudes. (e,f) The variation in the anomaly produced in mid-latitudes by an inductively magnetized sheet with varying dip. All profiles are oriented north–south and all bodies are assumed to be "2D," i.e. to have infinite strike length at right angles to the profile.

interpreters become familiar with, and understand, the anomalies typical of the latitudes in which they work than that they rely on processes that can significantly degrade their data.

Magnetic anomalies are caused by magnetite, pyrrhotite, and maghemite (loosely describable as a form of hematite with the crystal form of magnetite). Magnetite is by far the commonest. Ordinary hematite, the most abundant ore of iron, only rarely produces anomalies large enough to be detectable in conventional aeromagnetic surveys. Because all geologically important magnetic minerals lose their magnetic properties at about 600°C, a temperature reached near the base of the continental crust, local features on magnetic maps are virtually all of crustal origin. Magnetic field variations over sedimentary basins are often only a few nanotesla in amplitude, but changes of hundreds and even thousands of nanotesla are common in areas of exposed basement (see Fig. 4.4). The largest known anomalies reach to more than 150,000 nT, which is several times the strength of the Earth's normal field.

Because magnetite is a very common accessory mineral but tends to be concentrated in specific types of rock, magnetic maps contain a wealth of information about rock types and structural trends that can be interpreted qualitatively. Image processing techniques such as shaded relief maps are increasingly being used to emphasize features with pre-selected strikes or amplitudes. Quantitative interpretation involves estimating source depths, shapes, and magnetization intensities for individual anomalies, but it would seldom be useful, or even practicable, to do this for all the anomalies in an area of basement outcrop. Magnetization intensities are not directly related to any economically important parameter, even for magnetite ores, and are rarely of interest. Automatic methods such as Werner deconvolution (which matches observed anomalies to those produced by simple sources) and Euler deconvolution (which uses field gradients) generally focus on depth determination. Direct determination of gradients is becoming commonplace, especially in exploration for kimberlites. In airborne installations horizontal gradients can be measured using nose-and-tail or wingtip sensors.

7.4   GRAVITY METHOD

The gravity field at the surface of the Earth is influenced, to a very minor extent, by density variations in the underlying rocks. Rock densities range from less than $2.0 \, \text{Mg m}^{-3}$ for soft sediments and coals to more than $3.0 \, \text{Mg m}^{-3}$ for mafic and ultramafic rocks. Many ore minerals, particularly metal sulfides and oxides, are very much denser than the minerals that make up the bulk of most rocks, and orebodies are thus often denser than their surroundings. However, the actual effects are tiny, generally amounting even in the case of large massive sulfide deposits to less than 1 part per million of the Earth's total field (i.e. 1 mgal, or 10 of the SI *gravity units* or *g.u.* that are equal to $10^{-6} \, \text{m s}^{-2}$). Gravity meters must therefore be extremely sensitive, a requirement which to some extent conflicts with the need for them also to be rugged and field-worthy. They measure only gravity differences and are subject to drift, so that surveys involve repeated references to base stations. Manual instruments are relatively difficult to read, with even experienced observers needing about a minute for each reading, while the automatic instruments now becoming popular require a similar time to stabilize. Because of the slow rate of coverage, gravity surveys are more often used to follow up anomalies detected by other methods than to obtain systematic coverage of large areas. The potential of the systematic approach was, however, illustrated by the discovery of the Neves Corvo group of massive sulfide deposits (Fig. 7.5) following regional gravity surveys of the Portuguese pyrite belt on 100 and 200 meter grids (Leca 1990). Airborne gravity is technically difficult and, although available in a rather primitive form as early as 1980, is still not widely used. However, BHP-Billiton invested several million dollars in developing the Falcon airborne gradiometer system specifically for use in their own mineral exploration programs and has reported several successes since beginning operations in 1999. The Bell Geospace Full Tensor Gradiometer (FTG), which is now available commercially, gives comparable results, equivalent roughly to ground stations recorded on a 150 m grid, with 0.2 mgal noise.

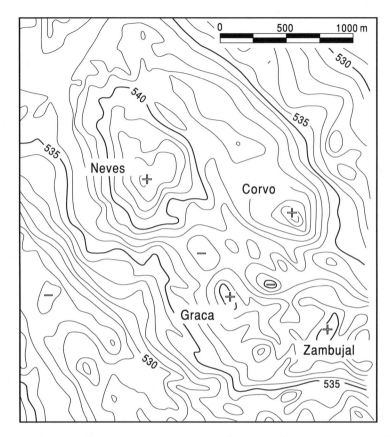

FIG. 7.5 Gravity anomalies over the Neves Corvo group of orebodies in the Portuguese pyrite belt. The main bodies, indicated by crosses, coincide with local culminations in a regional gravity high associated with high density strata. Under these circumstances it is difficult to separate the anomalies from their background for total mass or other analysis. (Data from Leca 1990.)

Gravity tends to increase towards the poles. The difference between the polar and equatorial fields at sea level amounts to about 0.5 % of the total field, and corrections for this effect are made using the International Gravity Formula (IGF). Gravity also varies with elevation. The 0.1 g.u. (0.01 mgal) effect of a height change of only 5 cm is detectable by modern gravity meters and corrections for height must be made on all surveys. Adding the free-air correction, which allows for height differences, and subtracting the Bouguer correction (which approximates the effect of the rock mass between the station and the reference surface by that of a uniform flat plate extending to infinity in all directions) to the latitude-corrected gravity produces a quantity known as the Bouguer anomaly or Bouguer gravity. Provided that the density of the topography has been accurately estimated, a Bouguer gravity map is usually a good guide to the effect of subsurface geology on the gravity field, although additional corrections are needed in rugged terrain. Although much can be achieved with careful processing (e.g. Leaman 1991), terrain corrections can be so large, and subject to such large errors, as to leave little chance of detecting the small anomalies produced by local exploration targets.

Geophysical interpretations are notoriously ambiguous but the gravity method does provide, at least in theory, a unique and unambiguous answer to one exploration question. If an anomaly is fully defined over the ground surface, the total gravitational flux it represents is proportional to the total excess mass of the source body. Any errors or uncertainties are due to the difficulty, under most circumstances, of defining background, and are not inherent in the method. The true total mass of the body is obtained by adding in the mass of an equivalent volume of country rock, so some knowledge of source and country-rock densities is needed.

Very much less can be deduced about either the shapes or the depths of sources. A variety of rule of thumb methods exist, all relying on the rough general relationship between the lateral extent of an anomaly and the depth of its source (e.g. Milsom 2002). A deeper body will, other things being equal, give rise to a broader (and flatter) anomaly. Strictly speaking, only homogeneous spherical bodies can be approximated by the point sources on which most of the rules are based, and the widths of real anomalies obviously depend on the widths of their sources, but rough guides are obtained to the depths of centers of mass in many cases. Also, the peaks of gravity anomalies are generally located directly above the causative bodies, which is not the case with many of the other geophysical methods. However, all interpretation is subject to a fundamental limitation because the field produced by a given body at a given depth can always also be produced by a laterally more extensive body at a shallower depth. No amount of more detailed survey or increased precision can remove this ambiguity.

Full quantitative interpretations are usually made by entering a geological model into a computer program that calculates the corresponding gravity field, and then modifying the model until there is an acceptable degree of fit between the observed and calculated fields. This process is known as *forward modeling*. Several packages based on algorithms published by Cady (1980) are available for modeling fields due to *2D* bodies (which have constant cross-section and infinite strike extent) and to $2^{1}/_{2}D$ bodies, which are limited in strike length. Full three-dimensional modeling of bodies of complex shape is now becoming more common as desktop computers become faster and memory sizes increase. The reliability of any interpretation, no matter how sophisticated the technique, depends, of course, on the validity of the input assumptions. For an example

of the use of modeling in gold exploration see section 14.5.4.

7.5 RADIOMETRICS

Natural radioactive decay produces alpha particles (consisting of two neutrons and two protons bound together), beta particles (high-energy electrons), and gamma rays (very high frequency electromagnetic waves which quantum theory allows us to treat as particles). Alpha and beta radiations are screened out by one or two centimeters of solid rock, and even a little transported soil may conceal the alpha and beta effects of mineralisation. Gamma rays are more useful in exploration but even they have ranges of only one or two meters in solid matter. The traditional picture of a radiometric survey is of the bearded prospector, with or without donkey, plodding through the desert and listening hopefully to clicking noises coming from a box on his hip. The Geiger counter he used was sensitive only to alpha particles and is now obsolete, but tradition dies hard. Modern gamma rays detectors (*scintillometers*) have dials or digital readouts, but ground survey instruments can usually also be set to "click" in areas of high radioactivity. For an example of a carborne survey see Fig. 4.8.

Radiometric methods are geophysical oddities because the measurements are of count rates, which are subject to statistical rules, rather than of fields with definite, even if diurnally variable, values. Survey procedures must strike a balance between the accuracy obtainable by observing for long periods at each station and the speed of coverage demanded by economics. Statistics can be improved by using very large detector crystals, and these are essential for airborne work, but their cost rises dramatically with size. The slow speeds needed to obtain statistically valid counts in the air generally necessitate helicopter installations.

Gamma ray energies can be measured by *spectrometers*, allowing different sources of radiation to be identified. Terrestrial radiation may come from the decay of ^{40}K, which makes up about 0.01% of naturally occurring potassium, or from thorium or uranium. Radiation arises not only from decay of these long-lived *primeval* elements but also from unstable

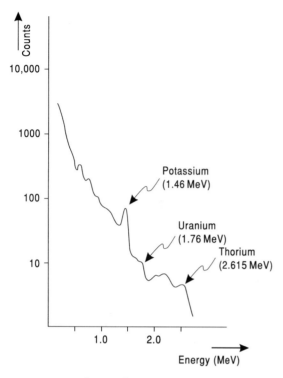

FIG. 7.6 Typical natural gamma-ray spectrum recorded at ground level. The peaks due to specific decay events are superimposed on a background of scattered radiation from cosmic rays and higher energy decays. Note the logarithmic vertical scale.

daughter isotopes of lighter elements, formed during decay and usually themselves unstable and producing further offspring. Gamma-ray *photons* with energies above 2.7 MeV can only be of extraterrestrial (usually solar) origin, but a 2.615 MeV signal is produced by a thorium daughter. The most prominent uranium (actually ^{214}Bi) signal is at 1.76 MeV and the single peak associated with potassium decay is at 1.46 MeV. These and other peaks are superimposed on a background of scattered radiation (Fig. 7.6), but the relative importance of the three main radioelements in any source can be estimated by analyzing spectra using rather simple techniques based on the count rates within windows centerd on these three energy levels. Corrections must be made for the effects of scattered thorium radiation in the uranium window and for the effect of both thorium and uranium in the potassium window. It is usual to also record total count but discrimination is important, as was shown by the discovery of the Yeeleerie uranium deposit in Western Australia in a salt lake where the radiometric anomaly had initially been attributed to potash in the evaporites. Interpretational difficulties are introduced in the search for uranium by the fact that ^{214}Bi lies below radon in the decay chain. Radon is a gas and the isotope concerned has a half-life of several days, allowing considerable dispersion from the primary source.

The unpopularity of nuclear power, and the availability of uranium from dismantled nuclear bombs, made exploration for uranium much less attractive, and the importance of radiometric methods declined accordingly. Much of the radiometric work being undertaken at present is for public health purposes and in such applications the monitoring of alpha particles from radon gas using *alpha cups* and *alpha cards* may be as useful as the detection of gamma rays. Gamma-ray surveys do, however, have geological applications in locating alteration zones in acid and intermediate intrusions and in the search for, and evaluation of, phosphate and some placer deposits. Another, and developing, application is in airborne soils mapping, since soils derived from different rocks have different radiometric signatures. False-color maps, produced by assigning each of the three primary colors to one of the three main radioactive source elements, can supplement magnetic maps as geological mapping tools and may reveal features such as old river channels that have exploration significance. The addition of a gamma-ray sensor (typically, a 256-channel spectrometer) to an aeromagnetic system, at an increase in cost per line kilometer of the order of 10–20%, is therefore becoming routine.

7.6 RESISTIVITY

The electrical properties of continuous media are characterized by their resistivities, which are the resistances of meter cubes. Rock resistivities vary widely but are normally within the range from 0.1 to 1000 ohm-meters. Because most minerals are insulators, these

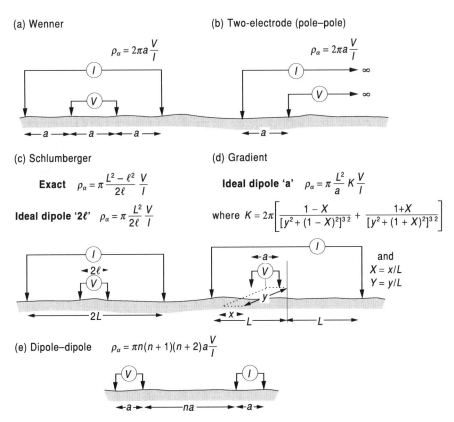

FIG. 7.7   Arrays and geometric conversion factors for resistivity and IP surveys: (a) Wenner array; (b) pole–pole array; (c) Schlumberger array; (d) gradient array; (e) dipole–dipole array.

resistivities are commonly controlled by rock porosity and by the salinity of the pore waters. However, clay minerals are electrically polarized and rocks containing them are highly conductive when even slightly moist.

A few minerals, notably graphite and the base metal sulfides (except sphalerite), conduct by electron flow and can reduce rock resistivity to very low values if present in significant amounts. Even so, straightforward measurement of direct-current resistivity is seldom used by itself in the search for base metals. More common uses are in estimating overburden thicknesses and in determining the extents of deposits of various bulk minerals. Lenses of clean, and therefore resistive, gravels can be found in clays and, conversely, china clay deposits can be found in granites or redeposited as ball clays in bedrock depressions.

The two fundamental requirements of any resistivity survey are the introduction of current and the measurement of voltage but, because of contact effects at electrodes, subsurface resistivity cannot be estimated using the same pair of electrodes for both purposes. Instead, two current and two voltage electrodes are used in arrays which are usually, but not necessarily, linear. For convenience, current is usually supplied through the outer electrodes but the geometrical factors of Fig. 7.7 can also be used to calculate the averaged or *apparent* resistivities if the two inner electrodes are used for this purpose.

Power may be provided by motor generators or rechargeable batteries. Generators and voltmeters may be separated or combined in single units that record resistance values directly. To reduce polarization effects at the electrodes, and also to compensate for any natural currents and voltages, directions of current flow are reversed periodically, usually at intervals of the order of a second. Voltage electrodes that do not polarize can be made by immersing copper rods in saturated copper sulfate solution contained in porous pots. Contact with the ground is made by solution that leaks through the bases of the pots. These electrodes are messy and inconvenient and are used only when, as in induced polarization surveys (section 7.8), they are absolutely essential.

For studying lateral variations in resistivity, the Wenner, pole–pole, and gradient arrays (Fig. 7.7a,b,d) are the most convenient. In Wenner traversing, the leading electrode is moved one electrode interval along the line for each new station and each following electrode is moved into the place vacated by its neighbor. With the gradient array, which requires very powerful current generators, the outer electrodes are fixed and far apart and the inner electrodes, separated by only a few meters, are moved together. The pole–pole array also has the advantage of requiring only two moving electrodes but the very long cables needed to link these to the fixed electrodes "at infinity" may be inconvenient.

Resistivity surveys are also used to investigate interfaces, such as water tables or bedrock surfaces, that are approximately horizontal. Current can be sent progressively deeper into the ground by moving electrodes further apart. Resistivities can thus be estimated for progressively deeper levels, although, inevitably, resolution decreases as electrode separations increase. The Wenner array is often used, as is the Schlumberger array (Fig. 7.7c), which is symmetrical but has outer electrodes very much further apart than the inner ones. All four electrodes must be moved when expanding the Wenner array but with the Schlumberger array, within limits, the inner electrodes can be left in the same positions. Traditionally, the results obtained with either array were interpreted by plotting apparent resistivity against array expansion on log–log paper and comparing the graphs with type curves. A single sheet of curves was sufficient if only two layers were present (Fig. 7.8), but a book was needed for three layers and a library would have been

needed for four! Multiple layers were therefore usually interpreted by matching successive curve segments using two-layer curves together with auxiliary curves that constrained the two-layer curve positions. This technique is now largely obsolete, having been replaced by interactive computer modeling, but the insights provided by type curves are still valuable. More sophisticated computer programs now allow the results obtained from multiple Wenner traverses at different electrode separations along single lines to be inverted to cross-sections approximating the actual subsurface distribution of resistivity. The fieldwork is much more time-consuming than simple single-expansion depth sounding, but the extra effort is almost always justified by the improved results.

7.7   SPONTANEOUS POLARIZATION (SP)

Natural currents and natural potentials exist and have exploration significance. In particular, sulfide orebodies may produce negative anomalies of several hundred millivolts. It was originally thought that these potentials were maintained by oxidation of the ore itself, but it is now generally agreed that the ore acts as a passive conductor, focussing currents produced by the oxidation-reduction reactions that take place across the water table. For a voltage to be produced, the conductor must straddle the water table and the method, although quick and simple, requiring only a high-impedance voltmeter, some cables, and a pair of nonpolarizing electrodes, is often rejected because economically exploitable bodies do not necessarily produce anomalies even when highly conductive.

7.8   INDUCED POLARIZATION (IP)

Flow of electric current in a rock mass can cause parts of it to become electrically polarized. The effect is almost negligible in sandstones, quite marked in clays, in which small pore spaces and electrically active surfaces impede ionic flow, and can be very strong at the surfaces of grains of electronic conductors such as graphite and metallic sulfides. If current

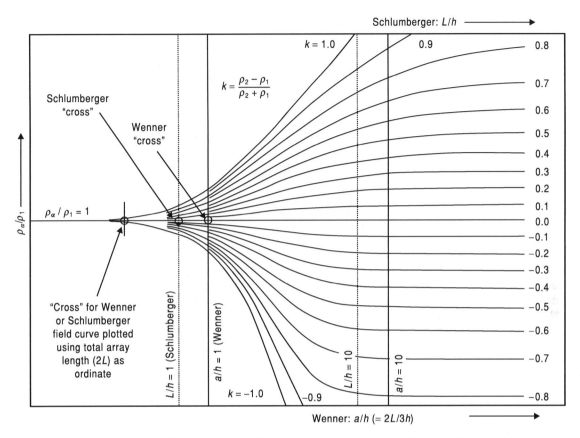

FIG. 7.8 Two-layer type curves for Wenner and Schlumberger depth sounding. Although the curves for these two arrays are not actually identical, the differences are so small as to be undetectable at this scale, and are less than the inevitable errors in measurement. The curves are plotted on 2 × 3 cycle log–log paper. For interpretation, field curves plotted on similar but transparent paper are laid over the type curves and then moved parallel to the axes until a "match" is obtained. The value ρ_1 of the upper layer resistivity is then given by the point where the $\rho_a/\rho_1 = 1$ line cuts the field curve vertical axis and the depth, h, to the interface by the point where the $a/h = 1$ line (Wenner) or $L/h = 1$ line (Schlumberger) cuts the field curve horizontal axis. The value ρ_2 of the lower layer resistivity is determined using the value of k corresponding to the best-matching type curve.

flow ceases, the polarization cells discharge, causing a brief flow of current in the reverse direction. The effect can be measured in several different ways, all of which can be illustrated by considering a simple square wave.

A square-wave voltage applied to the ground will cause currents to flow, but the current waveform will not be perfectly square and will, moreover, be different in different parts of the subsurface. Its shape in any given area can be monitored by observing the voltage developed

between two appropriately placed nonpolarizing electrodes. Attainment of the theoretical steady voltage, V_0, is delayed by polarization for some time after current begins to flow. Furthermore, when the applied voltage is terminated, the voltage drops rapidly not to zero but to a much lower value, V_p, and then decays slowly to zero (Fig. 7.9a). Time-domain IP results are recorded in terms of chargeability, defined in theory as the ratio of V_p to V_0, but in practice it is impossible to measure V_p and voltages

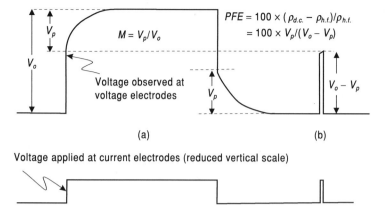

$$PFE = 100 \times (\rho_{d.c.} - \rho_{h.f.})/\rho_{h.f.}$$
$$= 100 \times V_p/(V_o - V_p)$$

$M = V_p/V_o$

Voltage observed at voltage electrodes

V_p

(a)

(b)

Voltage applied at current electrodes (reduced vertical scale)

FIG. 7.9 Idealized IP effects in (a) the time domain and (b) the frequency domain. Not to scale. The voltage applied at the current electrodes may be several orders greater than the voltage observed across the voltage electrodes, and V_p will usually be at least two orders of magnitude smaller than V_0.

are measured after one or more pre-determined time delays. The ratios of these to V_0 are measures of the polarization effect and are usually quoted in millivolts per volt. Some early instruments measured areas under the decay curves and results were quoted in milliseconds.

If, as in Fig. 7.9b, the voltage is applied for a very short time only, V_0 is never reached and, since resistance is proportional to voltage in all of the array formulae (Fig. 7.7), a low apparent resistivity is calculated. Calculations involving the relationship between this value and the direct-current resistivity are the basis of frequency domain IP. The difference between the two divided the by the high frequency resistivity and expressed as a percentage gives the most commonly used parameter, the percent frequency effect (PFE). In practice, because of the need to reverse the current at intervals to eliminate the effects of natural potentials (see section 7.5), the "direct current" resistivity is actually measured at a low alternating frequency, typically 0.25 Hz. Nor can very high frequencies be used without electromagnetic induction affecting the results, so the "high" frequency may be as low as 4 Hz. Because measurement of IP effects involves arbitrary choices (of time delays or of frequencies), it is not normally possible to relate results obtained with different makes of equipment quantitatively, and even instruments of the same make may record slightly different values in the same places.

The asymmetry of the observed voltage curve implies frequency-dependent phase dif-ferences between it and the primary signal. Advances in solid-state circuitry and precise crystal-controlled clocks now allow these differences to be measured and plotted as functions of frequency. This is phase IP.

Time, frequency, and phase techniques all have their supporters. Time-domain work tends to require dangerously high voltage and current levels, and correspondingly powerful and bulky generators. Frequency-domain equipment can be lighter and more portable but the measurements are more vulnerable to electromagnetic noise. Phase measurements, which span a range of frequencies, allow this noise (electromagnetic coupling) to be estimated and eliminated, but add complexity to field operations.

Coupling can be significant in IP work of any type and especially severe if cables carrying current pass close to cables connected to voltage electrodes. Keeping cables apart is easy with the dipole–dipole array (Fig. 7.7e), which is therefore very popular.

Low frequency resistivity is one of the parameters measured in frequency IP, and can be calculated in time-domain IP provided that the input current is recorded. IP surveys are thus also resistivity surveys. In the frequency-domain it is common practice to calculate a *metal factor* by dividing the PFE by the resistivity and, since this would be a rather small quantity, multiplying by a factor of the order of ten thousand. Metal factors emphasize regions that are both chargeable and conductive and can thus be useful for detecting massive

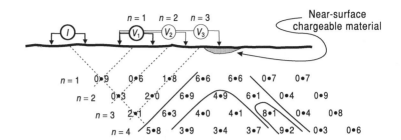

FIG. 7.10 An IP pseudo-section, showing a "pant's-leg" anomaly due to a shallow chargeable body.

sulfides, but may very effectively conceal orebodies of other types. A clear chargeability anomaly produced by disseminated copper ore can appear less attractive on a metal factor plot than an adjacent zone of conductive alteration. The moral is that resistivity and IP data should each be respected for the information that they individually provide and should not be confused into a single composite quantity.

Gradient arrays (Fig. 7.7d) are often used for IP reconnaissance, with results presented as profiles. Dipole–dipole results may also be presented in this way, with separate profiles for different values of the ratio (n) of the interdipole to intradipole distance. Increasing n increases the penetration (at the expense, inevitably, of resolution), and an alternative form of presentation known as the pseudo-section is often used (Fig. 7.10). Pseudo-sections give some indications of the depths of chargeable or conductive bodies but can be very misleading. Indications of dip are notoriously ambiguous and it is quite possible for the apparent dip of a body on a pseudo-section to be in the opposite direction to its actual dip. The classic "pant's leg" anomaly, shown in Fig. 7.10 and apparently caused by a body dipping both ways at 45 degrees from the surface, is usually due to a small source near to one specific dipole position. Every measurement that uses that position will be anomalous. Such problems are now addressed by the use of two-dimensional inversions similar to those used in resistivity work, which produce depth sections plotted against true depth scales. The reliability of the interpretations can be improved by using high-density acquisition along closely spaced lines. However, the fact remains that the arrays most commonly used for IP surveys, and especially

the dipole–dipole array, are poorly adapted to precise target location. Model studies can provide a guide to the position of a chargeable mass but it may still require several drillholes to actually find it.

IP is important in base metal exploration because it depends on the surface area of the conductive mineral grains rather than their connectivity and is therefore especially sensitive to disseminated mineralisation which may produce no resistivity anomaly. In theory, a solid mass of conductive sulfides would give a negligible IP response, but real massive conductive ores seem always to be sufficiently complex to respond well. Since both massive and disseminated deposits can be detected, IP is very widely used, even though it is rather slow, requires moderately large field crews, and is consequently relatively expensive. The need for actual ground contact also causes problems, especially in areas of lateritic or caliche overburden. IP also suffers from the drawbacks inherent in electrical methods, in responding to graphite and barren pyrite (the initials have, rather unkindly, sometimes been said to stand for "indicator of pyrite") but not to sphalerite. Galena also may produce little in the way of an anomaly, especially in "colloidal" deposits.

There is now much interest in the possibility of using phase/frequency plots to discriminate between different types of chargeable materials. It seems that curve shapes are dictated by grain size rather than actual mineral type but even this may allow sulfide mineralisation to be distinguished from graphite because of its generally smaller grains. Phase and multifrequency surveys are now commonplace, often deliberately extending to frequencies at which electromagnetic effects are dominant.

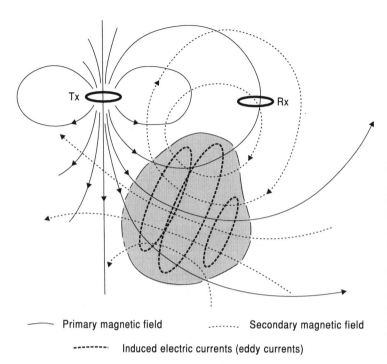

—— Primary magnetic field   ········· Secondary magnetic field

--------· Induced electric currents (eddy currents)

FIG. 7.11 Schematic of an electromagnetic prospecting system. The transmitter (Tx) and receiver (Rx) coils are horizontal and co-planar. They may also be described as "vertical dipoles," referring to the magnetic field produced. The system is said to be "maximum-coupled" because the primary field is at right angles to the plane of the receiver coil where it passes through it. In ground surveys this system is sometimes denoted by the Swedish term *Slingram*. The currents induced in the conductor generate an alternating magnetic field that opposes the primary (transmitted) field, and hence produces a negative anomaly maximum (see Fig. 7.1c). (Drawing based on Grant & West 1965.)

7.9 CONTINUOUS-WAVE ELECTROMAGNETICS

Electric currents produce magnetic fields. Conversely, voltages are induced in conductors that are exposed to changing magnetic fields, and currents, limited by circuit resistivities and self-inductances, will flow in any closed circuits present. These *induced* currents produce secondary magnetic fields. Electromagnetic surveys, which measure these fields, use either continuous, usually sinusoidal, waves (CWEM) or transients (TEM). Transmitters are usually small coils through which electric currents are passed, producing magnetic fields that decrease as the inverse cube of distance. Other methods use long, grounded, current-carrying wires or extended current loops with dimensions comparable to the areas being surveyed. Receivers are almost always small coils. Physical contact with the ground is not required and electromagnetic methods can therefore be used from aircraft. Figure 7.11 shows, in schematic form, a system of the type used in exploration for massive sulfide orebodies, using transmitter and receiver loops with their planes horizontal and axes vertical. Because

the magnetic field from the transmitter cuts the plane of the receiver coil at right angles, such systems are said to be *maximum-coupled*.

In CWEM surveys, alternating currents are passed through transmitter loops or wires at frequencies that usually lie between 200 and 4000 Hz. There will, in general, be phase differences between the primary and secondary fields, and anomalies are normally expressed in terms of the amplitudes of the secondary field components that are *in-phase* and in *phase-quadrature* (i.e. 90 degrees out of phase) with the primary field. For any given body, the amplitudes (response functions) of the in-phase and quadrature anomalies are determined by a *response parameter* that varies with conductivity and frequency. For a target consisting of a simple conducting loop the parameter is equal to the frequency multiplied by the self-inductance and divided by the resistance (Grant & West 1965). In Fig. 7.12 the response of such a loop to the system of Fig. 7.11 is shown plotted against response parameter. If the loop is made of relatively resistive material, induced currents are small, quadrature signals are also small (being, in the limit, proportional

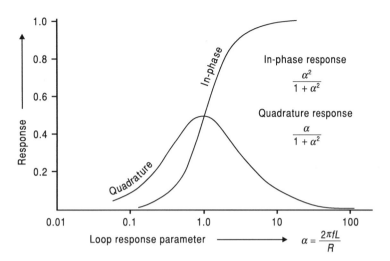

FIG. 7.12   Peak responses recorded by a "Slingram" horizontal-loop e.m. prospecting system over a vertical conducting loop. The size of the anomaly is determined by the response parameter, which is a function of the frequency, f, the self-inductance, L, and the resistance, R, of the loop. Natural conductors respond in more complicated but generally similar ways.

to frequency and inversely proportional to resistivity), and in-phase signals are even smaller. For a good conductor, on the other hand, the induced currents are large but almost in phase with the transmitted field. Quadrature signals are then again small but in-phase signals are large. The quadrature signal reaches a maximum at the point where the quadrature and in-phase responses are equal. Similar but more complex relationships apply to real geological bodies, complicated still further if these have significant magnetic permeability. It is evident that the in-phase/quadrature ratio, which for the loop target is equal to the response parameter, provides crucial information on conductivity. For steeply dipping tabular bodies the effects of conductivity and thickness are difficult to separate and results are often interpreted in terms of a *conductivity-thickness product*. In principle, readings at a single frequency are sufficient to determine this, but readings at several frequencies improve the chances of distinguishing signal from noise.

The depth of penetration achievable with an electromagnetic system is determined in most circumstances by the source–receiver spacing, but there is also an effect associated with the frequency-dependent attenuation of electromagnetic waves in conducting media. This is conveniently expressed in terms of the *skin depth* in which the field strength falls to about one-third of its surface value (Fig. 7.13). If re-

ceiver and transmitter are widely separated, penetration will be skin depth limited.

The induced currents circulating in each part of a conductor produce fields which act on every other part, and computed solutions, except in very simple cases, have to be obtained by successive approximations. Considerable computer power is needed, even for simple two-dimensional bodies, and an alternative is to use physical scale models. A model will be valid if the ratio of the skin depths in the model and in the real geology is the same as the ratio of their lateral dimensions. In principle the permeabilities and dielectric constants could be scaled, but if only the conductivities and lateral dimensions are varied, field instruments can be used in the laboratory, coupled to miniature receiver and transmitter loops. Aluminum sheets may be used to model conductive overburden and copper sheets to represent sulfide orebodies. Physical models are convenient for three-dimensional studies, being easily configured to represent the complicated combinations of orebodies and formational conductors found in the real environment.

In early airborne systems, transmitter coils were mounted on the aircraft and the receiver coils were towed in aerodynamic *birds*. Problems due to the almost random coupling of the receiver to the primary field were avoided by recording only the quadrature signals. This led to considerable arguments as to whether

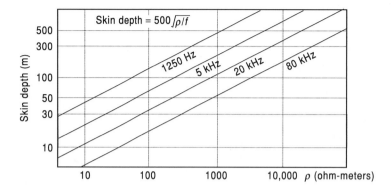

FIG. 7.13   Variation of skin depth with resistivity and frequency in homogenous media. The signal amplitude falls to about one third (actually $1/e = 1/2.718$) of its original value in one skin depth.

natural conductors existed of such quality that the quadrature anomalies would be negligible. The question never seems to have been formally answered, there being commercial vested interests on both sides, but it does seem to be generally agreed that some good targets may produce only very small phase shifts. In any case, as noted above, the in-phase to quadrature ratio is an important interpretation parameter. For CWEM surveys the transmitter and receiver coils are now rigidly mounted either on the aircraft frame or in a single large towed bird. Birds are now used almost exclusively from helicopters, as civil aviation authorities enforce increasingly stringent safety regulations on fixed wing aircraft.

Originally almost all airborne systems used the vertical-loop coaxial maximum-coupled coil configuration, but multi-coil systems are now general. Because the source–receiver separations in airborne systems are inevitably small compared to the distances to the conducting bodies (even though surveys with sensors only 30 m above the ground are becoming commonplace), anomalies generally amount to only a few parts per million of the primary field. Inevitably, there is some degree of flexure in even the most rigid of coil mountings and this produces noise of similar amplitude. In some systems flexure is monitored electronically and corrections are made automatically.

7.10   TRANSIENT ELECTROMAGNETICS (TEM)

One solution to the problem of variable source–receiver coupling is to work in the time-domain. A magnetic field can be suddenly collapsed by cutting off the current producing it, and the magnetic effects of the currents induced in ground conductors by this abrupt change can be observed at times when no primary field exists. In the case of a fixed-wing aircraft, the transmitter coil may extend from the tailplane to the nose via the wingtips (Fig. 7.2). The receiver coil may be mounted in a towed bird, the variation in coupling to the primary field being irrelevant since this field does not exist when measurements are being made. Helicopter TEM is becoming steadily more popular, with a number of new systems competing in the market. Induced currents attenuate quite rapidly in poorly conductive country rock but can persist for considerable periods in massive sulfide orebodies (and often in graphitic shales). Signals associated with these currents were originally observed at four delay-times only (Fig. 7.14), but the number of channels has steadily increased, up to 256 in some recent instruments.

The TEM method was first patented by Barringer Research of Toronto as the INduced PUlse Transient system (INPUT) and was used almost exclusively for airborne surveys. The success of INPUT (and the skills of its patent lawyers) delayed development of ground-based transient systems outside the communist bloc for some years, but by 1968 a Russian instrument was being used for surveys in regions of high-conductivity overburden around Kalgoorlie. The SIROTEM system was then developed in Australia and, at about the same time, the Crone PEM system was designed independently in North America. Subsequently

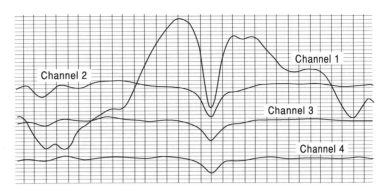

FIG. 7.14   INPUT Mk III transient response over the Timmins orebody. Poor quality, large volume surficial conductors cause large excursions in Channel 1 which die away rapidly at longer delay times. The orebody response persists through all four channels. Modern TEM systems may have as many as 256 channels but the basic principles remain unchanged. (Data from Barringer Research Inc., Toronto.)

another system, the EM37, was produced by Geonics in Canada. A different approach to transients is found in UTEM systems, which use transmitter currents with precisely triangular waveforms. In the absence of ground conductivity, the received UTEM signal, proportional to the time-derivative of the magnetic field, is a square-wave. In most TEM systems (but not UTEM), the transmitter loop can also be used to receive signals since measurements are made after the primary current has been terminated. Even if separate loops are used, the absence of primary field at measurement time allows the receiver coil to be placed within the transmitter loop, which is impractical in CWEM because of the very strong coupling to the primary field.

The Fourier theorem states that any transient wave can be regarded as the sum of a large number of sine and cosine waves, and TEM surveys are therefore equivalent to multi-frequency CWEM surveys. Although first developed almost entirely for the detection of massive sulfide ores, their multi-frequency character allows calculation of conductivity–depth plots that can be used in studies of porphyry systems, in detection of stratiform orebodies and kimberlites, and even in the development of water resources. Interpretations of TEM depth-soundings typically use one-dimensional "smooth-model" inversions.

Transient methods are also superficially similar to time-domain IP, and in IP also there is a time–frequency duality. The most obvious difference between the electromagnetic and IP methods is that currents are applied directly to the ground in IP surveys whereas in most

electromagnetic work they are induced. More fundamentally, IP systems either sample at delay times of from 0.1 to 2 seconds (the initial delay being introduced specifically to avoid electromagnetic noise) or work at frequencies at which electromagnetic effects are negligible. Modern IP units are designed to work over wide ranges of frequencies to obtain conductivity spectra, and so necessarily, and often deliberately, do record some electromagnetic effects, but it is quite possible to avoid working in regions, of either frequency or time-delay, where both effects are both significant.

7.11   REMOTE SOURCE METHODS (VLF AND MAGNETOTELLURICS)

Life would be much easier for EM field crews if they had only to make measurements and did not also have to transmit primary signals. Efforts have therefore been made to use virtually all the various existing forms of background electromagnetic radiation for geophysical purposes. The most useful are military transmissions in the 15–25 kHz range (termed very low frequency or *VLF* by radio engineers, although very high by geophysical standards) and natural radiation in the 10 Hz to 20 kHz range produced by thunderstorms (audio-frequency magnetotellurics or *AMT*). In both AMT and VLF work, the electromagnetic wavefronts are regarded as essentially planar, and in neither is it possible to measure phase differences between primary and secondary radiation. However, differences between the phases of various magnetic and electrical

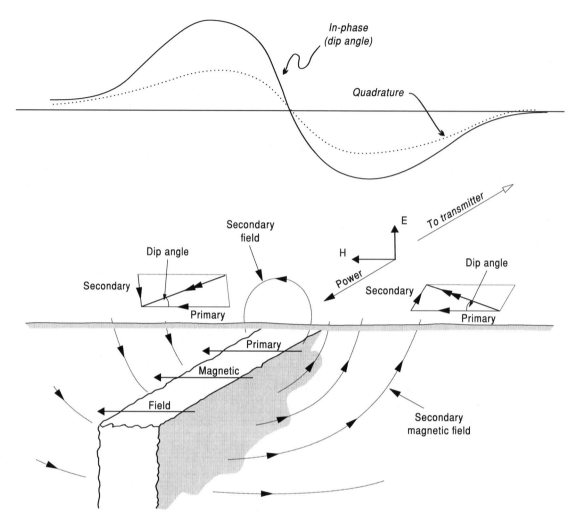

FIG. 7.15 Dip angle anomaly in the VLF magnetic field. Over homogenous ground the VLF magnetic vector is horizontal. The largest anomalies are produced when, as in this case, the conductor is elongated in the direction towards the VLF transmitter (which may be thousands of kilometers away).

components of the waves can be measured and used diagnostically.

When using VLF radiation, the magnetic components are generally the most important if the conductors dip steeply (Fig. 7.15), while the electrical components provide information on the conductivities of flat-lying near-surface layers. Because of the high frequencies used, even moderate conductors respond strongly provided they are large enough and oriented in directions that ensure good coupling to the

primary field, and the VLF method is especially useful in locating steeply dipping fracture zones.

Magnetotelluric signals offer a wide frequency range but poor consistency in source direction and power, and there has been increasing use of local artificial sources that transmit over the same frequency range. In controlled-source magnetotellurics (*CSAMT*), the source is usually a grounded wire several kilometers long. Ideally, it should be located

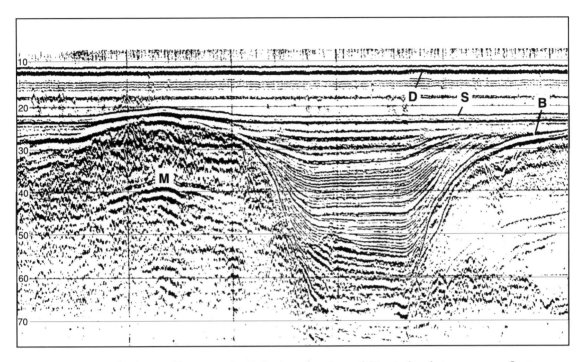

FIG. 7.16   Seismic reflection profile across a buried submarine channel. Vertical scale is two-way reflection time (TWT), in milliseconds. Ten milliseconds TWT is approximately equivalent to 7.5 m of water or 10 m of soft sediments. The reflector indicated by "S" is the sea floor and "B" is a strongly reflecting "basement" surface. "M" is a multiple of the seafloor, created by wavefronts that have made two trips between the reflector and the sea surface. "D" is created by the wave that has traveled directly through the water from the source to the receiver. (Reproduced with the permission of Dr D.E. Searle and the Geological Survey of Queensland.)

far enough from the survey area for plane-wave approximations to apply, since complex corrections are needed where this is not the case. Unfortunately, because distances in the AMT equations are not absolute but are expressed in terms of wavelengths, near-field corrections have to be applied to at least the low frequency, long wavelength data in almost all surveys. For an example of CSAMT use see Fig. 16.3.

7.12   SEISMIC METHODS

Seismic methods dominate oil industry geophysics but are comparatively little used in mineral exploration, partly because of their high cost but more especially because most orebodies in igneous and metamorphic rocks

lack coherent layering. One obvious application is in the search for offshore placers. Sub-bottom profiling using a sparker or boomer source and perhaps only a single hydrophone detector can allow bedrock depressions, and hence areas of possible heavy mineral accumulation, to be identified. Subsea resources of bulk minerals such as sands and gravels may be similarly evaluated. The images produced can be very striking (Fig. 7.16) but, being scaled vertically in two-way reflection time rather than depth, need some form of velocity control before they can be used quantitatively.

Where ores occur in sedimentary rocks that have been only gently folded or faulted, seismic surveys can be useful. However, reflection work onshore is slow and expensive because detectors (*geophones*) have to be positioned

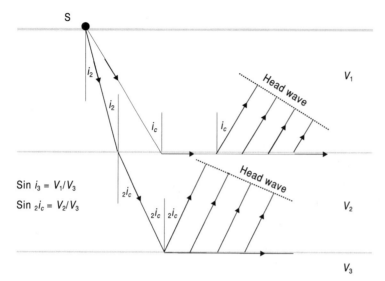

$$\text{Sin } i_3 = V_1/V_3$$

$$\text{Sin } _2i_c = V_2/V_3$$

FIG. 7.17 Principles of seismic refraction for a three layer case. If $V_1 < V_2 < V_3$, critical refraction can occur at both interfaces, producing planar wavefronts known as "head waves." Near to the shot point (S) the direct wave will arrive first, but it will soon be overtaken by the wave that travels via the first interface and, eventually, by the wave that travels via the second.

individually by hand and sources may need to be buried. Also, because of the variability of near-surface weathered layers, results require sophisticated processing and even then are usually less easy to interpret than marine data. The use of reflection in onshore exploration for solid minerals other than coal is consequently rare, although Witwatersrand gold reefs (see section 14.5.4), flat-lying kimberlite sills (see section 17.2), and some deep nickel sulfide bodies have all been investigated in this way (Eaton et al. 2003).

Refraction methods use simpler equipment and need less processing than reflection methods but can succeed only if the subsurface is regularly layered and velocity increases with depth at each interface. Generally, only the first arrivals of energy at each geophone are used, minimizing processing but limiting applications to areas with no more than four significant interfaces at any one locality. Depths to interfaces can be estimated because, except close to the shot point, where the direct route is the quickest, the first arrivals will have been critically refracted. Critically refracted rays can be pictured as traveling quite steeply from the surface to the refractor at the overburden velocity, then along the refractor at the velocity of the underlying layer, and then steeply, and

slowly, back to the surface (Fig. 7.17). Because horizontal velocities estimated from plots of distance against time are used in calculations as if they were vertical velocities, calibration against borehole or other subsurface data is desirable, the more so since an interface may be hidden by others if the velocity contrast or underlying layer thickness is small, and will be completely undetectable if velocity decreases with depth. Despite these limitations, there are many cases, especially in the assessment of deposits of industrial mineral, where refraction surveys can be useful.

7.13   GROUND RADAR

Ground penetrating radar (GPR) is a relatively new addition to the geophysical armoury, but reports of mineral exploration applications are beginning to appear (e.g. Francké & Yelf 2003). GPR results are displayed as images very similar to those used for seismic reflection work, but penetration is limited to a few tens of meters at the best, and may stop at the water table, which is usually a very strong reflector. In some cases the mapping of the water table, which is typically a surface at which the costs of extracting bulk minerals increase dramatic-

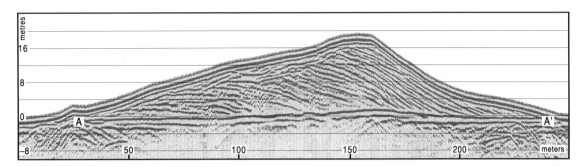

FIG. 7.18   Radar profile across a sand dune in Namibia. The subhorizontal reflector AA' is rockhead. Horizontal lines are at intervals of 50 ns TWT. Calculated time shifts (*static corrections*) have been applied to individual traces to adjust the record to the topographic profile. The vertical scale on the left shows elevations in meters relative to an off-dune ground-surface datum, assuming a radar wave velocity of 0.16 m ns^{-1}. The data were obtained using a 100 MHz radar signal, an antenna spacing of 1 m, and a 0.5 m step size. (Image reproduced by courtesy of Dr C. Bristow, Birkbeck College, University of London.)

ally, is, by itself, sufficient to justify the use of radar. GPR can also be used onshore to locate river paleochannels beneath overburden and to define favorable locations within them for placer deposits. Sedimentary layering can often be imaged within such channels, and also in sand dunes (Fig. 7.18), which may include beds rich in heavy minerals.

7.14   BOREHOLE GEOPHYSICS

Most geophysical techniques can be modified for use in boreholes but neither gravity nor magnetic logging is common and seismic velocity and radiometric logs, both very important in the oil industry, are little used in mineral exploration. In downhole resistivity logging, as in the surface equivalents, four electrodes must be used, but usually at least one is outside the hole. In *normal* logs, one current and one voltage electrode, a few tens of centimeters apart, are lowered downhole, with the other two *at infinity* on the surface. The limitations of direct current resistivity in base metal exploration apply downhole as much as on the surface and more useful results may be obtained with IP or electromagnetics. Electrode polarization can be reduced in downhole IP surveys by using lead (Pb) electrodes, and can be eliminated, with some practical difficulty, using porous pots. In electromagnetic work,

small coils are used for both transmitting and detecting signals, which may be continuous waves or transients. Transient (TEM) techniques are now generally considered the more effective, with a proven ability to detect "missed target" massive sulfides (Bishop & Lewis 1996, see Fig. 15.7).

When conductive ore has been intersected, the *mise-à-la-masse* method can be used. One current electrode is positioned down the hole and the other is placed on the surface well beyond the area of interest. If the ground were electrically homogenous (and its surface flat), equipotentials mapped at the surface using one fixed and one mobile voltage electrode would contour as circles centered above the downhole electrode. If, however, the downhole electrode is placed at an ore intersection, there are departures from this pattern indicating the strike direction and extent of continuous mineralisation. Equipotentials are actually diverted away from good conductors that are not in electrical contact with the electrode. *Mise-à-la-masse* is thus a powerful tool for investigating the continuity of mineralisation between intersections in a number of boreholes.

Because mineral exploration holes are normally cored completely, the mineral industry has been slower than the oil industry to recognize the advantages of geophysical logging. However, the high cost of drilling makes it essential to obtain the maximum possible information from each hole, and the electrical characteristics of the rocks surrounding a borehole can often be better guides to the significance of an ore intersection, or of the presence of ore nearby, than the information contained in a few centimeters of core.

7.15   GEOPHYSICS IN EXPLORATION PROGRAMS

A logically designed exploration program progresses through a number of stages, from regional reconnaissance to semi-detailed follow-up and thence to detailed evaluation. The geophysical component will tend to pass through the same stages, reconnaissance often being airborne and the final evaluation perhaps involving downhole techniques. The exact route followed will depend on the nature of the target, but a typical base metal or kimberlite exploration program would have, as its first stage, the assembling of all the geophysical data already available. Lease conditions may in some cases require the integration of certain types of survey data into national databases, and most countries have at least partial magnetic and gravity coverage and there may have been local surveys of these or other types. With gravity and magnetics it is usually worthwhile going beyond the published maps, which often contain errors, to the original data.

For sulfide ores or kimberlites, the next stage might involve a combined airborne magnetic and electromagnetic survey, using the smallest possible terrain clearance and very closely spaced lines. Semi-regional magnetic maps would be produced and, with luck, a number of interesting electromagnetic anomalies might be located. Some sources might be obvious and irrelevant (e.g. powerlines, pipelines, corrugated-iron barns) but any other anomalies would have to be checked on the ground, with magnetic and electromagnetic surveys to locate the anomaly precisely, geological mapping to determine rock type, and perhaps a gravity survey if density differences are probable between ore and country rock. Other electrical methods, such as IP and even resistivity, may also be used to define drilling targets. Ground geophysics might also be used in areas that had

not been shown to be anomalous from the air but which appeared geologically or geochemically promising. Once a drilling program has begun, not only can the new information be fed back into the geophysical interpretation but the holes can be used for electrical or radiometric logging or, if there are ore intersections, for *mise-à-la-masse*. Work of this sort may well continue into the development stage.

Not all geophysical exploration follows this pattern. Sometimes only one or two methods have any chance of success. Seismic techniques, whether reflection or refraction, tend to be used alone, being applicable where, as in the search for placer deposits, other methods fail, and virtually useless in dealing with the complex geologies encountered in hard rock mining. Lateral thinking has also produced some quite surprising specialized applications of geophysical methods, e.g. phosphate grades can in some cases be estimated by radiometric logging of uncored boreholes.

7.16   INTEGRATION OF GEOLOGICAL AND GEOPHYSICAL DATA

Making good use of data from a variety of very different sources is the most testing part of an exploration geologist's work, requiring theoretical knowledge, practical understanding, and a degree of flair and imagination. All these will be useless if information is not available in convenient and easily usable forms. The most effective way of recognizing correlations is to overlay one set of data on another (see sections 5.1.8. & 9.2), and in a paper-based exploration program decisions on map scales and map boundaries have to be made at an early stage and adhered to thereafter. Computer-based geographical information systems (GIS) offer greater flexibility, although most require a considerable investment of time and effort on the part of the users before their advantages can be realized (discussed in detail in section 9.2).

Presenting different types of data together on a single map is generally useful if they are shown as separate entities that can still be clearly distinguished. Combining data of different types before they are plotted can be dangerous. This approach is common in exploration geochemistry, but its extension to geophysics has not been notably successful. Displays of mathematically combined gravity and magnetic data have generally succeeded only in disguising important features while producing no new insights. Allowing computers to cross-correlate geophysics with geology or geochemistry is even less likely to give meaningful results. Correlation is essential but should be done in the brain of the interpreter.

Of the various possible correlations, those most likely to be useful in the early stages of an exploration program are between geology, air photographs, and magnetic maps. Structural dislocations such as faults and shears are likely to be indicated magnetically and the intersections of magnetic lineaments have been fruitful exploration targets in many areas. Gravity and radiometric data, along with infrared or radar imagery and soils geochemistry, may also help in obtaining the best possible understanding of the geology. Geophysics provides independent information on the third dimension (depth), and may be especially useful in preparing regional cross-sections.

The inherent ambiguity of all geophysical solutions makes it essential to use all possible additional information. For there to be even a chance of a model being correct, it must satisfy all the known geological and geophysical constraints. Consequently, it must also be continually refined, not only up to but also throughout any drilling program. If a worthwhile target has been defined but the hole has proved unsuccessful, it is not enough to simply blame geophysics. There is a reason for any failure. It may be obvious, as in the case of an IP anomaly produced by barren pyrite, but until the reason has been established, a valid target remains to be tested.

7.17 THE ROLE OF THE GEOLOGIST IN GEOPHYSICAL EXPLORATION

A mineral exploration program will usually be the responsibility of a small team, and whether this should include a geophysicist is itself a decision that requires geophysical advice. Factors to be considered are the types of targets sought (and therefore the types of methods

appropriate), the likely duration of the project and, of course, the availability of personnel. If a full-time geophysicist cannot be justified but some geophysical techniques are to be used, there may be company geophysicists who can allocate a percentage of their time. Alternatively, consultants can be retained and/or one or more of the team geologists can be given responsibility for day to day geophysical matters, including supervision of contractors and possibly the conduct of some surveys. It is usually best if one (and preferably only one) geologist is responsible for routine geophysics, but he or she must be able to call on more geophysically experienced help when needed and the freedom to seek such help must not be unduly restricted by short-term cost considerations. Geophysical surveys are often among the most expensive parts of an exploration program and much money can be wasted by a few apparently trivial, but wrong, decisions.

Project geologists should not be expected to decide, unaided, on methods, instrument purchases, and overall geophysical exploration strategy. In companies with their own geophysicists it would be unusual for them to have to do so, but where these are not available it is all too common for consultants to be called in only when things have already gone irretrievably wrong. The instruments purchased, and used, the contracts let, and the contractors selected may all have been totally unsuitable for the work required.

What, then, can reasonably be expected of a project geologist whose only experience of geophysics may be a few dimly remembered lectures on a university course several years in the past, possibly accompanied by some miserable days in the field pressing the keys on a proton magnetometer or trying to level a gravity meter? First, and perhaps most important, is a degree of humility and a willingness to seek help. Second, a variety of geophysical texts are required and a readiness to read them all (beware of relying on any single authority). Third, a realistic attitude is needed. Exploration programs have limited and usually predefined budgets, and geophysical instruments are neither magic wands that reveal everything about an area, nor pieces of useless electronic circuitry inflicted on hard-working geologists specifically to complicate their lives. In considering geophysics, geologists must ask themselves a number of questions. What information is needed? Is there a realistic expectation that geophysical methods will provide it? What information will they not provide? Is this information essential and can any incompleteness be tolerated? Is geophysics affordable and (even if the answer is yes) would the money be better spent on something else? Team geologists may not be able to answer all these questions themselves but they need to ask them, and to get answers. The necessary approaches can be illustrated by specific examples:

1 An exploration area is some hundreds of square kilometers in extent and is covered by regional aeromagnetic maps. Are these adequate or should a new, detailed survey be planned? (Seek geophysical advice, but not from a contractor who may be hoping for work.) Is it reasonable to suppose that the targets will be magnetic or that there will be critical structural or lithological information in the magnetic data? Will such data only be obtained by high sensitivity or gradient surveys or will a 1 nT magnetometer do? If aeromagnetic work is to be done, is there a case for adding other sensors (electromagnetic, radiometric), bearing in mind the inevitable increases in cost? Is the area topographically suitable for fixed-wing survey or will a helicopter have to be used? How much help will have to be bought in for contract supervision and interpretation, and can this be accommodated within the existing budget?

2 A lease of a few square kilometers is to be explored for massive sulfides. SP surveys have located some mineralized bodies in the region but have missed others. Bearing in mind that other, more expensive methods will also have to be used if the area is to be fully evaluated, is more SP worthwhile? Should it be kept in reserve (the equipment is cheap) in case there are field hands spare for a few days? Or, having gone to the trouble of gridding and pegging lines, should not every possible method be used everywhere?

Geophysical work will involve surveying, and may also require line clearing and cutting. At this point the geologist responsible for geophysics may well come into conflict with the geologist responsible for geochemistry, compli-

cating their lives even further. Geochemists, once beyond the reconnaissance stream sediment stage, also take samples along pegged lines, but like to be able to plot locations at regular intervals on planimetric base maps. Planimetrically regular spacing (secant-chaining) is also desirable for geophysical *point data* (gravity, magnetic, and SP), but in resistivity, IP, and electromagnetic surveys it is the actual distances between coils or electrodes that must be constant and these may differ significantly from plan distances. Whether lines are secant- or slope-chained will depend on how much, of what, is being measured, but this apparently trivial problem can be a major cause of friction between team members.

The final requirements of all exploration geologists, and particularly of those who involve themselves in geophysics, are serendipity and an awareness of the possible. Geophysical surveys always provide some information, even though it may not seem directly relevant at first, and many mineral deposits have been discovered by what may be considered luck but at least required geologists with their wits about them. In the 1960s, radiometric surveys were flown in the Solomon Islands primarily for detecting phosphates. A ground follow-up team sent to Rennel Island expected to find phosphatic rock, of which there was very little, but did recognize that the source of the anomaly was uraniferous bauxite, in mineable quantities.

7.18   DEALING WITH CONTRACTORS

Exploration geologists may be involved with geophysical contractors at many different stages, and will often need the support and advice of an experienced geophysicist who is committed to them and not to the contractor. The most important stage is when the contract is first being drafted. Geophysical consultants find it frustrating (but also sometimes very profitable) to be asked to sort out contracts, already written and agreed, which are either so loose that data of almost any quality must be accepted or which specify procedures that actually prevent useful data being acquired. Project geologists will share in the frustration but not in the profit!

Survey parameters such as line separation and line orientation, and all specifications relating to final reports and data displays, should be decided between the project geologists and their geophysical adviser. More technical questions, such as tolerances (navigational and instrumental), instrument settings, flying heights in airborne surveys, and data control and reduction procedures, should be left to the geophysicist. Geologists who set specifications without this sort of help, perhaps relying on the contractor's advice and goodwill, are taking grave risks. Goodwill is usually abundant during initial negotiations, and contractors will, after all, only stay in business if they keep their clients reasonably happy, but as deadlines draw closer, and profit margins come to look tighter, it becomes more and more tempting to rely on the letter of the contract.

Once a contract has been signed, the work has to be done. The day-to-day supervision can usually be left to the geologists on the spot, who will find a comprehensive and tightly written contract invaluable. The greater the detail in which procedures are specified, the easier it is to see if they are being followed and to monitor actual progress. Safety and the environment are becoming more important every year, and although the contract may (and should) specify that care for these are the responsibility of the contractor, the company commissioning the work will shoulder much of the blame if things go wrong. Random and unannounced visits to a field crew, especially early in the morning (are they getting up at a reasonable hour, without hangovers, and do they have precise and sensible plans for the day's work?) or at the end of the day (have they worked a full day and are their field notes for that day clear and intelligible?), are likely to be more useful than any number of formal meetings with the party chief. Evening visits are particularly useful, revealing whether data reductions and paperwork are keeping pace with the field operations and whether the evenings are being passed in an alcoholic haze.

A geophysical contractor can reasonably be expected to acquire data competently, with due regard to safety and environmental protection, reduce it correctly, and present results clearly. Many also offer interpretations, but these may be of limited value. Few contractors are likely to have anything approaching the knowledge of a specific area and its geological problems that is possessed by the members of an exploration team. Moreover, mining companies are often reluctant to provide contractors, who may next week be working for a competitor, with recently acquired data. Since such information can be vital for good interpretation, contractors' reports are often, through no fault of their own, bland and superficial. They may only be worth having if the method being used is so new or arcane that only the operating geophysicists actually understand it.

7.19 FURTHER READING

A number of textbooks cover the geophysical techniques used in mineral exploration in some detail. Of these, the most comprehensive is *Applied Geophysics* (Telford et al. 1990). A briefer but still excellent coverage is provided by *An Introduction to Geophysical Exploration* (Kearey et al. 2002), while *Principles of Applied Geophysics* (Parasnis 1996) has a strong mineral exploration bias, derived from the author's own background in Scandinavia. *Interpretation Theory in Applied Geophysics* (Grant & West 1965) is now hard to find but also hard to beat as a summary of the theoretical backgrounds to most geophysical methods. *Practical Geophysics II* (van Blaricom 1993) and *Geophysics and Geochemistry at the Millenium* (Gubins 1997) provide a wealth of largely North American case histories and practical discussions but both tend to disintegrate if taken into the field. A third edition of *Practical Geophysics* was produced in 1998, on CD-ROM only. Practical aspects of small scale ground-based geophysical surveys and some elementary interpretational rules are discussed in *Field Geophysics* (Milsom 2002). Valuable case histories are to be found in many journals, but especially in *Geophysics* and *Exploration Geophysics*, published by the Society of Exploration Geophysicists and the Australian Society of Exploration Geophysicists respectively. The same two organizations publish newsletter journals (*The Leading Edge* and *Preview*) that provide up-to-the-minute reviews and case histories. Three Australian compilations, on orebodies at Elura

(Emerson 1980), Woodlawn (Whitely 1981), and throughout Western Australia (Dentith et al. 1995), provide useful, although now somewhat dated, case histories where multiple geophysical techniques have been applied to specific base metal deposits. Hoover et al. (1992) provide similar information with a Western Hemisphere emphasis.

8

EXPLORATION GEOCHEMISTRY

CHARLES J. MOON

Geochemistry is now used in virtually every exploration program, if only to determine the grade of the material to be mined. However exploration geochemistry has evolved from its early origins in assaying, to using the chemistry of the environment surrounding a deposit in order to locate it. This particularly applies to the use of surficial material, such as soil, till, or vegetation, that can be used in areas where there is little outcrop. The object is to define a geochemical anomaly which distinguishes the deposit from enhancements in background and nonsignificant deposits. This chapter explains how geochemistry may be employed in the search for mineral deposits. Further details of the theory behind exploration geochemistry are given in the exhaustive but dated Rose et al. (1979) and Levinson (1980).

Exploration geologists are likely to be more directly involved in geochemistry than with geophysics which is usually conducted by contractors and supervised by specialist geophysicists. A geochemical program can be divided into the following phases:

1 Planning;
2 Sampling;
3 Chemical analysis;
4 Interpretation;
5 Follow-up.

The field geologist will probably carry out phases 1, 2, 4, and 5, while analysis is normally performed by a commercial laboratory.

8.1   PLANNING

The choice of the field survey technique and the analytical methods depends on the commodity sought and its location. In the same way as geological and grade–tonnage models are generated (seesection 4.1.3) modeling can be extended to include geochemical factors, summarized in Barton (1986). Thus the geologist will start with a knowledge of the elements associated with a particular deposit type, an idea of the economic size of the deposit to be sought, the mineralogical form of the elements, and the probable size of the elemental anomalies around it. The outline of a deposit is defined by economic criteria and the mineable material is surrounded by lower concentrations of the mined elements which are however substantially enriched compared with unmineralized rock. This area of enrichment is known as the *primary halo*, by analogy with the light surrounding the outline of the moon, and the process of enrichment as *primary dispersion*. In addition ore-forming processes concentrate or deplete elements other than those mined. For example, massive sulfide deposits often contain substantial arsenic and gold in addition to the copper, lead, and zinc for which they are mined. A summary of typical elemental associations is shown in Table 8.1.

The geologist's problem is then to adapt this knowledge of primary concentration to the exploration area. The geochemical response at the surface depends on the type of terrain and especially on the type of material covering the deposit as shown in Fig. 8.1. The response in an area of 2 m of residual overburden is very different from that of an area with 100 m deep cover, or if the overburden has been transported. Also elements behave differently in the near-surface environment from that in which the deposit formed. For example, in cases where copper,

TABLE 8.1 Elemental associations and associated elements (pathfinders) useful in exploration. (Largely from Rose et al. 1979 with some data from Beus & Grigorian 1977, p. 232, and Boyle 1974.)

Type of deposit	Major components	Associated elements
Magmatic deposits		
Chromite ores (Bushveld)	Cr	Ni, Fe, Mg
Layered magnetite (Bushveld)	Fe	V, Ti, P
Immiscible Cu–Ni–sulfide (Sudbury)	Cu, Ni, S	Pt, Co, As, Au
Pt–Ni–Cu in layered intrusion (Bushveld)	Pt, Ni, Cu	Sr, Co, S
Immiscible Fe–Ti–oxide (Allard Lake)	Fe, Ti	P
Nb–Ta carbonatite (Oka)	Nb, Ta	Na, Zr, P
Rare–metal pegmatite	Be, Li, Cs, Rb	B, U, Th, rare earths
Hydrothermal deposits		
Porphyry copper (Bingham)	Cu, S	Mo, Au, Ag, Re, As, Pb, Zn, K
Porphyry molybdenum (Climax)	Mo, S	W, Sn, F, Cu
Skarn–magnetite (Iron Springs)	Fe	Cu, Co, S
Skarn–Cu (Yerington)	Cu, Fe, S	Au, Ag
Skarn–Pb–Zn (Hanover)	Pb, Zn, S	Cu, Co
Skarn–W–Mo–Sn (Bishop)	W, Mo, Sn	F, S, Cu, Be, Bi
Base metal veins	Pb, Zn, Cu, S	Ag, Au, As, Sb, Mn
Sn–W greisens	Sn, W	Cu, Mo, Bi, Li, Rb, Si, Cs, Re, F, B
Sn–sulfide veins	Sn, S	Cu, Pb, Zn, Ag, Sb
Co–Ni–Ag veins (Cobalt)	Co, Ni, Ag, S	As, Sb, Bi, U
Epithermal precious metal	Au, Ag	Sb, As, Hg, Te, Se, S, Cu
Sediment hosted precious metal (Carlin)	Au, Ag	As, Sb, Hg, W
Vein gold (Archaean)	Au	As, Sb, W
Mercury	Hg, S	Sb, As
Uranium vein in granite	U	Mo, Pb, F
Unconformity associated uranium	U	Ni, Se, Au, Pd, As
Copper in basalt (L. Superior type)	Cu	Ag, As, S
Volcanic-associated massive sulfide Cu	Cu, S	Zn, Au
Volcanic-associated massive sulfide Zn–Cu–Pb	Zn, Pb, Cu, S	Ag, Ba, Au, As
Au–As rich Fe formation	Au, As, S	Sb
Mississippi Valley Pb–Zn	Zn, Pb, S	Ba, F, Cd, Cu, Ni, Co, Hg
Mississippi Valley fluorite	F	Ba, Pb, Zn
Sandstone-type U	U	Se, Mo, V, Cu, Pb
Red bed Cu	Cu, S	Ag, Pb
Sedimentary types		
Copper shale (Kupferschiefer)	Cu, S	Ag, Zn, Pb, Co, Ni, Cd, Hg
Copper sandstone	Cu, S	Ag, Co, Ni
Calcrete U	U	V

lead, and zinc are associated in volcanic-associated massive sulfide deposits, zinc is normally more mobile in the surface environment than copper and much more so than lead. Lead is more likely to be concentrated immediately over the deposit, as it is relatively insoluble, whereas zinc will move or disperse from the deposit. This process of movement away from the primary source is termed *secondary dispersion* and it can also be effected by mechanical movement of fragments under gravity, movement as a gas, or diffusion of the elements in the form of ions as well as movement in solution.

The background levels of an element in rocks and soils also have to be considered when trying to find secondary dispersion from deposits. All elements are present in every rock and soil sample; the concentration will depend on the mode of formation of the rock and the

Residual

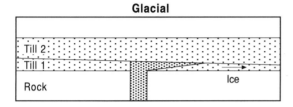

Glacial

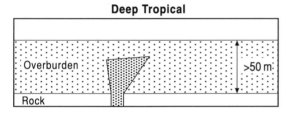

Deep Tropical

FIG. 8.1 Sketch showing dispersion through major types of overburden.

process forming the soil. Indications of background levels of elements in soils are given in Table 8.2, as are known lithologies with elevated concentrations which might provide spurious or non-significant anomalies during a survey. These background levels can be of use in preparing geochemical maps which can be used to infer lithology in areas of poor outcrop. The reader is advised to get some idea of the background variation over ordinary rock formations from a geochemical atlas, such as those for England and Wales (Webb et al. 1978), Alaska (Weaver et al. 1983), the former West Germany (Fauth et al. 1985), and Europe (Salmimen et al. 2004).

The basis of a geochemical program is a systematic sampling program (Thomson 1987) and thus decisions must be made in a cost-effective manner as to the material to be sampled, the density of sampling, and the analytical method to be employed. Cost/benefit ratios should be considered carefully as it may be that a slightly more expensive method will be the only effective technique. The material to be sampled will

largely be determined by the overburden conditions but the density and exact nature of the sample will be based either on previous experience in the area or, if possible, on an orientation survey.

8.1.1 Orientation surveys

One of the key aspects of planning is to evaluate which techniques are effective for the commodity sought and in the area of search. This is known as an orientation survey. The best orientation survey is that in which a variety of sampling methods is tested over a prospect or deposit of similar geology to the target and in similar topographical conditions to determine the method which yields the best results. A checklist for an orientation study is given below (Closs & Nichol 1989):
1 Clear understanding of target deposit type;
2 Understanding of surficial environment of the search area;
3 Nature of primary and secondary dispersion from the mineralisation;
4 Sample types available;
5 Sample collection procedures;
6 Sample size requirements;
7 Sample interval, orientation, and areal density;
8 Field observations required;
9 Sample preparation procedures;
10 Sample fraction for analysis;
11 Analytical method required;
12 Elemental suite to be analyzed;
13 Data format for interpretation.

The relatively small cost involved in undertaking an orientation survey compared with that of a major geochemical survey is always justified and whenever possible this approach should be used. However, if a physical orientation survey is not possible, then a thorough review of the available literature and discussion with geochemical experts is a reasonable alternative option. Of particular use in orientation studies are the models of dispersion produced for various parts of the world; Canadian Shield and Canadian Cordillera (Bradshaw 1975), Western USA (Lovering & McCarthy 1977), and Australia (Butt & Smith, 1980).

Particularly useful discussion on orientation procedures is provided by Thomson (1987) and Closs and Nichol (1989).

TABLE 8.2 Background concentration of trace elements and utility in geochemical exploration. (Largely after Rose et al. 1979, Levinson 1980.)

Element	Typical background concentration in soils (ppm)	Enriched lithologies	Surficial mobility	Use in exploration
Antimony	1		Low	Pathfinder
Arsenic	10	Ironstones	Mobile; Fe scavenged	Pathfinder especially for Au
Barium	300	Sandstones	Low often barite	Panned concentrations in VMS search
Beryllium	500 ppb	Granites	High if not in beryl	Occasional use
Boron	30	Granites	Moderate	For borates
Bismuth	500 ppb	Granites	Low	
Cadmium	100 ppb	Black shales	High	Zn deposits
Chromium	45	Ultramafics	Low	Chromite in panned concs.
Cobalt	10	Ultramafics	Moderate Mn scavenged	Widely used
Copper	15	Basic igneous	pH > 5 low, else moderate	Most surveys
Fluorine	300	Alkaline igneous	High	F deposits
Gold	1 ppb	Black shales	Low	Gold deposits
Lead	15	Sandstones	Low	Wide use
Lithium	20	Granites	Moderate	Tin deposits
Manganese	300	Volcanics	Moderate, high at acid pH	Scavenges Co, Zn, Ag
Mercury	50 ppb		High	Wide use as pathfinder
Molydenum	3	Black shale	Moderate to high pH > 10	Wide use
Nickel	17	Ultramafics	Low, scavenged	Wide use
Platinum	1 ppb	Ultramafics	Very low	Difficult to determine
Rare earths	La 30	Beach sands	Very low	
Selenium	300 ppb	Black shales	High	Little use
Silver	100 ppb?		High; Mn scavenged	Difficult to use
Tellurium	10 ppb	Acid intrusives	Low	Difficult to use
Thallium	200 ppb?		Low	Epithermal Au
Tin	10	Granites	Very low	Panned concentrates
Tungsten	1	Granites	Very low	Scheelite fluoresces in UV
Uranium	1	Phosphorites	Very high; organic scavenged	Determine in water
Vanadium	55	Ultramafics	Moderate?	Little use
Zinc	35	Black shales	High; scavenged by Mn	Most surveys
Zirconium	270	Alkaline igneous	Very low	Little use

After the orientation survey has been conducted the logistics of the major survey need to be planned. The reader should devise a checklist similar to that of Thomson (1987).

1 Hire field crew with appropriate experience and training;

2 Obtain base maps and devise simple sample numbering scheme;

3 Designate personnel for communication with the laboratory;

4 Arrange quality control of laboratory;

5 Arrange for data handling and interpretation;

6 Organize archive of samples and data;

7 Liaise with other project staff (e.g. geophysicists) and arrange reporting to management.

Reporting of geochemical surveys is important and readers should consult the short booklet by Bloom (2001). This provides a wealth of information as well as the basis for fulfilling legal requirements in Canada.

8.2 ANALYSIS

As the geologist generally sees little of the process of analysis, which is usually done at some distance from the exploration project, analytical data tend to be used uncritically. While most laboratories provide good quality data they are usually in business to make a profit and it is up to the geologist to monitor the quality of data produced and investigate the appropriateness of the analytical methods used.

8.2.1 Accuracy and precision

The critical question for the geologist is how reproducible the analysis is and how representative of the "correct" concentration the concentration is, as shown in Fig. 8.2. The reproducibility of an analysis is termed the precision and its relation to the expected or consensus value the accuracy. For most purposes in exploration geochemistry it is vitally important that an analysis is precise but the accuracy is generally not so crucial, although some indication of the accuracy is needed. At the evaluation stage the analyses must be precise and accurate. The measurement of accuracy and precision requires careful planning and an understanding of the theory involved.

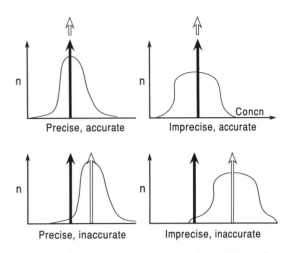

FIG. 8.2 Schematic representation of precision and accuracy assuming normal distribution of analytical error.

A number of schemes have been devised but the most comprehensive is that of Thompson (1982). Precision is measured by analyzing samples in duplicate whereas accuracy requires the analysis of a sample of known composition, a reference material. The use of duplicate samples means that precision is monitored across the whole range of sample compositions. Reference materials can be acquired commercially but they are exceedingly expensive (>US$100 per 250 g) and the usual practice is to develop in-house materials which are then calibrated against international reference materials. The in-house reference materials can be made by thoroughly mixing and grinding weakly anomalous soils from a variety of sites. The contents should be high enough to give some indication of accuracy in the anticipated range but not so high as to require special treatment (for copper, materials in the range 30–100 ppm are recommended). If commercial laboratories are used then the reference materials and duplicates should be included at random; suggested frequencies are 10% for duplicates and 4% for reference materials. If the analyses are in-house then checks can be made on the purity of reagents by running blank samples, i.e. chemicals with no sample. Samples should if possible be run in a different and random order to that in which they are collected. This enables the monitoring of systematic drift. In practice this is often not easy and it is not desirable to operate two numbering systems, so some arrangement should be made with the laboratories to randomize the samples within an analytical batch. The results can then easily be re-sorted by computer.

When laboratory results become available the data should be plotted batch by batch to examine the within- and between-batch effects. The effects observable within a batch are precision and systematic instrumental drift. Precision should be monitored by calculating the average and difference between duplicates on a chart such as is detailed by Thompson (1982). The precision for most applications should be less than about 20%, although older methods may reach 50%. If precision is worse than anticipated then the batch should be re-analyzed. Systematic drift can be monitored from the reference materials within a batch and the reagent blanks. Between-batch effects represent the deviation from the expected value of the reference material and should be monitored by control charts. Any batch with results from reference materials outside the mean ±2 standard deviations should be re-analyzed; commercial laboratories normally do this at no further charge. Long-term monitoring of drift can show some interesting effects. Coope (1991) was able to demonstrate the disruption caused by the moving of laboratories in a study of the gold reference materials used by Newmont Gold Inc.

8.2.2 Sample collection and preparation

Samples should be collected in nonmetallic containers to avoid contamination. Kraft paper bags are best suited for sampling soils and stream sediments because the bags retain their strength if the samples are wet and the samples can be oven dried without removing them from their bags. Thick gauge plastic or cloth are preferred for rock samples. All samples should be clearly labeled by pens containing nonmetallic ink.

Most sample preparation is carried out in the field, particularly when it involves the collection of soils and stream sediments. The aim of the sample preparation is to reduce the bulk of the samples and prepare them for shipment. Soils and stream sediments are generally dried either in the sun, in low temperature ovens, or freeze dried; the temperature should be below

65°C so that volatile elements such as mercury are not lost. Drying is generally followed by gentle disaggregation and sieving to obtain the desired size fraction. Care should be taken to avoid the use of metallic materials and to avoid carryover from highly mineralized to background samples. Preparation of rocks and vegetation is usually carried out in the laboratory and care should be taken in the selection of crushing materials. For example, in a rock geochemical program a company searching for volcanic-associated massive sulfides found manganese anomalies associated with a hard amphibolite. They were encouraged by this and took it as a sign of exhalative activity. Unfortunately further work showed that the manganese highs were related to pieces of manganese steel breaking off the jaw crusher and contaminating the amphibolite samples. Other less systematic variation can be caused by carryover from high grade samples, for example not cleaning small grains of mineralized vein material (e.g. 100,000 ppb Au) will cause significant anomalies when mixed with background (1 ppb Au) rock. Contamination can be eliminated by cleaning crushing equipment thoroughly between samples and by checking this by analyzing materials such as silica sand.

The Bre-X scandal, in which alluvial gold was added to drill pulps before sending the samples for analysis (see section 5.4), emphasizes the importance of recording the methods of sample preparation and controlling access to the samples.

8.2.3 Analytical methods

Most analysis is aimed at the determination of the elemental concentrations in a sample and usually of trace metals. At present it is impossible to analyze all elements simultaneously at the required levels, so some compromises have to be made (Fig. 8.3). In exploration for base metals it is usual to analyze for the elements sought, e.g. copper in the case of a copper deposit, and as many useful elements as possible at a limited extra cost. With modern techniques it is often possible to get 20–30 extra elements, including some that provide little extra information but a lot of extra data for interpretation. The major methods are as shown in Table 8.3 but for detail on the methods used

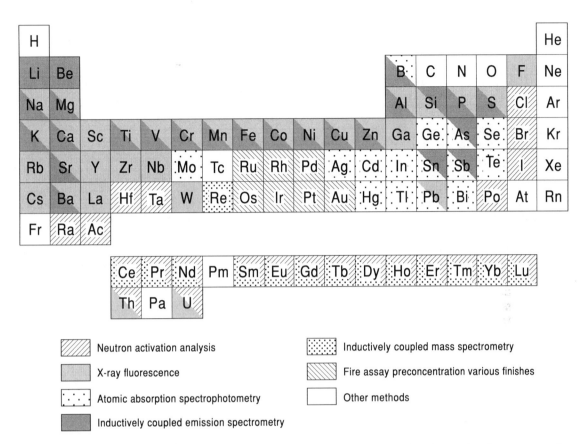

FIG. 8.3 Summary periodic table showing cost-effective methods of elemental analysis at levels encountered in background exploration samples.

Legend:
- Neutron activation analysis
- X-ray fluorescence
- Atomic absorption spectrophotometry
- Inductively coupled emission spectrometry
- Inductively coupled mass spectrometry
- Fire assay preconcentration various finishes
- Other methods

TABLE 8.3 Summary of the main methods used in exploration geochemistry.

Method	Capital cost $ (approx.)	Multielement	Precision	Sample type ($US)	Cost per	comments sample
Colorimetry	8000	No	Poor	Solution	2–10	Good for field use and W, Mo
Atomic absorption spectrophotometry	60,000	No	Good	Solution	1–5	Cheap and precise
X-ray fluorescence	200,000	Yes	Good	Solid	20	Good for refractory elements
ICP–ES	150,000	Yes	Good	Solution	10	Good for transition metals
ICP–MS	150,000– 1000,000	Yes	Good	Solution	10	Good for heavy metals. High resolution instruments very low detection limits

the reader should consult Fletcher (1981, 1987) and Thompson and Walsh (1989).

The differences between the methods shown are cost, the detection limits of analysis, speed of analysis, and the need to take material into solution. Most general analysis in developed countries is carried out by inductively coupled plasma emission spectrometry (ICP–ES), often in combination with inductively coupled plasma mass spectrometry (ICP–MS), or X-ray fluorescence (XRF). All three methods require highly sophisticated laboratories, pure chemicals, continuous, nonfluctuating power supplies, and readily available service personnel, features not always present in developing countries. In less sophisticated environments, high quality analysis can be provided by atomic absorption spectrophotometry (AAS), which was the most commonly used method in developed countries until about 1980. Another method which is widely used in industry is neutron activation analysis (NAA) but its use is restricted to countries with cheap nuclear reactor time, mainly Canada.

Precious metals (gold and platinum group elements) have been extremely difficult to determine accurately at background levels. The boom in precious metal exploration has, however, changed this and commercial laboratories are able to offer inexpensive gold analysis at geochemical levels (5 ppb to 1 ppm) using solvent extraction and AAS, ICP–ES or alternatively NAA on solid samples. For evaluation the method of fire assay is still without equal: in this the precious metals are extracted into a small button which is then separated from the slag and determined by AAS, ICP–ES, or ICP–MS. The analysis of precious metals is different from most major elements and base metals in that large subsamples are preferred to overcome the occurrence of gold as discrete grains. Typically 30 or 50 g are taken in contrast to 0.25–1 g for base metals; in Australia, 8 kg are often leached with cyanide to provide better sampling statistics (further details in section 16.1.2).

Elements which occur as anionic species are generally difficult to measure, especially the chloride, bromide, and iodide ions which serve as some of the ore transporting ligands. Although some of these elements can be determined by ICP–ES or XRF, the most useful method is ion chromatography.

Isotopic analysis is not yet widely used in exploration although some pilot studies, such as that of Gulson (1986), have been carried out. The main reason for this is the difficulty and cost of analysis. Although high resolution ICP sourced mass spectrometers are finding their way into commercial laboratories and are the main hope for cheap analysis, they are not yet routine.

The choice of analytical method will aim at optimizing contrast of the main target element. For example, it is little use determining the total amount of nickel in an ultramafic rock when the majority of nickel is in olivine and the target sought is nickel sulfides. It would be better in this case to choose a reagent which will mainly extract nickel from sulfides and little from olivine. In soils and stream sediments, optimum contrast for base metals is normally obtained by a strong acid (e.g. nitric + hydrochloric acids) attack that does not dissolve all silicates, and an ICP–ES or AAS finish. Most rock analysis uses total analysis by XRF or by ICP–ES/ICP–MS following a fusion or nitric–perchloric–hydrofluoric acid attack.

8.3   INTERPRETATION

Once the analytical data have been received from the laboratory and checked for precision and accuracy, the question of how the data is treated and interpreted needs to be addressed. As the data are likely to be multi-element and there are likely to be a large number of samples this will involve the use of statistical analysis on a computer. It is recommended that the data are received from the analyst either in the form of a floppy disk, CD-ROM, or over the Internet. Re-entering data into a computer from a paper copy is expensive and almost certain to introduce major errors. Normally the data will be transferred into an electronic database (see section 9.1) to allow easy access or, in the case of small data sets stored in a spreadsheet.

8.3.1   Statistics

The object of geochemical exploration is to define significant anomalies. In the simplest case these are the highest values of the element sought but they could be an elemental association reflecting hydrothermal alteration or even element depletion. Anomalies are defined by statistically grouping data and comparing these with geology and sampling information. Normally this grouping will be undertaken by computer and a wide variety of statistical packages are available; for micro-computers, one of MINITAB, SYSTAT, and STATISTICA is recommended at a cost of $US200–500 each.

The best means of statistically grouping data is graphical examination using histograms and box plots (Howarth 1984, Garrett 1989). This is coupled with description using measures of central tendency (mean or median) and of stat-istical dispersion (usually standard deviation). It would be expected that if data are homo-genous then they will form a continuous normal or, more likely, log-normal distribu-tion but if the data fall into several groups then they will be multi-modal as shown in the example in Fig. 8.4. These are a set of copper

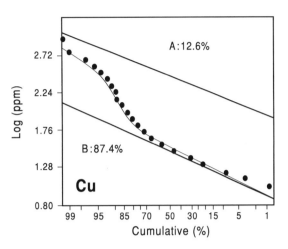

FIG. 8.5 Probability plot of the Daisy Creek copper data. The data have been subdivided into two log normal groups.

determinations of soil samples from the Daisy Creek area of Montana and have been described in detail by Sinclair (1991). The histogram shows a break (dip) in the highly skewed data at 90 and 210 ppm. This grouping can be well seen by plotting the data with a probability scale on the x axis (Fig. 8.5); log-normal dis-tributions form straight lines and multi-modal groups form straight lines separated by curves (for full details see Sinclair 1976). The Daisy Creek data show thresholds (taken at the upper limit, 99%, of the lower subgroups) which can be set at 100 ppm to divide the data into two groups and at 71 and 128 ppm if three groups are used. The relationship of the groups to other elemental data should then be examined by plotting and calculating the correlation matrix for the data set. In the Daisy Creek example there is a strong correlation (0.765) between copper and silver (Fig. 8.6) which re-flects the close primary association between the two elements, whereas the correlation between copper and lead is low (0.12) reflect-ing the spatial displacement of galena veins from chalcopyrite. If geological or sampling data are coded it will be possible to compare these groups with that information.

The spatial distribution of the data must then be examined by plotting elemental data using the intervals that delineate the groups or by plotting the data by percentiles (percentage of data sorted into ascending order); 50, 75, 90,

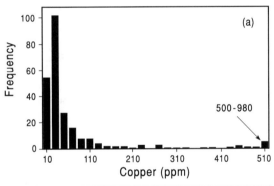

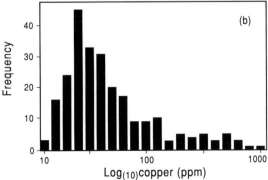

FIG. 8.4 Arithmetic and log$_{10}$ transformed histograms of Daisy Creek copper data. Note the positive skew of the arithmetic histogram. For further discussion see text and Sinclair 1991.

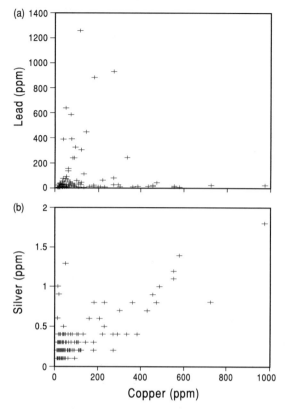

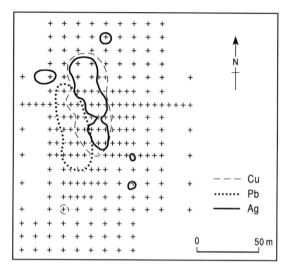

FIG. 8.7   Contour plot of Daisy Creek data showing the highest anomalous populations for copper (>128 ppm), lead (>69 ppm), and silver (>0.55 ppm). The relative size and relative positions of the anomalous areas reflect the primary zoning in the underlying prospect. (From Sinclair 1991.)

FIG. 8.6   Scatter plots of Daisy Creek data showing (a) the strong correlation of copper and silver and (b) the lack of correlation of lead and copper.

95 are recommended for elements that are enriched (see Fig. 9.8).

Geologists are used to thinking in terms of maps and the most useful end product to compare geochemical data with geology and geophysics is to summarize the data in map form using a gridding or GIS package such as Geosoft Oasis Montaj, ArcView, or MapInfo (see section 9.2). However, some care should be taken with the preparation of these maps as it is extremely easy to prejudice interpretation and many maps in the literature do not reflect the true meaning of the data but merely the easiest way of representing it. If the data reflect the chemistry of an area, such as a catchment, it is best if the whole area is shaded with an appropriate color or tone, rather than contouring (e.g. Figs 8.8, 9.8). Data which essentially represent a point can be plotted by posting the value at that point or if there are a lot of data

by representing them by symbols, which are easier to interpret. Typical symbols are given by Howarth (1982). A more usual method of presentation if the samples were collected on a grid is to present the data as a contour plot using the intervals from a histogram (Figs 8.7, 8.4). Plots are best colored from blue to green to yellow to red – cold to hot, as an indication of low to high concentrations.

The combined statistical analysis and element mapping is used to define areas of interest which need to be investigated in detail, using detailed geology, topography sampling information, and, probably, a return to the site. These areas can be classified as to their suggested origin, seepage, and precipitation from local groundwaters, from transported material, or probable residual anomalies. The aim must be to provide a rational explanation for the chemistry of all the areas of interest and not merely to say that they are of unknown origin.

8.4   RECONNAISSANCE TECHNIQUES

The application of these techniques has been discussed in section 4.2.2 but the actual

TABLE 8.4   Application of the main exploration geochemical methods.

	Planning scale	Drainage sediments	Lake sediments	Water	Soil	Deep overburden	Rock	Vegetation	Gas
Reconnaissance	1:10,000 to 1:100,000	Common	Glacial	Arid	Common	Laterite Glacial	Good outcrop	Forests	
Detailed	1:2500 to 1:10,000	Common			Common			Forests	Arid
Drilling	>1:1000						Common		

Common, commonly used. Other techniques useful in areas noted.

method to be used depends on the origin of the overburden. If it is residual, i.e. derived from the rock underneath, the problem is relatively straightforward but if the overburden is exotic, i.e. transported, then different methods must be used. A summary of the application of various geochemical techniques is given in Table 8.4.

8.4.1   Stream sediment sampling

The most widely used reconnaissance technique in residual areas undergoing active weathering is stream sediment sampling, the object of which is to obtain a sample that is representative of the catchment area of the stream sampled. The active sediment in the bed of a river forms as a result of the passage of elements in solution and in particulate form past the sampling point. Thus a sample can be regarded as an integral of the element fluxes. The simplicity of the method allows the rapid evaluation of areas at relatively low cost. Interpretation of stream sediment data is carried out by comparing the elemental concentrations of catchments, as there is only a poorly defined relationship between the chemistry of the stream sediment and the chemistry of the catchment from which it is derived. The simple mass balance method put forward by Hawkes (1976), that relates the chemistry of anomalous soils to that of stream sediments, works in tropical areas of intense weathering but requires modification by topographical and elemental factors in other areas (Solovov 1987, Moon 1999).

In stream sediment sampling the whole stream sediment or a particular grain size or mineralogical fraction of the sediment, such as a heavy mineral concentrate, can be collected. In temperate terrains maximum anomaly/

background contrast for trace metals is obtained in the fine fraction of the sediments as this contains the majority of the organic material, clays, and iron and manganese oxides. The coarser fractions including pebbles are generally of more local origin and depleted in trace elements. Usually the size cut-off is taken at 80 mesh (<177 μm). However, the grain size giving best contrast should be determined by an orientation survey. For base metal analysis and geochemical mapping a 0.5 kg sample is sufficient but a much larger sample is required for gold analysis due to the very erratic distribution of gold particles. A number of authors (Gunn 1989, Hawkins 1991, Akçay et al. 1996) have collected 8–10 kg of –2 mm material and carefully subsampled this using methods such as automated splitting (section 10.1.4).

The most common sample collection method is to take a grab sample of active stream sediment at the chosen location. This is best achieved by taking a number of subsamples over 20–30 m along the stream and at depths of 10–15 cm to avoid excessive iron and manganese oxides. Care must be taken, if a fine fraction is required, that sufficient sample is collected. In addition contamination should be avoided by sampling upstream from roads, farms, factories, and galvanized fences. Any old mine workings or adits should be carefully noted as the signature from these will mask natural anomalies. Most surveys use quite a dense sample spacing such as 1 km^{-2} but this should be determined by an orientation survey. The main mistake to avoid is sampling large rivers in which the signature from a deposit is diluted.

An alternative approach used by geological surveys, for example the British Geological

Survey, is to sieve the samples in the field using a minimum of water. The advantages of this are that it is certain that enough fine fraction is collected and samples are easier to carry in remote areas. It also seems that the samples are more representative but the cost efficiency of this in small surveys remains to be demonstrated. A full review of the use of regional geochemical surveys is given in Hale and Plant (1994) and useful discussion in Fletcher (1997).

An example of a stream sediment survey is shown in Fig. 8.8. The object was to define further mineralisation in an area of Besshi style massive sulfide mineralisation and the copper data is shown in the figure. The area sampled consists of interbedded units of amphibolite and mica schist, within which a deposit had been drilled at point A. The area suffers very high rainfall and much precipitation of iron and manganese; as a result dispersion trains are very short, of the order of 200 m, as shown by an orientation survey. Thus the regional survey was conducted at very close-spaced intervals. However, the deposit is clearly detected and other prospects were defined along strike from the deposit (point B) and disseminated chalcopyrite was found at C.

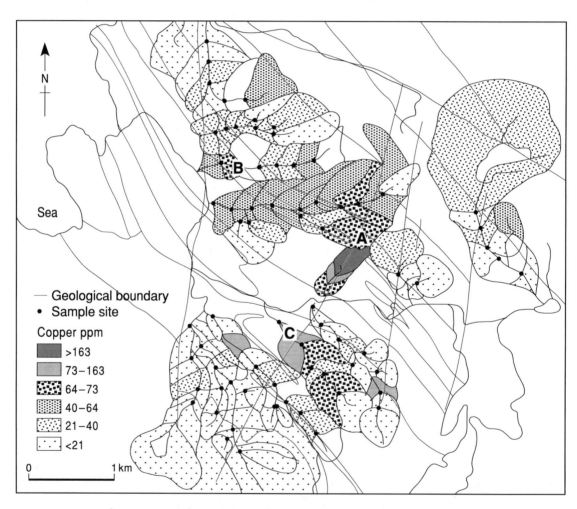

FIG. 8.8   Stream sediment survey of part of the Gairloch area, Wester Ross, NW Scotland. The survey, which was designed to follow up known mineralisation (A), detected further anomalies along strike (B) and disseminated mineralisation at C. The analytical method was ICP–ES following a HNO_3–$HClO_4$ digestion.

Heavy mineral concentrates have been widely used as a means of combating excessive dilution and of enhancing weak signals. The method is essentially a quantification of the gold panning method, which separates grains on the basis of density differences. Panning in water usually separates discrete minerals with a density of greater than 3. Besides precious metals, panning will detect gossanous fragments enriched in metals, secondary ore minerals such as anglesite, and insoluble minerals such as cassiterite, zircon, cinnabar, baryte, and most gemstones, including diamond. The mobility of each heavy mineral will depend on its stability in water, for example sulfides can only be panned close to their source in temperate environments whereas diamonds will survive transport for thousands of kilometers.

The samples collected are usually analyzed or the number of heavy mineral grains are counted. The examination of the concentrate can prove very useful in remote areas where laboratory turnaround is slow and the cost of revisiting the area high, as it will be possible to locate areas for immediate follow-up. The major problem is that panning is still something of an art and must be practised for a few days before the sampler is proficient. The diagram and photograph in Figs 8.9 and 8.10 show panning procedures. Useful checks on panning technique can be made by trial runs in areas of known gold or by adding a known number of lead shot to the pan and checking their recovery. It is usual to start from a known sample size and finish with a concentrate of known mass. However, the differences in panning technique mean that comparisons between different surveys are usually impossible.

8.4.2 Lake sediments

In the glaciated areas of northern Canada and Scandinavia access to rivers is difficult on foot but the numerous small lakes provide an ideal reconnaissance sampling medium as they are accessible from the air. A sample is taken by dropping a heavy sampler into the lake sediment and retrieving it. The sampling density is highly varied and similar to stream sediments. Productivity is of the order of ten samples per flying hour. Essentially data are interpreted in a similar way to that from stream sediments and corrections should be made for organic matter. Lake sediment is only useful if glacial material is locally derived and is ineffective in areas of glaciolacustrine material such as parts of Manitoba. Further details are contained in the reviews by Coker et al. (1979) and Davenport et al. (1997).

8.4.3 Overburden geochemistry

In areas of residual overburden, overburden sampling is generally employed as a follow-up to stream sediment surveys but may be used as a primary survey for small licence blocks or more particularly in areas of exotic overburden. Perhaps the key feature is that it is only employed when land has been acquired. The results are generally plotted at scales from 1:10,000 to 1:1000.

The method of sampling depends on the nature of the overburden; if the chemistry of the near-surface soils reflects that at depth then it is safe to use the cheap option of sampling soil. If not, then samples of deep overburden must be taken. The main areas where surface soils do not reflect the chemistry at depth are glaciated areas, where the overburden has been transported from another area, in areas of wind-blown sand, and in areas of lateritic weathering where most trace metals have been removed from the near-surface layers.

Surface soil sampling

The simplest sampling scheme is to take near-surface soil samples. The major problem is to decide which layer of the soil to sample, as the differences between the layers are often greater than that between sites. The type of soil reflects the surface processes but in general the most effective samples are from a zone at around 30 cm depth formed by the downward movement of clays, organic material, and iron oxides, the B horizon indicated in Fig. 8.11, or the near-surface organic material (A horizon). This downward movement of material is responsible for the depletion and concentration of trace elements causing variations that may be greater than that over mineralisation. It is essential that the characteristics of the soil sampled are recorded and that an attempt is

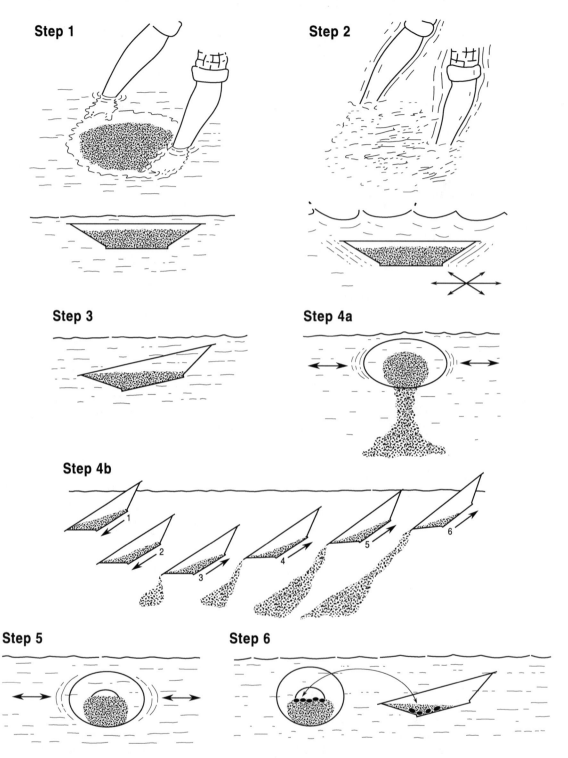

FIG. 8.9   The stages of panning a concentrate from sediment. Often the sediment is sieved to ≤2 mm although this may eliminate any large nugget. Note the importance of mixing the sediment well so that the heavy minerals stay at the bottom of the pan. For an experienced panner the operation will take about 20 minutes. (Modified from Goldspear 1987.)

FIG. 8.10 Panned concentrate sampling in northern Pakistan. The water is glacial so edge of river sampling is essential.

made to sample a consistent horizon. If different horizons are sampled anomalies will reflect this as shown in Fig. 8.12.

The usual method of soil collection in temperate terrains is to use a soil hand auger, as shown in Fig. 8.13. This allows sampling to a depth of the order of 1 m, although normally samples are taken from around 30 cm and masses of around 100–200 g collected for base metal exploration. In other climatic terrains, particularly where the surface is hard or where large samples (500 g to 2 kg) are required for gold analysis, then small pits can be dug. The area of influence of a soil sample is relatively small and should be determined during an orientation survey. The spacing is dependent on the size of the primary halo expected to

occur across the target and the type of overburden, but a rule of thumb is to have at least two anomalous samples per line if a target is cut. Spacing in the search for veins may be as little as 5 m between samples but 300 m between lines, but for more regularly distributed disseminated deposits may be as much as 100 m by 100 m. Sample spacing may also be dictated by topography: in flat areas or where the topography is subdued then rectilinear grids are the ideal choice, but in mountainous areas ridge and spur sampling may be the only reasonable choice.

A typical example of overburden sampling is shown in Figs 8.14a–c. The area is on the eastern side of the Leinster Granite in Ireland and the target was tin–tungsten mineralisation

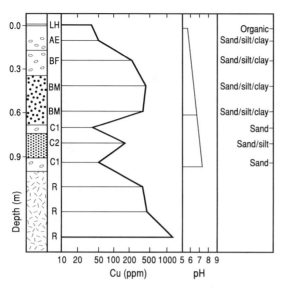

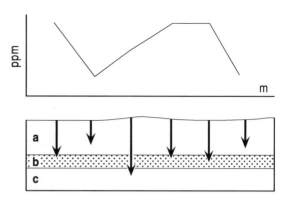

FIG. 8.11 Soil profile showing the effect of sample texture and soil horizons on copper content. Note the bedrock copper content of 0.15% Cu. Letters refer to soil horizons and subdivisions, from the top L, A, B, C, and R (rock). (From Hoffman 1987.)

FIG. 8.12 Effect of sampling differing soil horizons with very different metals contents during a soil traverse.

FIG. 8.13 Typical surface soil sampling in temperate terrains using hand augers over the Coed-y-Brenin copper prospect, Wales.

(a)

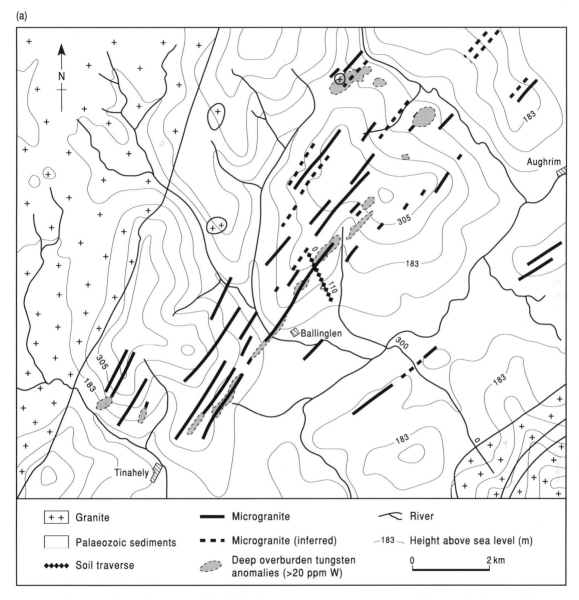

Granite

Palaeozoic sediments

Soil traverse

Microgranite

Microgranite (inferred)

Deep overburden tungsten anomalies (>20 ppm W)

River

—183— Height above sea level (m)

0   2 km

FIG. 8.14   (a) Regional geochemistry of the Ballinglen tungsten prospects, Co. Wicklow, Ireland. Detailed dispersion through overburden and soil response are shown in b and c. (From Steiger & Bowden 1982.)

associated with microgranite dykes, which are also significantly enriched in arsenic and lesser copper and bismuth. Regional base of overburden sampling (usually 2–5 m depth), which was undertaken as the area has been glaciated, clearly delineates the mineralized dykes (Fig. 8.14a). Surface soil sampling was also effective in delineating the dykes using arsenic and to some extent copper and Fig. 8.14b demonstrates the movement of elements through the overburden. Tungsten is very immobile and merely moves down slope under the influence of gravity whereas copper and arsenic are more mobile and move further down slope. The

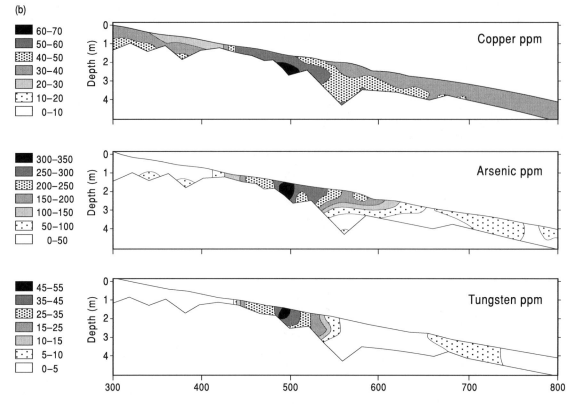

FIG. 8.14   (b) Dispersion through overburden along the soil traverse shown in Fig. 8.14(a). Note the different mobilities of tungsten, copper, and arsenic.

density of soil samples on the surface traverse is clearly greater than needed. The reader may care to consider the maximum sample spacing necessary to detect two anomalous samples in the traverse.

A special form of residual overburden is found in lateritic areas such as in Amazonian Brazil or Western Australia where deep weathering has removed trace elements from the near-surface environment and geochemical signatures are often very weak. One technique that has been widely used is the examination of gossans. As discussed in section 5.1.6, gossans have relic textures which allow the geologist to predict the primary sulfide textures present at depth. In addition the trace element signature of the primary mineralisation is preserved by immobile elements. For example, it is difficult to discriminate gossans overlying nickel deposits from other iron-rich rocks on the basis of their nickel contents, as nickel is mobile, but possible if their multi-element signatures are used (e.g. Ni, Cr, Cu, Zn, Mo, Mn) or the Pt, Pd, Ir content (Travis et al. 1976, Moeskops 1977, Smith 1977).

Drilling to fresh rock, often 50–60 m, has been widely used to obtain a reasonable signature despite the costs involved. Mazzucchelli (1989) suggests that costs are of the order of $A10 m^{-1} and could easily reach $A1M km^{-2} for detailed grids. He recommends surface sampling to outline areas for follow-up work.

Transported overburden

In areas of transported overburden, such as glaciated terrains of the Canadian Shield, sandy deserts, or gravel-covered areas in the Andes, sampling problems are severe and solutions to them expensive.

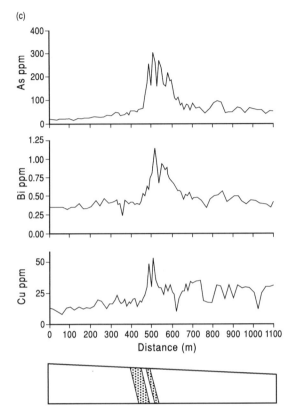

FIG. 8.14   (c) Soil traverse across the mineralized dyke. Note the sharp cut-off upslope and dispersion downslope.

In glaciated terrains overburden rarely reflects the underlying bedrock and seepages of elements are only present where the overburden is less than about 5 m thick. In addition the overburden can be stratified with material of differing origins at different depths. If the mineralisation is distinctive, it is often possible to use boulder tracing to follow the boulders back to the apex of the boulder fan, as in the case of boulders with visible gold, sulfides, or radioactive material. Generally however the chemistry of the tills must be examined and basal tills which are usually of local origin sampled. Figure 8.15 shows a typical glacial fan in Nova Scotia. Usually basal till sampling can only be accomplished by drilling; the most common methods are light percussion drills with flow-through samplers or heavier reverse circulation or sonic drills.

Lightweight drills are cheaper and easier to operate but results are often ambiguous, as it is not easy to differentiate the base of overburden from striking a boulder. The use of heavy equipment in most glaciated areas is restricted to the winter when the ground is frozen. The difficulties of finding the source of glacial dispersion trains for small targets at Lac de Gras, Canada, are discussed in section 17.2.

In sandy deserts water is scarce, most movement is mechanical, and most fine material is windblown. Thus the −80 mesh fraction of overburden is enriched in windblown material and is of little use. In such areas either a coarse fraction (e.g. 2–6 mm), reflecting locally derived material, or the clay fraction reflecting elements moved in solution is used (Carver et al. 1987). One of the most successful uses of this approach has been in the exploration for kimberlites in central Botswana. Figure 8.16 shows the result of a regional sampling program which discovered kimberlite pipes in the Jwaneng area. Samples were taken on a 0.5 km grid, heavy minerals separated from the +0.42 mm fraction and the number of kimberlite indicator minerals, such as picroilmenite, counted. The anomalies shown are displaced from the suboutcrop probably due to transport by the prevailing northeasterly winds.

Recent studies in Chile and Canada have examined the use of various weak extractions to maximize the signature of deposits covered by gravel and till. Figure 8.17 shows the result of a survey over the Gaby Sur porphyry copper deposit. The weak extractions, including deionized water, show high contrast anomalies at the edge of the deposit, probably generated by the pumping of groundwater leaching the deposit (Cameron et al. 2004).

8.4.4 Hydrogeochemistry

Hydrogeochemistry uses water as a sampling medium. Although water is the most widely available material for geochemistry, its use is restricted to very specific circumstances. The reasons for this are that not all elements show equal dissolution rates, indeed many are insoluble, contents of trace elements are very low and have been difficult to measure until recently, being highly dependent on the weather

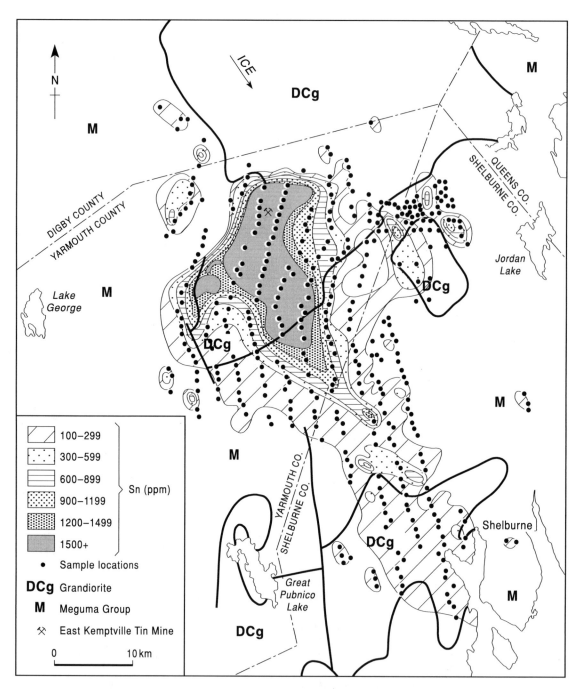

FIG. 8.15 Glacial dispersion train from the East Kemptville tin deposit, Nova Scotia, Canada. Basal till samples. (From Rogers et al. 1990.)

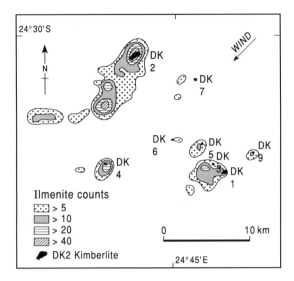

FIG. 8.16 Dispersion of ilmenite around the Jwaneng kimberlites, Botswana. Ilmenite is one of the heavy minerals used to locate kimberlites. (From Lock 1985.)

and easily contaminated by human activity. In general where surface waters are present it is far more reliable and easier to take a stream sediment sample. The possible exception is when exploring for fluorite as fluoride is easily measured in the field using a portable single ion electrode. Where hydrogeochemistry becomes useful is in exploration of arid areas with poor outcrop. In this terrain water wells are often drilled for irrigation and these wells tap deep aquifers that can be used to explore in the subsurface. This approach has been used in uranium exploration although it has not been very successful as the uranium concentration is dependent on the age of groundwater, the amount of evaporation, and its source, which are extremely difficult to determine. For good discussions with case studies see Taufen (1997) and Cameron et al. (2004).

8.4.5 Gases

Gases are potentially an attractive medium to sample as gases can diffuse through thick overburden; in practice most surveys have met with discouraging results, although there has been a recent resurgence in interest due to more robust sampling methods. A number of gases

have been used of which mercury has been the most successful. Mercury is the only metallic element which forms a vapor at room temperature and it is widely present in sulfide deposits, particularly volcanic-associated base metal deposits. The gas radon is generated during the decay of uranium and has been widely used with some success. More recently the enrichment of carbon dioxide and depletion of oxygen caused by weathering of sulfide deposits has been tested, particularly in the western USA (Lovell & Reid 1989).

In general results have been disappointing because of the large variations in gas concentration (partial pressure) caused by changes in environmental conditions, particularly changes in barometric pressure and rainfall. The more successful radon and mercury surveys have sought to overcome these variations by integrating measurements over weeks or months. Gaseous methods work best in arid areas as diffusion in temperate climates tends to be overshadowed by movement in groundwater. The need to "see through" thick overburden coupled with improvements in analytical and sampling methodologies has prompted a reassessment of gas geochemistry and the technique is likely to be used more in the future.

8.4.6 Vegetation

Vegetation is used in two ways in exploration geochemistry. Firstly the presence, absence or condition of a particular plant or species can indicate the presence of mineralisation or a particular rock type, and is known as *geobotany*. Secondly the elemental content of a particular plant has been measured, this is known as *biogeochemistry*. Biogeochemistry has been used more widely than geobotany and has found particular application in the forest regions of northern Canada and Siberia where surface sampling is difficult, but it should only be used with caution. A comprehensive treatment of the subject can be found in Brooks (1983) and the biogeochemical part has been revised in Dunn (2001).

Geobotany

One of the pleasurable, although all too infrequent, parts of exploration is to identify

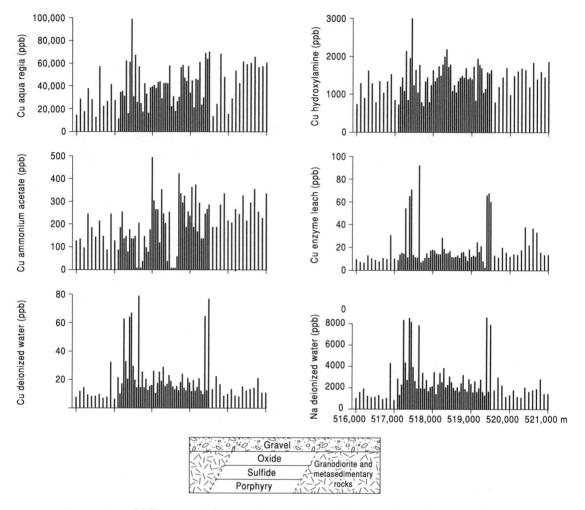

FIG. 8.17   Comparison of different partial extractions on soil samples across the Gaby Sur porphyry copper deposit, Chile. (From Cameron et al. 2004.)

plants, in particular flowers associated with mineralisation. The most famous of these is the small mauve copper flower of the Zambian Copper Belt, *Becium homblei*. In general this plant requires a soil copper content of 50–1600 ppm Cu to thrive, conditions that are poisonous to most plants (Reedman 1979). This type of plant is known as an indicator plant. Unfortunately recent research has demonstrated that most indicator plants will flourish under other conditions and are not very reliable.

A more reliable indicator of metal in soils is the stunting of plants or yellowing (chlorosis).

This condition is particularly amenable to remote sensing from satellites. Most gardeners appreciate that plants have favorite soil conditions and it is possible to make an estimate of bedrock based on the distribution of plant species. Unfortunately this method tends to be rather expensive and is little used.

Biogeochemistry

Plants require most trace elements for their survival and take these through their roots, transpiring any residues from their leaves and concentrating most trace elements in their recent growth. Unfortunately for the geochemist the rate of uptake and concentration of elements is highly dependent on the species and the season. In general sampling is conducted on one plant species and one part of the plant, usually first and second year leaves or twigs. The variation in concentration with season is so severe that sampling must be restricted in time. In general around 500 g of sample are collected and ashed prior to analysis.

The main advantage of biogeochemistry is that tree roots often tap relatively deep water tables which in glacial areas can be below the transported material and representative of bedrock. A similar effect can be obtained by sampling the near-surface humic horizons which largely reflect decayed leaves and twigs from the surrounding trees (Curtin et al. 1974).

8.5 FOLLOW-UP SAMPLING

Once an anomaly has been found during reconnaissance sampling and a possible source identified, it is necessary to define that source by more detailed sampling, by highlighting areas of elemental enrichment, and eliminating background areas until the anomaly is explained and a bedrock source, hopefully a drilling target, proven. If the reconnaissance phase was stream or lake sampling, this will probably involve overburden sampling of the catchments. In the case of soil or overburden sampling it will mean increasing the density of sampling until the source of an anomaly is found, proving the source by deep overburden sampling, and sampling rocks at depth or at surface when there is outcrop. A typical source can be seen in Fig. 8.14. The regional tungsten anomaly was followed up by deep overburden sampling such as that shown in Fig. 8.14b and the suboutcrop of the dyke defined. The dip of the dyke could be estimated from the occurrence of disseminated sulfides, which would also respond to induced polarization, and a drilling target outlined.

8.5.1 Rock geochemistry

Rock sampling is included in the techniques for follow-up because although it has been applied with some success in regional re-

connaissance, it is really in detailed work, where there is good outcrop or where there is drill core, that this technique becomes most effective.

On a regional basis the most successful applications have been in the delineation of mineralized felsic plutons and of exhalative horizons. Full details are provided in reviews by Govett (1983, 1989) and Franklin (1997). Plutons mineralized in copper and tungsten are usually enriched in these elements but invariably show high variability within a pluton. Tin mineralisation is associated with highly evolved and altered intrusives and these are easily delineated by plotting on a K–Rb versus Rb–Sr and Mg–Li diagram or examing the geochemistry of minerals, such as micas. It is recommended that 30–40 samples are taken from each pluton to eliminate local effects. Studies in the modern oceans indicate haloes of the order of kilometers around black smoker fields and these have been detected around a number of terrestrial massive sulfide deposits, notably by the manganese content in the exhalative horizon.

Some care needs to be taken in the collection of rock samples. In general 1 kg samples are sufficient for base metal exploration but precise precious metal determination requires larger samples, perhaps as large as 5 kg, if the gold is present as discrete grains. Surface weathering products should be removed with a steel brush. Rock geochemistry depends on multi-element interpretation and computer-based interpretation with careful subdivision of samples on the basis of lithology.

Mine scale uses

One of the major applications of rock geochemistry is in determining the sense of top and bottom of prospects and detecting alteration that is not obvious from visual examination. Geochemists from the former Soviet Union have been particularly active in developing an overall zonation sequence which they have used to construct multiplicative indications of younging for a variety of deposit types (Govett 1983, 1989). Rock geochemistry has been widely applied in the search for volcanic-associated massive sulfide deposits in Canada, especially in combination with downhole electromagnetic methods (see section 15.3.8). In particular rock geochemistry has detected the cone-shaped alteration zones beneath them, which are enriched in Mg and depleted in Na, Ca and sometimes in K, and recognized distal exhalite horizons which are enriched in more mobile elements such as arsenic, antimony, and barium (for a good case history see Severin et al. 1989). Another major application is in the exploration of sulfide-rich epithermal gold deposits, which often show element zonation (Heitt et al. 2003). The more volatile elements such as As, Sb and Hg are concentrated above the major gold zones whereas any copper is at depth. The successful application of these methods on a prospect demands a detailed knowledge of the subsurface geology.

8.6   SUMMARY

Exploration geologists are more likely to be directly involved in geochemical surveys than in geophysical ones and they should therefore have considerable knowledge of this technique and its application in mineral exploration. A geochemical program naturally commences with a planning stage which involves choice of the appropriate field survey and analytical methods suitable for use for a particular area and the commodity sought. Individual areas will present very different problems, e.g. areas with deep or shallow overburdens, and therefore orientation surveys (section 8.1.1) are very desirable. If possible such a survey should be one in which a variety of sampling methods are tested over a deposit of similar geology to that of the target and in similar topographical conditions to determine which produces the best results. This done, the logistics of the major survey can be planned.

The analysis of samples collected during geochemical surveys is generally carried out by contract companies and the quality of the data they produce must be monitored, and for this reason the geologist in charge must be familiar with sample collection and preparation, analytical methods, and the statistical interpretation of the data obtained.

The material which will be sampled depends upon whether the overburden is residual or transported. The most widely used technique in residual areas suffering active weathering is stream sediment sampling which is a relatively simple method and allows the rapid evaluation of large areas at low cost. Panning to collect heavy mineral concentrates can be carried out at the same time, thus enhancing the value of this method. In glaciated areas of difficult access lake sediment rather than stream sediment surveys may be used.

Overburden geochemistry is often a follow-up to stream sediment surveys in areas of residual and transported overburden. The method of sampling is very varied and dictated by the nature of the overburden. When the overburden is transported (glaciated terrains or sandy deserts) sampling problems are severe and solutions to them expensive.

The use of hydrogeochemistry, gases, and vegetation is restricted but rock geochemistry is increasing in importance both in surface and underground work as well as by utilizing drill core samples.

8.7   FURTHER READING

The literature on exploration geochemistry is large and reasonably accessible. *Geochemistry in Mineral Exploration* by Rose et al. (1979) remains the best starting point. A good general update on soil sampling is provided by Fletcher et al. (1987) with reviews of most other geochemical techniques in Gubins (1997). The *Handbook of Exploration Geochemistry* series, although expensive, provides comprehensive coverage in the volumes so far published: chemical analysis (Fletcher 1981), statistics and data analysis (Howarth 1982), rock geochemistry (Govett 1983), lateritic areas (Butt & Zeegers 1992), arctic areas (Kauranne et al. 1992), and stream sediments (Hale & Plant 1994). Readers are strongly advised to try the practical problems in Levinson et al. (1987).

Discussion of techniques above can be found in *Explore*, the newsletter of the Association of Exploration (now Applied) Geochemists which also publishes the more formal *Geochemistry: Exploration, Environment, Analysis*. The reader will also find articles of interest in *Journal of Geochemical Exploration*, *Applied Geochemistry*, and *Exploration and Mining Geology*.

9

MINERAL EXPLORATION DATA

CHARLES J. MOON AND MICHAEL K.G. WHATELEY

One of the major developments in mineral exploration has been the increased use of computerized data management. This has been used to handle the flow of the large amounts of data generated by modern instrumentation as well as to speed up and improve decision making. This chapter details some of the techniques used to integrate data sets and to visualize this integration. Two types of computer packages have evolved to handle exploration and development data: (i) Geographical Informations Systems (GIS) for early stage exploration data, usually generic software developed for other nongeologic applications, discussed in section 9.2, and (ii) mining-specific packages designed to enable mine planning and resource calculations, discussed in section 9.3.

What must be emphasized is that the quality of data is all important. The old adage "rubbish in and rubbish out" unfortunately still applies. It is essential that all data should be carefully checked before interpretation, and the best times to do this are during entry of the data into the database and when the data are collected. A clear record should also be maintained of the origin of the data and when and who edited the data. These data about data are known as metadata.

9.1 DATA CAPTURE AND STORAGE

9.1.1 Theory

In order to integrate data, they must be available in an appropriate digital form. If the data are in paper form they require conversion, a process known as digitization. Even if data are derived as output from digital instruments, such as airborne magnetometers or downhole loggers, the data may need conversion to a different format.

Computers do not know how to classify geological objects so a format for storing data must be defined. This format will be determined by the type, relationship, attributes, geometry, and quality of data objects (for further details see Bonham-Carter 1994). An example of simple geological map data based on Fig. 9.1 is: (i) type – geological unit; (ii) relationship – contacts; (iii) attributes – age, lithology; (iv) shape – polygon. Two main components can be separated out for all data types: (i) a spatial component dependent on location (e.g. sample location for a point sample with x, $y \pm z$ components); and (ii) an attribute component not dependent on location but linked to the spatial component by a unique identifier (e.g. sample number or drillhole number and depth). Figure 9.1 shows typical geological objects with differing dimensions: lithological units as polygons, samples as points, faults as lines, structures as points with orientation, drillholes as points (vertical) or lines (inclined).

9.1.2 Spatial data models

There are two major methods of representing spatial data, raster and vector. In the vector model the spatial element of the data is represented by a series of coordinates, whereas in the raster model space is divided into regular pixels, usually square. Each model has advantages and disadvantages summarized in Table 9.1 but the

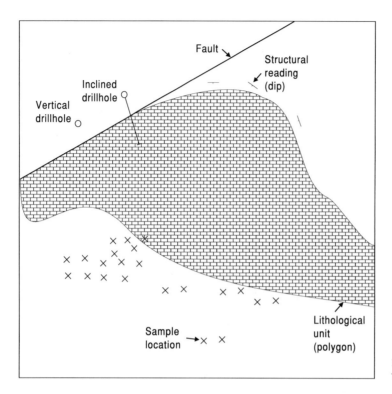

FIG. 9.1 Typical geological objects with differing dimensions.

TABLE 9.1 Typical geological features and their representation and quality.

	Relationship	Typical attributes	Geometry	Typical quality
Geological unit	Contacts	Lithology, age, name	Polygon	±25 m
Fault	Contacts	Type, name	Line	±25 m
Structural reading		Type, orientations	Point	±5 m
Sample location		Number, lithology, assay	Point	±5 m
Inclined drillhole		Number, date, driller	Line	±0.1 m
Vertical drillhole		Number, date, driller	Point	±0.1 m
Drillhole sample		Depth, lithology, assay	Point or line (interval)	±0.5 m

key factors in deciding on a format are resolution and amount of storage required. Figure 9.2 shows a simple geological map in vector and raster format. The raster method is commonly used for remote sensing and discussed in section 6.2.2, whereas the vector method is used for drillholes and geological mapping. Most modern systems allow for integration of the two different types as well as conversion from one model to another, although raster to vector conversion is much more difficult than that from vector to raster.

In a simple (two-dimensional) vector model, points are represented by x and y coordinates, lines as a series of connected points (known as vertices), and polygons as a series of connected lines or strings. This simple model for polygons is known as a spaghetti model and is that adopted by computer aided drawing (CAD) packages. For more complex querying and modeling of polygons, the relationship between adjoining polygons must be established and the entire space of the study area subdivided. This is known as the topological model. In this

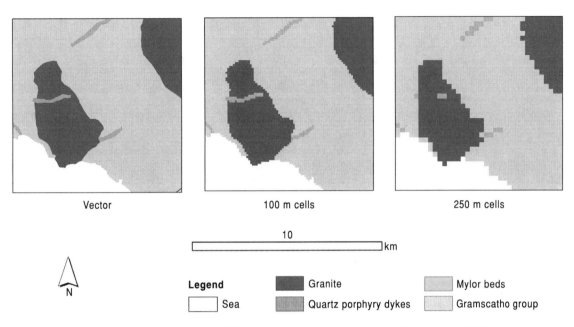

Vector   100 m cells   250 m cells

10
└───┘ km

△
N

Legend   ▨ Granite   ▨ Mylor beds

☐ Sea   ▨ Quartz porphyry dykes   ☐ Gramscatho group

FIG. 9.2   Vector and raster representation of lithologies from a geological map. The area in SW Cornwall, England and the original map was compiled at 1:250,000 scale based largely on a map of the British Geological Survey. Note the impact of using different cell sizes in a raster map. The lithology of the rasterized cell is taken from the lithology with the largest area within the cell.

model, polygons are formed by the use of software as a mesh of lines, often known as arcs, that meet at nodes. Another variation of this model often used for height data is that of the triangular irregular networks (TIN) and is used to visualize digital elevation surfaces or construct digital terrain models (DTM). The TIN model is similar to the polygons used in ore resource and reserve calculation (see section 10.5).

9.1.3   Storage methods

The simplest solution for data storage is that of the flat file method in which each point has associated *x, y* (and *z*) coordinates, as well as attributes. This is the familiar style of an accounting ledger or table and implemented electronically in a spreadsheet, such as Lotus 1-2-3 and Microsoft Excel. In this format (Table 9.2) attributes are stored in columns or fields, and rows, that are known as tuples. The features of this type of storage are that: (i) all the data are represented in the table; (ii) any cell in the table must have a single value, replicate samples require additional tuples; (iii) no

TABLE 9.2   Advantages and disadvantages of raster and vector formats. (After Bernhardsen 1992.)

	Raster	**Vector**
Data collection	Rapid	Slow
Data volume	Large	Small
Data structure	Simple	Complex
Graphical treatment	Average	Good
Geometrical accuracy	Low	High
Area analysis	Good	Average
Generalization	Simple	Complex

duplicate tuples are allowed; and (iv) tuples can be rearranged without changing the nature of the relation (Bonham-Carter 1994). The flat file method is however an inefficient way to store data as a minor change, for example, a change to the name of a lithology in Table 9.3 (which is part of the table associated with Figs 9.4 & 9.5), requires a global search and change of all examples. Data are more efficiently stored and edited in a relational database in which the data are stored as a series of tables linked by unique keys, such as sample numbers. The flat file

TABLE 9.3   Part of the data associated with Fig. 9.3 stored as a flat file. Note that a change (e.g. name of geological unit) would require editing of all instances.

Area	Perimeter	Modgeol_	Unit	Lithology	Age
12,983	711	2	Basic volcanics	basic volcanics	Middle Devonian
2648,177	8049	3	Upper Devonian slates	slates	Upper Devonian
14,476	866	4	Basic volcanics	basic volcanics	Middle Devonian
977,190	10,015	5	Sea	none	None
7062,567	27,324	6	Upper Devonian slates	slates	Upper Devonian
69,818,680	231,598	7	Mid Devonian slates	slates	Middle Devonian
16,556	598	8	Dolerite	dolerite	Upper Devonian
11,447,985	51,973	9	Basic volcanics	basic volcanics	Middle Devonian
415,067	5493	10	Mid Devonian slates	slates	Middle Devonian
8805	566	11	Basic volcanics	basic volcanics	Middle Devonian
6650	468	12	Mid Devonian limestone	limestone	Middle Devonian
860,230	5309	13	Basic volcanics	basic volcanics	Middle Devonian
9946	437	14	Dolerite	dolerite	Upper Devonian
47,900	958	15	Dolerite	dolerite	Upper Devonian
36,817	980	16	Dolerite	dolerite	Upper Devonian
15,053	492	17	Dolerite	dolerite	Upper Devonian
297,739	3253	18	Basic volcanics	basic volcanics	Middle Devonian
39,071	972	19	Dolerite	dolerite	Upper Devonian
156	86	20	Sea	none	None

format can be converted to a relational database by a process known as normalization. Table 9.4 shows an example of this process, a unit number (finalgeol) which is a simplified version of modgeol, lithology number (lith#) and age# have been added as the first step in this process. The next step is (Table 9.5) to break Table 9.3 into its components. The first table of Table 9.5 link the polygons with a finalgeol. The second links finalgeol and with lith# and age# (Table 9.5b). The third part of Table 9.5 links lith# with lithology name (Table 9.5c) and the fourth age# with age (Table 9.5d). This decomposition into their normalized form allows easy editing, e.g. an error in the formation name can be corrected with a single edit.

The flat file and relational database systems are mature technologies and used extensively for 2D GIS and geometric models. It is however becoming increasingly difficult for them to manage 3D information with complex topological relationships. Difficulties arise in converting the complex data types used in portraying these objects being represented into relational tables with links. This problem is overcome in an object-oriented (OO) database approach in which conversion is unnecessary as the database stores the objects directly, complete with topological and other information links. For this reason, OO database systems are being increasing used in Internet information delivery where complex multimedia objects need to be retrieved and displayed rapidly (FracSIS 2002).

9.1.4   Corporate solutions

As large amounts of money are invested in collecting the data, it is crucial that the data are safely archived and made available to those who need them as easily as possible. Integrity of data is paramount for any mining or exploration company, both from a technical and legal viewpoint (acQuire 2004). However this integrity has often been lacking in the past and many organizations have had poor systems giving rise to inconsistencies, lost data, and errors. Increasingly, in the wake of incidents such as the Bre-X fraud (see section 5.4), both industry and government departments require higher levels of reporting standards. Relational databases provide the means by which data can be stored with correct quality control procedures and retrieved in a secure environment.

TABLE 9.4   Table with lith# and age# inserted to allow the table to be normalized as shown in Table 9.5.

Area	Perimeter	Modgeol_	Finalgeol	Unit	Lith#	Lithology	Age#	Age
12,983	711	2	6	Basic volcanics	6	basic volcanics	2	Middle Devonian
2648,177	8049	3	8	Upper Devonian slates	4	slates	3	Upper Devonian
14,476	866	4	6	Basic volcanics	6	basic volcanics	2	Middle Devonian
977,190	10,015	5	10	Sea	0	none	0	None
7062,567	27,324	6	8	Upper Devonian slates	4	slates	3	Upper Devonian
69,818,680	231,598	7	5	Mid Devonian slates	4	slates	2	Middle Devonian
16,556	598	8	9	Dolerite	8	dolerite	3	Upper Devonian
11,447,985	51,973	9	6	Basic volcanics	6	basic volcanics	2	Middle Devonian
415,067	5493	10	5	Mid Devonian slates	4	slates	2	Middle Devonian
8805	566	11	6	Basic volcanics	6	basic volcanics	2	Middle Devonian
6650	468	12	7	Mid Devonian limestone	5	limestone	2	Middle Devonian
860,230	5309	13	6	Basic volcanics	6	basic volcanics	2	Middle Devonian
9946	437	14	9	Dolerite	8	dolerite	3	Upper Devonian
47,900	958	15	9	Dolerite	8	dolerite	3	Upper Devonian
36,817	980	16	9	Dolerite	8	dolerite	3	Upper Devonian
15,053	492	17	9	Dolerite	8	dolerite	3	Upper Devonian
297,739	3253	18	6	Basic volcanics	6	basic volcanics	2	Middle Devonian
39,071	972	19	9	Dolerite	8	dolerite	3	Upper Devonian
156	86	20	10	Sea	0	none	0	None

(a)

Area	Perimeter	Modgeol_	Finalgeol
12,983	711	2	6
2,648,177	8049	3	8
14,476	866	4	6
977,190	10,015	5	10
7,062,567	27,324	6	8
69,818,680	231,598	7	5
16,556	598	8	9
11,447,985	51,973	9	6
415,067	5493	10	5
8805	566	11	6
6650	468	12	7
860,230	5309	13	6
9946	437	14	9
47,900	958	15	9
36,817	980	16	9
15,053	492	17	9
297,739	3253	18	6
39,071	972	19	9
156	86	20	10

TABLE 9.5 Normalized form of Table 9.4. Note the linkages between tables and how a change in lithology or age can be more easily edited.

(b)

Finalgeol	Unit	Lith#	Age#
1	Dartmouth Group	1	1
2	Meadfoot Group	2	1
3	Acid volcanics	7	1
4	Staddon Formation	3	1
5	Mid Devonian slates	4	2
6	Basic volcanics	6	2
7	Mid Devonian limestone	5	2
8	Upper Devonian slates	4	3
9	Dolerite	8	3
10	Sea	0	0

(c)

Lith#	Lithology
0	None
1	sandstone and shale
2	shale and sandstone
3	sandstone
4	Slate
5	limestone
6	basic volcanics
7	acid volcanics
8	Dolerite

(d)

Age#	Age
0	None
1	Lower Devonian
2	Middle Devonian
3	Upper Devonian

Many proprietary technical software products provide such storage facilities, for example acQuire (acQuire 2004) provides such a solution for storage and reporting of data that also interfaces with files in text formats such as csv, dif, txt (tab delimited and fixed width formats), as well as numerous proprietary formats.

The strategy for collection and evaluation (checking) of data (Walters 1999) is often a matter of company procedure. Most errors are gross and can be easily filtered out. Each geologist and mining or processing engineer knows what the database should contain in terms of ranges, values, and units. It is a simple matter of setting up the validation tables to check that the data conform to the ranges, values, and units expected. A simple example would be ensuring that the dip of drillholes is between 0 and −90 degrees for surface drilling.

Documentation for the database is essential including the source and method of data entry as well as their format. It is essential to maintain adequate, long-term archiving and access. For contingencies, it should be possible to add fields and columns to the database should it be necessary to add information at a later stage. The database should also be capable of manipulating all data and the ability to evaluate and review the data is essential.

Normally a single, maintained, flexible and secure storage source is recommended. This means that only the most recent copy of the database is being used for evaluation at the mine site with older versions archived regularly. Multiple access to the database, even from other sites, can be achieved through web technology. The additional advantage of such web technology is that different sets of data, each containing information about a particular item such as a borehole, can be available through a single search interface (Whateley 2002). The ultimate long-term storage however remains paper and a hard copy should be regularly filed of all important data sets, e.g. drillhole logs and assays.

9.1.5 Data capture

The capture of attribute data is usually relatively straightforward but spatial data is often more difficult. Most field data are now generated in digital form but any data only available on paper will require digitization.

Field data capture

Capture of data in the field using computers is becoming routine often using ruggedized computers that can be easily transported. At present this approach is widely used for routine tasks such as sample collection and core logging but is less well suited to geological mapping. However this situation is changing rapidly with the advent of inexpensive portable digital assistants using a stylus for input and better quality displays.

Digitization

Spatial data are normally digitized either by scanning a map or by using a digitizing table. In the digitizing table method a point or line is entered by tracing the position of a puck over it relative to fine wires within the table. The position of the point or line is then converted by software into the original map coordinates from the position measured on the table. Tracing lines is very laborious and maps often have to be simplified or linework traced to avoid confusion during the digitizing process. The scanning method has become much easier with the advent of inexpensive scanners. In this method a georeferenced image is traced on a computer screen using a mouse or puck. When the map or scan has been digitized all linework should be carefully edited. Generally this editing is laborious and more prone to errors than the original digitization.

Attribute data

Attribute data can be captured by typing handwritten data or by scanning data that is already typewritten and relatively clean. The scanned images are then converted into characters for storage using optical character recognition software. This software is however not perfect and the resulting characters should be carefully checked for errors.

9.1.6 Data sources

Sources of digital data have become widespread and much less expensive over the last 10 years. In particular there are digital topographical (ESRI 2004, Globe 2004), Landsat (Geocover 2004, GLCF 2004) and geological map coverages of the world, albeit at small scales (GSC 1995, Arcatlas 1996). In North America and Australia most geological, as well as airborne geophysical data, are available from the national or provincial geological surveys for the cost of reproduction or freely across the Internet. The reader is advised to make a search of the Internet using Google or a similar search engine to find up to date information.

9.1.7 Coordinate systems and projections

Geologists have in the past generally managed to avoid dealing with different coordinate systems in any detail, as the areas they were dealing with were small. The advent of GPS and computerized data management has changed this. The plotting of real world data on a flat surface is known as projection and is the result of the need to visualize data as a flat surface when the shape of the earth is best approximated by a spheroid, a flattened sphere. For small areas the distortion is not important but for larger areas there will be a compromise between preserving area and distance relationships. For example, the well-known Mercator projection emphasizes Europe at the expense of Africa. The scale of the data also governs the choice of projection. For maps of scales larger than 1:250,000, either a national grid or a Universal Transverse Mercator (UTM) grid is generally used. In the latter projection, the earth is divided into segments of 6 degrees longitude with a value of 500,000 m E given to the central meridian of longitude and a northing origin of 0 m at the equator, if north of the equator, or large number, often 10,000,000 m, if south of the equator.

There are a variety of different values in use for the ellipsoid that approximates the shape of the earth, known as the datum. The most commonly used datum for GPS work is World Geodetic System (WGS 1984) but the datum used on the map must be carefully checked, as the use of different datums can change coordinates by up to 1500 m. The reader is advised to read about the problems in more detail in texts such as Longley et al. (2001) and Snyder (1987).

9.2 DATA INTEGRATION AND GEOGRAPHICAL INFORMATION SYSTEMS

One of the major advances in technology at the early exploration stage has been the ability to integrate data easily. This has been driven by the development of Geographical Information Systems (GIS). Although GIS is usually defined to include data storage, its main function is to allow the easy integration of data and output, usually in the form of maps (Longley et al. 1999, 2001). Its development has been designed for a wide range of applications involving spatial data, including the design of optimum siting of pizza delivery sites and monitoring the spread of disease, but the generic commercial systems are applicable (with limitations) to mineral exploration data. The dominant commercial systems, at the time of writing, are ArcView and ArcGIS (ESRI) with 35% of total market share and they have wide use in geological applications; MapInfo has a much smaller overall market share but a wide following in the geological world, particularly in Australia as add-on programs have been developed for mineral exploration. Current commercial systems allow the display of both vector and raster data with varying degrees of querying and modeling facilities. These systems are well suited to 2D data but, at the time of writing, only partly usable for 3D drill data including drillholes and underground sampling.

In addition to the complex (and expensive) fullblown GIS systems there are a number of simpler software packages, such as Oasis Montaj (Geosoft 2004), Interdex (Visidata 2004), and Micromine (Micromine 2004), specifically designed for mineral exploration. Some have features not easily accessed in GIS packages, such as the gridding facilities for geophysical data and integrated data management in Oasis Montaj.

9.2.1 Arrangement of data using layers

The basis for the integration of data in a GIS is their use as a series of layers. This method is really an extension of the old light table method in which maps were overlaid and the result viewed by shining a light through them. Combining layers is much easier then splitting them apart and a good rule is to build layers from the simplest components. For example, rather than having a topographical layer it is better to have separate layers for roads, field boundaries, buildings, and rivers. In the case of a conventional geological map separate layers would be generated for geological units (polygons), structural readings (points), and faults (lines).

9.2.2 South Devon example

Perhaps the best way of explaining the methods of handling data in a GIS is to use a worked example. This example is typical of early stage exploration and refers to an area in south Devon, England. Prior to the mid 1980s this area was not thought prospective for gold and no systematic exploration had been undertaken. The British Geological Survey had long been active in the area and had investigated the area of basic volcanics in the north of Fig. 9.3 for its base metal potential. The initial orientation samples derived from acid volcanics (266,000E, 49,000N, British National Grid) returned small grains of gold in a panned concentrate and it was decide to sample for gold. Gold was also known from a small prospect at Hopes Nose, Torquay, approximately 30 km to the northeast of the discovery site.

The area discussed is on the southern flank of one of the major early Permian granite intrusives, the Dartmoor Granite. This granite intrudes an area generally of low grade marginal marine and fluvial sediments with interbedded volcanics of Devonian age that were deformed during the late Carboniferous. The Devonian sediments (and originally the granite) are unconformably overlain by red-bed sandstones and fine sediments of Permo-Triassic age. Marine sediments of Jurassic and Cretaceous age probably originally covered the area but have been removed by Tertiary erosion.

Simple display and querying: geological map

The basis for geological interpretation in the area is a digitized version of the 1:50,000 scale

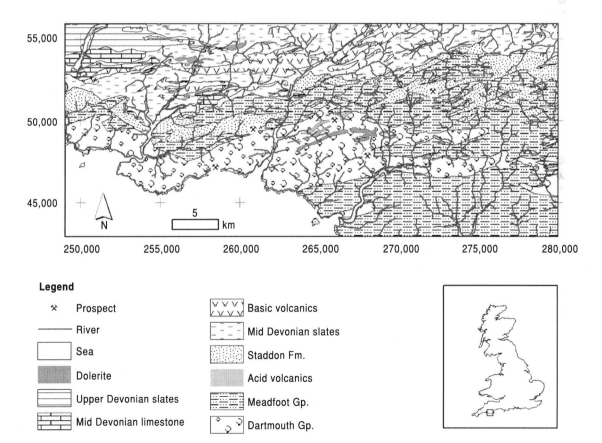

FIG. 9.3 Geological map (in ArcGIS) and index map. This map shows lithologies and no structure. (Largely after Ussher 1912.)

published geological mapping (Ussher 1912). The most important component is the distribution of lithologies that were outlined by generating polygons. Each polygon has a unique number which is linked in a table to the lithological number and lithological name in the key (Fig. 9.4, Tables 9.3 & 9.4). In contrast to conventional maps it is easy to select individual units; in this case (Fig. 9.5) the acid and basic volcanics have been chosen.

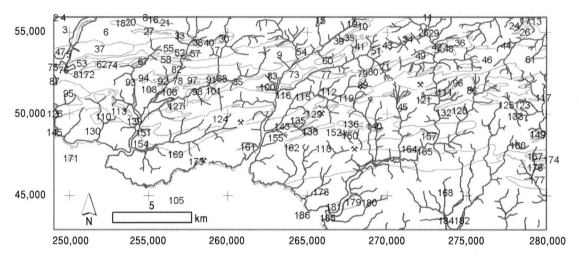

FIG. 9.4   Geological map of Fig. 9.3 showing polygon numbers.

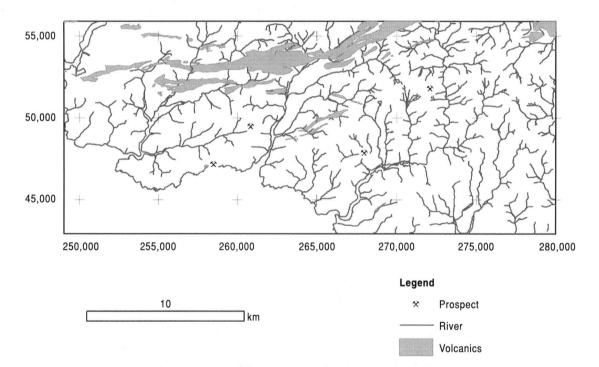

Legend

✕   Prospect

———   River

▨   Volcanics

FIG. 9.5   Geological map with only acid and basic volcanics selected. Compare this map with Fig. 9.3.

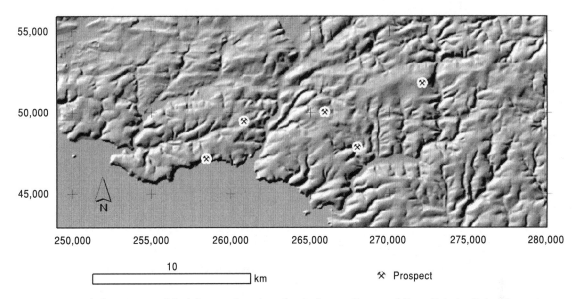

FIG. 9.6   Digital elevation model of the area based on the Ordnance Survey of Great Britain digital data collected at 50 m intervals. The image is hillshaded with sun elevation of 45 degrees from 315 degrees so SW–NE striking structures are highlighted. (Source: Digimap, University of Edinburgh.)

Remote sensing data

As the geological mapping was quite old (Ussher 1912) and the oucrop poor, it was possible to check the mapping with Landsat, air photography, and digital elevation models. Figure 9.6 shows a hill shaded image of the digital elevation model. A number of features can be detected by comparison with Fig. 9.3, particularly the high ground underlain by the Staddon Formation, a coarse sandstone unit. Some mapped structures are obvious, a mapped fault at 266,000E, 51,000N and the strong N–S lineament at ~274,000E. Landsat images (Fig. 9.7) are difficult to interpret as the area is intensely farmed in small fields. Vegetation is dominant and many of the dark gray areas of Fig. 9.7 are woodland. In spite of this it was possible to extract lineaments, which are shown in Fig. 9.10.

Geochemical data

One of the key data sets in the area is the result of a panned concentrate sampling exercise (Leake et al. 1992). Samples were analyzed by X-ray fluorescence for trace elements and for gold by solvent extraction and atomic absorp-tion spectrometry. Gold geochemistry is the most obvious indicator of area for follow-up. Gold show a strongly skewed distribution with a mean of 273 ppb, median of 10 ppb, and maximum of 5700 ppb Au (Fig. 9.8). The soft-ware used offers four options to divide the data: (i) quantiles – e.g. 20, 40, 60, 80, 100 percentiles in this case 5, 5, 24, 200, 5700 ppb; (ii) equal intervals – 1140, 2280, 3420, 4560, 5700 ppb; (iii) natural breaks (based on breaks in the histogram) 215, 708, 1540, 3100 ppb; (iv) mean and standard deviations – 273, 1060, 1846, 2633 ppb. As the distribution is so highly skewed and 40% of the concentrations are less than the detection limit 10 ppb (set to 5 ppb), in this case a percentile method based on the higher part of the distribution was used: 50, 75, 90, 95 percentiles rounded to familiar numbers, 10, 150, 500, 1500 (Fig. 9.8). This map picks out catchments with detectable gold as well as differentiating those which have very high gold concentrations. These were followed up and gold grains were discovered in soil at the locations marked by the prospect symbols.

If the log-transformed data are plotted on a probability scale (Fig. 9.9), a threshold of 50 pbb might be selected. This simple subdivision is shown in Fig. 9.9, although it is, at least in the

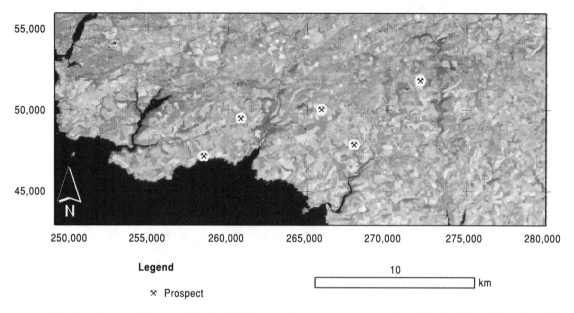

FIG. 9.7 Landsat image of the area. Black and white version of a color composite of bands 7, 5, and 4. The dark grey areas are mainly trees in river valleys and the black area is sea.

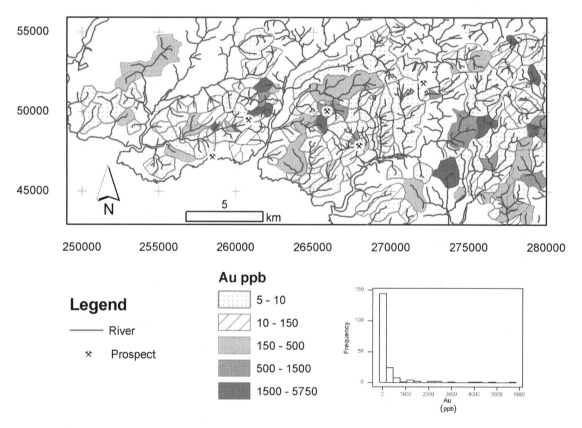

FIG. 9.8 Panned concentrate gold data in ppb. Intervals are approximately 50, 75, 90, and 95 percentiles, as discussed in the text. (Data courtesy British Geological Survey but converted into their catchments (drainage basins) by C.L. Wang.)

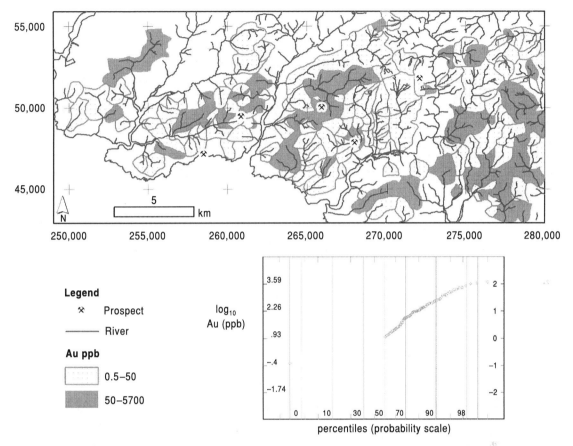

FIG. 9.9   Gold data of Fig. 9.8 replotted using a simple threshold of 50 ppb ($\log_{10} = 1.7$ of the lower plot). The $\log_{10}$–probability plot of the Au data was made in Interdex software and should be compared with the histogram of Fig. 9.8.

opinion of the writers, inferior to the multiple class map of Fig. 9.8.

The panned concentrates were analyzed for a variety of elements in addition to gold and much more information can be deduced from the other elements. One surprise was the high concentrations of tin, present as cassiterite, as the area is distant from the well-known tin mineralisation of Dartmoor and Cornwall. These high tin concentrations probably reflect heavy mineral concentration on an erosion surface of unknown, although probably Permian and Miocene, ages.

Data overlay and buffering

Overlaying data is one of the key methods in GIS. A typical use of the data overlay function is to analyze the control of lineaments on the distribution of the known gold soil anomalies. In the study area two major trends of lineaments, NW–SE and NE–SW, have been recognized. Possible lineaments following these orientations were derived from an analysis of a Landsat TM image and digitized into two coverages (Fig. 9.10). The relation of lineaments with gold mineralisation can be analyzed by calculating the distance of the gold occurrences from the lineaments. In this example there does not seem to be a definitive relationship between gold occurrences and a particular set of lineaments.

Lineaments are often inaccurately defined and it more practical to define zones of influence. The ability to generate corridors of a given width around features or buffer around

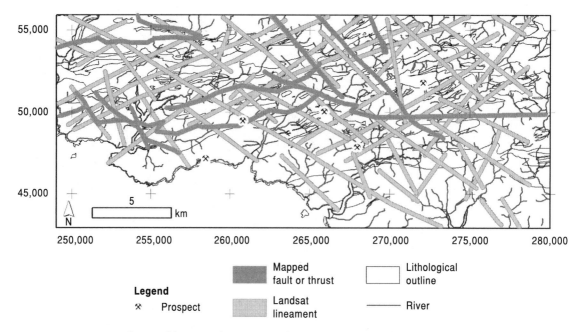

55,000

50,000

45,000

250,000   255,000   260,000   265,000   270,000   275,000   280,000

Legend

⚹   Prospect

Mapped fault or thrust

Landsat lineament

Lithological outline

River

FIG. 9.10   Lineaments digitized from Landsat coverage by C.L. Wang and buffered at 250 m for clarity. Known gold occurrences and mapped faults are added for comparison.

structures and geological contacts is therefore a major advance. In Fig. 9.10 buffers of 250 m are shown around the NW–SE lineaments.

Boolean methods

Once the layers to be investigated have been overlaid they can be used to define areas for follow-up. The simplest method of selecting areas is to use straightforward Boolean algebra. In this approach data from layers are selected and combined with other layers depending on the prejudices of the geologist (Fig. 9.11). An example in the south Devon data is to select catchments which are underlain by acid volcanics and have panned concentrate gold >1000 ppb Au based on the premise that the acid volcanics are the most favorable rock for gold deposits. The statements used by the computer are "geology = acid volcanics and Au > 1000 ppb". The database is then scanned and each unit is either true (1) or false (0) for the statement. If dolerite is considered to be of similar favorability to acid volcanics then the statements "(geology = acid volcanics) and (Au > 1000 ppb)" will meet these requirements

(Fig. 9.12). Although the Boolean approach can be very effective and is simple, it assumes that there is a sharp break in attributes, for example, gold in panned concentrates is significant at 1001 ppb but not at 999 ppb, and in the simple use of the method that each layer is of equal importance. In the index overlay variety of this method, weights are assigned to the different layers, depending on their perceived importance.

Complex methods

Although the Boolean approach can be effective, and is easily understood, it is difficult to detect more subtle patterns, to use very large and complex data sets, or to apply complicated rules. More complex methods have been widely applied to mineral exploration data but are highly dependent on software. Originally the programs were expensive and difficult to use but now many methods are available through the use of add-on software to Arcview and Mapinfo. Of particular note is the Arc-SDM software developed by the US and Canadian federal geological surveys and available freely

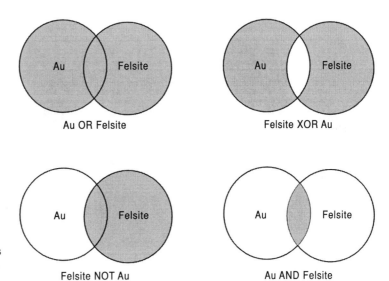

FIG. 9.11 Boolean algebra. The circles are diagrammatic representations of two sets, felsite and Au. The possible combinations are shown. XOR is for felsite or Au but not where they overlap.

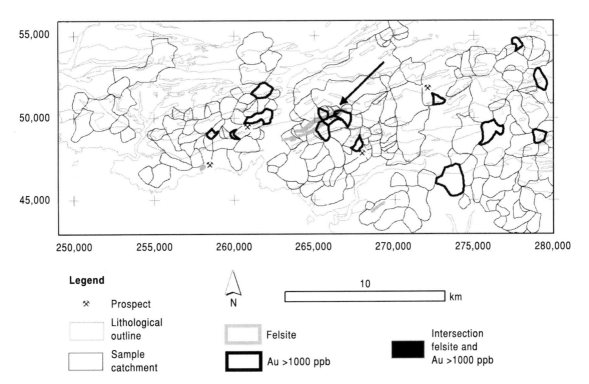

FIG. 9.12 Simple Boolean search in south Devon data for the presence of acid volcanics (felsite) and gold in catchment >1000 ppb, based on the maps of Figs. 9.3 and 9.8. The selected lithological units and catchments are shown by outlining and the very small area which meets both these criteria is in black and arrowed.

on the Internet (Arc-SDM 2001). The differing complex methods can be divided into two main approaches: (i) data driven, without a preconceived model; (ii) model driven, dependent on the type of model used. Although the methods differ, their aim is to map the probability of finding mineralisation.

Data-driven approaches

The main data driven techniques are (1) weights of evidence and (2) logistic regression, both of which require the location of mineralisation in at least part of the area to be known, in order to use as a training set for use in the remainder of the area. The reader can find further details in Bonham-Carter (1994) and at the Arc-SDM website (Arc-SDM 2001).

1 Weights of evidence was pioneered in mineral exploration use by Bonham-Carter et al. (1988). This method uses Bayesian statistics to compare the distribution of the known mineralisation with geological patterns that are suspected to be important controls.

The first stage in using the method is to calculate the probability of deposit occurrence calculated without knowing anything of the test pattern (prior probability). The distribution of mineralisation in the training area is then compared with geological patterns in turn. The calculations for each pattern give two weights: (i) a positive weight indicating how important the pattern is for indicating an occurrence; and (ii) a negative value indicating how important the absence of the pattern is for indicating no occurrence. The weights for the different patterns are then examined to see if they are significant and combined to calculate an overall posterior probability. This is then used to calculate the probability of mineralisation based on the geological patterns.

In the example of the data from south Devon, the summary weights for different geological controls are shown in Table 9.6. The most important patterns using the known gold occurrences are the occurrence of felsite and dolerite which can be seen to dominate the final probability map (Fig. 9.13).

2 Logistic regression uses a generalized regression equation to predict the occurrence of mineralisation. Logistic regression is distinct from the more familiar least squares regression in

that it uses presence or absence of geological patterns, rather than a continuous variable. The significance of the various geological controls is tested and a model built up.

In the south Devon example (Fig. 9.14) using drainage catchments containing more than 1000 ppb Au generated a probability p of:

$$\text{Logit}(p) = -3.81 + 1.47X5 - 1.74X11 + 2.04X2X5 + 3.07X3X4$$

where X3X4 is combined felsite and Dartmouth Group occurrence, X5 Meadfoot Group occurrence, X2X5 Meadfoot Group and basic volcanics, and X11 faults. In this model, the felsite and Dartmouth Group factor is the most important and faults are a less important control.

Model-driven approaches

A number of model-driven approaches have been used which develop on the index overlay theme: (1) fuzzy logic, (2) Dempster Shafer methods, and (3) neural networks.

1 Fuzzy logic. This has probably been the most widely used technique. In the fuzzy logic method variables are converted from raw data to probability using a user-defined function. For example, if arsenic soil geochemistry is suspected as a pathfinder for gold mineralisation, then low As concentrations (say <100 ppm As) will be given a probability of 0 and high concentrations (say >250 ppm As) a probability of 1. Intermediate concentrations will be given a probability of between 0 and 1 depending on the function applied (Fig. 9.15). The variables are then combined using algebra similar to the Boolean functions but including sum and product functions. The latter two functions are often combined together using a gamma-function to generate an overall probability from individual layers (Bonham-Carter 1994, Knox-Robinson 2000). Knox-Robinson (2000) advocates the use of methods that allow the visualization of the degree of support of geological evidence. An example of the use of these techniques on Pb-Zn deposits in Western Australia can be found in D'Ercole et al. (2000). In the south Devon example, a fuzzy "or" has been used to combine layers which were estimated to indicate the occurrence of

TABLE 9.6 Calculated weights and contrasts for patterns in generating a probability map. The mineralisation pattern was known gold occurrences as 1 km² cells.

Patterns	W_j^+	W_j^-	C
Landsat lineaments NE–SW 250 m buffer	−0.222	0.025	−0.248
Landsat lineaments NW–SE 250 m buffer	0.247	−0.039	0.285
Landsat lineaments NW–SE 500 m buffer	0.298	−0.119	0.417
Landsat lineaments NW–SE 750 m buffer	0.315	−0.245	0.560
Landsat lineaments NW–SE 1 km buffer	**0.259**	**−0.339**	**0.598**
Landsat lineaments NW–SE 1.25 km buffer	0.143	−0.276	0.419
Landsat lineaments NW–SE 1.5 km buffer	0.052	−0.145	0.197
Landsat lineaments NW–SE 1.75 km buffer	0.002	−0.009	0.011
Landsat lineaments NW–SE 2 km buffer	−0.003	0.023	−0.027
Felsite	**3.739**	**−0.059**	**3.797**
Dartmouth Slate	**1.050**	**−0.618**	**1.668**
Anomalous catchments with Au > 5 ppb	**0.844**	**−0.689**	**1.533**
Contact (Dart – Mead) 2 km buffer	**0.625**	**−0.741**	**1.366**
Dolerite	**1.184**	**−0.017**	**1.201**
Thrust 250 m buffer	**0.784**	**−0.062**	**0.846**
Contact (Mead – Stad) 750 m buffer	**0.548**	**−0.199**	**0.746**
Contact (Mead – Mdev) 750 m buffer	**0.605**	**−0.084**	**0.689**

For each lineament set a series of buffers were calculated and an example of the results for different buffers of the NW–SE lineaments is shown. Otherwise only the final results are shown. The significant features (shown in bold) were used to calculate the final probability.
W_j^+, positive weight; W_j^-, negative weight; C, contrast.
Dart, Dartmouth Slate; Mead, Meadfoot Group; Stad, Staddon Grits; Mdev, Middle Devonian Slates.

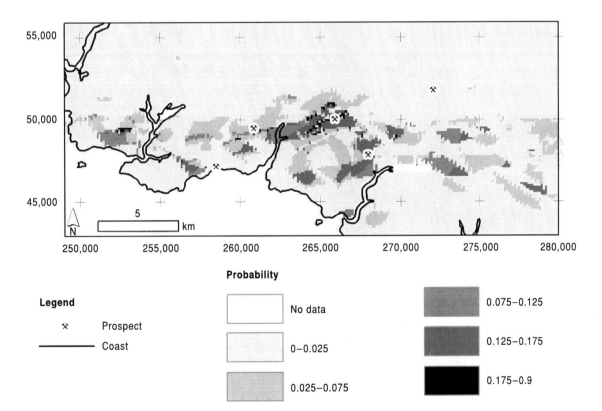

FIG. 9.13 Posterior probability map resulting from a weights of evidence analysis of geological and geochemical data discussed in the text. Note the general good correlation with known occurrences with the exception of that at ~2,72000E, 52,000N, which is probably of different origin. (Source: Wang 1995.)

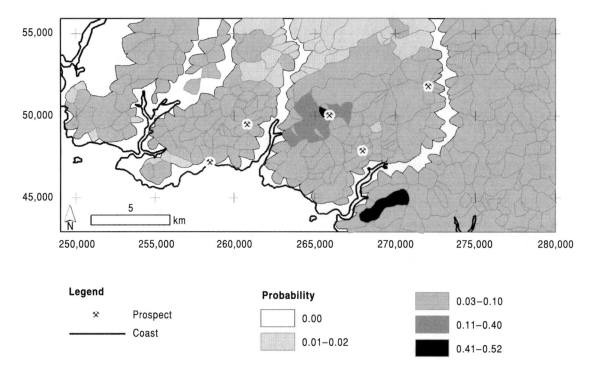

FIG. 9.14 Probability map based on logistic regression of Au panned concentrate values >1 ppm against geology and structure. Note the clustering of high probabilities in the center of the map and the prospective catchment at ~27,000E, 45,000N. (Source: Wang 1995.)

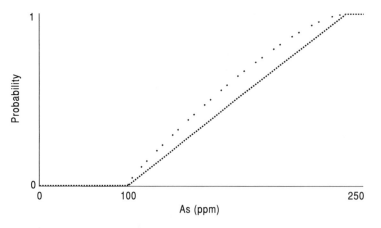

FIG. 9.15 Relationship between arsenic geochemistry and probability of indicating mineralisation in a hypothetical example, in which <100 ppm has no association with mineralisation and >250 ppm is associated with all gold shows. A straight line function is dashed and another function is shown in dots.

mineralisation, as the "or" includes most of the control broad areas that are defined within which there are higher probability zones (Fig. 9.16).

2 Dempster Shafer belief functions. These allow the input of some idea of the degree of confidence in the evidence used. In practice these can be difficult to apply but a good example of an application in Canada is found in Moon (1990).

3 Neural networks. This is a method in which software is trained by using the absence or presence of mineral deposits in large mixed data sets to find predictive geological factors.

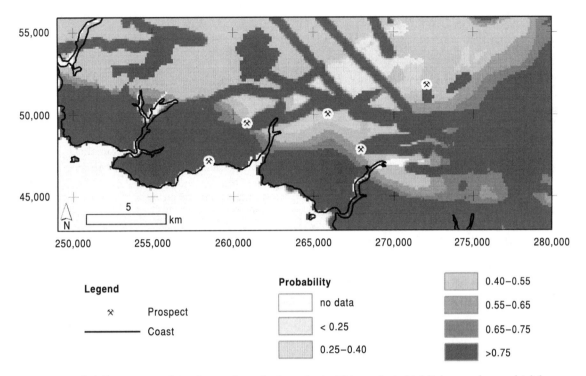

FIG. 9.16   Probability map resulting from a fuzzy logic analysis. This analysis highlights any layer which has possible association with gold mineralisation and is the result of a fuzzy "or." This should include all occurrences. (Source: Wang 1995.)

Brown et al. (2000) compared the method with fuzzy logic and weights of evidence on data from an area in New South Wales and found that it gave superior results. They claim that although neural networks require a training set the number of occurrences required is not as onerous as that of the weights of evidence method.

There have been a number of other studies of using complex methods in integrating data and the reader should examine some of these and decide which approach is suitable. A good starting point is the volume of Gubins (1997). The paper by Chinn and Ascough (1997) details the effort of Noranda geologists in the Bathurst area of Canada. They used a fuzzy logic approach to define areas for follow-up but concluded that the main advantage of computer-based methods was to get their deposit experts to sit down together and put their experience on paper. The methods are therefore not a substitute for excellent geology but a complement.

9.3   INTEGRATION WITH RESOURCE CALCULATION AND MINE PLANNING SOFTWARE

Most resource calculations performed in exploration, in feasibility studies, or in routine mine grade control and scheduling use a specialized computer package (or packages) that deals with 3D data. At the early stages of exploration, the key features of the package will be to input drill information and relate this to surface features, as discussed in Chapter 5 and section 10.4. When a resource is being calculated, the ability of the package to model the shape of geological units and calculate volumes and tonnages becomes more important. Subsequently in feasibility studies (see section 11.4.4) the capacity to design underground or surface workings and schedule production is crucial.

The package chosen will depend on the finance available and the nature of the operation, as there are packages specifically designed for both open cut and underground mining. The selection of such technical software is often an emotive issue and often depends upon a person's previous exposure to and their familiarity with the software, or its impressive graphics capability. Unfortunately software is rarely selected for the relevance to the operation of the functions that it provides. There are merits and weaknesses in each package. In addition, improvements keep appearing and a different decision may be reached if the selection exercise was repeated at a different moment in time.

Most operations have selected geological and mine modeling software that suits their budget, and geological and mining complexity. Occasionally, a mining house will standardize on one software package, e.g. BHP Billiton uses Vulcan software. The software modeling systems currently available include Surpac and Minex from the Surpac Minex Group, Maptek's Vulcan, Mincom, Datamine, Mintec's MineSight, and Gemcom. Other operations may have task-specific proprietary software, such as Isatis (Geovariances 2001), used for resource and/or reserve calculations, or Whittle or Earthworks products for strategic mine planning.

9.4   FURTHER READING

This is a rapidly developing area and there is not a single source covering the field. There are many good texts covering the use of GIS but probably the best starting point is the general textbook of Longley et al. (2001) which is a cut-down version of the comprehensive Longley et al. (1999). *Geographic Information Systems for Geoscientists* by Bonham-Carter (1994) remains the only text aimed at geologists, although the field has advanced significantly since the book was published.

The use of mining-specific software is not well covered in the formal literature although much information is available on the Internet. For Spanish speakers the book by Bustillo and Lopez (1997) provides a well illustrated guide to exploration geology and mine design.

10

EVALUATION TECHNIQUES

MICHAEL K.G. WHATELEY AND BARRY C. SCOTT

As an exploration geologist you will be expected to be familiar with mineral deposit geology, to understand the implications of extraction on the hydrology of the mineral deposit area, to recognize the importance of collecting geotechnical data as strata control problems may have considerable impact on mine viability, to be able to propose suitable mining methods, and to assess the economic viability of a deposit. To be able to cope with all these tasks you will need to be technically competent particularly in sample collection, computing, and mineral resource evaluation. The basis of all geological evaluation is the sample. Poor sample collection results in unreliable evaluation.

This chapter will present some of the methods that are used in the field to obtain representative samples of the mineralized rock that will enable the geologist to undertake mineral deposit evaluation. This includes the various drilling techniques, pitting and trenching as well as face and stope sampling. Once these data have been collected they are evaluated to determine how representative they are of the whole deposit. This is achieved using statistics, a subject covered in the first part of the chapter.

The geologist prepares and evaluates the deposit's geological and assay (grade) data. In an early phase of a project, global resource estimates using classical methods will normally be adopted. Later, as more data become available, local estimates are calculated using geostatistical methods (often computer based). An outline of these evaluation methods is also presented.

Finally, to ensure that the maximum amount of information is derived from drill core (other than the obvious grade, thickness, rock type, etc.) a basic geotechnical and hydrogeological outline is given. This information may be used in assessing possible dilution of the ore upon mining, or may identify strata control or serious water problems which may occur during mining.

10.1 SAMPLING

10.1.1 Introduction to statistical concepts

Population and sample

Sampling is a scientific, selective process applied to a large mass or group (a *population*, as defined by the investigator) in order to reduce its bulk for interpretation purposes. This is achieved by identifying a component part (a *sample*) which reflects the characteristics of the parent population within acceptable limits of *accuracy*, *precision*, and *cost effectiveness*. In the minerals industry the average grade of a tonnage of mineralized rock (the population) is estimated by taking samples which are either a few kilograms or tonnes in weight. These samples are reduced to a few grams (*the assay portion*) which are analyzed for elements of interest.

Results of analyzed samples plotted as a frequency curve are a pictorial representation of their distribution (Fig. 10.1). Distributions have characteristics such as mid-points and other measures which indicate the spread

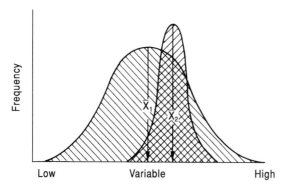

FIG. 10.1   Two normal distributions. Distribution 1 has a lower arithmetic mean ($\bar{x}_1$) than distribution 2 but higher variance (i.e. a wider spread). Distribution 2 has the reverse, a higher mean ($\bar{x}_2$) but a lower variance.

of the values and their symmetry. These are *parameters* if they describe a population and *statistics* if they refer to samples.

In any study the investigator wishes to know the parameters of the population (i.e. the "true" values) but these cannot be established unless the population is taken as the sample. This is normally not possible as the population is usually several hundred thousands or millions of tonnes of mineralized rock. A *best estimate* of these parameters can be made from sampling the population and from the statistics of these samples. Indeed a population can be regarded as a collection of potential samples, probably several million or more, waiting to be collected.

It is fundamental in sampling that samples are representative, at all times, of the population. If they are not the results are incorrect. The failure of some mineral ventures, and losses recorded in the trading of mineral commodities, can be traced to unacceptable sampling procedures due to confusion between taking samples that are representive of the deposit being evaluated, and specimens whose degree of representation is not known.

Homogeneity and heterogeneity

Homogeneity is the property that defines a population whose constituent units are strictly identical with one another. Heterogeneity is the reverse condition. Sampling of the former material can be completed by taking any group of these units such as the most accessible frac-

tion. Mineralisation is composed of dissimilar constituents and is heterogeneous – it is rarely (if ever) homogenous – and correct sampling of such material has to ensure that all constituent units of the population have a uniform probability of being selected to form the sample, and the integrity of the sample is respected (see later). This is the concept of *random sampling.*

Normal and asymmetrical distributions

This is a brief introduction to the subject of distributions. The reader is referred to standard texts such as Issaks and Srivastava (1989) for more details.

Normal distribution

In a normal distribution the distribution curve is always symmetrical and bell shaped (Fig. 10.2). By definition, the mean of a normal distribution is its mid-point and the areas under the curve on either side of this value are equal. Another characteristic of this distribution, or curve, is the spread or dispersion of values about the mean which is measured by the variance, or the square root of the variance called the standard deviation. Variance (σ^2) is the average squared deviation of all possible values from the population mean:

$$\sigma^2 = \frac{\Sigma(x_i - \bar{x})^2}{n}$$

where σ^2 = population variance, x_i = any sample value, $\bar{x}$ = population mean, n = number of samples.

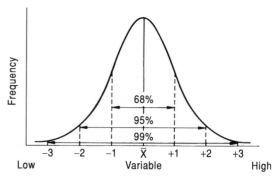

FIG. 10.2   The normal distribution: the variable has a continuous and symmetrical distribution about the mean. The curve shows areas limited and occupied by successive standard deviations.

Variance units are the square of the units of the original observation. A small variance indicates that observations are clustered tightly about the arithmetic mean whilst a large variance shows that they are scattered widely about the mean, and that their central clustering is weak. A useful property of a normal distribution is that within any specified range, areas under its curve can be exactly calculated. For example, slightly over two-thirds (68%) of all observations are within one standard deviation on either side of the arithmetic mean and 95% of all values are within ±2 (actually 1.96) standard deviations from the mean (Fig. 10.2). The mean of observations, or average, is their total sum divided by the number of observations, n.

Asymmetrical distributions
Much of the data used in geology have an asymmetrical rather than a normal distribution. Usually such distributions have a preponderance of low values with a long tail of high values. These data have a positive skew. Measures of such populations include the mode which is the value occurring with the greatest frequency (i.e. the highest probability), the median which is the value midway in the frequency distribution which divides the area below the distribution curve into two equal parts, and the mean which is the arithmetic average of all values. In asymmetrical distributions the median lies between the mode and the mean (Fig. 10.3); in normal curves these three measures coincide.

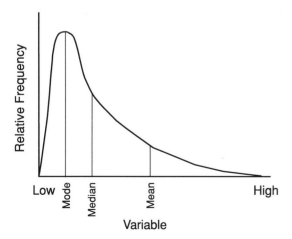

FIG. 10.3 Asymmetrical distribution, positively skewed (to the right).

In asymmetrical distributions there is no comparable relationship between standard deviation (or variance) and the area under the distribution curve, as in a normal counterpart. Consequently, mathematical transformations have been used whereby skewed data are transformed to normal. Perhaps the commonest is the log normal transformation where the natural log of all values is used as the distribution and many geological data approximate to this type of distribution; however, this method should be used with caution. Geovariances (2001) recommend the use of gaussian transformation particularly for variable transformation in the geostatistical conditional simulation process (see section 10.4.3) and for use in nonlinear geostatisical techniques such as disjunctive kriging and uniform conditioning (Deutsch 2002). An alternative approach is to use nonparametric statistics.

Parametric and nonparametric statistics
Parametric statistics, discussed above, specify conditions regarding the nature of the population being sampled. The major concern is that the values must have a normal distribution but we know that many geological data do not have this characteristic. Nonparametric statistics, however, are independent of this requirement and thus relevant to the type of statistical testing necessary in mineral exploration. Routine nonparametric statistical tests are available as computer programs, e.g. Henley (1981).

Point and interval estimates of a sample

Point estimates are the arithmetic mean ($\bar{x}$), variance (S^2), and sample size (n). The point estimate $\bar{x}$ provides a best estimate of the population mean ($\bar{X}$) but by itself it is usually wrong and contains no information as to the size of this error. This is contained in the variance (S^2). A quantitative measure of this, however, is obtained from a two-sided interval estimate based on the square root of the variance, the standard deviation (S).

Two-sided interval estimates
If $\bar{x}$ is the mean of n random samples taken from a normal distribution with population mean $\bar{X}$ and variance S^2, then at a 95% probability $\bar{x}$ is not more than 2 standard deviations (S) larger or smaller than $\bar{X}$. In other words, there is a

95% chance that the population mean is in this calculated interval and five chances in a hundred that it is outside. The relationship is:

$$\bar{x} + \frac{1.96\sigma}{\sqrt{n}} > \bar{X} > \bar{X} - \frac{1.96\sigma}{\sqrt{n}}$$

This method is correct but useless because we do not know σ and the expression becomes:

$$\bar{x} + t_{2.5\%}\frac{S}{\sqrt{n}} > \bar{X} > \bar{x} - t_{2.5\%}\frac{S}{\sqrt{n}}$$

where σ is replaced by its best estimate S, and the normal distribution is replaced by a distribution which has a wider spread than the normal case. In this t distribution (Davis 2003) the shape of its distribution curve varies according to the number of samples (n) used to estimate the population parameters (Table 10.1). As n increases, however, the curve resembles that of the normal case and for all practical purposes for more than 50 samples the curves are identical. Examples of the use of this formula are given in Table 10.2 and Fig. 10.4.

The width (i.e. spread) of a two-sided interval estimate about the sample arithmetic mean depends on the term:

$$t = \frac{S}{\sqrt{n}}$$

TABLE 10.2 Two-sided and one-sided confidence intervals at 95% probability.

	Two-sided	One-sided
$\bar{x}$	10.55% Zn	10.55% Zn
S	3.60%	3.60%
n	121	121
$\sqrt{n}$	11	11
$(S/\sqrt{n})$	0.33	0.33
$t_{2.5\%}$	1.96	NA
$t_{5\%}$	NA	1.65
$t(S/\sqrt{n})$	0.65	0.54
Upper limit	11.20%	NA
Lower limit	9.90%	10.01%

1. The total number of samples needed to halve the standard error of the mean $(S/\sqrt{n})$, and thus the range of the confidence interval, is 484 $[= (2 \times 11)^2]$.
2. With a one-sided confidence interval all the risk is placed into one side of the interval. Consequently the calculated limit is closer to the x statistic than to the corresponding limit for a two-sided interval. In the above example the upper limit could have been calculated instead of the lower.
3. See Table 10.1 for t values and Fig. 10.4 for a graphical representation.
NA, not appropriate.

TABLE 10.1 Critical values of t for different sample numbers and confidence levels.

Number of samples (n)	Confidence levels			
	80%	90%	95%	99%
1	3.08	6.3	12.71	31.82
5	1.48	2.02	2.57	3.37
10	1.37	1.81	2.23	2.76
25	1.32	1.71	2.06	2.49
40	1.30	1.68	2.02	2.42
60	1.30	1.67	2.00	2.39
120	1.29	1.66	1.98	2.36
>120	1.28	1.65	1.96	2.33

The number of samples (n) is referred to as "the degrees of freedom."

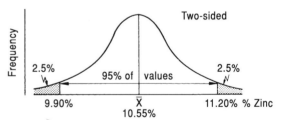

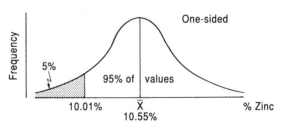

FIG. 10.4   Graphical representation of a two-sided and one-sided confidence interval at 95% confidence level. See Table 10.2 for calculations.

TABLE 10.3 Effect of changing the confidence level on the range of a two-sided confidence interval. The same zinc mineralisation as in Table 10.2 is used. Grade $(x) = 10.55\%$ zinc, standard deviation $(S) = 3.60\%$, number of samples $(n) = 121$.

	Confidence level			
	50%	**80%**	**90%**	**95%**
Percentile of t	$t_{25\%}$	$t_{10\%}$	$t_{5\%}$	$t_{2.5\%}$
Value of t	0.68	1.28	1.65	1.96
$t(S/\sqrt{n})$	0.22	0.42	0.54	0.65
Upper limit (U1)	10.77%	10.98%	11.10%	11.20%
Lower limit (U2)	10.33%	10.12%	10.00%	9.90%
Range (U1–U2)	0.44%	0.76%	1.10%	1.30%

1. The range becomes narrower as the confidence level decreases, and the risk of the population mean being outside this interval correspondingly decreases. At the 50% level this risk is 1 in 2, but at the 95% level it is 1 in 20.
2. See Table 10.1 for values of t.

The magnitude, and width, of this term is reduced by:

1 Choosing a lower confidence interval (a lower t value); an example of this approach is given in Table 10.3. The disadvantage of this is that as the t value decreases the probability (i.e. chance) of the population mean being outside the calculated confidence interval increases and usually a probability of 95% is taken as a useful compromise.

2 Increasing the number of samples (n), but the relationship is an inverse square root factor. Thus the sample size, n, must be increased to $4n$ to reduce the factor $1/n$ by half, and to $16n$ to reduce the factor to one quarter, clearly the question of cost effectiveness arises.

3 Decreasing the size of the total variance, or total error $[S^2(TE)]$, of the sampling scheme. In this sense sampling error refers to variance (S^2), or standard deviation (S). A small standard deviation indicates a narrower interval about the arithmetic mean and the smaller this interval the closer the sampling mean is to the population mean. In other words, the accuracy of the sampling mean as a best estimate of the population mean is improved by reducing the variance. This is achieved by correct sampling procedures, as discussed below.

One-sided interval estimates
In this approach only one side of the estimate is calculated, either an upper or lower limit, then at 95% probability:

$$U_L = \bar{x} - t_{5\%}\frac{S}{\sqrt{n}} < \bar{X}(U_L = \text{lower limit})$$

$$U_u = \bar{x} - t_{5\%}\frac{S}{\sqrt{n}} > \bar{X}(U_U = \text{upper limit})$$

It is usually more important to set U_L, so that mined rock is above a certain cut-off grade. With a one-sided estimate all the risk has been placed into one side of the interval. Consequently, the calculated limit for a one-sided estimate is closer to the $\bar{X}$ statistic than it is to the corresponding limit for a two-sided interval. Examples are given in Table 10.2 and Fig. 10.4. Through interval estimates, the variability (S), sample size (n), and the mean $(\bar{X})$ are incorporated into a single quantitative statement.

10.1.2 Sampling error

Sampling consists of three main steps: (i) the actual extraction of the sample(s) from the *in situ* material comprising the population; (ii) the preparation of the assay portion which involves a mass reduction from a few kilograms (or tonnes) to a few grams for chemical analysis; and (iii) the analysis of the assay portion. Gy (1992) and Pitard (1993) explain in some detail the formula, usually known as Gy's formula, to control the variances (errors) induced during sampling.

Global estimation error [S²(GE)]

Global error (GE) includes all errors in sample extraction and preparation (total sampling error or *TE*), and analysis errors (*AE*).

$$S^2(GE) = S^2(TE) + S^2(AE)$$

The *TE* includes selecting the assay portion for analysis while the analytical error proper excludes this step. Gy (1992) maintains that analysts usually complete their work in the laboratory with great accuracy and precision. In this strict sense, associated analytical error is negligible when compared with the variance of taking the assay portion. Consequently, for practical purposes $S^2(AE) \approx 0$ and from this $S^2(GE) \approx S^2(TE)$.

Total sampling error [S²(TE)]

The total sampling error (TE) includes all extraction and preparation errors. It includes errors of selection (*SE*) and preparation (*PE*): the former are inherent in both extraction and preparation while the latter are restricted to mechanical processes used in sample reduction such as crushing, etc.:

$$S^2(TE) = S^2(SE) + S^2(PE)$$

Selection errors (SE) are minimized by:
1 A correct definition of the number of point increments, or samples. Sample spacing is best defined by geostatistics, described in section 10.4.
2 A correct definition of the area sampled. Sampling of mineral deposits presents a three-dimensional problem but for practical purposes they are sampled as two-dimensional objects which equate to the surface into which samples are cut. The long axis of the sample preferably should be perpendicular to the dip of the mineralisation, or at least at an angle to it, but not parallel. The cut in plan is ideally a circle, sometimes a square or rectangle, and in length should penetrate the full length of the sampled area.
3 A correct extraction and collection of the material delimited by the above cut. Usually this presents a problem with the planes of weakness in rocks, and their variation in hard-

ness. Following from (2) and (3) an optimum sample is provided by diamond drilling which, with 100% core recovery, cuts almost perfect cylindrical lengths of rock the full length of the sampled area.
4 A correct preparation of the assay portion from the original sample, as described later.

Selection errors are of two main categories, which arise from:
4.1 the inherent heterogeneity of the sample and
4.2 the selective nature of the sample extraction and its subsequent reduction in mass and size to the assay portion.

The first category comprises the *fundamental error* [$S^2(FE)$] and is the irreducible minimum of the total sampling error and can be estimated from the sampling model of Gy (1992) (section 10.1.4). Gy (1992) separates those of the second category into seven different types, which are estimated from measurements (i.e. mass and grain size) from each stage of a sample reduction system.

Preparation errors [$S^2(PE)$] are related to processes such as weighing, drying, crushing, grinding, whose purpose is to bring successive samples into the form required by the next selection stage, and ultimate analysis. Sources of possible error are:
1 Alteration of the sample's chemical composition by overheating during drying as, for example, with coal and sulfides of mercury, arsenic, antimony, and bismuth.
2 Alteration of the sample's physical condition. If the grain size of the sample is important this can be changed by drying and careless handling.
3 Once a sample is collected, precautions must be taken to avoid losses during preparation of the assay portion. Losing material always introduces error (i.e. increases the variance) because the various sized fractions differ in grade and it is usually the finer grained material which is lost. Losses can be checked by weighing samples and rejects at each reduction stage.
4 Sample contamination must be avoided: sample containers and preparation circuits must be clean and free from foreign material. Dust produced by the size reduction of other samples should be excluded. Accidental sources of added material include steel chips

from hammers, and that derived from the crushing and grinding equipment.

5 Unintentional sampling errors must be avoided such as mislabeling and the mixing of fractions of different samples.

In essence preparation errors result from the careless treatment of samples and can be caused by the use of inadequate crushing and grinding equipment, insufficient drying capacity, and inappropriate sample splitting.

Obviously, PE values can be estimated only from the results of an operating sample reduction system and, like SE values, cannot be estimated directly for either preliminary calculations or the design of such a system. Fortunately, the totality of both can be estimated from the relationship:

$$S^2(TE) \le 2S^2(FE)$$

where the variance of the total sampling error (TE) does not exceed twice that of the fundamental error (FE) (Gy 1992).

Summary

Sampling consists of three stages:

1 *Extraction* of the original sample from *in situ* material (i.e. the population), including its delimitation and collection.

2 *Preparation* which involves a reduction in both mass and grain size of the original sample to an assay portion for chemical analysis.

3 *Chemical analysis* of the assay portion.

Each of the above three stages is discussed in turn.

10.1.3 Sample extraction

Random sampling

Correct sampling technique requires a random selection of each sample from the population. At an operating mine, it is possible that sampling is nonrandom since it is completed from mine development which is planned on a systematic basis. Provided, however, that the mineral particles which form the population are themselves randomly distributed this objection is overcome. Many mineral deposits display trends and variations in the spatial distribution of their grade variables and the effect of this on the concept of their random distribution is a matter of debate.

Sample volume–variance relationship

Preferably samples should be of equal volume, based on similar cross-sectional area. If volumes are unequal there must be no correlation between this and their analytical results. Although the mean of observations based on large volumes of rock should not be expected to be different from those based on small volumes (e.g. larger or smaller diameter drill core), the corresponding variance of samples from the larger volumes might be expected to be smaller. This is because the variability of the sample depends both on the variance and the number of samples taken (n):

$$\text{variability} \propto \frac{S^2}{n}$$

and instinctively we feel that there should be a similar relationship with the volume of the sample taken. Indeed, it has been advanced that:

$$\text{variability} \propto \frac{S^2}{\text{volume}}$$

It appears in practice, however, that there is no such simple, direct, relationship. This can be tested by splitting samples in half (e.g. mineralized parts of a drill core), analyzing them separately, and then combining the results as if the whole core sample had been tested. If the values from the two halves are statistically independent values from a single population, then the ratio of the variance of the values for the whole sample to the average values for the two halves should be 0.5. Results show a calculated ratio of 0.9 or greater so that little appears to be gained by analyzing the whole sample rather than half of it.

If the sample volume–variance relationship were important it would be desirable to analyze all portions of a core from a drillhole rather than follow the common practice of splitting the core longitudinally to save a half as a record and use half for analysis. In sampling, however, this relationship should always be considered and, if possible, tested. This is particularly

Length	Pb + Zn%	SG	Length × Grade	Length × SG	Length × Grade × SG
1.5	8.7	3.03	13.05	4.55	39.54
2.0	15.3	3.27	30.60	6.54	100.06
1.5	18.1	3.37	27.15	5.06	91.50
1.0	7.6	2.99	7.6	2.99	22.72
1.0	5.1	2.91	5.1	2.91	14.84
2.0	14.9	3.26	29.80	6.52	97.15
1.5	8.2	3.01	12.30	4.51	37.02
1.0	12.3	3.16	12.30	3.16	38.87
11.5			137.90	36.24	441.70

TABLE 10.4 Sample density–volume relationship.

Average grade by volume % = 137.90 / 11.5 = 12.0%.
Average grade by weight % = 441.70 / 36.24 = 12.2%.

relevant if the valuable constituents have a low grade and are erratically distributed (e.g. disseminated gold and PGE values) and little is known of the variance. In such cases it is sensible either to slice the entire core longitudinally saving 10% as a record and sending 90% for analysis, or collect a larger core.

This topic was discussed by Sutherland and Dale (1984) in relation to the minimum sample size for sampling placer deposits, such as alluvial diamonds, and they concluded that the major factor affecting variance was sample volume. Annels (1991) explains that the standard error of the mean, expressed as:

$$\text{Standard error of the mean} = \frac{\sqrt{S^2}}{n}$$

decreases as the number of samples increase. Taking larger samples will have a similar effect, because they will contain more "point" samples than a smaller volume of samples. When undertaking resource estimation, semivariograms (section 10.4.3) should be constructed for each sample set. These will demonstrate whether the variances are proportional or whether there is a more fundamental difference between the sample sets.

Sample density–volume relationship
In making reserve calculations, grades and tonnages are determined by using linear %, area %, and volume % relationships. Analyt-

ical data are expressed in terms of weight and unless the weight/volume ratio is unity, or varies within a narrow range, average grades calculated by volume % will be in error. This problem is illustrated in Table 10.4 which is an example of galena and sphalerite mineralisation in limestone where the average grade calculated using volume percentage is lower than that calculated from weight percentage.

Sampling procedures

Introduction
Variance can be reduced by using well planned sampling procedures with as thorough and meticulous extraction of samples as possible. The sample location is surveyed to the nearest survey point and cleaned by either chipping or washing from the rock face all oxidized and introduced material so as to expose fresh rock. A sampling team usually consists of two people, the senior of whom maintains the sampling notebook. At each location a visual description of the rock is made, noting the main rock types and minerals present, percentage of potential valuable mineral(s), rock alteration, other points of interest, and who is taking the sample. Obviously each sample has a unique number – a safe practice is to use numbered aluminum sample tags with one placed inside the plastic sample bag and the other attached by thin wire about its neck. Any collection device, whether a scoop, shovel, or

pipe, must obey the three times the maximum particle size rule, i.e. the width, length, and depth must be at least three times the maximum particle size. This is defined as the largest screen size which retains 5% of the material. There are three hand sampling techniques, namely channel, chip, and grab sampling.

Channel sampling
In mineral exploration, a channel is cut in an outcrop, usually the same diameter as the core being collected, to maintain the volume–variance relationships. It is cut using a hammer and chisel or a circular saw, across the strike of the mineralisation (see Fig. 5.3). As the material is cut it is allowed to fall to the floor on to a plastic sheet or sample tray from which it is collected and bagged. Samples are normally 0.5–5 kg in weight, mostly 1–2 kg, and each is rarely taken over 2 m or so in length, but should match the core sample length. There is no point in oversampling and sample spacing perhaps can be determined from the range (a) (section 10.4.1) derived from a study of a semi-variogram.

A mineral deposit can often be resolved into distinct and separate types of mineralisation. Sampling these different types as separate entities rather than as one large sample can reduce natural variation and thus the variance, and keep sample weights to a minimum. This *stratified sampling* is also of value where the separate types require different mineral processing techniques due to variation in either grain size or mineralogy. Such sample data are then more useful for mine planning. Such samples can be recombined statistically into one composite result for the whole of the mineralisation.

In well-jointed or well-bedded rocks collection by hammering the rock face presents difficulties in that adjacent nonsample material will fall with the material being cut but this must be rejected. Additionally, it is important that a representative collection of rock masses is collected. Frequently, mineral values of interest occur in altered rocks which may well be more friable than adjacent, harder, unmineralized ground, and overcollection of this easily cut material will provide a biased sample.

Channel sampling provides the best technique of delimiting and extracting a sample and, consequently, provides the smallest pos-

sible contribution to the total error. The two techniques which follow are less rigorous, and more error prone, but less expensive.

Chip sampling
Chip samples are obtained by collecting rock particles chipped from a surface, either along a line or over an area. In an established mine, rock chips from blastholes are sampled using scoops, channels, or pipes pushed into the heap. With a large database of chip samples, statistical correlation between core and chip samples may establish a correction factor for chip sampling results and thus reduce error. Chip sampling is used also as an inexpensive reconnaissance tool to see if mineralisation is of sufficient interest to warrant the more expensive channel sampling.

Grab sampling
In this case the samples of mineralized rock are not taken in place, as are channel and chip samples, but consist of already broken material. Representative handfuls or shovelfuls of broken rock are picked at random at some convenient location and these form the sample. It is a low cost and rapid method and best used where the mineralized rock has a low variance and mineralisation and waste break into particles of about the same size. It is particularly useful as a means of quality control of mineralisation at strategic sampling points such as stope outlets and in an open pit (Annels 1991).

Comparison of methods
Bingham Canyon mine in the United States and the Palabora Mine in South Africa conducted a series of tests whereby the entire product from blasthole drilling was collected in 30-degree segments from the full 360 degrees (C.P. Fish, personal communication). One blasthole may produce between 2 and 3 t. Each segment was sampled by the traditional means described above, and compared to the remaining material in the segment after crushing to 3 mm and reducing in a 16-segment rotary-divider. The results showed:
1 No segment ever gave the same assay as the whole mass. The spatial variation was great and there was no consistency in the way the material dispersed.
2 Subsample analyses gave higher analytical results than the 100% mass. Scoop samples

tend to collect the finer particles as they are invariably taken from the top part of any heap. The larger, usually less mineralized particles roll down to the base of the perimeter and are not sampled. When creating a channel, material will always collapse into the void. Thus, the top of the heap is sampled. Similarly, in an upward scoop, the scoop is usually full before it gets anywhere near the top of the channel. When forcing a pipe sampler into a pile of material, the pipe preferentially collects the top 6 inches (15 cm) of the material and then blocks.

3 Grab sample analyses also gave higher results than the 100% sample.

10.1.4   Sample preparation

The ultimate purpose of sampling is to estimate the content of valuable constituents in the samples (the sample mean and variance) and from this to infer their content in the population. Chemical analysis is completed on a few grams of material (the assay portion) selected from several kilograms or tonnes of sample(s). Consequently, after collection, sampling involves a systematic reduction of mass and grain size as an inevitable prerequisite to analysis. This reduction is of the order of 1000 times with a kilogram sample and 1,000,000 with a tonne sample.

Samples are reduced by crushing and grinding and the resultant finer grained material is split, or separated, into discrete mass components for further reduction (Geelhoed & Glass 2001). A sample reduction system is a sequence of stages that progressively substitutes a series of smaller and smaller samples from the original until the assay portion is obtained for chemical analysis. Such a system is essentially an alternation of size and mass reduction stages with each stage generating a new sample and a sample reject, and its own sampling error. For this reduction Gy (1992) established a relationship between the sample particle size, mass and sampling error (Box 10.1):

BOX 10.1   Gy sampling reduction formula.

Gy (1992) is believed to be the first to devise a relationship between the sample mass (M), its particle size (d) and the variance of the sampling error (S^2). This variance is that of the fundamental error (FE): a best estimate of the total sampling error (TE) is obtained by doubling the FE.

$$M \geq \frac{Cd^3}{S^2}$$

where M is the minimum sample mass in *grams* and d is the particle size of the coarsest top 5% of the sample in *centimeters*. Cumulative size analyses are rarely available and from a practical viewpoint d is the size of fragments that can be visually separated out as the coarsest of the batch. S^2 is the fundamental variance and C is a heterogeneity constant characteristic of the material being sampled and $= c\beta fg$, where:

$$c = \text{a mineralogical constitution factor} = \frac{(1 - a_L)}{a_L}[(1 - a_L)\delta A + \delta G.a_L]$$

a_L is the amount of mineral of interest as a *fraction*,
δA is the specific gravity of the mineral of interest in g cm^{-3},
δG is the specific gravity of the gangue,
β is a factor which represents the degree of liberation of the mineral of interest. It varies from 0.0 (no liberation) to 1.0 (perfect liberation) but practically it is seldom less than 0.1. If the liberation size (d_{lib}) is not known it is safe to use β equal to 1.0. If the liberation size is known $d \geq d_{lib}$ then $\beta = (d_{lib}/d)^{0.5}$ which is less than 1. In practice there is no unique liberation size but rather a size range.
f is a fragment shape factor and it is assumed in the formula that the general shape is spherical in which case f equals 0.5,
g is a size dispersion factor and cannot be disassociated from d. Practically g extends from 0.20 to 0.75 with the narrower the range of particle sizes the higher the value of g but, with the definition of d as above, g equals 0.25.

$$M = \frac{Cd^3}{S^2}$$

M is the minimum sample mass in *grams*, d is the size of the coarsest top 5% of the sample in *centimeters*, C is a heterogeneity constant characteristic of the material being sampled, and S^2 is the variance. Consequently, in reducing the mass of a sample there is a cube relationship between this reduction and its particle size if the variance is to remain constant.

In practice the observed variance of a sampling system is larger than that calculated from the above model. This is because the observed variance comprises the total sampling error (*TE*) while the above model calculates a particular component of this totality, the fundamental error (*FE*). This important point has already been referred to, where $S^2(TE) = 2S^2(FE)$.

Design of a sample reduction system

The choice of suitable crushing, grinding, drying, and splitting equipment is critical and systems are best designed and installed by those with appropriate experience. The design of a reduction system is often neglected but it is of prime importance particularly in dealing with low grade and/or fine-grained mineralisation. In particular, equipment may have to handle wet, sticky material with a high clay content which will choke normal comminution equipment. Also the creation of dust when processing dry material is an environmental and health hazard and may also contaminate other samples. Dust has to be controlled by adequate ventilation and appropriate isolation.

In any reduction system the most sensitive pieces of equipment are the crushers and grinders. Each such item works efficiently within a limited range of weight throughput and size reduction and there has to be a realistic assessment in the planning stage of the expected number and weight of samples to be processed each hour or shift. The use of inadequate comminution equipment at best may introduce excessive preparation errors and at the worst can close the entire system.

The calculation of *TE* from twice the *FE* provides a safe estimate (see previously) from which a flexible reduction system can be designed. When it is in operation its actual total error can be calculated and compared with the initial assumption (see later).

Graphical solution of sample reduction

Introduction
Sample preparation is essentially an alternation of size and mass reduction within the parameters of the Gy formula given above which provides for a control of the variance. Such reduction is best shown graphically with log d (top particle size) contrasted with log M (sample mass), where a horizontal line represents a size reduction and a vertical line a mass reduction (Fig. 10.5).

These reductions have to be completed within an acceptable variance which can be taken as a relative standard deviation of less than 5%. Ideally what is required is a safety line which divides the above graph into two parts, so that on one side of the line all reduction would be safe (i.e. with an acceptable variance) while on the other side unacceptable errors occur. Gy (1992) provides a mathematical expression for such a line as:

$$Mo = Kd^3$$

Empirically Gy (1992) suggests a value of 125,000 for K but emphasizes that for low concentrations of valuable constituents (<0.01%) a higher K value should be used of up to 250,000. Consequently:

$$Mo = 125{,}000\, d^3 \text{ or } \log Mo = \log K + 3 \log d$$

This relationship on log–log graph paper (Fig. 10.5) is a straight line, with a slope of 3.0, which divides the graph into two areas.
1 To the left of this safety line Mo is greater than M and, consequently, from the Gy formula the fundamental sampling variance So^2 is less than S^2, and is acceptable.
2 On the line, $Mo = M$ and $So^2 = S^2$, which is acceptable. All points on the safety line correspond to a constant fundamental variance.
3 To the right of the line Mo is less than M and the sampling variance is too high ($So^2 > S^2$) and unacceptable and either Mo has to be increased or d decreased to place the reduction point to the left of the safety line.

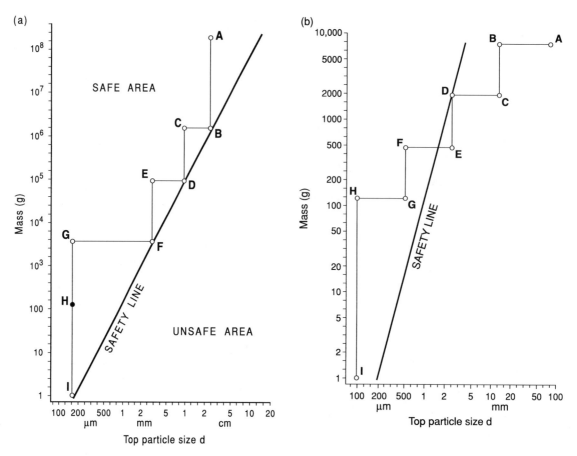

FIG. 10.5   Graphical representation of a sample reduction scheme. Fig. 10.5a is an example of a safe scheme while Fig. 10.5b is unsafe. For explanation see text.

Equal variance lines

The above allocates equal variance to each sampling stage but this may not always be an optimum practice. Usually the larger the mass (M) to be sampled the larger the top particle size (d) and the higher its reduction cost. From the Gy sampling formula, however, the necessity for particle size reduction at any stage can be reduced by allocating a higher proportion of the total variance to that particular stage. With the mass (M) remaining constant this allows a coarser top particle size. Consequently it is preferable to allocate a maximum portion of the variance to the samples with the coarsest particle sizes, and less to later successive samples of finer grain size. Gy (1992) suggests that one half of the total variance is allocated to the first sampling stage, one fourth to the next, etc.

As previously, a safe estimate of the total sampling variance is calculated by:

$$S^2(FE) = \frac{Cd^3}{M}$$

Using

$$M = Kd^3$$

then

$$S^2(FE) = \frac{Cd^3}{M} = \frac{C}{K}$$

and

$$So(TE) = 2\frac{C}{K}$$

where C is the heterogeneity constant (Box 10.1) and K equals 125,000.

From this a series of equal variance lines can be drawn with decreasing values of K, such as $K/2$, $K/4$, etc., which correspond to increasing variances. For each stage, knowing M and the increased value of $So^2(FE)$ the related coarser grain size (d) can be calculated from the Gy formula.

Variation of the safety line
The heterogeneity constant (C) contains the liberation factor β (Box 10.1). The safety lines above are calculated on the basis that this factor is at its maximum value of 1.0 which is a safe assumption but only correct when the top particle size is smaller than or equal to the liberation size (d_{lib}) of the valuable components. If d is larger than d_{lib} then β becomes $[d_{lib}/d]^{0.5}$, which is less than 1.0.

When the liberation size of the mineral of interest is not known the value of 1.0 is used as a safe procedure. This provides for the continuously straight equal variance lines as described above, with a slope of 3.0. When the liberation size is known, the above formula can be used with a β value of less than 1.0 but approaching this value as the grain size d is reduced. In this case each sampling stage has a unique safety line slope from about 2.5 increasing to 3.0 as d approaches d_{lib}, with the less steep sections in the coarser grain sizes. François-Bongarçon and Gy (2002) point out that there are still common errors being made when applying Gy's formula. They describe the main variable, the liberation size, and illustrate the importance of modeling it with a worked example.

Splitting of samples during reduction

In each stage after reduction in grain size the sample mass is correspondingly reduced by splitting. The sampling described in this text is concerned with small batches of less than 40 t and usually less than 1 t which are stationary in the sense that they are not moving streams of material on, say, a conveyor belt. In exploration, stationary samples include chip samples produced by percussion or reverse circulation drilling (section 10.4).

In the field it may be difficult to calculate the weight of sample required. In such cases, collect a minimum of 1000 particles. This will seem a trivial amount when collecting, for instance, a froth flotation product, but plants have accurate sampling systems built into the process system. It becomes important when dealing with larger particles, such as RC or blasthole rock chips, rocks in stockpiles, or on trucks. Invariably the amount required will be larger than anticipated, but don't forget if the primary sample is unrepresentative, everything else is a waste of time.

Hand or mechanical shovel splitting
Splitting samples is best completed by the process of alternative shovelling which minimizes the splitting and preparation error contribution to the total error. The sample is spread either on a sheet of thick plastic or sheet iron which is on a flat, clean, and smooth floor which can easily be swept. The material is piled into a cone (step 1) which is then flattened into a circular cake (step 2) and the first step is repeated three times for maximum mixing of the sample. Studies have shown that rolling on a cloth or plastic sheet does not necessarily cause mixing. If too slow the material just slides, and if too fast, there is a loss of the finer particles thereby causing bias. From the final cone, shovelfuls of material are taken consecutively around its circumference and placed alternatively into two separate piles. In this way the original cone, and sample, is split into two separate but equal smaller cones and the choice as to which is the new sample is decided by the toss of a coin and not by the sampler. Further weight reduction is continued in exactly the same manner on the new sample. Alternatively, the cone may be quartered with a shovel and each quarter is further reduced in the same way. This technique is referred to as cone and quartering.

The width of the hand, or small mechanical, shovel should be at least three times the width of the diameter of the coarsest sample particle size which provides for a maximum possible size of about less than 100 mm with the former and less than 500 mm with the latter tool. Shovels should always be overflowing and working with them partially filled is not a good practice. Obviously, towards the end of a splitting operation with finer grain sizes a smaller sized tool has to be used.

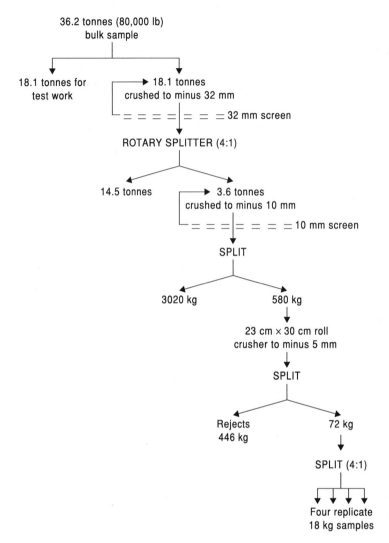

FIG. 10.6   Reduction of 36.2 tonne (80,000 lb) bulk sample by successive crushing and splitting. This involves a concomitant reduction in grain size from +15 cm to 0.5 cm. (After Springett 1983b.)

In order to provide an adequate number of increments, shovel capacity should be between one thirtieth to one fiftieth of the original sample mass to provide some 30–50 shovelfuls, or increments, divided equally between the two splits. In a final, fine-grained 100 g sample the required asssay portion can be obtained by repeated alternative "shovelling" with a spatula.

An example of the reduction of a bulk sample for chemical analysis is given in Figs 10.6 and 10.7. The example is from a large, low grade, disseminated gold deposit with a grade of about 1 g t^{-1}. Due to the high specific gravity of gold

and the broad range of gold particle size (i.e. d_{lib} in Gy's formula) in the micrometer range, devising an adequate subsampling plant was difficult. The most problematical stage was the reduction of the initial bulk of 36.2-tonne samples. This was effected by splitting each such sample into two equal portions with one half (18.1 t) for sampling purposes and the other for heap leaching gold tests (Fig. 10.6). Figure 10.7 illustrates a flow sheet for reduction of 18 kg portions, which were treated in the same fashion as drilling samples, to four replicate samples each of 5 g at a grain size of 0.01 mm (Springett 1983a,b).

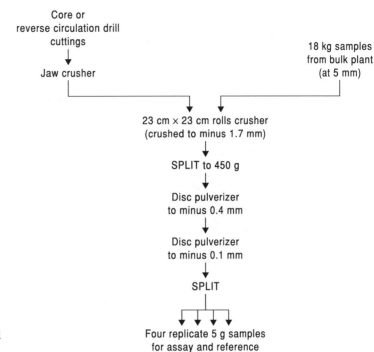

FIG. 10.7   Reduction of 18 kg (40 lb) samples at 0.5 cm size from bulk sampling plant (Fig. 10.5), and drill material, to 5 g samples at −0.01 cm size suitable for chemical analysis. (After Springett 1983b.)

Instrument splitters

The Jones Riffle is the most satisfactory of stationary instrument splitters (Fig. 10.8) which divide a dry sample into two equal parts. It comprises an even number (between 12 and 20) of equally sized chutes of a slope between 45 and 60 degrees where adjacent chutes discharge on opposite sides of the instrument. For correct implementation, the material to be split is evenly spread on a rectangular scoop designed to fit the splitter width, and discharged slowly and evenly into the chutes. This prevents loss of fines due to overenthusiastic pouring. To prevent obstruction the chutes should be at least twice the diameter of the coarsest particle. With repeated splitting it is better practice to implement an even rather than an odd number of splitting stages, and to alternate the choice of right and left split samples. Asymmetrical splitters are now available which split samples on a 70:30 or 80:20 basis (Christison 2002). This speeds up sample reduction and, with fewer separations, reduces any errors. A sample splitter may be used at the drill site especially when percussion or reverse circulation drilling is in progress and samples

are collected at regular intervals. A 1 m section of 102 mm diameter hole produces approximately 21 kg of rock chips.

Rotary splitters capable of separating laboratory size (a few kilograms) or bulk samples (a few tonnes) are also available. They use centrifugal force to split a sample into four or six containers.

The use of several riffle splitters in series, each one directly receiving the split from the preceding stage, is poor practice which can introduce sampling errors.

Comparison of calculated and observed total sampling errors

A best estimate of the total sampling error (*TE*) is obtained from the fundamental error (*FE*):

$$S^2(TE) = 2S^2(FE)$$

This estimate is used to design a sample reduction system from which, when in operation, an observed $S^2(FE)$ can be obtained by sampling each reduction stage to establish the sample mass (*M*) and the maximum particle size (*d*),

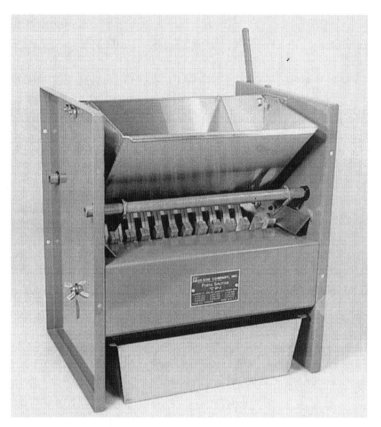

FIG. 10.8   A Jones sample splitter.
For explanation see text.

using Gy's formula. This is then compared with the previously estimated total error, on the following basis:

1 When the observed total error is smaller than or of the same order of magnitude as the estimated total error nothing should be changed in the reduction system. This provides a safety margin against the future wear of the equipment which will deliver progressively coarser material for which the sampling variance increases.

2 When the observed total error is larger than the estimated total error the former has to be reduced and particularly at the first (or coarsest) reduction stage where the variance is likely to be the major component of $S^2(FE)$. This can be achieved by retaining the same number of sampling stages but increasing the weight (i.e. the mass) treated at each separate stage and increasing the number of stages but with the same sample weight at each stage. Usually the latter choice is preferable as the

crushing–grinding equipment does not have to be enlarged, as in the first option.

The total sampling error of any sample reduction system should be checked at regular intervals, say every 6 months depending upon throughput and use, to ensure that it is working satisfactorily.

10.1.5   Preparation of samples containing native gold and platinum

With native precious metals a problem arises from the significant difference between the density of an associated gangue (about 2.6) and that of the metals (110.0 to 21.5). Gold may be diluted with silver and copper to a density in the order of 15.0 but even at this level its density is twice that of galena, the densest of all common sulfide minerals, and six times that of quartz. This contrast means that when such valuable constituents are liberated from their gangue, flakes and nuggets will segregate rapidly under gravity and homogenization of such material is difficult (Austr. Inst. Geoscientists 1987, 1988, Royle 1995).

Calculation of the variance of the fundamental error

Under such circumstances, and with the very low concentrations of precious metals ($\approx$1 ppm), Gy (1992) suggests a modification of his formula for calculating the fundamental error.

Normally, the factor $C = c\beta fg$ (Box 10.1) but for gold and platinum Gy (1992) recommends a factor of:

$$C = \frac{2.4d^3}{a_L} \text{ and } S^2 = \frac{C}{M}$$

where a_L is the concentration of the valuable constituent expressed as a fraction (i.e. 1 ppm = 1×10^{-6}).

An example of the use of this formula is given in Box 10.2.

10.1.6 Analytical errors

According to Gy (1992) analytical errors (AE), in the strict sense, are negligible and errors which are attributed to analytical laboratories arise from the selection of the assay portion and not, on the whole, from actual analytical techniques (section 10.1.2). Errors occur, however, and vary from laboratory to laboratory. An example is given in Table 10.5 where an artificial sample was made from gold in silica sand, at 2.74 g t^{-1} and was sent to four different laboratories for analysis (Springett 1983b). The following conclusions can be drawn:

Laboratory 1 Results are both inaccurate and imprecise.
Laboratory 2 Results are accurate but imprecise.
Laboratory 3 Results are inaccurate but precise.
Laboratory 4 Results are accurate and precise.
Obviously only the results from Laboratory 4 are acceptable.

The scope of the control of error in chemical analysis is beyond this chapter but these errors must be considered (see section 8.2). External control is exercised by submitting duplicate samples and reference materials of similar matrices to the unknowns (see section 8.2.1). It is essential that the geologist is aware of the internal control procedures of the laboratory. These usually employ the analysis of replicate

BOX 10.2 Variance of fundamental error [$S^2(FE)$] with liberated gold mineralization.

Gold mineralization contains 1 g t^{-1} ($a_L = 10^{-6}$) and is liberated from its gangue (i.e. $\beta = 1$) with the coarsest particles about 2 mm in size ($d = 0.2$ cm.). Under such circumstances Gy (1992) recommends a modification of his standard formula for the calculation of the fundamental variance. Normally the heterogeneity factor C equals $c\beta fg$ (Box 10.1), but for free gold he recommends $C = 2.4d^3/A_L$ and $S^2(FE) = C/M$.

With a 1 tonne sample ($M = 10^6$ g), $C = 2.4(0.2)^3/10^{-6} = 19,200$ and $S^2(FE) = 1.92 \times 10^{-2}$, $S = 1.4 \times 10^{-1}$, or 14%. If a normal distribution is assumed the 95% confidence level double-sided interval is $\pm28\%$, which is a low precision. If, say, a 10% precision is required at 95% confidence then a larger sample (M) is needed. At this level $S(FE)$ equals 5×10^{-2}, and $S^2(FE)$ equals 25×10^{-4}; then

$$M = C/S^2(FE) = 19,200/25 \times 10^{-4} = 768 \times 10^4 \text{ g or 7.68 tonnes.}$$

The treatment of such large samples presents a problem, but Gy (1992) recommends the use of *stratified sampling*. In this process the sample is reduced to a grain size of a few millimeters and then concentrated on a shaking table, such as the Wilfley table (Wills 1985). This produces a high density concentrate that contains all the coarse gold, an intermediate fraction, and low density tailings which is largely gangue. The intermediate fraction is recycled on the table while the tailings are sampled again on a 1% sampling ratio. In this way the 8 tonne sample is reduced to about 80 kg which is then processed as in Figs 10.6 and 10.7.

TABLE 10.5 Errors in assays on a prepared sample of 2.74 grams per tonne gold. This is an example where the population parameters are known since the material was artificially prepared. The results of the four different sets of five identical samples (A to E), each analyzed separately at four different laboratories (1 to 4), are given below. (After Springett 1983b.)

	Laboratory (g t^{-1} gold)			
	1	2	3	4
Sample				
A	5.72	2.74	4.39	2.81
B	4.18	2.40	4.01	2.78
C	5.66	2.06	3.81	2.67
D	4.42	3.43	4.01	2.67
E	3.26	2.06	3.81	2.71
Mean	4.65	2.54	4.01	2.73
S^2	0.87	0.26	0.04	0.003
S	0.93	0.51	0.21	0.06

1. The range of values at the 95% confidence level (which is the mean $\pm 2S$) for laboratory 1 is 2.79–6.51 g t^{-1}. This range does not contain the population mean (2.74 g t^{-1}) and the variance is high. For laboratory 2 the range is 1.52 to 3.56 g t^{-1} which does contain the population mean, but the variance is high. At laboratory 3 the range is 3.59–4.43 g t^{-1}. This does not contain the population mean, although the variance is acceptable. Laboratory 4 results have a range of 2.61–2.85 g t^{-1}, which contains the population mean and has a low variance.
2. Only the results from laboratory 4 are acceptable.

samples (e.g. every, say, tenth sample is repeated) and by introducing artificially prepared reference materials (e.g. every 30th or 40th sample). Some laboratories are found to perform relatively poorly and a rigorous program of interlaboratory checking by the use of blanks, duplicates and prepared materials is important (Reimann 1989).

10.1.7 Sample acquisition

Before sampling

Pre-sampling procedures consist of gaining as much familiarity as possible with the geology and possible problems before visiting the location to be tested. Obviously if the sampling is at an operating mine these problems are already familiar and the following remarks mainly apply to work at remote prospects.

As much background information as possible should be collected from previous publications and reports. A preliminary description of the exercise serves as a source of stimulation that will help to isolate points on which further thought is needed. The tasks to be completed have to be briefly but adequately described.

Logistical factors are important. How much time is available for the exercise? Can the site be revisited or is this the only opportunity for sampling? What budget restraints are there? Is assistance necessary and if so of what type – surveyors, drivers, cooks, laborers, etc.? Decide on the objectives and how it is proposed to achieve them; are they cost-effective?

What is the purpose of the sampling? Is it purely qualitative, where it is necessary to prove the presence or absence of certain minerals or chemical elements, or is a statistically valid estimate of the grade of (say) a base metal prospect required? Is the study preliminary (i.e. a return visit is possible) or final (i.e. no chance of return)? Certainly, before carrying out extensive systematic sampling at various locations, an orientation survey should be completed to ascertain the main characteristics of the prospective target population – a preliminary mean, standard deviation, geochemical threshold, and probability of error (see also section 8.3). Finally, check that the sampling procedure is sound and adequate and that the projected sample size (weight and number) is large enough for the purpose but not too large. Carefully assess the cost-effectiveness of the proposed exercise.

Safety

Safety aspects are paramount. Before visiting a new site or abandoned mines, samplers should undertake a risk assessment (Grayson 2001). This will consist of hazard identification and risk assessment, followed by developing procedures to manage the risk and control the hazards. Well-designed risk management programs will also bring health aspects into consideration. Effective assessment of hazards required good analytical tools and judgment. Ask for experienced help when starting out in the field. A formalized safety code will be available in all exploration offices and must be read and understood. Samplers should rarely, if ever, work alone and visits to mine workings (surface or underground) should be accompanied. Visits to abandoned mine workings need particular care and the workings should be assessed fully for potential weak rock conditions, weak roof support, water extent, noxious gases, and other hazards before sampling commences.

After sampling

As an example, suppose the calculated mean of a prospect from 25 samples ($n = 25$) is 5.72% zinc with a standard deviation of 0.35% zinc. At a 95% probability level a two-sided confidence interval is:

$$5.72\% + t_{2.5\%}\frac{0.35\%}{\sqrt{25}} > \bar{X} > 5.72\% - t_{2.5\%}\frac{0.35\%}{\sqrt{25}}$$

$$= 5.72\% + 2.06\frac{0.35\%}{5} > \bar{X}$$

$$> 5.72\% - 2.06\frac{0.35\%}{5}$$

$$= 5.72\% \pm 0.15\% \text{ at a 95\% confidence level}$$

That is, there are 19 chances in 20 of the population mean (i.e. the true grade) being within this range, and one chance in 20 of it being outside.

If we are dissatisfied with this result the remedy, in theory, is simple. If we wish a narrower confidence interval the site can be revisited and additional samples taken (i.e. increase n), but remember the square root relationship in the above calculation. That is, if the interval is to be halved (to become 5.72% ±0.075%), then the total number of samples (n) required is not 50 but 100 – an additional 75 samples have to be taken. Alternatively, the value of the t statistic can be reduced (Table 10.3). At an 80% confidence level the interval is:

$$5.72\% + 1.32\frac{0.35\%}{5} > \bar{X} > 5.72\% - 1.32\frac{0.35\%}{5}$$

$$= 5.72 \pm 0.09\%.$$

In other words there are 16 chances in 20 of the population mean being within this narrower range. As it is narrower the probability of it being outside the range is higher – four chances in 20 or quadruple the previous calculation.

Commonsense must be used in the interpretation of the results of any sampling program and particularly with the problem of re-sampling in order to increase the n statistic. As an example, a zinc prospect 200 km from the nearest road or railway is sampled and preliminary estimates are 2.1% zinc with an estimated overall reserve tonnage of about 200,000 t. There is no point in sharpening this grade estimate by taking more samples as the low tonnage, approximate grade and location mean that this deposit is of little economic interest. There is nothing to be gained by extra work which may demonstrate that a better estimate is 2.3% zinc.

10.1.8 Summary

In a perfect world we would all like to follow exactly the sampling theories of Pierre Gy and Francis Pitard. In practice a company does not always have the equipment in place, or the opportunity or the skill to implement them, particularly when operating in the field. It is important therefore to pick the relevant important sampling theory facts and use them at the basic level in a way that an operator can understand. The person with the shovel does not wish to see a complicated mathmatical formula. An explanation of what is important can yield real dividends. It is relatively easy to get from poor sampling to good sampling, but very difficuilt (and the cost may be exponential) to get from good to perfect.

Sample acquisition and preparation methods must be clearly defined, easy to implement and monitor, bias free, as precise and accurate as possible, and cost effective. This can be achieved by:
1 Making a thorough analysis of the material with attention to the particle size distribution and the composition of the particles in each size class. These details are required to establish an optimal sampling strategy.
2 Ensuring that the batch-to-sample size ratio and the sample-to-maximum particle size ratio are sufficiently large.
3 Determining the optimum sample size by considering the economics of the entire process. In some cases, the sample size is determined by

minimizing the cost of collecting the samples when compared with re-sampling.

10.2   PITTING AND TRENCHING

In areas where soil cover is thin, the location and testing of bedrock mineralisation is made relatively straightforward by the examination and sampling of outcrops. However in locations of thick cover such testing may involve a deep sampling program by pitting, trenching, or drilling. Pitting to depths of up to 30 m is feasible and, with trenching, forms the simplest and least expensive method of deep sampling but is much more costly below the water table. For safety purposes, all pits and trenches are filled in when evaluation work is completed. Drilling penetrates to greater depth but is more expensive and requires specialized equipment and expertise that may be supplied by a contractor. Despite their relatively shallow depth, pits and trenches have some distinct advantages over drilling in that detailed geological logging can be carried out, and large and, if necessary, undisturbed samples collected.

10.2.1   Pitting

In areas where the ground is wet, or labor is expensive, pits are best dug with a mechanical excavator. Pits dug to depths of 3–4 m are common and with large equipment excavation to 6 m can be achieved. In wet, soft ground any pit deeper than 1 m is dangerous and boarding must be used. Diggers excavate rapidly and pits 3–4 m deep can be dug, logged, sampled, and re-filled within an hour. In tropical regions, thick lateritic soil forms ideal conditions for pitting and, provided the soil is dry, vertical pits to 30 m depth can be safely excavated. Two laborers are used and with a 1 m square pit, using simple local equipment, advances of up to 2 m per day down to 10 m depth are possible, with half that rate for depths from 10 to 20 m, and half again to 30 m depth.

10.2.2   Trenching

Trenching is usually completed at right angles to the general strike to test and sample over long lengths, as across a mineralized zone.

Excavation can be either by hand, mechanical digger, or by bulldozer on sloping ground. Excavated depths of up to 4 m are common.

10.3   DRILLING

10.3.1   Auger drilling

Augers are hand-held or truck-mounted drills, which have rods with spiral flights to bring soft material to the surface. They are used particularly to sample placer deposits. Power augers are particularly useful for deep sampling in easily penetrable material where pitting is not practicable (Barrett 1987). They vary in size from those used to dig fence post holes to large, truck-mounted rigs capable of reaching depths of up to 60 m, but depths of less than 30 m are more common. Hole diameters are from 5 to 15 cm in the larger units, although holes 1 m in diameter were drilled to evaluate the Argyle diamond deposit in Australia. In soft ground augering is rapid and sampling procedures need to be well organized to cope with the material continuously brought to the surface by the spiralling action of the auger. Considerable care is required to minimize cross-contamination between samples. Augers are light drills and are incapable of penetrating either hard ground or boulders. For this purpose, and holes deeper than about 60 m, heavier equipment is necessary and this is described in the next section.

10.3.2   Other drilling

For anyone interested in understanding the subsurface, drilling is the most frequently used technology. The various methods of drilling serve different purposes at various stages of an exploration program (Annels 1991). The Australian Drilling Industry Training Committee (1997) gives a comprehensive account of methods, applications, and safety issues. Early on when budgets are low, inexpensive drilling is required. The disadvantage of cheaper methods, such as augering, rotary or percussion drilling, is that the quality of sampling is poor with considerable mixing of different levels in the hole. Later, more expensive, but quality samples are usually collected using reverse circulation or diamond core drilling as shown in the following table:

Purpose	Quality	Type
Reconnaissance and exploration	Low	Auger, rotary and percussion (chips)
	High	Reverse circulation (chips)
Resource or reserve evaluation	High	Reverse circulation coring in soft formations
		Diamond or sonic core drilling in hard formations

Fundamentally, drilling is concerned with making a small diameter hole (usually less than 1 m and in mineral exploration usually only a few centimeters) in a particular geological target, which may be several hundred meters distant, to recover a representative sample. Among the available methods rotary, percussion, reverse circulation, and diamond core drilling are the most important.

Rotary drilling

Rotary drilling is a noncoring method and is unequalled for drilling through soft to medium hard rocks such as limestone, chalk, or mudstone. A typical rotary bit is the tricone or roller rock bit that is tipped with tungsten carbide insets. Rock chips are flushed to the surface by the drilling fluid for examination and advances of up to 100 m per hour are possible. This type of drilling is typically used in the oil industry with large diameter holes (>20 cm) to depths of several thousand meters with the extensive use of drilling muds to lift the rock chips to the surface. The large size of the equipment presents a mobility problem.

Percussion drilling

In percussion drilling, a hammer unit driven by compressed air imparts a series of short rapid blows to the drill rods or bit and at the same time imparts a rotary motion. The drills vary in size from small hand-held units (as used in road repair work) to large truck-mounted rigs capable of drilling large diameter holes to several hundred meters depth (see Fig. 11.3). The units can be divided broadly into two types:

1 *Down-the-hole hammer drills*. The hammer unit is lowered into the hole attached to the lower end of the drill rods to operate a noncoring, tungsten carbide-tipped, drill bit. Holes with diameters of up to 20 cm and penetration depths of up to 200 m are possible, but depths of 100–150 m are more usual. Drill cuttings are flushed to the surface by compressed air and a regular and efficient source of this is vital for the successful operation of this technique. As the cuttings come to the surface they can be related to the depth of the hole. However, such direct correlation is not always reliable as holes may not be cased and material may fall from higher levels. In wet ground the lifting effect of compressed air is rapidly dissipated but special foaming agents are available to alleviate this drawback. Rigs are usually truck or track mounted and very mobile. With the latter slopes of up to 30 degrees can be negotiated and the ability to drill on slopes steeper than 25 degrees partly eliminates the need to work on access roads.

2 *Top hammer drills*. As the name suggests the hammer unit, driven by compressed air, is at the top of the drill stem and the energy to the noncoring drill bit is imparted through the drill rods. These are usually lighter units than down-the-hole hammer drills. They are used for holes up to 10 cm diameter and depths of up to 100 m, but more usually 20 m. Most only use light air compressors and this restricts drilling depths to, at the most, only a few meters below a water table. It becomes impossible for the pressurized air to blow the heavy wet rock sludge to the surface. Usually they are mounted on either light trucks or tractors.

General remarks

Percussion drilling is a rapid and cheap method but suffers from the great disadvantage of not providing the precise location of samples, as is the case in diamond drilling. However, costs are one third to one half of those for diamond drilling and this technique has proved particularly useful in evaluating deposits which present more of a sampling problem than a geological one, e.g. a porphyry copper. There is a rapid penetration rate of around a meter a minute and it is possible to drill 150–200 m in an 8-hour shift. With such penetration rates and several machines, several hundred samples can be collected each day. As each 1.5 m of a 10 cm diameter hole is likely to produce about 20–30 kg of rock chips and dust, sample collection and examination requires a high degree of organization. Like all compressed air equipment these drills are noisy in operation.

Reverse circulation

The reverse circulation (RC) drilling technique has been in use in exploration since the mid-seventies and can be used in unconsolidated sediments such as alluvial deposits or for drilling rock. Both air and water can be used as the drill flushing medium and both cuttings or core can be recovered. The technique employs a double-walled string of drill rods (Fig. 10.9), with either a compressed air driven percussion hammer or a rotating tungsten carbide coring bit at the cutting end of the string. The medium is supplied to the cutting bit between the twin-walled drill rods and returned to the surface up the center of the rods. In the case of percussion drilling the rock chippings are also transported to the surface up the center of the rods and from there, via a flexible pipe, to a cyclone where they are deposited in a sample collection container (Fig. 10.9).

The advantages of using this method to collect rock chippings, rather than auger, rotary or percussion drilling, are that the entire sample is collected, the method is extremely quick (up to 40 m per hour can be drilled) and there is very little contamination. The specialized rods, the need for a compressor and additional equipment makes this a more expensive drilling technique than auger or percussion drilling, but the additional costs are outweighed by the higher quality of sample collection. The dual nature of some RC rigs (chips and core) means that high quality core can be taken through the zone of interest without the need to mobilize a second (core) rig, thus reducing the overall drilling costs.

Diamond core drilling

The sample is cut from the target by a diamond-armoured or impregnated bit. This produces a cylinder of rock that is recovered from the inner tube of a core barrel. The bit and core barrel are connected to the surface by a continuous length (string) of steel or aluminum alloy rods, which allow the bit plus core barrel to be lowered into the hole, and pulled to the surface. They also transmit the rotary cutting motion to the diamond bit from the surface diesel power unit, and appropriate pressure to its cutting edge (Fig. 10.10 & Table 10.6).

1 *Drill bits.* Drill bits are classified as either impregnated or surface-set. The former consist of fine-grained synthetic or industrial grade diamonds within a metallic cement while the latter have individual diamonds, sized by their number per carat. In general, impregnated bits are suitable for tough compact rocks such as chert, while surface-set varieties with large individual diamonds are used for softer rocks such as limestone (Fig. 10.11).

Diamond bits will penetrate any rock in time but because of their high cost and the need to maximize core advance and core recovery with minimum bit wear, the choice of bit requires considerable experience and judgment. Second-hand (i.e. used) surface-set bits also have a diamond salvage value and consequently are not used to destruction. Drill bit diameters are classified with either a letter code (American practice) or in millimeters (European practice) (Table 10.7).

2 *Core barrels.* As the cylinder of rock (the core) is cut by the circular motion of the drill bit it is forced up into the core barrel by the advancing drill rods. Core barrels are classified by the length of core they contain. They are usually 1.5–3.0 m in length but can be as long as 6 m. They are normally double-tubed in the sense that in order to improve core recovery an inner core barrel is independent of the motion of the drill rods and does not rotate. Triple-tubed barrels can be used in poor ground and for collecting undisturbed samples for geotechnical analysis.

Previously, to recover core the barrel had to be removed from the hole by pulling the entire length of drill rods to the surface, a time-consuming process. Wire-line drilling (Q series core) is now standard practice; in this method the barrel is pulled to the surface inside the connecting drill rods using a thin steel cable. This has the advantage of saving time but often

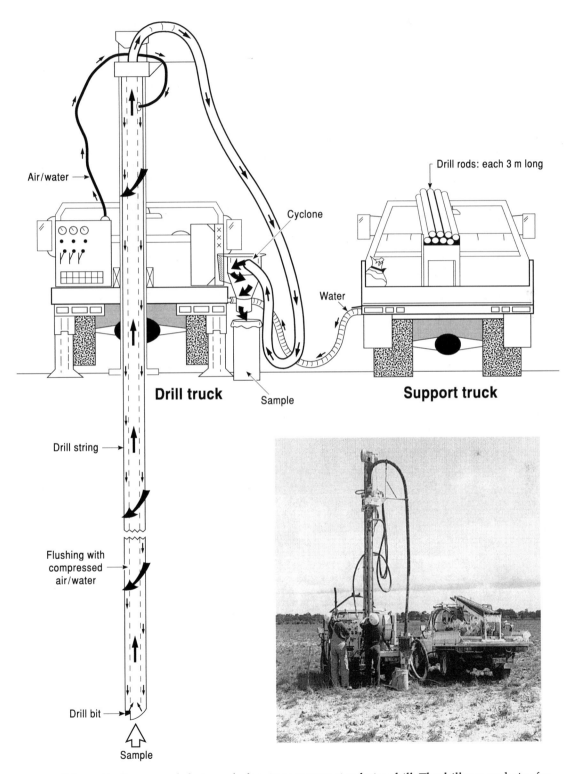

Air/water

Cyclone

Drill rods: each 3 m long

Water

Drill truck

Sample

Support truck

Drill string

Flushing with
compressed
air/water

Drill bit

Sample

FIG. 10.9   Schematic diagram and photograph showing a reverse circulation drill. The drill was exploring for mineral sands in Western Australia (White 1989).

FIG. 10.10   A diamond drill. From left to right the components are the power, control, drill, and water pump units. (Reproduced by permission of Diamond Boart Craelius Ltd.)

Drill unit weight	890 kg	Length of holes:
Power unit weight	790 kg	AQ (48 mm) 425 m
Motor rating	40 kW at 2200 rpm^{-1}	BQ (60 mm) 350 m
Flush pump flow	76 liters min^{-1}	NQ (76 mm) 250 m
Flush pump pressure	5 Mpa	

TABLE 10.6   Diamond drilling rigs. In mineral exploration most drill holes are less than 450 m in length. A typical drill rig for this purpose is shown in Fig. 10.10 but many models are in use. Some of the specifications for the drill in Fig. 10.10 are given in this table.

The drill is a fully hydraulic rig. It is suitable for both underground and surface units and is powered by either an air, diesel or electric motor. To a certain extent penetration rates can be increased by an increase in pressure on the drill bit. This is regulated by hydraulic rams in the drill unit at surface and transmitted to the bit by the drill rods. A limit is reached when all the diamond chips in the bit's surface have been pressed as far as possible into the rock. Higher penetration rates can then only be achieved by increasing the bit rotation speed. Excessively high bit pressure, with low rotation speed, results in abnormally high wear of the bit. Conversely, low pressure with high rotation results in low penetration and polishing (i.e. blunting) of the diamonds.

In the above unit, drill water would flush through a 60-mm-diameter hole at about 50 cm s^{-1}. In a 350-m-length hole return water would take about 12 minutes to pass from the drill bit to the surface. The ascending velocity of the fluid should be greater than the settling velocity of the largest rock particles generated by the cutting and grinding action of the bit.

FIG. 10.11   On the left-hand side are several types of surface-set diamond bits, on the right are bits with tungsten carbide insets and at the back are impregnated bits. (Reproduced by permission of Diamond Boart Craelius Ltd.)

necessitates having a smaller diameter core (Table 10.7).

3 *Circulating medium.* Usually water is circulated down the inside of the drill rods, washing over the cutting surface of the drill bit, and returning to surface through the narrow space between the outside of the rods and the wall of the drillhole. The purpose of this action is to lubricate the bit, cool it, and remove crushed and ground rock fragments from the bit surface. In friable and soft rock this action may flush away part of the sample and this can sometimes be alleviated with the addition of specialized additives. Water may be used in combination with various clays or chemicals, which, in addition to the above functions, have added uses such as sealing the rock face of the drillhole. The science and use of such drilling fluids is a subject in itself and is much developed in the oil industry which drills significantly larger diameter and deeper holes than are required in mineral exploration

Size	Hole diameter (mm)	Core diameter (mm)	Core area (% of hole area)
METRIC SIZES			
Conventional drilling			
36T	36.3	21.7	36
56T2	56.3	41.7	55
66T2	66.3	51.7	61
Wire-line drilling			
56ST	56.3	35.3	39
AMERICAN SIZES			
Conventional drilling			
AX	48	30.1	39
BX	59.9	42.0	49
NX	75.7	54.7	52
Wire-line drilling			
AQ	48	27	31
BQ	60	36.5	36
NQ	75.8	47.6	39
HQ	96.1	63.5	44

TABLE 10.7   A selection of core, drillhole and drill bit sizes.

The hole diameter is the diameter of rock cut by the drill bit. Part of this is removed by the cutting and grinding action of the bit and water flushed to the surface, while the central part is retained as core in the core barrel.

(Australian Drilling Industry Training Committee (1997).

4 *Casing.* Cylindrical casing is used to seal the rock face of the hole. It provides a steel tube in which the drill string can operate in safety and prevents loss of drill strings caused by rock collapse and either loss or influx of water. Casing and drill bits are sized so that the next lower size (i.e. smaller diameter) will pass through the larger size in which the hole had been previously drilled.

5 *Speed and cost of drilling.* Drilling machines used in mineral exploration usually have a capacity of up to 2000 m, exceptionally to 6000 m (see section 14.4) and may be inclined from horizontal to vertical. The rate of advance depends upon the type of drill rig, the bit, the hole diameter (usually the larger the hole diameter the slower the advance), the depth of hole (the deeper the hole the slower the advance), the rock type being drilled (hard or soft, friable, well jointed, etc.), and the skill of the driller.

The driller must decide on the best volume of drilling fluid to be flushed over the drill bit, the right pressure to apply at its cutting surface

(this can be adjusted using hydraulic rams at the surface), the right number of revolutions per minute, and the correct choice of drill bit. Until recently there was no direct method of knowing what operating conditions existed at the drill bit, which are the critical factors if drilling efficiency is to be optimized. Recent developments in "measurements while drilling (MWD)" have overcome this and allow a variety of direct measurements such as temperature, pressure at the bit face, and rate of water flow, to be made (Dickinson et al. 1986, Prain 1989, Schunnesson & Holme 1997). Another technique being developed is a retractable bit that can be raised, replaced, and lowered through the string of drill rods. At present when a worn bit requires replacement the complete drill string has to be withdrawn, which is a time-consuming process (Jenner 1986, Morris 1986). However, modern rigs with automated rod handling can pull the string and unscrew the rods at very high speed.

Drilling advance rates of up to 10 m an hour are possible but depend a great deal upon the skill of the driller and rock conditions. Costs vary from $US50 to 80 a meter for holes up to

300 m long and from $US70 to 150 a meter for lengths up to 1000 m in accessible areas. Standard references for diamond drilling are Australian Drilling Industry Training Committee (1997), Cummings and Wickland (1985), the *Australian Drillers Guide* (Eggington 1985) and *The Management of Drilling Projects* (1981).

Sonic core drilling

Sonic drilling retrieves core but without the contamination caused by drilling muds. Sonic drilling applies the principle of harmonics to drill and case a borehole (Potts 2003). It uses a variable-frequency drill head to transmit vibration energy through the drill pipe and core barrel to allow continuous core sampling to take place. Sonic drilling can penetrate overburden, fine sand, boulders, and hard rock. It can collect samples up to 254 mm in diameter and can drill up to 200 m vertically or in inclined holes. The big advantage of sonic drilling is that uncontaminated, undisturbed samples can be collected because no air, water, or other drilling medium is used. Sonic drilling can realize 100% core recovery, even in glacial till, clay, sands and gravels as well as hard rock. It offers rapid penetration, reduced on-site costs and minimal environmental impact. In the later case the clean up and waste disposal costs are significantly lower.

10.3.3 Logging of drillhole samples

Information from drillholes comes from the following main sources: rock, core, or chips; down-the-hole geophysical equipment; instruments inside the hole (see MWD above); and performance of the drilling machinery. In this section we are only concerned with geological logging, but the geologist on site at a drill location must be familiar with all sources of information. The collection of geotechnical data from core is discussed in section 10.6.

Geological logging

Effective *core recovery* is essential – that is, the length or volume (weight) of sample recovered divided by the length or volume (weight) drilled expressed as a percentage. If recovery is less than 85–90% the value of the core is doubtful as mineralized and altered rock zones are frequently most friable and the first to be ground away and lost during drilling. The core is not then representative of the rock drilled, it is not a true sample, and it is probably misleading (Fig. 10.12).

Core drilling

Often, initial, rapid core logging is done at the drill site. This information is used to decide whether the hole is to be either continued or abandoned. Wetted core is more easily examined, using either a hand lens or a binocular microscope. Most organizations have a standard procedure for core logging and a standard terminology to describe geological features. Field data loggers are now used to gather company standardized digital data, which are downloaded to the central database upon return to the field base or office (see section 9.1). Onions and Tweedie (1992) discuss the time and costs saved using this integrated approach to data gathering, storage, and processing.

Once the initial logging at a drill site is complete, the core is moved to a field base, where a more detailed examination of the core takes place at a later date. Nevertheless, the main structural features should be recorded (fracture spacing and orientation) and a lithological description (colour, texture, mineralogy, rock alteration, and rock name) with other details such as core recovery and the location of excessive core loss (when say >5%). The description should be systematic and as quantitative as possible; qualitative descriptions should be avoided. These data are plotted on graphical core logs (see Fig. 5.13) and used as an aid in interpreting the geology of the current and next holes to be drilled.

Core is stored in slotted wooden, plastic or metal boxes short enough to allow two persons to lift and stack them easily (Fig. 10.13; see Fig. 13.10). Core is collected for a variety of purposes other than geological description, e.g. metallurgical testing and assaying. For these latter purposes the core is measured into appropriate lengths (remembering the principle of stratified sampling) and divided or split into two equal halves either by a diamond saw or a mechanical splitter. Half the core is sent for assay or other investigations whilst the other

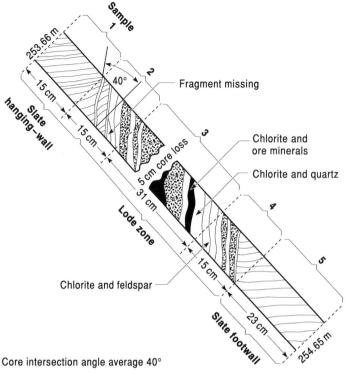

Core intersection angle average 40°

Sample number	Depth		Recovery		True width	%Sn	cm%
	from(m)	to(m)	(m)	(%)	(cm)		
1	253.66	253.81	0.15	100	12	0.03	0.36
2	253.81	253.96	0.15	100	12	0.56	6.72
3	253.96	254.27	0.25	81	20	4.75	95.00
4	254.27	254.42	0.15	100	12	1.27	15.24
5	254.42	254.65	0.23	100	18	NIL	–

$$\text{Weighted average lode value} = \frac{116.96\,\text{cm\%}}{44\,\text{cm (true thickness)}} = 2.66\%\,\text{Sn}$$

at a weighted average core recovery of 91%

FIG. 10.12 Typical intersection of a tin-bearing vein showing sampling intervals and the uncertainty introduced by incomplete core recovery. (After Walsham 1967.)

half is returned to the core box for record purposes. Obviously structural features have to be recorded before splitting and a good practice is to photograph wet core, box by box, before logging it, to produce a permanent photographic record (Fig. 13.10).

When core from coal seams is to be sampled or the samples are to be collected for geotechnical analysis, the core should be sealed as soon as the core leaves the core barrel. This is to prevent loss of moisture, which can adversely

affect the measurement of relative density of the core (Preston & Sander 1993) or alter the nature of the material.

Rock chips and dust ("sludge") can be collected during core drilling; they represent the rock cut away by the diamond drill bit. Drilling with air circulation in relatively shallow holes (as in most percussion drilling) delivers cuttings to the surface within a minute or so. However, with core drilling, water circulation, and longer holes, there is an appreciable

FIG. 10.13 Diamond drill core being examined and stored in a wooden core box. Note the excellent core recovery. (Reproduced by permission of Diamond Boart Craelius Ltd.)

time lag before cuttings reach the surface. It is estimated that from a depth of 1000 m cuttings can take 20–30 minutes to reach the surface, with the inherent danger of differential settlement in the column of rising water due to differences in mineral and rock-specific gravities and shape. Consequently rock sludge is rarely examined during core drilling.

Obtaining core is expensive so it is sensible to retain it for future examination. However, adequate and long-term storage involves time, space, and expense, but the value of the contained information is important particularly as during mining some drill locations may be permanently lost.

Noncore drilling
In noncore drilling the chips and dust are usually collected at 1- to 2-m intervals, dried and separately bagged at the drill site. After washing they are relatively easy to examine with the use of a hand lens and binocular microscope. Samples can be panned so as to recover a heavy mineral concentrate. It is a good practice to

sprinkle, and glue, a sample of rock chips and panned concentrates from each sample interval on to a board so that a continuous visual representation of the hole can be made. Again, descriptions must be systematic and quantitative.

10.3.4 Drilling contracts

Drilling can be carried out with either in-house (company) equipment or it can be contracted out to specialist drilling companies. In the latter case the conditions of drilling, amount of work required, and the cost will be specified in a written contract. The purpose of drilling is to safely obtain a representative sample of the target mineralisation in a cost-effective manner. Once the mining company has been assured, in writing, of the drilling company's safe working practices, the choice of drilling equipment is crucial and much depends upon the experience of the project manager. Unless the drilling conditions are well known test work should, if possible, be carried out to compare different drilling methods before any large-scale program begins (Box 10.3).

The main items of cost in a contract are as follows:

1 Mobilization and transport of equipment to the drilling site. This can vary from movement along a major road by truck to transport by helicopter.
2 Setting up at each site and movement between successive borehole locations. Again costs can vary greatly depending on distance and terrain.
3 A basic cost per meter of hole drilled.
4 Optional items costed individually, e.g. cementing holes, casing holes, surveying.
5 Demobilization and return of equipment to driller's depot.

All items of cost should be detailed in the contract. There may be a difference of interest between the driller and the company representative on site, normally a geologist, who is there to see that drilling proceeds according to the company's plan and the contract. The driller may be paid by distance drilled during each shift whilst the geologist in charge is more concerned with adequate core recovery and that the hole is proceeding to its desired target. The company representative usually signs documents at the completion of each shift where progress and problems are described and it is on the basis of these documents that

BOX 10.3   Comparative sampling results from two different drilling methods. (after Springett 1983b.)

If drilling conditions are not known a program of test work to compare and contrast appropriate drilling procedures may be desirable before any large scale sampling program is commenced, as in the following example.

Silver mineralization 12.2 m thick consisting of fine-grained dolomite and quartz with minor chalcopyrite, galena, and sphalerite occurs in limestone. The silver is principally acanthite but with some native silver. It has been tested by diamond and percussion drilling, and also by mining from a test shaft.

Diamond drilling: BQ core, diameter 36.5 mm with samples every 1 m over the total length of 12.2 m; core recovery was 93%. The core was split for assay and the total weight of the overall sample was 10 kg. The weighted assay average was 62 g t^{-1} silver. The main concern in diamond drilling was the possibility of high grade, friable silver mineralization being ground by the diamond bit and washed away by the circulating water.

Percussion drilling: With a 1.6-cm-diameter hole samples were collected every 1 m; material recovery was 87%. All the recovered material was taken for the analysis and totalled 255 kg. The weighted assay average was 86 g t^{-1}. Whilst this method was cheaper than core drilling, there was the possibility that the high specific gravity silver minerals might sink in the column of water in the bore hole and not be fully recovered.

Bulk sample: From a 1.1-m square test shaft all the material was taken for assay, totalling 51.8 t. The weighted assay average was 78 g t^{-1}.

Accepting the bulk sample result as the best estimate of silver grade the cheaper percussion drilling method can be used to sample the deposit rather than the more expensive core drilling.

the final cash payment is made. Consequently, it is important for this representative to be thoroughly familiar with the contract and the problems that can arise at a drilling site, particularly if in a remote location.

The client (i.e. the company) is at liberty to fix specifications such as, say, a plus 90% core recovery, a vertical hole with less than 5 degrees deviation. It is then at the discretion of the drilling company as to whether or not these requirements are accepted, but if they are and the conditions are not met the failure is remedied at the expense of the contractor.

10.3.5 Borehole surveying

Drills are usually unpredictable wanderers. In section holes drilled at a low angle to rock structures tend to follow that structure while holes at a higher angle tend to become perpendicular to the feature. A similar behavior can be expected from alternations of hard and soft rock (e.g. layers of chert in chalk). Steep holes are likely to flatten and all holes tend to spiral counter clockwise with the rotation of the drill rods. However, wandering also depends upon the pressure on the drilling bit and its rate of revolution, so that two drillholes in similar geological situations may take different paths. Holes with a length of 1000 m have been found to be several 100 m off course – while individual lengths of drill rod are essentially inflexible, a drill string several hundred meters in length will bend and curve. Since drillholes are used for sampling at depth, if the orientation of the hole is not known the location of the sample is similarly unknown, and geological projections based on these samples will be in error.

Borehole surveys are conducted as a matter of routine in all holes and there are a variety of instruments for this purpose. They most commonly indicate the direction (azimuth) and inclination of the hole at selected intervals, commonly every 100 m, using a small magnetic or gyroscopic compass. The calculation of corrected positions at successive depths is a straightforward mathematical procedure, if one knows the location of the top of the hole (X, Y, and Z co-ordinates) and the initial hole inclination.

Deliberate deflection can be achieved either by wedging holes (say, 1 degree per wedge)

or using a hydraulically driven downhole motor, which acts as a directional control device (Cooper & Sternberg 1988, Dawson & Tokle 1999). This relatively new directional-drilling tool was originally designed for the oil industry. The downhole motor is powered by the circulating fluids passed down the center of the rod string, which drives the bit without drill-rod rotation. A spring deflection shoe located just above the bit acts as a directional control device and delivers a constant side pressure to the drillhole wall forcing the bit to move (and deflect) in the opposite direction. The amount of spring loading in the shoe is preset before the unit is lowered into the hole and, obviously, progress is monitored by closely spaced surveys. The deflection unit is a special tool and is not used in normal drilling.

10.3.6 Drillhole patterns

The pattern of drilling in an exploration program depends primarily upon the intended use of the data. In reconnaissance work where the geology is poorly known the first holes may be isolated from each other and drilled for geological orientation purposes. In exploration for sedimentary uranium deposits, coal and borates, holes may be drilled up to 10 km apart to locate sedimentary formations of interest and obtain structural data. Later holes may be a few kilometers apart whilst drilling a specific target calls for "fences" or lines of holes, at a close spacing of 100–200 m. However, the location and spacing of holes in the last eventuality are guided by detailed geological mapping, geochemical, geophysical, and geostatistical results (see sections 5.2 and 10.4.1).

Generally, a systematic grid of drillholes (and samples) taken normal to the expected mineralized zone is a preferred pattern. This provides a good statistical coverage and geological cross sections can be made with a minimum of projection. Such fences may well be 200–400 m apart with individual holes spaced at 100–200 m intervals, which allows room for systematic infill drilling if the results justify extra work. The inclination of individual holes is obviously important and generally they should be drilled at right angles to the expected average dip of the mineralisation.

Before drilling a hole it is recommended that a section be drawn along its projected length, allowing for hole deviation if this is possible. As the hole progresses the section is modified. Drilling is an expensive and a time-consuming part of mineral exploration. The objective is to drill a precise number of holes, within budget, safely, and provide the exact number of intersections needed to demonstrate the grade and tonnage (dimensions) of the mineralisation at an appropriate level of accuracy and precision. Unfortunately, such optimization is rarely, if ever, possible. At second best an iterative procedure is followed whereby a series of successively closer spaced drilling programs are completed with each followed by a re-assessment of all existing data – a process of successive approximations. The main problem is that in the calculation of mineral resources the zone of influence of each sample is not known until a minimum amount of work is completed. If two adjacent samples (taking a drillhole as a sample) cannot be correlated at an acceptable confidence level, then neither has an acceptable zone of influence in the intervening space and further sampling is necessary (see section 10.4.1). Conversely where adjacent samples show appropriate correlation further sampling is not required.

There are several methods of assigning zones of influence to either successive samples or individual boreholes: the mean-square (dn) successive difference test, the use of correlation coefficients (R^2) or geostatistics using the range (a) of a semi-variogram. However, a relatively large nugget effect (section 10.4.3) in the semi-variogram indicates an element of uncertainty that would call for a reduction in spacing. Such an effect is apparent in the evaluation of low-grade disseminated gold deposits where drilling may be required on a ±50 m grid basis.

10.4   MINERAL RESOURCE AND ORE RESERVE ESTIMATION, GRADE CALCULATIONS

During exploration and initial evaluation of a base metal, industrial mineral, or coal deposit, the principal emphasis is placed upon its geology and the estimation of the quality and quantity of the resources present. Data are collected from several sampling programs as described above, including trenching, pitting, percussion drilling, reverse circulation drilling, and diamond drilling, as well as during trial mining. The object of this sampling is to provide a mineral inventory. Once the mining engineers and financial analysts have established that the mineral can be mined at a profit, or in the case of industrial minerals can be marketed at a profit, it can then be referred to as an ore reserve.

The stock market controversies that surrounded the discovery of the Poseidon nickel deposits in Western Australia in the 1960s led the Minerals Council of Australia to establish a committee to resolve these issues. When the Australasian Institute of Mining and Metallurgy joined it, the resulting committee was called the Australasian Joint Ore Reserves Committee (JORC). Other controversies, including that surrounding the Bre-X deposit in Indonesia in the 1990s (see section 5.4), prompted Australian and north American stock exchanges to adopt rigorous procedures for companies to report their resources. The leader in determining these procedures has been JORC, which has defined the JORC Code (Australasian Joint Ore Reserves Committee 2003) that has been taken up by a number of companies and stock exchanges, and forms the basis for most international reporting systems.

As early as 1976 the United States Geological Survey (USGS 1976) published definitions of reserves and resources for mineral and coal deposits. These publications take into account the increasing degree of confidence in the resources and the financial feasibility of mining them.

Between 1972 and 1989, a number of reports were issued by JORC which made recommendations on public reporting, and resource and reserve classification. These gradually developed the principles now incorporated in the JORC Code. In 2004, JORC published its latest version of the code.

The JORC Code is currently used as a model for reporting codes of other countries with appropriate modifications to reflect local conditions and regulatory systems. Examples include the USA (SME 2004), South Africa (SAIMM 2004), and UK (IOM3 2004). Agreement has also been reached with the United

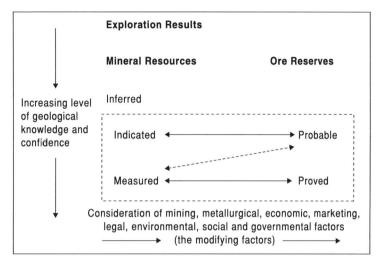

FIG. 10.14 A JORC Box used as an early basis for classifying mineral resources and reserves. It takes into account the increasing degree of confidence in the resources and the financial feasibility of mining them. (From Australasian Joint Ore Reserves Committee 2003.)

Nations (UNCE 2001) on a common set of terms to describe resources and reserves, and an international set of definitions and guidelines.

10.4.1 The JORC Code

The JORC code sets out a system for classifying tonnage and grade estimates as either ore reserves or mineral resources, and of subdividing these into categories that reflect different levels of confidence (Fig. 10.14). The purpose of this is purely for public reporting. JORC does not regulate how estimates should be done, nor does it attempt to quantify the amount of data needed for each mineral resource category.

A Competent Person(s) makes these decisions based on direct knowledge of the deposit in question. The criteria used may be entirely subjective or they may involve some quantitative measures, or mixtures of both. There are no prescribed methods that can be applied "off the shelf." It is very important that the selected methods of distinguishing resource classes should be geologically sensible in the context of the deposit being studied.

Stephenson (2000) explained that the JORC Code has been operating successfully for over 10 years and, together with complementary developments in stock exchange listing rules, has brought about substantially improved standards of public reporting by Australasian mining and exploration companies. The success of the Code is based upon its early adoption in full by the Australian and New Zealand Stock Exchanges, and the ability and willingness of the mining industry to bring Competent Persons to account when necessary.

The JORC Code is designed for reporting of mineral resources, which it defines as a concentration or occurrence of material of intrinsic economic interest in or on the Earth's crust in such form and quantity that there are reasonable prospects for eventual economic extraction. The location, quantity, grade, geological characteristics, and continuity of a Mineral Resource are known, estimated, or interpreted from specific geological evidence and knowledge, by a Competent Person appointed by the Company. Mineral Resources are subdivided, in order of increasing geological confidence, into Inferred, Indicated, and Measured categories. Portions of a deposit that do not have reasonable prospects for eventual economic extraction must not be included in a Mineral Resource.

A mineral resource is not an inventory of all mineralisation drilled or sampled, regardless of cut-off grade, likely mining dimensions, location, or continuity. It is a realistic inventory of mineralisation, which, under assumed and justifiable technical and economic conditions, might, in whole or in part, become economically extractable.

JORC defines ore and reserves as the economically mineable part of a measured or indicated mineral resource. Reserves include diluting materials and allowances for mining losses. Reserves are usually assessed during feasibility studies and include consideration of and modification by realistically assumed mining, metallurgical, economic, marketing, legal, environmental, social and governmental factors. These assessments demonstrate at the time of reporting that extraction could reasonably be justified. Ore reserves are subdivided in order of increasing confidence into probable and proved ore reserves. A probable ore reserve is the economically mineable part of an indicated, and in some circumstances measured, mineral resource. A probable ore reserve is the economically mineable part of a measured mineral resource.

Geologists are interested in the amount and quality of *in situ* mineralisation within a geologically defined envelope. *In situ* mineral resources tend to be slightly greater than the ore reserves. Discussions of the subject are given by the Australasian Joint Ore Reserves Committee (JORC 2003), Whateley and Harvey (1994), and Taylor (1989).

Before resource calculations can proceed, a study of the mineralisation envelope is required. This envelope can often be defined by readily identifiable geological boundaries. In some cases inferences and projections must be made and borne in mind when assessing confidence in the resources. Some deposits can only be delineated by selecting an assay cut-off to define geographically and quantitatively the potential mineralized limits (see Chapter 16, where an assay cut-off is used to define the outline of the Trinity Silver Mine and where waste blocks are identified within the deposit for separate disposal). Initially this is highly subjective. In later stages when confidence is higher, the ore limits will be carefully calculated from conceptual mining, metallurgical, cost, and marketing data (see section 11.2).

Geologists must provide the basis for investment decisions in mining using the data provided from the above sampling. It is their responsibility to ensure that the database that they are using is valid (see section 9.1). Once the geological and assay (grade) cut-offs (Lane 1988, King 2000) have been established, usu-

ally in discussions with the mining engineers, it is possible to start the evaluation procedure.

It is therefore important to understand the cut-off grade theory and grade tonnage curves.

10.4.2 Cut-off grade theory

Geologists will be required to understand the concept of cut-off grade theory (Lane 1988, King 2000), because they will be required to identify only the material above the lowest grade that will be included in the potentially economic part of the deposit. This is defined as the cut-off grade, which separates the mineralized rock from barren or low grade rock. Deposits with only one valuable metal, such as gold, would normally use grade $(g\ t^{-1})$ to rank the resource. In a simple process, the highest grade material is processed to recover the gold and the low grade material sent to a waste dump.

The resource potential will therefore be determined by the cut-off grade. It is common practice to calculate the resource tonnage at a series of cut-off grades. This information is plotted on a grade–tonnage graph (Fig. 10.15). Eventually the material above cut-off will be used to develop a mine plan and the blocks within the mine will be scheduled for extraction. The schedule will be affected by location and distribution of ore in respect to topography and elevation, mineral types, physical characteristics, and grade–tonnage distribution, and direct operating expenses associated with mining, processing, and converting the commodity into a saleable form. The ultimate aim is to calculate the net present value (NPV; see section 11.5.3).

All this is based upon the tonnage and grade calculations made by the geologist. Although cut-off grades are used to classify material into streams for processing, stockpiling, or being sent to waste dumps, Fig. 10.16 shows how errors in predicting properties of the resource may result in misclassification of this material (Wellmer 1998). The ellipse represents a band of confidence between the estimated and actual values (typically 95% confidence band). The same cut-off has been applied to both axes. It is apparent that there are two areas where material is classified correctly and two areas of misclassification. Misclassified waste has a

(a)

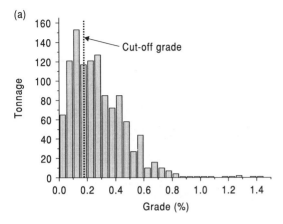

(b)

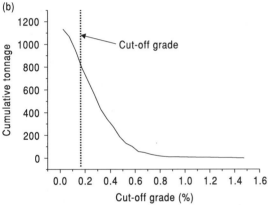

FIG. 10.15   Grade–tonnage graphs. (a) Resource tonnage frequency distribution at different grades. (b) Resource tonnage cumulative frequency distribution at different grades.

grade below the cut-off grade but is estimated to have a grade above cut-off. This misclassified waste will be mined and processed as ore. Likewise, misclassified ore is classified as waste and dumped without extracting any metal, with the subsequent loss of revenue. The importance of making as good an estimate as possible cannot be overemphasized. Some of the methods available to make these estimates are described below.

10.4.3   Tonnage calculation methods

Classical statistical methods

The use of classical statistical methods in resource calculations is generally restricted to

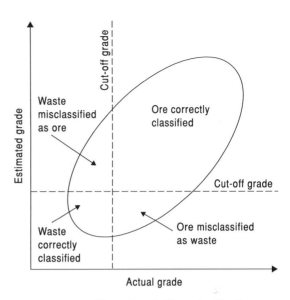

FIG. 10.16   A confidence band ellipse showing how errors in predicting properties of the resource may result in misclassification of this material (Wellmer 1998). The ellipse represents a band of confidence between the estimated and actual values (typically 95% confidence band).

a global estimation of volume or grade within the mineralisation envelope. The simplest statistical estimate involves calculating the mean of a series of values, which gives an average value within the geologically defined area. Calculation of the variance will give a measure of the error of estimation of the mean (Davis 2003). When using these nonspatial statistical methods it is important that the samples are independent of one another, i.e. random. If sample locations were chosen because of some assumption or geological knowledge of the deposit then the results may contain bias. Separate calculations must also be made of different geological areas to ensure that the results are meaningful.

The data are usually plotted on frequency distribution graphs (histograms) (see Figs 8.4, 10.1 & 16.10) and scatter diagrams (correlation graphs) (see Fig. 8.6). The distributions are usually found to approximate to a gaussian or to a log–normal distribution. Some geostatistical techniques only operate in gaussian space. Techniques exist that enable geologists to transform the variable data into gaussian space

(Geovariances 2001). Cumulative frequency distributions can also be plotted on normal probability paper and a straight line on such a plot represents a normal distribution. Significant departures from a straight line may indicate the presence of more than one grade zone and thus more than one population within the data. Each population should be treated separately (c.f. section 8.3.1).

Conventional methods

In calculating the resource or reserve potential of a deposit, one formula, or a variation of it, is used throughout, namely:

$$T = A \times Th \times BD$$

where T = tonnage (in tonnes), A = area of influence on a plan or section in km², Th = thickness of the deposit within the area of influence in meters, BD = bulk density. This includes the volume of the pore spaces. It is obtained by laboratory measurement of field samples.

The area of influence is derived from a plan or section of the geologically defined deposit. The conventional methods commonly used for obtaining these areas are: thickness contours (constructed manually), polygons, triangles, cross-sections, or a random stratified grid (Fig. 10.17). Popoff (1966) outlined the principles and conventional methods of resource calculation in some detail. The choice of method depends upon the shape, dimensions and complexity of the mineral deposit, and the type, dimensions and pattern of spacing of the sampling information. These methods have various drawbacks that relate to the assumptions on which they are based, especially the area of influence of the sampling data, and generally do not take into account any correlation of mineralisation between sample points nor quantify any error of estimation.

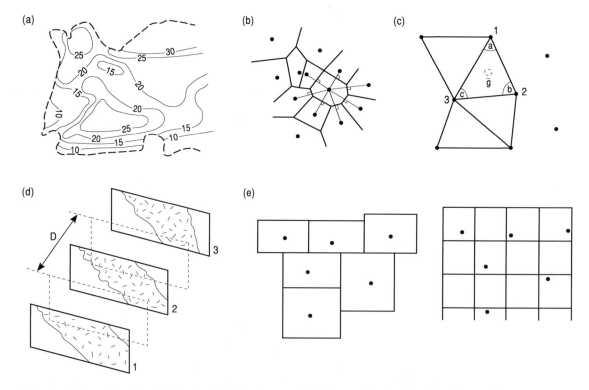

FIG. 10.17 Conventional methods of estimating the area of influence of a sample using (a) isopachs, (b) polygons, (c) triangles, (d) cross-sections, or (e) a random stratified grid. (From Whateley 1992.)

Large errors in estimation of thickness (or grade) can therefore be made when assuming that the thickness or grade of a block is equal to the thickness or grade of a single sample point about which the block has been drawn. Care should be taken when using the cross-sectional method that the appropriate formula is used for computing volume between sections, as significant errors in tonnage estimation can be introduced where the area and shapes on adjacent sections vary considerably. Wire frame modeling, in technical software packages such as Datamine, offers the user the ability to construct a series of sections between which the user can link the areas of interest with a wire frame. The package calculates the volume, limiting some of the errors introduced by manual methods. Manual contouring methods are less prone to this estimation error by virtue of the fact that contours are constructed by linear interpolation, thereby smoothing the data irregularities.

The inverse power of the distance $(1/D^n)$ method can also be used to calculate resources, usually using a geological modeling software package. The deposit is divided into a series of regular blocks within the geologically defined boundary, and the available data are used to calculate the thickness value for the center of each block. Near sample points are given greater weighting than points further away. The weighted average value for each block is calculated using the following general formula:

$$Th = \frac{\Sigma(V \times f)}{\Sigma f}$$

where Th = the thickness at any block center, V = known thickness value at a sample point, $f = 1/D^n$ weighting function, n = the power to which the distance (D) is raised.

The geologist has to decide how many of the available data are to be used. This is normally decided by choosing a distance factor for the search area (usually the distance, a, derived from the range of a semi-variogram), the power factor which should be employed (normally D^2), and how many sample points should be used to calculate the center point for each block. The time needed to calculate many

hundreds of block values on a complex deposit requires the aid of a computer. This method begins to take the spatial distribution of data points into account in the calculations.

Most of these methods have found application in the mining industry, some with considerable success. The results from one method can be cross checked using one of the other methods.

Geostatistical methods

Semi-variograms

The variables in a deposit (grade, thickness, etc.) are a function of the geological environment. Changing geological and structural conditions results in variations in grade or quality and thickness between deposits and even within one deposit. However, samples that are taken close together tend to reflect the same geological conditions, and have similar thickness and quality. As the sample distance increases, the similarity, or degree of correlation, decreases, until at some distance there will be no correlation. Usually the thickness variable correlates over a greater distance than coal quality variables (Whateley 1991, 1992) or grade.

Geostatistical methods quantify this concept of spatial variability within a deposit and display it in the form of a semi-variogram (Box 10.4). Once a semi-variogram has been calculated, it must be interpreted by fitting a "model" to it. This model will help to identify the characteristics of the deposit. An example of a model semi-variogram is shown in Fig. 10.22. The model fitted to the experimental data has the mathematical form shown in the two equations below. It is known as a spherical scheme model and is the most common type used, although other types do exist (Journel & Huijbregts 1978, David 1988, Isaaks & Srivastava 1989, Annels 1991).

There is often a discontinuity near the origin, and this is called the nugget variance or nugget effect (C_0). This is generally attributable to differences in sample values over very small distances, e.g. two halves of core, and can include inaccuracies in sampling and assaying, and the associated random errors (see section 10.1.2). C_0 is added where a nugget effect exists:

BOX 10.4   Calculating the semi-variogram.

It is assumed that the variability between two samples depends upon the distance between them and their relative orientation. By definition this variability (semi-variance or $\gamma(h)$) is represented as half the average squared difference between samples that are a given separation or lag distance (h) apart (see the equation below).

In the example (Fig. 10.18), the sample grid shows thickness values of a stratiform mineral deposit or coal seam at 100 m centers. Suppose we call this distance between the samples (in the first instance 100 m) and their orientation, h. We can calculate the semi-variance from the difference between samples at different distances (lags) and in different orientations, e.g. at 200 m (Fig. 10.22) and 300 m (Fig. 10.23), etc. These examples are all shown with an E–W orientation, but the principle equally applies with N–S, NE–SW, etc. directions. We calculate the semi-variance of the distances for as many different values of h as possible using the following formula:

$$\gamma(h) = \frac{\sum_{i=1}^{n(h)} (z_{(i)} - z_{(z-h)})^2}{2n(h)}$$

where $\gamma(h)$ = semi-variance, $n(h)$ = number of pairs used in the calculation, z = grade, thickness, or whatever, (i) = the position of one sample in the pair, $(i+h)$ = the position of the other sample in the pair, h meters away from $z(i)$.

Using the above equation above, the $\gamma(h)$ values calculated for lag 1 ($h = 100$ m) are as follows:

$(1.11 - 1.16)^2 + (1.16 - 1.08)^2 + (1.08 - 1.03)^2 +$
$(1.03 - 1.00)^2 + (1.19 - 1.12)^2 + (1.12 - 1.03)^2 +$
$(1.08 - 1.05)^2 + (1.05 - 1.03)^2 + (1.03 - 0.97)^2 +$
$(0.97 - 1.00)^2 + (1.00 - 0.92)^2 + (0.92 - 0.94)^2 +$
$(1.05 - 1.03)^2 + (0.97 - 1.03)^2 + (1.03 - 1.00)^2 +$
$(1.00 - 0.95)^2 = 0.0481$
$(100) = 0.0481/[2 \times 16] = 0.0015 \ (m)^2$

Using the same formula the $\gamma(h)$ values for lag 1 to lag 6 ($h = 600$ m) are as follows:

$(100) = 0.0015, (200) = 0.0037, (300) = 0.0036,$
$(400) = 0.0055, (500) = 0.0108, (600) = 0.0104.$

The values thus calculated are presented in a graphical form as a semi-variogram. The horizontal axis shows the distance between the pairs, while the vertical axis displays the values of $\gamma(h)$ (Fig. 10.21).

$$\gamma(h) = C_0 + C \left(\frac{3h}{2a} - \frac{1(h)^3}{2(a)^3} \right) \quad \text{(for } h < a\text{)}$$

$$\gamma(h) = C_0 + C \quad \text{(for } h > a\text{)}$$

The semi-variogram starts at 0 on both axes (Fig. 10.21). At zero separation ($h = 0$) there should be no variance. Even at relatively close spacings there are small differences. Variability increases with separation distance (h). This is seen on the semi-variogram where a rapid rate of change in variability is marked by a steep gradient until a point where the rate of change decreases and the gradient is 0. Beyond this point sample values are independent and have variability equal to the theoretical variance of sample values (Fluor Mining and Metals Inc. 1978, Journel & Huijbregts 1978, Isaaks & Srivastava 1989). This variability is termed the sill ($C + C_0$ on Fig. 10.22) of the semi-variogram, and the point at which this sill value is reached is termed the range (a) of the semi-variogram.

The range, a, on Fig. 10.22 is 500 m. One can see that if the drill spacing is greater than 500 m then the data from each sample point would not show any correlation. They would in

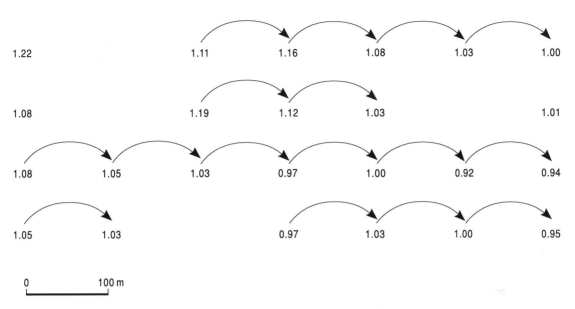

FIG. 10.18   A portion of a borehole grid with arrows used to identify all the pairs 100 m apart in an E–W direction. (From Whateley 1992.)

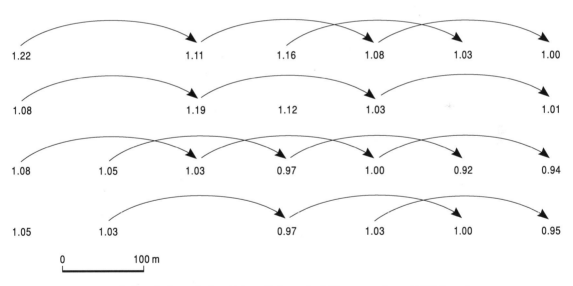

FIG. 10.19   A portion of a borehole grid identifying all the pairs 200 m apart in an E–W direction. (From Whateley 1992.)

effect be random sample points. If, however, one were to drill at a spacing less than 500 m, then the semi-variogram shows us that the data will, to a greater or lesser extent (depending on their distance apart), show correlation. Holes 100 m apart will show less variance than holes 400 m apart. Industry standard is to accept a sample spacing at two thirds of the range (e.g. 350 m of 500 m).

Semi-variograms with a reasonably high nugget variance and where semi-variance also increases rapidly over short distances suggest

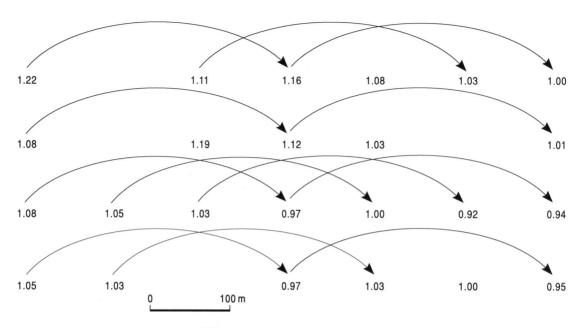

FIG. 10.20   A portion of a borehole grid illustrating some of the pairs 300 m apart in an E–W direction. (From Whateley 1992.)

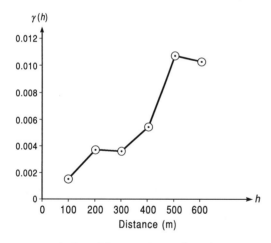

FIG. 10.21   A plot of the experimental semi-variogram constructed from the data derived from the portion of the borehole grid shown in Figs 10.18–10.20. The calculations are shown in the text.

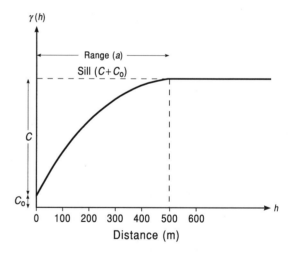

FIG. 10.22   The idealized shape of a semi-variogram (spherical model). (From Whateley 1992.)

that there is a large degree of variability between the samples. When h is greater than or equal to a then the data show no correlation and this gives a random semi-variogram. This can be interpreted to mean that sample spacing is too great.

It is important to note that the rate of increase in variability with distance may depend on the direction of a vector separating the samples. In a coal deposit through which a channel passes, the variability in the coal quality and thickness in a direction normal to the

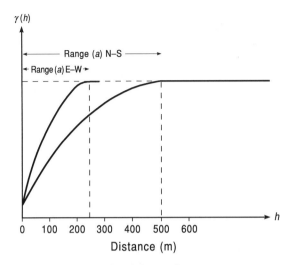

FIG. 10.23 The idealized shape of a semi-variogram (spherical model) illustrating the concept of anisotropy within a deposit. (From Whateley 1992.)

channel (i.e. across any "zoning") will increase faster than the variability parallel to the channel (Fig. 10.23). The different semi-variograms in the two directions will have different ranges of influence and as such represent anisotropy or "hidden" structure within the deposit.

The squared differences between samples are prone to fluctuate widely, but theoretical reasoning and practical experience have shown that a minimum of 30 evenly distributed sample points are necessary to obtain a reliable semi-variogram (Whitchurch et al. 1987). This means that in the early stages of an exploration program a semi-variogram may be difficult to calculate. In this case typical semi-variogram models from similar deposits can be used as a guide to the sample spacing, until sufficient samples have been collected from the deposit under investigation. This is analogous to using the genetic model to guide exploration.

Kriging
Kriging is a geostatistical estimation technique used to estimate the value of a point or a block as a linear combination of the available, normally distributed samples in or near that block. It considers sample variability in assigning sample weights. This is done with reference to the semi-variogram. Kriging is also

capable of producing a measurement of the expected error of the estimate (see kriging variance below). The principles of kriging and types of kriging are discussed by Clark (1979), Journel and Huijbregts (1978), David (1988), Isaaks and Srivastava (1989), Annels (1991), and Deutsch (2002) amongst others. The kriging weights are calculated such that they minimize estimation variance by making extensive use of the semi-variogram. The kriging estimator can be considered as unbiased under the constraint that the sum of the weights is one. These weights are then used to estimate the values of thickness, elevation, grade, etc. for a series of blocks.

Kriging gives more weight to closer samples than more distant ones, it addresses clustering, continuity and anisotropy are considered, the geometry of the data points and the character of the variable being estimated are taken into account, and the size of the blocks being estimated is also assessed.

Having estimated a reliable set of regular data values, these values can be contoured as can the corresponding estimation variances (see below). Areas with relatively high estimation variances can be studied to see whether there are any data errors, or whether additional drilling is required to reduce the estimation variance.

Cross-validation
Cross-validation is a geostatistical technique used to verify the kriging procedure, to ensure that the weights being used do minimize the estimation variance. Each of the known points is estimated by kriging using between four and twelve of the closest data points. In this case both the known value (Z) and the estimated value ($Z^\star$) are known, and the experimental error, $Z - Z^\star$ can be calculated, as well as the theoretical estimation variance (σ_e^2). By plotting Z against $Z^\star$, it is possible to test the semi-variogram model used to undertake the cross-validation procedure. If the values show a high correlation coefficient (R^2 near 1), we can assume that the variables selected from the semi-variogram provide the best parameters for the solution of the system of linear equations used in the kriging matrices to provide the weighting factors for the block estimate and estimation variance calculations (Journel & Huijbregts 1978).

Estimation variance (σ_e^2)

No matter which method is used to estimate the properties of a deposit from sample data, whether it is manual contouring or computer methods, some errors will always be introduced. The advantage of geostatistical methods over conventional contouring methods is their ability to quantify the potential size of such errors. The size of the error depends on factors such as the size of the block being estimated (Whitchurch et al. 1987), the continuity of the data, and the distance of the point (block) being estimated from the sample. The value to be estimated ($Z^\star$) generally differs from its estimator (Z) because there is an implicit error of estimation in $Z - Z^\star$ (Journel & Huijbregts 1978). The expected squared difference between Z and $Z^\star$ is known as the estimation variance (σ_e^2):

$$\sigma_e^2 = E([Z - Z^\star]^2)$$

with E = expected value.

The estimation variance can be derived from the semi-variogram:

$$2\gamma(h) = E(z_{(i)} - z_{(i+h)})^2$$

By rewriting this equation in terms of the semi-variogram, as described by Journel and Huijbregts (1978), the estimation variance can then be calculated.

Knudsen (1988) explained that this calculation takes into account the main factors affecting the reliability of an estimate. It measures the closeness of the samples to the grid node being estimated. As the samples get farther from the grid node, this term increases, and the magnitude of the estimation variance will increase. It also takes into account the size of the block being estimated. As the size of the block increases the estimation variance will decrease. In addition it measures the spatial relationship of the samples to each other. If the samples are clustered then the reliability of the estimate will decrease. The variability of the data also affects the reliability of the estimate, but since the continuity (C) is measured by the semi-variogram this factor is already taken into account. The estimation variance is a linear function of the weights (see "Kriging" above) established by kriging (Journel & Huijbregts 1978). Kriging determines the optimal set of weights that minimize the variance. These weighting factors are also used to estimate the value of a point or block. As it is rare to know the actual value (Z) for a point or block it is not possible to calculate $Z - Z^\star$ (c.f. cross-validation above).

In order to minimize the estimation variance, it is essential that only the nearest samples to the block being estimated are be used in the estimation process.

Kriging variance (σ_k^2)

Kriging variance is a function of the distance of the samples used to estimate the block value. A block, which is estimated from a near set of samples, will have a lower σ_k^2 than one that is estimated using samples some distance away. Once the kriging weights for each block have been calculated (see "Kriging" above) the variances can be calculated. Once σ_k^2 is determined it is possible to calculate the precision with which we know the various properties of the deposit we are investigating, by obtaining confidence limits (σ_e) for critical parameters. Knudsen and Kim (1978) have shown that the errors have a normal distribution. This allows the 95% confidence limit, ±2(σ_e), to be calculated (sections 8.3.1 & 10.1):

$$\sqrt{\sigma_e^2} = \sigma_e$$

95% confidence limit $= Z^\star \pm 2(\sigma_e)$.

Estimation variance and confidence limit are extremely useful for estimating the reliability of a series of block estimates or a set of contours (Whateley 1991). An acceptable confidence limit is set and areas that fall outside this parameter are considered unreliable and should have additional sample data collected. Thus the optimal location of additional holes is determined. The confidence limits for critical parameters such as Au grade or ash content of coal can be obtained. Areas which have unacceptable variations may require additional drilling, or in a mining situation it may lead to a change in the mine plan or a change in the storage and blending strategy.

10.4.4 Grade calculation methods

Conventional methods

Often borehole chips or cores, or channels, are sampled on a significantly smaller interval than a bench height in an open pit or a stope width in an underground mine. It is the geologist's responsibility to calculate a composite grade or quality value over a predetermined interval from the samples taken from the cores or channels. In an open pit the predetermined interval is often a bench height. In some instances where selective mining is to take place, such as in a coal deposit, the sample composites are calculated for coal and waste separately (see Chapter 13 for a more detailed description of sampling and weighted average quality calculations). In an underground vein mine, the composite usually represents the minimum mining width. Sometimes a thickness × grade accumulation is used to calculate a minimum mining width (e.g. section 14.6).

The general formula given above for calculating the weighted average thickness value for the $1/D^n$ block model, also applies when calculating weighted average grade or quality values. The weighting function changes though, and three variables, length, specific gravity (SG), and area are used. For example when compositing several core samples for a particular bench in a proposed open pit, length of sample is used. This is particularly important when the geologist sampling the core has done so using geological criteria to choose sample lengths and samples are of different lengths (e.g. Fig. 10.12). Thus:

$$\bar{G} = \frac{\sum L \times G}{\sum L}$$

where $\bar{G}$ = weighted average grade of each borehole on that bench, G = grade of each core sample, L = length of sample.

When samples are to be composited from, say, core, and the rocks are of significantly different densities, then the weighting factor should have bulk density (BD) included. Thus:

$$\bar{G} = \frac{\sum BD \times L \times G}{\sum BD \times L}$$

Area of influence can also be used as part of the weighting function. For example, to calculate the weighted average grade of a stratiform deposit which is to be assessed by polygons of different sizes, the area of each polygon is used as the weighting function. Thus:

$$\bar{G} = \frac{\sum A \times G}{\sum A}$$

where A = area of influence of each polygon.

Computational methods

In large ore bodies, or deposits which have a large amount of data, such as Trinity Silver Mine (see Chapter 16), computers are used to manipulate the data. Commercial software packages are available (e.g. MineSight, Datamine, see section 9.3) which can perform all the conventional methods described above. Computers offer speed and repeatability, as well as the chance to vary parameters to undertake sensitivity analyses. In addition computers can use exactly the same procedures as described for geostatistical mineral resource estimation methods. As well as using thickness to calculate volume and (with SG) tonnage, grade or quality can be estimated along with the estimation error.

10.4.5 Software simulation

Geostatistical tools may be used for spatial data analysis, to model the heterogeneity of mineral deposits, to assess the uncertainty (risk) in the estimation process, and to use this information in the decision process (Deutsch 2002). The fundamental problem exploration geologists have is to create a model from limited sample data. Methods to mitigate this, such as manual contouring or kriging, are described above. Once a project appears to have economic potential, engineers become involved in the evaluation process. They may wish to predict grade changes as the deposit is "mined" in a manner that mimics the proposed mining method. This can be achieved using conditional simulation.

Normally, sample points will support three estimated (kriged) points. Thus, on a 250 m drilling grid, block size would be 80 m. Mineralized rock with a density of 2.6 t m^{-3} in an 80×80 m block with a 15 m bench will provide 250 kt of material. This is usually too coarse to provide an estimate of local variability. This is a situation where conditional simulation can be used because it allows the estimator to generate values on a very fine grid. All calculations are performed using gaussian transformed (normalized) data (Deutsch 2002). The properties of the resultant simulations are such that they replicate the statistical and geostatistical properties of the input data, as well as honoring the input data. The simulated values are controlled, or conditioned, by drill-hole samples, which gives rise to the name of this technique (Deutsch & Journel 1998, Deutsch 2002).

In coal deposits, some coal quality variables, such as ash and calorific value, have large numbers of data points. There are often fewer sulfur or trace element data points (Whateley 2002). This creates estimation problems and difficulties in estimating life of mine (LOM) coal quality variability. This lack of data for some of the layers within a model presents problems to the estimator for the application of the method. This is particularly true for sulfur values within some plies when less than 30 samples of hard data are available for each layer. Adapting the conditional simulation technique, by using the more abundant, but less reliable whole seam data (soft data) to augment the ply data, can mitigate this problem. Conditional simulation models thus generated still need to be carefully examined and verified. Once they are found to satisfactorily replicate the geological, statistical (Fig. 10.24), and geostatistical characteristics of raw data, the simulated quality variability can be used for predicting LOM coal feed (Fig. 10.25).

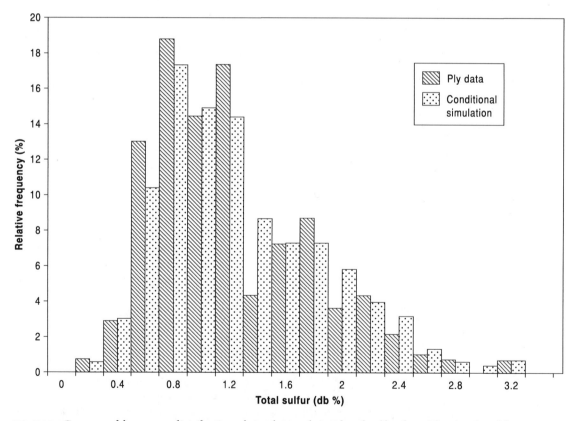

FIG. 10.24 Compound frequency distribution of raw data and simulated sulfur data. The simulated data are found to satisfactorily replicate the geological, statistical, and geostatistical characteristics of raw data.

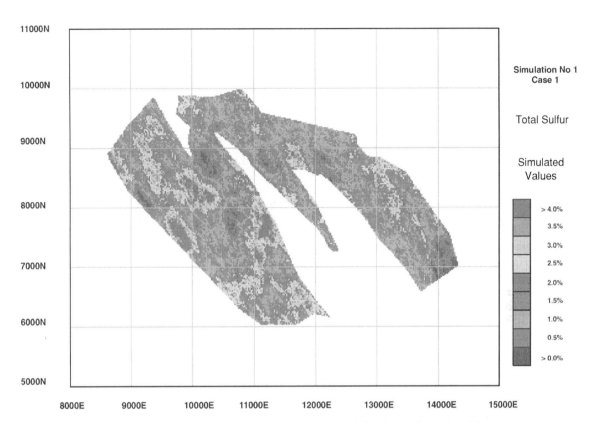

FIG. 10.25   Plot of the simulated sulfur variable, used for predicting life-of-mine (LOM) coal feed.

From 100 simulations the probability distribution of a grade variable can be derived. Prediction of grade variability is then possible.

10.5   GEOTECHNICAL DATA

During the sampling and evaluation of a mineral deposit, the exploration geologist is in an ideal position to initiate the geotechnical investigation of that deposit. Usually, geotechnical and mining engineers are invited on to the investigating team after the initial drilling is complete. They will need to know some of the properties of the rocks, such as strength, degree of weathering, and the nature of the discontinuities within the rock, in order to begin designing the rock-related components of the engineering scheme (haulages, underground chambers, pit slopes). It is important to investigate some of the geotechnical properties of the overburden and of the rocks in which the mineralisation is found during the exploration

phase of a project. The engineers must decide whether it is feasible to mine the potential ore, as this information is required for the financial analyses which may follow (see Chapter 11).

Geotechnical investigations essentially cover three aspects: soil, rock, and subsurface water. A good field guide for the description of these is given by West (1991). Soil refers to the uncemented material at or near the surface, and can include weathered or broken rock. Soil is mainly investigated for civil engineering purposes such as foundations of buildings. Rock is the solid material which is excavated during mining to reach and extract the ore. Subsurface water is part of the soil and the rock, and is found in the pore spaces and in joints, fractures and fissures.

There are two forms in which the data can be collected: (i) mainly subjective data obtained by logging core and the sides of excavations (Table 10.8), and (ii) quantitative data obtained by measuring rock and joint strengths and also virgin rock stress, etc., either in the laboratory

TABLE 10.8   A summary of the description and classification of rock masses for engineering purposes. (Adapted from Engineering Group Working Party 1977, Brown 1981.)

Descriptive indices of rock material
Consideration should be given to rock type, color, grain size, texture and fabric, weathering (condition), alteration, and strength

Rock condition (weathering)

Fresh	No visible signs of rock material weathering
Slightly weathered	Discoloration indicates weathering of rock material and discontinuity surfaces
Weathered	Half the rock material is decomposed or disaggregated. All rock material is discolored or stained
Highly weathered	More than half the rock material is decomposed or disaggregated. Discolored rock material is present as a discontinuous framework. Shows severe loss of strength and can be excavated with a hammer
Completely weathered	Most rock material is decomposed or disaggregated to a soil with only fragments of rock remaining

Rock material strength

Extremely strong compressive	Rock can only be chipped with geological hammer. Approximate unconfined strength >250 MPa
Very strong	Very hard rock which breaks with difficulty. Rock rings under hammer. Approximate unconfined compressive strenth 100–250 MPa
Strong	Hard rock. Hand specimen breaks with firm blows of a hammer, 50–100 MPa
Medium	Specimen cannot be scraped or peeled with a strong pocket knife, but can be fractured with a single firm blow of a hammer, 25–50 MPa
Weak	Soft rock which can be peeled by a pocket knife with difficulty, and indentations made up to 5 mm with sharp end of hammer, 5–25 MPa
Very weak	Very soft rock. Rock is friable, can be peeled with a knife and can be broken by finger pressure, 1–5.0 Mpa

Descriptive indices of discontinuities
Consideration should be given to type, number of discontinuities, location and orientation, frequency of spacing between discontinuities, separation or aperture of discontinuity surfaces, persistence and extent, infilling, and the nature of the surfaces

Rock quality designation

Sound RQD >100	Rock material has no joints or cracks. Size range >100 cm
Fissured RQD 90–100	Rock material has random joints. Cores break along these joints. Lightly broken. Size range 30–100 cm
Jointed RQD 50–90	Rock material consists of intact rock fragments separated from each other by joints. Broken. Size range 10–30 cm
Fractured RQD 1–50	Rock fragments separated by very close joints. Core lengths less than twice NX (Table 10.7) core diameter. Very broken. Size range 2.5–10 cm
Shattered RQD <1	Rock material is of gravel size or smaller. Extremely broken. Size range less than 2.5 cm

$$\text{RQD in per cent} = \frac{\text{Length of core in pieces 10 cm and longer}}{\text{Length of core run}} \times 100$$

RQD diagnostic description

>90%	Excellent
75–90	Good
50–75	Fair
25–50	Poor
<25%	Very poor

or *in situ*. Subjective data helps with the decision making fairly early in a project, but these data will always need quantification. It is therefore necessary to study and record the direction and properties of joints, cleavage, cleats, bedding fissures, and faults, and to study and record the mechanical properties (mechanics of deformation and fracture under load), petrology, and fabric of the rock between the discontinuities.

It is clear from the above that structural information is extremely important in mine planning. In the initial phases of a program most of the structural data come from core. The geologist can log fracture spacing, attitude, and fracture infill at the rig site. A subjective description of the quality of the core can be given using Rock Quality Designation (RQD) (Table 10.8). Pieces of core greater than 10 cm are measured and their length summed. This is divided by the total length of the core run and expressed as a percentage. A low percentage means a poor rock while a high percentage means a good quality rock.

Bieniawski (1976) developed a geomechanical classification scheme using five criteria: strength of the intact rock, RQD, joint spacing, conditions of the joints, and ground water conditions. He assigned a rating value to each of these variables, and by summing the values of the ratings determined for the individual properties he obtained an overall rock mass rating (RMR). Determination of the RMR of an unsupported excavation, for example, can be used to determine the stand up time of that excavation.

In order to provide information that will assist the geotechnical and mining engineers it is necessary to understand the stress–strain characteristics of the rocks. This is especially important in relation to the mining method that is proposed for the rocks, the way in which the rocks will respond in the long term to support (in underground mining), or in deep open pits how deformation characteristics affect the stability of a slope. As these rock properties become better understood, it is possible for the geotechnical and mining engineers to improve the design of underground mines, the way in which excavations are supported, the efficiency of mining machinery, and the safety of the mine environment.

10.5.1 Quantitative assessment of rock

There are numerous good textbooks which cover the properties of rocks from an engineering point of view (Krynine & Judd 1957, Goodman 1976, 1989, Farmer 1983, Brady & Brown 1985), the latter reference being particularly relevant to underground mining. The main properties which are used in the engineering classification of rocks for their quantitative assessment are porosity, permeability, specific gravity (bulk density), durability and slakability, sonic velocity, and rock strength.

Porosity and permeability are important in the assessment of subsurface water. Specific gravity is required to determine the mass of the rock. Durability and slakability tests reveal what affect alternate wetting and drying will have on surface or near-surface exposures. Under conditions of seasonal humidity some rocks have been known to disaggregate completely (Obert & Duval 1967). Sonic velocity tests give an indication of the fracture intensity of the rock. Probably the most important property of rock is its strength.

Rock strength

The fundamental parameters that define rock strength are stress and strain, and the relationship between stress and strain. Stress (σ) is a force acting on a unit area. It may be hydrostatic when the force is equal in all directions, compressional when the force is directed towards a plane, tensional when directed away from a plane, torsional (twisting), or shear stress (τ) when the forces are directed towards each other but not necessarily in line. Strain (ε) is the response of a material to stress by producing a deformation, i.e. a change in shape, length, or volume. The linear relationship between stress and strain is known as Hooke's law and the constant (E) connecting them is Young's modulus. It is also called the modulus of elasticity, and gives a measure of stiffness.

The strength of a rock sample can be tested (usually in the laboratory) in several ways. Unconfined (uniaxial) compressive strength is measured when a cylinder (usually core) of rock is loaded to a point of failure. The load immediately prior to failure indicates the rock's strength. Tensile strength is measured by

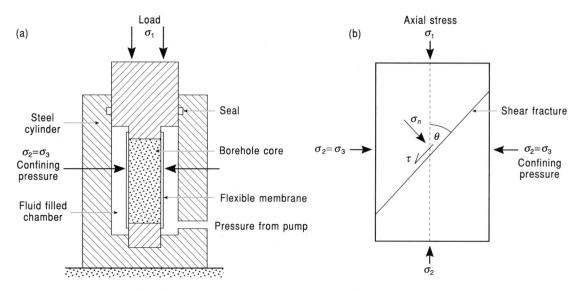

FIG. 10.26   (a) Schematic diagram of the elements of triaxial test equipment. (b) Diagram illustrating shear failure on a plane, an example of the results of the triaxial test on a piece of core. σ_1 is the major principal compressive stress, $\sigma_2 = \sigma_3 = $ intermediate and minor principal stresses respectively, σ_n is the normal stress perpendicular to the fracture, and τ is the shear stress parallel to the fracture.

"pulling" a cylinder of rock apart. The tensile strength of rock is generally less than one tenth of its compressive strength. Shear strength is measured as a component of failure in a triaxial compressive shear test. A cylinder of rock is placed inside a jacketed cylinder in a fluid-filled chamber. Lateral as well as vertical pressure is applied and the rock is tested to destruction (Fig. 10.26). The strength is that measured immediately prior to failure.

Failure

Observations on rocks in the field, in mines, and in open pits indicate that rocks which on some occasions appear strong and brittle may, under other circumstances, display plastic flow. Rocks which become highly loaded may yield to the point of fracture and collapse. A rock is said to be permanently strained when it has been deformed beyond its elastic limit.

With increasing load in a uniaxial (unconfined) compressive test, the slope of a stress–strain curve displays a linear elastic response (Fig. 10.27). Microfracturing within the rock occurs before the rock breaks, at which point (the yield point) the stress–strain curve starts

to show inelastic behavior followed by a rapid loss in load with increasing strain. If the load is released before the yield point is reached, the rock will return to its original form. Brittle rocks will shear and ductile rocks will flow, both types of behavior resulting in irreversible changes.

The results of the uniaxial (unconfined) compressive tests described above give a measure of rock strength and are convenient to make. The conditions that the rock was in prior to being removed by drilling (or mining) were somewhat different, because the rock had been surrounded (confined) by the rest of the rock mass. When a rock is confined its compressive strength increases. The triaxial compressive strength test gives a better measure of the effects of confining pressure at depth (Fig. 10.26). Core that has been subjected to triaxial testing, fractures in a characteristic way (Fig. 10.26b). Several failure criteria have been developed for rocks, the simplest being represented by Coulomb's shear strength criterion that can be developed on a fracture plane such as the one shown in Fig. 10.26b and is represented by:

$$\tau = \sigma_n \tan \theta + c$$

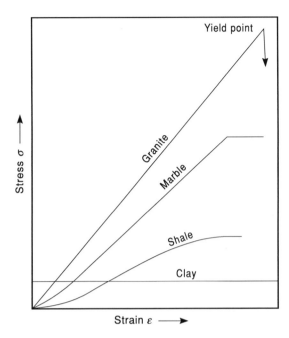

FIG. 10.27 A simplified stress–strain diagram illustrating the idealized deformational behavior of some rock types. That of granite represents an idealized perfectly elastic material. Clay is a typical material that behaves as a plastic material which will not deform until a certain stress is achieved. Marble is an example of an elastoplastic material and shale represents a ductile material where stress is not proportional to strain. (Modified after Peters 1987.)

where τ = total shearing resistance, σ_n = normal stress acting at right angles to the failure plane, c = cohesive strength, $\tan \theta$ = coefficient of internal friction where θ = angle between stress fracture induced and σ_1. The theory behind this equation is explained by Farmer (1983) and Brady and Brown (1985).

Triaxial test results are used to construct a Mohr diagram (Fig. 10.28). The symbol σ_3 represents lateral confining pressures in successive tests, while σ_1 represents the axial load required to break the rock. Circles are drawn using the distances between the different values of σ_3 and σ_1 as the diameter. The common tangent to these circles is known as the Coulomb strength envelope. The angle θ can be determined from the diagram, rather than having to measure the fracture angle on the

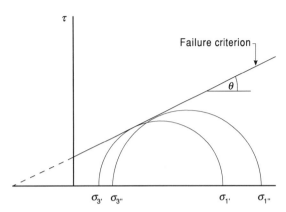

FIG. 10.28 An example of a Mohr diagram.

tested rock specimen. A value for c can also be determined from the diagram.

The expected confining pressures and axial loads that might be encountered in a mine can be calculated. Any sample whose test results plot as a Mohr diagram that falls below the tangent should not fail. A sample which failed in a test with lower confining stresses but similar axial stresses, would give rise to a circle which would cut the tangent suggesting that the critical stresses have been exceeded and failure might occur (Peters 1987). This does not take into account failure that might take place along discontinuities, nor does it consider the effects of anisotropic stress systems in the rock mass or changes that will occur with time.

The effect of discontinuities is probably more important than being able to measure the rock strength. The discontinuities may be weak and could result in failure at strengths below that of the rock itself.

In situ stress determination

Most *in situ* stress determinations are made before mining takes place usually from a borehole drilled from the surface or from an adit. Two common procedures are employed in the determination of *in situ* stress. The first is based on the measurement of deformation on the borehole wall induced by overcoring (Fig. 10.29) and the second by measuring the component of pressure in a borehole or slot needed to balance the *in situ* stress. In the overcoring system, a strain gauge is fixed to the borehole wall

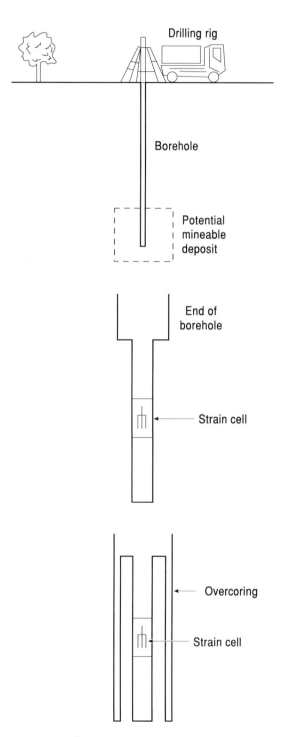

FIG. 10.29   A diagrammatic representation of the principle of overcoring.

using a suitable epoxy resin. Overcoring with a core barrel of larger size causes stress relief in the vicinity of the strain cell which induces changes on the strains registered in the gauges (Brady & Brown 1985). These measurements are used to help in mine design.

Additional tests are carried out during mining using various types of strain gauges. Some mines, e.g. Kidd Creek, have their own rock mechanics, backfill, and mine research departments whose rock mechanics instruments are regarded as an indispensable tool for detecting and predicting ground instability (Thiann 1983, Hannington & Barrie 1999). Hoek and Brown (1980) and Hoek and Bray (1977) give excellent examples of the use of rock mechanics in both underground and surface mining respectively.

10.5.2   Hydrogeology

During an exploration program it is important to note where the water table is encountered in a borehole because if one were to mine this deposit later, then it is essential to know where the water is, how much water there is, and the water pressure throughout the proposed mine. Useful tests that can be performed on boreholes include measuring the draw down of the water level by pumping, and upon ceasing pumping measuring the rate of recovery. Packer tests in boreholes are used to obtain a quantitative estimate of the contribution a particular bed or joint may make to the water inflow at a site (Price 1985). A good description of the basic fieldwork necessary to understand the hydrogeology of an area is given by Brassington (1988). Two properties of the rocks that can be measured are the porosity and the permeability. Porosity describes how much water a rock can hold in the voids between grains and in joints, etc. Permeability is a measure of how easily that water can flow through and out of the rock. The flow of water is described empirically by Darcy's law, and the relative ease of the flow is called the hydraulic conductivity. Darcy's law is a simple formula:

$$v = Ki$$

where v = the specific discharge or the velocity of laminar flow of water through a porous medium (the critical Reynold's number at

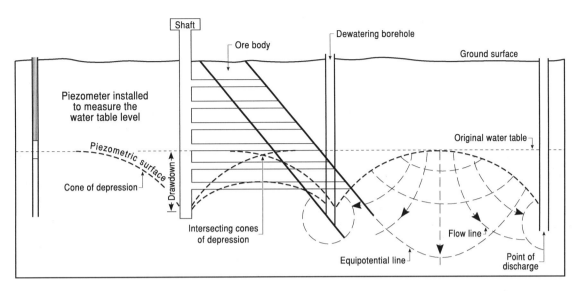

FIG. 10.30   Diagram to illustrate dewatering cones and a piezometric surface. The shaft and dewatering borehole are sited close enough to each other for their cones of depression to intersect and produce a lowering of the water table which effectively results in the dewatering of part of the mine. The flow net illustrates the flow of underground water through rocks of uniform permeability. Flow takes place towards the points of discharge, which in this case are pumps at the bottom of the shaft and dewatering boreholes. (Modified after McLean & Gribble 1985.)

which turbulence development is seldom reached), $K =$ hydraulic conductivity, $i =$ hydraulic gradient (the rate at which the head of water changes laterally with distance).

The hydraulic conductivity can be measured in boreholes by undertaking pumping tests. The head of water at a point (e.g. a borehole) within a rock mass is the level to which the water rises when the borehole is drilled. The tube inserted in a borehole to measure this water level is called a piezometer. Several boreholes may intersect points on a piezometric surface (Fig. 10.30), and the maximum tilt (i) of this surface from the horizontal defines the hydraulic gradient (McLean & Gribble 1985).

Groundwater moves slowly under the influence of gravity from areas of high water table to low areas (Fig. 10.30). The pumps lower the piezometric surface in the proximity of the borehole creating a difference in pressure, with the pressure much less at the point of discharge (the pumps) than in the areas between the pumps. This results in the water flowing in curved paths towards the pumps.

Rocks which are permeable and allow water to flow freely through the joints, fractures, and pores are known as aquifers. Rocks which have a low permeability and are impervious and do not allow water to flow through them are known as aquicludes.

Having found an orebody, finding water for the processing plant is often the next concern. A plant requires a considerable volume of water and it can happen that the amount of water available may make or break a project.

10.6   SUMMARY

Mineralisation is composed of dissimilar constituents, is heterogenous, and much mineralisation data may have an asymmetrical rather than a normal distribution. Simple statistical sampling methods used for investigating homogenous populations may therefore give a highly misleading result if used to interpret mineralisation data where in many cases a sample can represent only a minuscule proportion of the whole. The pitfalls and problems that this may cause have been discussed and emphasis placed on the need to pay careful attention to sampling error.

Sample extraction must be random and preferably of equal volume, but as the samples may amount to several kilograms or even tonnes in weight there must be a systematic reduction of mass and grain size before analysis can take place. Sample reducing systems require thoughtful design and must be carefully cleaned between each operation. Gold- and platinum-bearing samples present special problems because their high density may lead to their partial segregation from lighter gangue material during sample preparation.

Samples may be acquired by pitting and trenching and by various types of drilling. If coring methods are used then effective core recovery is essential and the core logging must be most methodically performed. Drilling contracts should be carefully thought out and based on sound knowledge of the drilling conditions in the area concerned. Drillholes deviate from a straight line and must be surveyed. Drilling patterns must be carefully planned and the geologist in charge ready to modify them as results come to hand.

The analytical data derived from the above investigations are used in mineral resource and ore reserve estimations, and for grade calculations. Classical statistical methods have only a limited application and geostatistical methods are normally used. The exploration sample collection work provides a great opportunity to initiate geotechnical investigations. The collection of geotechnical data will provide mining engineers with essential information for the design of open pit and underground workings. It will also provide a starting point from which the hydrogeology of the target area can be elucidated.

10.7   FURTHER READING

Two textbooks which cover statistical analysis of geological data are the eminently readable Davis (2003) and Swan and Sandilands (1995), while Isaaks and Srivastava (1989) give a very readable account of geostatistics. The mathematical details of geostatistical estimations are elaborated by Journel and Huijbregts (1978).

Annels (1991) covers this subject as well as drilling, sampling, and resource estimation. Discussions of the problems of resource definition are covered by the Australasian Joint Ore Reserves Committee (the JORC code: JORC 2003), Annels (1991), and Whateley and Harvey (1994). The JORC website hosts a number of technical papers that explain the JORC code, its history and implementation. A very good source for case histories is the volume of Edwards (2001).

An introductory text for geotechnical investigations is written by West (1991), while Hoek and Brown (1980) and Hoek and Bray (1977) are still the most sought after books that describe the application of rock mechanics to underground and surface mines respectively. Brassington (1988) gives a good description of the basic fieldwork necessary to understand the hydrogeology of an area.

10.8   APPENDIX: USE OF GY'S SAMPLING FORMULA

1 Introduction

These are worked examples of the calculation of the total variance (*TE*) of sphalerite and chalcopyrite mineralisation and the two-sided confidence of that mineralisation. These are followed by a worked example of how to calculate the mass of a sample to be taken knowing the confidence limits.

Using the short form of the Gy formula for the fundamental variance (section 10.1.4):

$$S_o^2(FE) = C/K$$

Taking K as 125,000:

$$S_o^2(FE) = C/125,000$$

and the variance of the total error (*TE*) is estimated as twice that of the fundamental error (*FE*). It is assumed that the liberation factor β is constant and equals unity, which is a safe assumption.

2 Calculation of total variance (*TE*) and double-sided confidence interval: sphalerite.

A prospect contains sphalerite mineralisation of mean grade 7.0% zinc in a gangue of density $2.7 \, \text{kg m}^{-3}$.

Sphalerite contains 67% zinc with a specific gravity of 4.0 kg m^{-3}.

(a) Fundamental variance of each sampling stage (see also Box 10.1). Critical content (a_L) = 7% zinc = 10.5% sphalerite as a fraction = 0.105.

Constitution factor (c):

$$= \frac{(1 - 0.105)}{0.105}[(1 - 0.105)4.0 + (0.105)2.7]$$

$$= 8.52(3.58 + 0.28) = 32.89, \text{ say } 33.$$

Heterogeneity constant (C)

$$= c\beta fg = 33 \times 1 \times 0.5 \times 0.5 = 4.13.$$

Fundamental variance:

$$= \frac{C}{K} = 24.38/125,000 = 33 \times 10^{-6}$$

(b) Total variance (TE) of the sampling scheme. At each reduction stage the total variance is equal to double the fundamental variance. The complete sampling scheme consists of four reduction stages (Fig. 10.5a) each ending on the safety line (see text). Consequently, the total variance equals eight times (2×4) the fundamental variance as calculated above.

Total variance $(TE) = 8 \times (33 \times 10^{-6}) = 264 \times 10^{-6}$.

$$\text{Relative standard deviation} = \sqrt{(264 \times 10^{-6})}$$
$$= 16.2 \times 10^{-6}$$

Absolute standard deviation as % of mineral = 10.5% sphalerite $\times$ 1.6% = 0.17% sphalerite. Absolute standard deviation as % of metal = 0.17% sphalerite $\times$ 67% = 0.11% zinc. The confidence level at 95% probability is $\pm$2 SD = $\pm$2(0.11%) = 0.22% zinc. The two-sided confidence level is then **7% ± 0.22 zinc** at 95% probability, which is an acceptable result. Generally, a relative standard deviation of <5% is acceptable.

3 Calculation of total variance (TE) and double-sided confidence interval: chalcopyrite mineralisation.

A prospect contains chalcopyrite mineralisation of grade 0.7% copper $(\bar{x})$ in gangue of density 2.7 kg m^{-3}.

Chalcopyrite contains 33% copper with a density of 4.2 kg m^{-3}.

(a) Fundamental variance of each sampling stage.

Critical content (a_L) = 0.7% = 2.1% chalcopyrite as a fraction = 0.021.

Constitution factor (c)

$$= \frac{1 - 0.021}{0.021}[(1 - 0.105)4.0 + (0.105)2.7]$$

$$= 46.62(4.11 + 0.06) = 194.4, \text{ say } 195.$$

Heterogeneity constant $(C) = c\beta fg = 195 \times 1 \times 0.5 \times 0.25 = 24.38$.

Fundamental variance:

$$= \frac{C}{K} = 24.38/125,000 = 195 \times 10^{-3}.$$

(b) Total variance (TE) of the sampling scheme. As above but, suppose, the system consists of five reduction stages each ending on the safety line. The total error variance now equals 10 times (2×5) the fundamental variance.

$$\text{Total variance } (TE) = 10 \times (195 \times 10^{-6})$$
$$= 1950 \times 10^{-6}$$

Relative standard deviation = $(1950 \times 10^{-6})^{1/2}$ = 44×10^{-3} = 4.4%.

Absolute standard deviation as % of mineral = 2.1% chalcopyrite $\times$ 4.4% = 0.09% chalcopyrite.

Absolute standard deviation as % of metal = 0.09% $\times$ 33% = 0.03% copper.

The confidence at 95% probability is $\pm$2 SD = $\pm$2(0.03%) = 0.06% copper.

The two-sided confidence level is then **0.7% ± 0.06%** copper at 95% probability, which is an acceptable result. Generally a relative standard deviation of <5% is acceptable.

4 Calculation of the weight of a sample (M) to be taken knowing the confidence interval.

(a) Introduction (see section 10.1.4)

$$M = Cd^3/S^2(FE)$$

(b) Sphalerite mineralisation.

Sphalerite mineralisation assays 7% zinc and the confidence level required is ± 0.2% zinc at

95% probability. The sphalerite is liberated from its gangue at a particle size of 150 µm: the gangue density is 2.7 kg m^{-3}. Sampling is undertaken during crushing when the top particle size (d) is 25 mm.

Critical content (a_t) = 7% zinc = 10.5% sphalerite = 0.105 as a fraction.

Constitution factor (c) as in previous example = 33.

Liberation factor (β) = (0.015/2.5)0.5 = (0.006)0.5 = 0.077.

Heterogeneity constant (C) = $c\beta fg$ = 33 × 0.077 × 0.5 × 0.25 = 0.32.

Top size (d) = (2.5)^3 = 15.62.

Cd^3 = 0.32 × 15.62 = 5.0.

Fundamental variance (FE).

Relative standard deviation $2S = \dfrac{0.2\%}{7}$. Therefore $S = 0.014$.

Mass of sample = $\dfrac{Cd^3}{S^2(FE)} = \dfrac{5.0}{(0.014)^2} = 25.5$ kg.

This is the sample weight to constrain the fundamental variance. For the total variance a sample weight of twice this quantity is required [i.e. $S^2(TE) \leq 2S^2(FE)$], or **51 kg**.

To illustrate the benefits of sampling material when it is in its most finely divided state, assume that the same limit (±0.2% zinc) is required when sampling aftergrinding to its liberation size when d is 0.015 cm and β equals 1.0:

$$C = 33 \times 1 \times 0.5 \times 0.25 = 4.13$$

$$d^3 = (0.015)^3 = 3.3 \times 10^{-6}$$

$$Cd^3 = 13.63 \times 10^{-6} \text{ and}$$

M = 6.9 grams, say 7 g

11

PROJECT EVALUATION

BARRY C. SCOTT AND MICHAEL K.G. WHATELEY

11.1 INTRODUCTION

It is said that mineralisation is found, orebodies are defined, and mines are made. Previous chapters have outlined how mineralized rock is located by a correct application of geology, geophysics, and geochemistry. Having located such mineralisation its definition follows and this leads to a calculation of mineral resources where a grade and tonnage is demonstrated at an appropriate level of accuracy and precision (see Chapters 3 and 10). The question now becomes how does a volume of mineralized rock become designated as a resource and by what process is this resource selected to become an ore reserve (see section 10.4) for a producing mine? What steps need to be taken to ensure that there is a logical progression that moves the project with the initial unconnected exploration data to a final reserve estimate that meets the requirements of potential investors and bankers? A mine will come into existence if it produces and sells something of value. What value does a certain tonnage of mineralized rock have? What is meant by value in this context? How is this value calculated and evaluated? These questions are considered in this chapter.

11.2 VALUE OF MINERALISATION

Legislation in a number of countries now requires directors of listed companies to comply with a corporate governance code. In the UK, the Turnbull report (Turnbull 1999) recommends a systematic approach to the identifica-tion, evaluation, and management of signific-ant risks to a company. In a mining company this requirement includes technical issues such as determination of the ore reserves, selection of appropriate mining and processing methods, as well as financial, social, environmental, and reclamation aspects of the project. Similarly there is a range of business issues to be add-ressed, such as the project's ownership struc-ture, permitting, marketing, and government relations and their interactions. The value of mineralisation is a function of these aspects and several other factors, which are listed below.

11.2.1 Mine life and production rate

Mine life

Other things being equal the greater the ton-nage of mineable mineralisation the longer the mine life and the greater its value. It follows that:

$$\text{Mine life in years} = \frac{\text{total ore reserve}}{\text{average annual production}}$$

Factors relating to mine life include the following:

Legal limitations
Exploitation of the mineralisation may be under the constraints of a lease granted by its former owner such as the government of the state concerned. If the lease has an expiry date then the probability of obtaining an extension has to be seriously considered.

Market forecast
What will be the future demand and value for the minerals produced? This important factor is considered separately below.

Political forecast
If a lease agreement has been negotiated with the local government how long is this government likely to remain in power? Will a new government wish to either cancel or modify this agreement? (See also section 1.5.4.)

Financial return
The development of a mine is a form of capital investment. The main concern is for this investment to produce a suitable level of return. The optimization of this financial return will probably decide the number of years of production. As a very general rule the minimum life of a mine should be 7–10 years.

Production rate

A commonly used guide for the Average Annual Production Rate (AAPR) is:

$$AAPR = \frac{(ore\ reserve)^{0.75}}{6.5}$$

This is a broad generalization but for ore reserves of 20 Mt, the AAPR is 1.5 Mt a^{-1}, and for 10 Mt it is 0.9 Mt a^{-1}. Factors relevant to the annual production rate include the following:

Type of mining
The two basic types of mining are either underground (Fig. 11.1) or from the surface (open pit), (Fig. 11.2). As production from the former can be more than three times the cost of the latter, underground extraction may require at least three times the content of valuable components in the mineralisation to meet this higher cost. This in turn implies a lower tonnage, but a higher grade of ore reserve and a lower annual production rate.

A basic characteristic of open pit mining is the surface removal of large tonnages of waste rock (overburden) in order to expose the ore for extraction (Fig. 11.2). This removal of waste rock and ore usually proceeds simultaneously and the resultant stripping ratio of tonnage of waste removed to tonnage of ore is a funda-

mental economic specification. With a stripping ratio of 4:1, 4 t of waste has to be removed for every 1 t of ore recovered; if the mining costs were $US 2 t^{-1} then the extraction cost per tonne of ore is $US10 t^{-1}. Conversely, if, say, the stripping ratio were 8:1 at the same mining cost then the extraction cost per tonne of ore would be $US 18 t^{-1}.

1 Open pit mining costs vary from about $US1 to $US3 t^{-1} of rock extracted. An economy of scale is apparent in that the higher charge is for production levels of lower than 10,000 tonnes per day (tpd), while the lower cost refers to outputs exceeding 40,000 tpd of ore plus waste rock.

2 Underground mining is generally on a smaller scale than open pit mining. An output of 10,000 tpd would be modest for an open pit but large for an underground mine. This generally lower level of production is in part related to the more selective nature of underground extraction in that large tonnages of waste rock do not have to be removed in order to expose and extract ore. This is exemplified in underground block caving. Northparkes Cu/Au mine in Australia is one of the most productive underground mines in the world, producing 5 Mt of ore per year (Chadwick 2002). Block caving is designed as a low maintenance working environment with minimal operating costs and high productivities.

3 Underground mining costs vary greatly depending upon the mining method employed. In large orebodies where production is mechanized, costs per tonne of ore extracted can be comparable with those of an open pit. At the other end of the scale, in narrow veins using labor-intensive methods, costs can be as high as $US30–40 t^{-1}.

4 Generally underground mining costs are higher than those at the surface. However, as the depth of an open pit increases, the ratio of waste to ore mined becomes greater and more waste rock per tonne of ore is extracted (Fig. 11.2). This has the effect of increasing the overall extraction cost until eventually it can equal underground mining costs.

Scale of operations
The economies of scale mean that usually the larger the annual output the lower the production cost per tonne of mineralized rock mined.

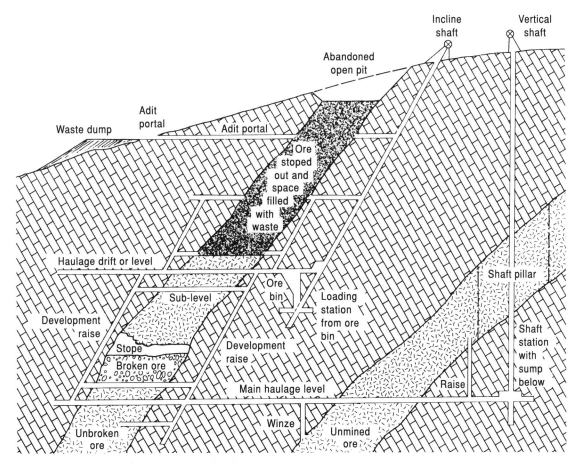

FIG. 11.1 Mining terminology. Ore was first extracted at the mineralized outcrop using an open pit. Later an adit was driven into the hillside to intersect the ore at a lower level. An inclined shaft was sunk to mine from still deeper levels and, eventually, a vertical shaft was developed. Ore is developed underground by driving main haulage and access drifts at various levels and connecting them by raises from which sublevels are developed. It is mined from the lower sublevel by extracting the mineralisation upwards to form a stope. Broken ore can be left in a stope to form a working platform for the miners and to support the walls against collapse (shrinkage stoping). Alternatively, it can be withdrawn and the stope left as a void (open stoping) or the resultant space can be filled with fine-grained waste rock material brought from surface (cut-and-fill stoping). Ore between the main haulage and sublevel is left as a remnant to support the former access until the level is abandoned. (From Barnes, 1988, *Ores and Minerals*, Open University Press, with permission.)

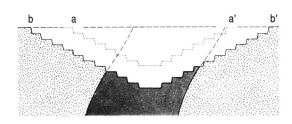

FIG. 11.2 Development of an open pit mine. During the early stages of development (a–a') more ore is removed than waste rock and the waste to ore ratio is 0.7:1. That is, for every tonne of ore removed 0.7 tonnes of waste has also to be taken. The waste rock is extracted in order to maintain a stable slope on the sides of the open pit, normally between 30 and 45 degrees. Obviously the greater the slope the less the waste rock to be taken, and the lower the waste to ore ratio. As the pit becomes deeper the ratio will increase; at stage b–b' it is 1.6:1. (From Barnes 1988, *Ores and Minerals*, Open University Press, with permission.)

Obviously there is an upper limit to this general principle otherwise most mines would be worked out in their first year of production!

Accessibility of the mineralisation
How accessible is the mineralisation? There are limitations on open pit and shaft depths, and shaft size, which restrict output. What type of mining extraction is to be followed? Underground methods vary considerably in their nature and scope depending upon the geology of the mineralisation and the host rocks. Some methods are suitable for large-scale production in wide zones of mineralisation whilst others are designed for small workings in narrow veins.

Capital expenditure limitations
There may well be restrictions on the availability of capital for developing and equipping the mine. Consequently, it may not be possible to achieve the maximum desirable rate of annual production despite the presence of adequate and accessible ore reserves.

11.2.2   Quality of mineralisation

Grade is but one facet of quality, albeit an important one. Other factors such as mineralogy and grain size enter into the definition (see section 2.2). Commodities like coal, iron ore, bulk materials, etc., can often be sold as a run-of-mine (ROM) product that does not require processing. However, most base and precious metal ores usually require some method of processing to separate waste rock and gangue from the valuable constituents (Wills 1997). Concentration plants vary from low cost (separating sand from gravel) to high cost, complex plants as in the separation of lead and zinc sulfides (Fig. 11.3).

11.2.3   Location of the mineralisation

The valuable products have to be taken to market for sale, and supplies for the mining operation follow the same route in reverse. Local availability of power, water supply, and skilled labor must be considered, and local housing, educational and recreational facilities for the workforce. Clearly mineralisation adjacent to existing facilities has greater value than that in a remote, inhospitable location.

11.2.4   Market conditions

The sale of minerals extracted from a mine is usually its *only* income. From this revenue the mining company has to pay back the capital borrowed to develop the mine and pay the related interest, production and processing costs, transport of the product to market, product marketing, dividends to shareholders, and local and national taxes (Fig. 11.4).

Can the product be sold and at what price? It cannot be assumed that demand is growing sufficiently fast to absorb all feasible new production. There has to be a reasonable assessment of future demand and price over the life of the mine, or at least its first 10 years, whichever is the shorter. The product price (see section 1.2.3) is the most important single variable in a feasibility study and yet the most difficult to predict. Historic price trends (Kelly et al. 2001, LME 2003) over the last 20 years show a declining market price for most base metals and energy materials (Fig. 11.5; see Figs 1.4–1.6, 1.8).

Many mineral products are sold on commodity exchanges, in London and New York, where a price is set each working day and considerable fluctuations occur (Figs 1.4–1.6). Most mine owners have little or no control over these prices, which are vital for the well being of producing mines. Market principles suggest that during a period of low prices production costs should be minimized and revenue maximized. One way of achieving this would be to adjust the cut-off grade to extract the maximum possible tonnage of easily accessible mineralisation at the highest grade. Alternatively during periods of higher price, lower grade material could be extracted.

Present day mineral and metal prices are available in publications such as the *Mining Journal*, *Engineering and Mining Journal*, and *Industrial Minerals*, and websites such as the London Metal Exchange (LME 2003) which summarize average weekly and monthly commodity prices, as well as providing annual reviews.

11.2.5   Economic climate

The prediction of general business sentiment, the demand for products, and inflation and

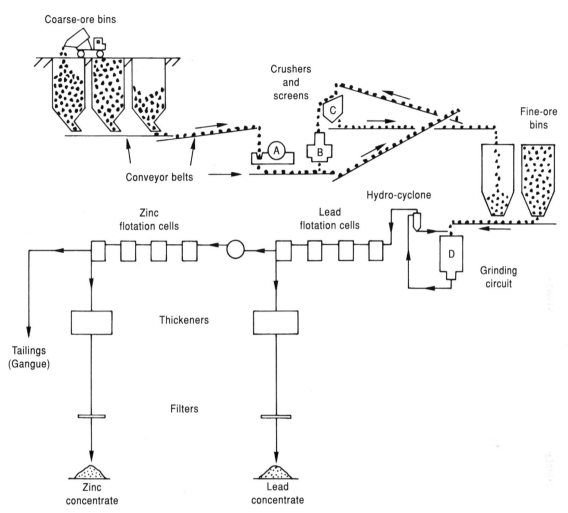

FIG. 11.3   Separation of lead and zinc sulfides. The ore has an average grade of 9.3% zinc and 2.0% lead, as sphalerite and galena respectively. At the main site the broken ore at 15 cm size is crushed via a primary (A) and a secondary crusher (B) to less than 40 mm size. The former crusher is in open circuit in that no material is returned for re-crushing. The secondary, finer, crusher is in a closed circuit and the crushed rock passes over screens (C) on which only material less than 40 mm size passes through to the fine-ore bins. Oversize is returned to the crusher B. The crushed ore is then ground, in water, to less than 0.075 mm, which is the effective liberation size of the contained sulfides, in the grinding circuit D. This is in a closed circuit with hydrocyclones, which separate oversize material (>0.075 mm) and return it for further grinding and size reduction, in D. The undersize is now a finely ground powder suspended in water and forms a slurry (about 40% solids) which can be pumped for treatment in the flotation units, or cells. In flotation the surface properties of the liberated sulfides are used whereby air bubbles become attached to individual particles of sphalerite which float to the surface of the cells and are skimmed off as a froth, whilst all other minerals and rock particles sink. Each sulfide has different surface properties depending upon the pH of the slurry and added chemicals. In this way sulfides are separated from their gangue, and from each other. A sphalerite concentrate is produced with, say, 54% zinc, and galena concentrate with 65% lead. The separation process is reasonably efficient and recovers 91% of the contained zinc values but only 77% of the lead. The waste rock, or tailings in the form of a fine slurry, contains about 0.2% zinc and 0.9% lead. (From Evans 1993 and Wills 1988.)

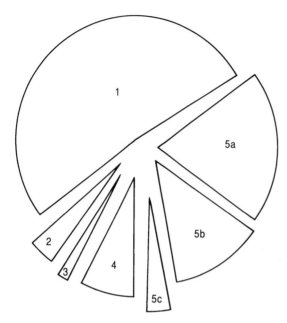

DISTRIBUTION OF INCOME

1 Wages, salaries and benefits to employees
2 Taxes and imposts to governments
3 Interest to providers of loan capital
4 Dividends to shareholders
5 Reinvested in R.G.C. Group
   a) depreciation
   b) exploration and evaluation
   c) retained earnings

FIG. 11.4   Distribution of cash flow in 1984 by
Renison Goldfields Consolidated Ltd. (After
*Consolidated Gold Fields PLC Annual Report
1985*.)

exchange rates over the life of the mine are crucial. Apart from rising equipment and labor costs, inflation and exchange rates may have disastrous effects if they give rise to higher interest rates on capital borrowed to develop the mine or on the value of local currency.

As far as mining companies are concerned economic cycles are doubly important because in times of recession, and inflation, they suffer falling mineral prices, rising interest rates on loans, but operating costs continue to rise. In the late 1980s the world economy entered a period of recession. This led to a dramatic fall in the copper price (Fig. 11.5) to a low of $US0.74 lb^{-1} in November 1993. Many producers incurred heavy financial losses on their operations. The reverse, however, happened in the mid 1990s when booming economies took the copper prices to $US1.47 lb^{-1} in July 1995. Conversely, in the late 1990s and early 2000s, a new recession commenced with depressed mineral prices and low profitability – in November 2001 the copper price fell to $US0.60 lb^{-1}, although it has risen sharply since then with sharp increase in demand from China. The overall long-term trend is downward (Fig. 11.5), which means that producers must always look for ways to reduce their costs.

11.2.6   Political stability of host country

Fundamentally a company developing a mineral property wishes to have its invested capital

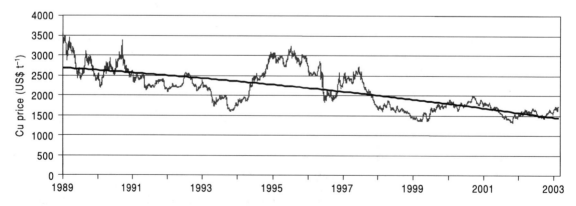

FIG. 11.5   Copper price trends on the London Metal Exchange, 1989–2003. Compare with long term trends corrected for inflation in Fig. 1.5. (After LME 2003.)

returned plus a healthy profit otherwise the investment will be taken elsewhere. Other necessary and associated objectives include safety and health of employees, efficient mineral extraction, care for the environment, and payment of local and national taxes. The government of the host country may have a different order of priorities, such as continuity of employment, social objectives, and taxes to central government for national social policies. It gives a low order of priority to dividends to shareholders and repayment of loans that were used to develop the mine. In addition an inflow of foreign investment is sometimes disliked with the thought of foreign shareholders controlling a national resource of the country. Thus it is essential to operate within a framework of acceptable law and, if necessary, a mutually agreed contract between developer and host. This may require delicate negotiation and it has to produce a true partnership otherwise the agreement will not last, particularly if there is a change of government or leadership. There are known deposits with measured resources that have no value because of the impossibility of achieving such a satisfactory agreement (see also section 1.4.4). Investment in mineral development almost ceased in Zimbabwe in the late 1990s and early 2000s because of Government policies relating to ownership, inflation, and fiscal policy.

11.2.7 Sustainable development, health and safety factors

The circumstances under which mining developments are viewed and judged have changed over the last few years. Government endorsement is no longer the only external factor in the approval process. Assessment of the social, environmental, health, and safety implications of project development now form an integral part of all applications in the mining lease approval process. Potential concerns of local communities as well as other interested parties such as non-governmental organizations (NGOs), shareholders, and investors can have significant influence on a project's viability as well as its design and scope (see section 4.3).

The long-term implications for sustainable development must also be considered. Sustainable development is development that meets the needs of the present without compromising the ability of future generations to meet their own needs (Brundtland 1987). It involves integrating economic activity with environmental integrity and social concerns (MMSD 2002). It is here that the mining industry has the opportunity to enhance the contribution that mining and metals can make to social and economic development (Walker & Howard 2002).

Frequently, Environmental Impact Assessments (EIA) are a requirement for permitting or regulatory approval with public consultation an underlying condition (Environment Agency 2002). It is at this time that concern to the NGOs can be addressed. The range of biophysical impacts to be considered may be quite exhaustive and include effects on air and water quality, flora and fauna, noise levels, climate, and hydrological systems. The reporting format can be quite prescriptive, demanding careful attention to content, procedures, and protocol (see section 17.2).

Typically, an EIA would explain how the company will mitigate issues such as blasting vibrations, dust, noise, atmospheric and water pollution, and increased traffic density. With modern equipment and design most, if not all, of these can be overcome but at a cost. This cost is, in effect, an added operating charge which may be one of the factors that has an influence upon the value of a potential mine (see also section 1.4.2).

In comparison with environmental concerns, legal requirements for consideration of social impacts may be limited to general demographic and economic concerns. The understanding of social risk is relatively new. Appropriate consideration needs to be given to the social elements. This could include changes that a proposed development would create in social relationships, community, people's quality and way of life, ritual, political/economic processes, attitudes, and values. Normally this is combined with the EIA in the form of a Social and Environmental Impact Assessment (S&EIA) report. It would include the sustainable development, health and safety aspects of a project. They also have to be managed to ensure that new projects and expansions can take place effectively. They provide an essential contribution throughout the mining cycle from the conceptual study to well beyond mine closure. They encompass both human and bio-physical considerations and they must be fully integrated with engineering, financial and other aspects of a project. These reports also have to consider the closure costs. Closure planning is now considered early in a mine life and is one of the factors that has a bearing on the value of the mineralisation. Decommissioning, reha-bilitation and sustainable development all contribute to closure costs.

11.2.8   Government controls

Any potential mine will have to operate within a national mining law and a fiscal policy enforced by an inspectorate. Fiscal policy is usually of prime importance for it decides items such as taxation of profits, import duties on equipment, and transfer of dividends and capital abroad if foreign loans and investors are involved. If this transfer cannot be guaranteed then the value of the proposed mine, and the mineralisation, will be considerably less (see also section 1.5.3).

11.2.9   Trade union policy

It is a foolish investor who does not consult the local trade unions before starting a new mining project. Union disputes can stop a mine as effectively as machinery breakdowns and con-tinual strikes and disputes over working prac-tices can make a mine unprofitable. In other words industrial disputes can change ore to mineralized rock of no value. In some countries trade unions are, in effect, a branch of the gov-ernment and if a contract is agreed between host and developer this will include a trade union agreement.

11.2.10   Value of mineralisation – a summary

Mineralisation has value if a saleable product can be produced from it. "Value" is a financial concept and is related to the several factors discussed above rather than to just grade and tonnage. The main factors which enhance the value of mineralisation are an increase in tonnage, grade, mineral recovery, product sales price, and political stability, together with decreases in costs of mining, ore beneficiation, transport to market, capital, and taxes.

Of course not all these factors are under the direct control of a mining company wishing to develop mineralisation into ore reserves. Changes in the way in which global business operates, and the impact of ever more sophist-icated and immediate means of communica-tions, have brought large mining companies under increasing scrutiny by, and pressure from, a wide range of interests. The demand is for these companies to demonstrate, through action, that they understand their responsibil-ity towards the communities in which their businesses are or will be located. As a result, issues that integrate economic activity with environmental integrity and social concerns have become central to the planning and man-agement of sustainable development good practice (MMSD 2002).

11.3   CASH FLOW

The factors in section 11.2 can be expressed in terms of revenue and cost (income and expend-iture) of a proposed mining operation. All rel-evant items are brought together in an annual summary which is referred to as a *cash flow*. In any particular mineral project:

Cash flow

= cash into the project (revenue) minus cash leaving the project (cost)

= (revenue) – (mining cost + ore beneficiation costs + transport cost + sales cost + capital costs + interest payments + taxes)

Cash flow (Fig. 11.6 & Box 11.1) is calculated on a yearly basis over either a 10-year period or the expected life of the mine, whichever is the shorter. It is a financial model of all relevant factors considered in section 11.2. It is the pro-cess by which the value of a tonne of mineral-ized rock is calculated to determine whether this tonnage is ore. Such mineralisation only has value by virtue of its ability to produce a series of annual positive cash flows (revenue exceeding expenditure) over a term of years. Another way of expressing this is to say that the value of a tonnage of mineralized rock is the value in today's terms of all future annual

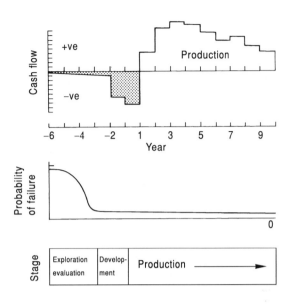

FIG. 11.6   The cash flow of a mineral project over 15 years. During the first 4 years the mineralisation is evaluated and over the next 2 years the mine is built at a total capital cost of $US250M. Over these 6 years the cash flow is negative in that there is no revenue generated. In the first year of production (Year 1) revenue commences and continues during the life of the mine. From this revenue is paid all the mine operating costs and repayment of the capital loans, plus interest, which were used to develop the mining operation. The lenders of capital require their money back within the first 3–5 years of production. Also shown is the probability of failure of the project, and its main stages. Obviously the size of the database increases with time and the probability of failure decreases correspondingly.

BOX 11.1   Cash flow calculation.

From the results of an order of magnitude feasibility study on an occurrence of sphalerite and galena mineralisation, the first 3 years of a 10-year cash flow are given below. The plan is to mine the mineralisation underground and produce galena and sphalerite concentrates at the main site. These would be sold to a lead and zinc smelter for metal production.

Calculations are for year end; interest payments on borrowed capital to build the project are excluded. Units are $US1000 and 1000 tonnes.

(a) Total initial capital (CAPEX) to establish the mine is $140,600. Construction will take 2 years: years −1 and −2.
(b) Ore production is 2400 t yr^{-1} but only 1600 t yr^{-1} in the first year of production (year 1). Ore grade is 9.3% zinc and 2.0% lead.
(c) Ore treatment is as in Fig. 11.3.
(d) Revenue: the selling price of zinc is taken as $0.43 lb^{-1} and for lead as $0.21 lb^{-1}. Translating this value to ore gives a revenue of $46.6 per tonne.
(e) Operating costs per tonne:

Mining	$7.25
Mineral processing	8.10
Overheads	4.35
	$19.70 per tonne ore

The cost includes transporting the sphalerite and galena concentrates to a local port, and loading charges. The concentrates are sold to the smelter at the port and it is then the smelter company's responsibility to ship the concentrates to their plant.
(f) There is a royalty payment to the previous mineral owners of 4.5% of gross revenue less production cost.
(g) Calculation f is made before the payment of tax.

cash flows arising from mine production. How these values are determined is discussed in section 11.5.

The basic data for the financial model, or cash flow, is collected in a series of studies, which are considered next. Some geologists may reject this involvement with finance. However, it has to be remembered that while all acknowledge the expertise of mine making (i.e. transposing mineralized rock into ore) which lies with geologists and other engineers, the ultimate success or failure of a project is measured in financial terms. Decisions are made with whatever geological information is available and it is the geologist's responsibility to see that as much critical information as possible is taken into account. Many geological assumptions are included in financial decisions regarding the development of a mineral property and geologists should participate fully in these financial deliberations.

11.4   STUDY DEFINITIONS

A mining company may be exploring for new deposits ("greenfields") or evaluating a mine extension ("brownfields") possibly as a result of the discovery of additional ore, an increase in commodity price, a change in the mining method, or the introduction of different technology. For each project technical and economic studies are required to determine if the project is valuable to the company concerned. Initially, especially during exploration, the amount of information available is limited and this constrains the range and detail possible in studies. As more data are collected across a range of disciplines, including geology, mining, mineral processing, marketing, etc., the studies become more extensive and eventually "home in" on the single project with the most value. The end result is a series of study stages that in most cases involve increasing levels of detail and expenditure.

The definition of each study stage is important and this is part of the job of the competent person (see section 10.4.1). In most cases study stages are defined by a set of objectives at the start, a series of work programs designed to achieve these objectives, and a decision point at which the project may progress to the next

stage, pause while more data are collected or perhaps be discarded in favor of a more attractive investment opportunity. Naturally, the detail of the content of the program will vary according to the mineral type, geological and metallurgical complexity, location, environmental considerations, etc., so that there is no ready-made recipe that applies in all situations.

Reports at the end of each stage describe technical issues such as geological, mining engineering, mineral processing and extractive metallurgy, infrastructure, environmental, social, financial, commercial and legal aspects of the project (Goode et al. 1991, Smith 1991, Barnes 1997, Northcote 1998, White 2001). Similarly there is a range of business issues to be addressed, such as the project's ownership structure, permitting, marketing and government relations. Business and technical issues also interact, e.g. when a particular product specification has to be achieved to enable market penetration.

From these inputs capital and operating costs are derived to prepare estimated cash flows from which the value of the mineralisation in question is determined. The value provides a base on which decisions can be made, including strategic or political, technical, financial, and social aspects. The project moves progressively from an exploration target, where "back of the envelope" estimates are made, to a feasibility study from which there is sufficient confidence in the information for a financial institute to lend the capital that will be required to bring the mine on stream.

Estimates may be ranked according to the information available and the extent and depth of study completed. A ranking proposed here (N. Weatherstone, personal communication) ranges from conceptual through order-of-magnitude and pre-feasibility to feasibility studies. An example for geological studies is shown in Fig. 11.7. A conceptual study report may only comprise a single sheet compiled in a few hours or days, whilst a feasibility study may involve a team of geologists, engineers, and other specialists for several months or years. A decision point is reached after each stage, which marks a significant point in the project development. The project only continues if there are sufficient indications of significant tonnage and grade, mining and

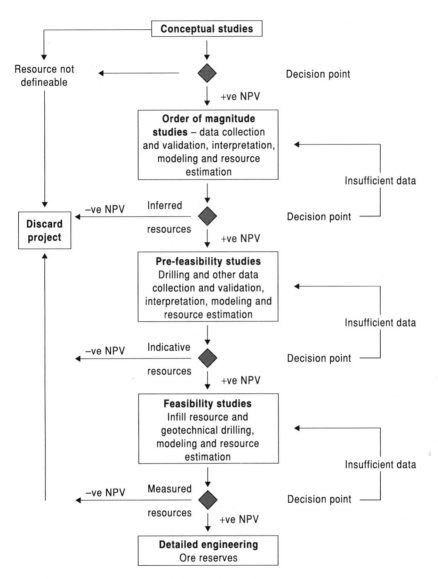

FIG. 11.7   Flow chart showing the range of feasibility studies and associated decision points. (After N. Weatherstone personal communication.)

processing are feasible, and indications are that it is economically viable.

It is recognized in the minerals industry that the chances of the successful establishment of a new mine are low. In other words, the conversion of mineralized rock into an ore reserve is a high-risk venture. A large and successful Canadian mining company investigated 1000 mineral projects over a 40-year period. Of these only 78 warranted major work, only 18

were brought into production but 11 did not return the company's capital investment. Only seven were viable from the thousand projects considered.

Consideration has to be given to the readership of these studies as there are different levels of understanding regarding the details of a project and differences of emphasis are always possible. Broadly is it intended for internal or external (outside the company) use? If the former, is it for an operational or an administrative group? External readers could be a government department, prospective investors, or banks that may loan money for the project.

11.4.1 Conceptual study

Most successful exploration programs start at this study level. A conceptual study establishes the likely presence of a resource. This estimate may be only preliminary, but is the first step to defining an inferred resource as classified by the JORC code. It is based upon the interpretation of the size, shape and grade of the potential resource and the possible mining and processing methods. The resource estimate may only be based upon reference to deposits of similar genesis and size and may only be an office-based evaluation. There might be sufficient indications of tonnage and grade to encourage further work. It is important to identify potential issues, such as access, topographical, or mining lease constraints.

During the study similar deposits elsewhere will be used as the basis to establish whether their mining or processing methods are technically feasible on your deposit. Infrastructure and engineering requirements will be costed to ±35%, although estimates will be too unreliable to attach confidence limits. Estimates can be calculated from empirical factors and comparison with similar operations, or in a mine extension, current and historical statistics and costs are immediately relevant and the accuracy of such estimates may be within ±10%. Such projects also follow the study definitions process.

For evaluation of exploration projects, much information is taken from sources such as the publications of state and national bureaux of mines and from reports of projects in the mining literature. Such examples and much useful data are included in O'Hara (1980), Mular (1982), Bureau of Mines (USA) Cost Estimating System Handbook (1987), Camm (1991), Goode et al. (1991), Smith (1991), Craig Smith (1992), and references such as the *Annual Review Reference Manual and Buyers Guide* of the *Canadian Mining Journal*. Websites that provide current information include that of Minecost (Minecost 2004) and Western Mine Engineering (Westernmine 2004). Major social and/or

environmental sensitivities will need to be identified. The project should indicate its potential to produce enhanced value (NPV, see section 11.4.3) when compared to other potential investments.

11.4.2 Order-of-magnitude study

An order-of-magnitude study is normally a fairly low cost assessment. However, the cost of order-of magnitude studies is increasing over the years, as companies have to address risks earlier in the process. The study is based upon limited drilling and other sample collection to establish whether there could be a viable project that would justify the cost of progressing to a pre-feasibility study. The results define the presence of sufficient inferred resources to warrant further work. Mineralogical studies at this stage will identify undesirable elements and other possible metallurgical issues. These resources should define sufficient tonnage above a given cut-off grade to enable engineers to determine possible mining options and production rates and thus the preliminary size of mining and processing equipment. Social and environmental baseline studies will be initiated. Preliminary capital and operating costs can be established to somewhere in the region of ±30%. Financial modeling of options and sensitivities will be undertaken.

11.4.3 Pre-feasibility study

The aim of the pre-feasibility study is to evaluate the various options and possible combinations of technical and business issues, to assess the sensitivity of the project to changes in the individual parameters, and to rank various scenarios prior to selecting the most likely for further, more detailed study.

The pre-feasibility study incorporates major sampling and test work programs. Upon completion of a pre-feasibility study, geological confidence is such that it should be possible to publicly declare ore reserves (from measured and indicated resources), and any other mineral resources that may become mineable in the future with further study. Similarly, the mining method and production rate will have been selected. Extensive mineral processing and metallurgical test work, possibly using a pilot plant, will have demonstrated that a product can be extracted using a viable process. The mine infrastructure and labor requirements, and the impact of the potential mine on the environment will have been evaluated. Detailed evaluation of capital and operating costs can be established to ±25%. These figures will be used, along with a preliminary mine schedule, in a financial and commercial evaluation. A number of sensitivities will have been assessed, including alternative methods of mining and processing the mineralisation, and the effects of various levels of output. The integration of social and environmental aspects is most important in the consideration of options for major project decisions. Findings from the social and environmental baseline studies are used as the basis to predict and evaluate the likely effects or impacts, whether negative or positive. The option that demonstrates the highest value with acceptable (lowest) risk will be selected as demonstrably viable. Such a study can cost several hundred thousand dollars.

11.4.4 Feasibility study

The aim of the feasibility study is to confirm and maximize the value of the preferred technical and business options identified in the pre-feasibility study stage.

During the feasibility study the environmental and other statutory approvals processes, commenced during the pre-feasibility study, will continue but at an accelerated pace. Consultations and negotiations with local community groups, landowners, and other interested parties will proceed to the point of basic agreement. Social and environmental impact assessments will be required that meet statutory and company requirements. The timing of formal submissions for permitting and other statutory processes needs careful consideration. Submission too early may prematurely commit the project to conditions which may constrain scope and compromise value.

Sufficient sample collection and test work has taken place during a feasibility study for more of the resource estimate to be reported in the measured category. On large projects several million dollars may have been spent to bring the project to feasibility study level. Sensitivity analyses will have been run to assess the major factors that may have an impact upon the reserve estimate. This will help quantify the risk associated with the reserves, which at this stage will fall within the company's acceptable risk category. If the company is approaching a financial institute to borrow the capital that will be required to bring the mine on stream, these risk factors will be assessed in detail. Often, financial institutes use independent consultants to audit the resource and reserve estimates. The reserves will be based upon a final mine design. This design will provide the information to enable the company to request bids for detailed engineering studies. Similarly, a final process flow sheet will have been developed. Costs will have been developed to ±15%.

11.4.5 Detailed engineering

Once the decision is taken to proceed to detailed engineering, costs will be developed to ±10%. The project is almost certain to progress to a mine. This is the final stage before major capital expenditure is incurred and involves producing detailed plans and work schedules so that orders can be placed for equipment and the mine development can begin. These investigations are thorough, penetrating, and usually completed by outside consultant companies who specialize in this type of work. Such a study usually takes several months or years, depending upon the initial database, and can cost several million dollars for a large project.

11.5 MINERAL PROJECT VALUATION AND SELECTION CRITERIA

Previous sections have considered the value of mineralisation in absolute terms (i.e. a sum of money expressed as a cash flow) and this is the basis of mineral project evaluation. Another aspect of value is its use in a relative sense, that is the ranking of a group of similar projects, or options within one project (King 2000). This relative value is based upon ranking each project's absolute value and is particularly useful when a company has several projects competing for funds. Any company has finite managerial and financial resources and a choice has to be made between two or more alternatives based on some tangible measurement of economic value or return. There are four main techniques, discussed below, all of which are based on the annual cash flow for the project in question.

11.5.1 Payback

This is a simple method which ranks mineral projects in order of their value by the number of years of production required to recover the initial capital investment from the project cash flow. In Box 11.2 the projects are ranked in order of least time of payback: B, D, C, and A. The method serves as a preliminary screening process but it is inadequate as a selection criterion for it fails to consider earnings after payback and does not take into account the time value of money (section 11.5.3). It is useful in areas of political instability where the recovery of the initial investment within a short period of time is particularly important and here project B would take precedence.

BOX 11.2   Mineral project evaluation and selection criteria.

Mineralisation at four similar prospects has been investigated and an order of magnitude feasibility study has been completed on each. At each location it is thought that a mine can be constructed in one year. Production commences in year 1. Which, if any, are worth retaining? Cash flow values are in arbitrary units.

		Year			
		A	B	C	D
Initial investment (CAPEX)	−1	(3100)	(2225)	(2350)	(2100)
Operating margin					
1		500	2000	150	1100
2		500	1000	450	900
3		1000	500	1000	1150
4		2000	500	3400	950
Payback (years)		3.6	1.2	3.2	2.1
Operating margin / CAPEX		1.3	1.8	2.1	1.9
NPV at 15% discount		−560	+835	+656	+780
DCF ROR		11%	40%	21%	30%

The value of the mineralisation at prospect A does not reach the minimum DCF return of 15% and is discarded. The remaining three meet this requirement and can be ranked in value:

Criteria	Most favorable	Middle	Least favorable
Payback	B	D	C
Operating margin / CAPEX	C	D	B
NPV	B	D	C
DCF ROR	B	D	C

The mineralisation at prospect B has the greatest value although it is not the lowest cost to develop, but it has the great virtue of generating a large cash flow at the beginning of the project. Prospect A, the least favorable, has the largest CAPEX and a low cash flow in years 1 and 2. At a discount rate of 15% the mineralisation at prospect B is worth 835 units, C is worth 656 units, and D 780 units. On the basis of this study prospect B is the one to be retained.

11.5.2 Operating margin to initial investment ratio

There are several variations on this theme but it produces a dimensionless number which indicates the amount of cash flow generated per dollar invested. It is an indication as to how financially safe the investment is and the larger the number the higher the value of the mineralisation. It is easy to calculate and considers the whole life of the project compared with payback which considers only the first few years. All factors in the cash flow (such as revenue, operating cost, taxes, etc.) are taken into consideration and all affect the ratio number. However, as in payback, the time value of money is not taken into consideration. As a simple example consider spending $100 today to receive $300 in 3 years, or spending $100 today to receive $400 in 10 years. The increase in the ratio overfavors the latter yet most would prefer the former. Using this approach the projects in Box 11.2 can be ranked in order of value as C, D, B, and A.

11.5.3 Techniques using the time value of money

These techniques are commonly used in the financial evaluation of mineralisation and mineral projects. The opening theme is that money has a time value (Wanless 1982). Disregarding inflation, money is worth more today than it will be at some future date because it can be put to work over that period. If a dollar were invested today at an interest rate of 15% compounded annually, it would amount to $1.00 \times (1 + 0.15)$ or $1.15 after 1 year and $1.00 \times (1.15)^5 = $2.01 after 5 years. This is the compound interest formula:

$$S = P(1 + i)n$$

where S is the sum of money after n periods of interest payment, i is the interest rate, and P the initial investment value. In this example we can say that the 5-year future value of the $1 invested today at 15% is $2.01. Reversing this viewpoint from the calculation of future values to one of value today, what is the value today of $2.01($S$) if this is a single cash flow occurring in 5 years' time at the accepted rate of interest of 15%? The answer is obviously $1.00 and the present value (P) is a variation of the compound interest formula above:

$$P = \frac{S}{(1+n)^n} = \frac{\$2.01}{(1+0.15)^5} = 2.01(0.497) = \$1.00$$

This expression is the present value discount factor, more usually termed the discount factor, and tables of calculated factors are readily available (Table 11.1). From the formula, and the table, it is evident that this factor decreases with increasing interest rates and number of

TABLE 11.1   A selection of present value discount factors.

		Years						
		1	**2**	**3**	**4**	**5**	**10**	**15**
Discount rate	5%	0.952	0.907	0.864	0.823	0.784	0.614	0.481
	10%	0.909	0.826	0.751	0.683	0.621	0.386	0.240
	15%	0.870	0.756	0.658	0.572	0.498	0.247	0.123
	20%	0.833	0.644	0.579	0.482	0.402	0.162	0.065

For explanation see text.
The table demonstrates the decrease in the magnitude of the discount factors with increasing years and increasing discount rate, governed by the basic formula:

$$\text{discount factor} = \frac{1}{(1 - i)n}$$

where i = interest or discount rate (as a fraction) and n = number of years.

years. From this two important characteristics in valuing mineralisation and mineral projects develop. Firstly, as discount factors are highest in the early years, the discounted value of any project is enhanced by generating high cash flows at the beginning of the project. That is, the ability to extract higher grade and more accessible (lower mining cost) ore at the beginning of a project can significantly enhance its value. This is the declining cut-off grade theory. Not all mineralisation is amenable to this arrangement but if this can be achieved it will provide a higher value than a comparable location in which this is not possible. Secondly, discount factors decrease with time and by convention and convenience cash flows are not calculated beyond usually a 10-year interval as their contribution to the value becomes minimal. This has the added advantage of limiting future predictions. In cash flow calculations it is difficult to predict with some degree of accuracy what is to happen next year – several years ahead is in the realm of speculation although best estimates have to be made.

Thus the value of money today is not the same as money received at some future date. This concept is concerned only with the interest that money can earn over this future intervening period and is not concerned with inflation. In this calculation money received at some time in the future is assumed to have the same purchasing power as money received today; it has constant purchasing power and is referred to as *constant money*. The effect of inflation on project value, however, is important and is considered in section 11.7.

Net present value (NPV)

The second formula above is used to determine the present value of the expected cash flow from a project at an agreed rate of interest. The cash flow from each year is discounted (i.e. multiplied by the factor from Table 11.1 appropriate to the year and interest rate). A summation of all these yearly present values over the review period provides the present value (PV) of the evaluation model. From this PV the initial capital investment is deducted to give the net present value (NPV). Some cases are given in Box 11.2.

The NPV method includes three indicators regarding the value of a mineral project, provided that the net present value summation is a positive sum:

1 the initial capital investment is returned;
2 the financial return on this investment is the specified interest rate;
3 the net present value summation provides a bonus payment which is sometimes called the acquisition value of the mineralisation concerned. Projects can be ranked according to the size of this bonus provided that the same interest rate is used throughout (Box 11.2). The reason why project B is superior to projects D and C is that the cash flows in project B are larger in the earlier years and the initial investment is returned sooner.

The selection of an appropriate interest rate is critical in the application of NPV as a valuation and ranking technique, as this discounts the cash flow and determines the net present value. A minimum rate is equal to the cost of initial capital used to develop the project. Additional factors such as market conditions, tax environment, political stability, payback period, etc. have to be considered over the proposed life of the project. This is a matter of company policy and usually the interest rate for discounting varies from 5% to 15% *above* the interest rate of the required initial capital investment. In times of high interest rates this discount rate is particularly onerous.

Discounted cash flow rate of return (DCF ROR)

This technique is a special case of NPV where the interest rate chosen is that which will exactly discount the future cash flows of a project to a present value equal to the initial capital investment (i.e. the NPV is zero). This DCF return is employed for screening and ranking alternative projects and is commonly used in industry. Since this discount rate is not known at the beginning of a calculation an iterative process has to be used which is ideally suited for computer processing. A very approximate first estimate can be obtained by dividing the total initial capital expenditure by the average annual cash flow, and dividing the result into 0.7, but it is dependent upon the shape of the overall cash flow (Box 11.2). Also over a narrow

BOX 11.3 Sensitivity analysis.

This example is taken from a feasibility study completed on a coal mine in western Canada. In the study it was apparent that two of the critical factors were variation in production cost and in tonnage of coal sold as these could both have a serious effect on the cash flow. These factors were varied and expressed as DCF ROR:

VARIATION IN MINE PRODUCTION COST

Base case	+5%	+10%	−5%	−10%
26.6%	22.3%	18.5%	31.5%	36.9%

VARIATION IN TONNAGE OF COAL PRODUCED SOLD

Base case	+5%	+10%	−5%	−10%
26.6%	33.9%	42.9%	20.2%	14.5%

Obviously the lower the tonnage produced the lower the revenue and the lower the DCF ROR, etc. None of the variations bring the DCF ROR to less than the hurdle of 15%. Results are best presented graphically.

range of discount rates the DCF ROR value correlates as a straight line (Box 11.3). Consequently by calculating a series (3–4) of positive and negative NPVs at selected discount rates the DCF rate can be seen graphically at the point where NPV equals zero.

Generally, mining companies finance mineralisation for development which has a value equal to or exceeding a DCF rate of return of 15% after tax or 20% before tax has been deducted.

Comment

These quantitative economic modeling techniques have contributed much to an improvement in the process of investment decision making. During the 1980s, however, there was an increasing concern with the totality of mineral projects and a greater emphasis on aspects such as business risks associated with rapidly rising capital requirements due to inflation, the unpredictability of future economic conditions, sophisticated financing arrangements, and host government attitudes.

Some mining organizations have used comparative production cost ranking to reduce investment risk. They require that the production cost to market for a new project is in the lower quartile of all major primary producers of that commodity. This is based on the premise that if the commodity prices fall, and there is a corresponding reduction of project revenue, then their operation will be protected by a cushion of other and higher cost producers, who, it is assumed, will be forced to reduce production at an early date thus stabilizing the commodity price. It is worth noting, however, that a project with such a low production cost would have a favorable investment rate of return (either NPV or DCF) and would be ranked highly by these techniques.

The NPV method of valuation requires management to specify their desired rate of return and indicates the excess or deficiency in the cash flow above or below this rate while DCF provides the project's rate of return and does not consider the desired rate. As a comparison between the two is said that NPV provides a conservative ranking of projects compared with DCF.

Long versus short run considerations have to be assessed. As an example compare two occurrences of mineralisation, one with a 15-year life and a 15% DCF and another with a 5-year life and a 20% DCF. If the short life, high DCF project is selected, management must consider the business opportunity at the end of the project. What is the probability of finding and developing a series of such short life, high return mineral deposits? The long-term future of the company may be better served by choosing the mineral project with the lower DCF. Lastly, there is the condition of selecting a series of small projects with high DCF. Small projects often require the same management time and attention as larger ones. With a finite amount of time its effective use will tend towards large projects with, perhaps, lower DCF rather than a series of smaller schemes.

In summary, the objective of valuation is to summarize and convey to management in a single determinant a quantitative summary of the value of a mineral deposit. Clearly there is no single perfect method of assessing this and good management uses every available relevant method and is aware of their inherent weaknesses and strengths. In the final analysis in the evaluation of mineral projects there is no substitute for sound managerial judgement.

11.6   RISK

Risk pervades our entire life and the way we act. It can be described in two ways – either with qualitative expressions (it's a sure thing, we have a fair chance of discovery in the Upper Palaeozoic, etc.) or in a quantitative sense using probability. A probability of 1.0 means that the event *will* occur while 0.0 means that it will *never* happen; negative probabilities do not exist. A probability of 0.31 means that there are 31 chances in 100 that the event will occur and 69 in 100 that it will not. What is an acceptable risk? Risk is very much a personal assessment. For instance many people would not work in an underground mine because it is seen as being dangerous, yet every day accept higher risks such as dying from fire in the home or in a road accident (Rothschild 1978).

In the previous section four methods of analyzing cash flows in the valuation of mineralisation were presented. The projected cash flows were assumed to occur (i.e. probability = 1.0) and they did not include quantitative statements of risk: they were treated as decisions under certainty. However the valuation of mineralisation involves the introduction of many factors (grade, tonnage, mining cost, transport cost, etc.) which are not constants (i.e. there

is not a single, unique value) but variables. Consequently mineral valuation comprises a series of uncertain decisions and the risk of these decisions being wrong can be assessed.

11.6.1   Risk analysis

Qualitative assessment

Adjustment of discount rates for NPV and DCF
A commonly used overall method of allowing for risk is to use an abnormally high discount rate in the valuation:

> Safe long term rate + risk premium rate
> = risk rate

For instance if the accepted minimum valuation rate is 15% DCF and there is a perceived risk, an added rate is included to bring this minimum to, say, 23%. This is the simplest method of allowing for risk but also the least satisfactory as the added rate is a matter of personal judgement; its determination with any degree of accuracy is impossible.

Adjustment of costs
A danger here is that of overadjusting. Consider a mineral valuation where as a safety factor it is decided to increase the best estimate of operating costs, to reduce ore grade and selling prices, and apply a high discount rate as an additional risk allowance. It will have to be a remarkable project to survive such treatment.

A better approach is to calculate a *base case* from the most accurate information available. Risk factors are then added in a *sensitivity analysis* (Box 11.3) by considering the variable components one at a time while all others are kept constant. In any evaluation certain components have a greater effect upon the size of the cash flow (and hence value) than others and the purpose of a sensitivity analysis is to identify them so that further investigations can improve their reliability (i.e. make them less risky). Commonly grade is varied incrementally, as are other components such as extraction cost, mineral recovery, etc.

The most significant component is revenue which is dependent upon the future sales price of the product and this is the most difficult to forecast. Capital and operating costs only move upwards but are not likely to increase at a rate exceeding 15% a year. However selling prices, and particularly for those commodities sold on exchanges, may decrease or increase and in extreme instances can either double or halve in a year (Figs 1.5 & 1.6). Forecasting commodity price trends is a specialist operation and there are groups who provide this service on a consultancy basis, such as Brook Hunt, London. If no reliable price trend predictions are available a break-even price can be calculated from a cash flow at a required rate of return – this is the price at which operating and financing costs can be met and the mine can continue in production. Once this minimum price is known, the possibility of its being maintained over a standard 10-year period can be assessed.

These sensitivity results are best appreciated graphically with variation in the single component concerned plotted against the related DCF ROR (Box 11.3).

Quantitative assessment

Discussion on components used in the calculation of a cash flow has not included any quantitative statement of risk. Each component has been thought of as if it were a constant with a probability of occurring of 1.0. Most of these components are variables, a concept readily apparent from considering the grade of a deposit which is an average value within its own distribution and probability of occurrence. The same principle applies to estimates of mining cost, recovery, selling price, etc.

Normally a spreadsheet model will only reveal a single outcome, generally the most likely or average case. The use of simulation enables the user to automatically analyze the effects of varying inputs on the modeled system. Probability of occurrence can be computed in a spreadsheet, by randomly generating values for uncertain variables many times over. The result is a probability distribution of the uncertain variable, or a simulated model. One type of spreadsheet simulation is referred to as Monte Carlo simulation (Decisioneering 2003, Lumina 2003).

It is possible to use these component distributions in the calculation of a cash flow with their probability of occurrence, instead of fixed values with no risk element. The probability distribution of DCF or NPV can be calculated and also the probability of a certain rate of return (Wanless 1982, Goode et al. 1991, Hatton 1994b, Decisioneering 2003).

11.7 INFLATION

Mining is among the most capital intensive of industries, and ranks near the top of the industrial sectors in this respect. This is related to the complex technology of modern production systems; high investment in major infrastructures such as town sites, railways and ports, and expenditure on social and environmental aspects. To offset this trend projects became larger (i.e. higher production) to benefit from an economy of scale which would either control or reduce increasing capital (and production) costs per unit of production. Generally, however, the capital cost of mineral projects is rising significantly. This inherent increase is exacerbated by inflation. As an example, an inflation rate of 7% a year for 5 years will increase estimates by 40% in year 5. The impact of high inflation rates on capital (and operating costs) for projects with a long pre-production period of several years can be considerable.

Related to inflation, monetary exchange rates are an important item to consider in mine evaluation. When a company operates a mine under one currency but sells its products in another, changes in exchange rates between them can have serious consequences for cash flows.

When inflation was in low single figures its effect on the valuation of mineral projects was negligible and could be ignored but with higher rates its effect on calculated rates of return may be considerable. The value of money, measured by what it will buy, becomes progressively less as time passes. The overall rate of increase in price (or fall in the value of money) is measured by a gross national product deflator and is the weighted average of the various price increases within a particular national economy. There are other measures of inflation (Camm 1991) which may be more appropriate in particular circumstances such as retail and wholesale price indices (Economagic 2003), the indices of capital goods in various categories, and wage indices for a country, an industry, or a particular skill.

The differential inflation (or escalation) which exists between the various components

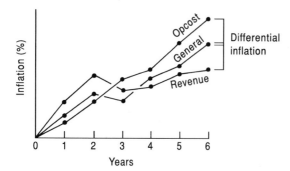

FIG. 11.8   Differential inflation. Revenue is increasing at a slower rate than general inflation, while the reverse is true with mine operating costs (opcost). There is a positive differential inflation with the former and negative with the latter.

in the cash flow is an important factor. While inflation may be assumed always to be a positive factor, escalation can be either positive or negative as the particular component moves at a rate either faster or slower than the measure of general inflation (Fig. 11.8).

11.7.1   Constant and current money

Cash flows may be calculated in either *constant* or *current* money terms. Constant money is assumed to have the same purchasing power throughout the valuation period and thus one year's money may be compared directly with that of any other. This is the basis on which DCF and NPV are calculated, as above and in Boxes 11.1–11.3.

In current money the figures used in calculations are adjusted for the anticipated rate of inflation each year and are, in essence, the figures which are expected to be entered into the books of account for that year and represent the net available profit or loss. Money from different years is not directly comparable and DCF and NPV derived from this type of cash flow are misleading.

If the rate of inflation is the same for capital and operating costs, and revenue (i.e. there is no escalation), and there are no tax or other financial complications (which is seldom, if ever, the case), the DCF yield may be correctly estimated by completing the calculation in constant money terms, without regard to infla-

tion. This is because, if the components which form the cash flow are inflated at the inflation rate and the resultant flow is then deflated for DCF and NPV calculations at the same rate, this will produce a flow identical with that produced by ignoring inflation altogether.

Opinions differ on the treatment of inflation in mineral project evaluation but a safe method is to use it as a type of sensitivity analysis on the base case. Separate specific cost and revenue items, which correlate well with published cost indices, can be used over a limited period of, say, 4 years ahead, and thus account can be taken of differential inflation over this period. Beyond this, separate predictions are unrealistic and it is advisable to use only a single, uniform rate for the complete cash flow for the remainder of the project duration. The base case cash flow prediction is in constant money whilst the inflation model will be in current money.

In most projects the greater part of the finance to build a mine is obtained from banks. Banks test the ability of a project cash flow to repay these loans, and related interest payments, by applying their own estimates of future inflation rates to the company base case. Many mineral companies then prefer to prepare cash flows for their base case, and sensitivity analysis, in constant money terms and leave the preparations of inflated flows in current money to the loan-making banks (Gentry 1988).

11.8   MINERAL PROJECT FINANCE

The development of mineralisation into a producing mine requires a skilful combination of financial and technical expertise (Box 11.4). For the purpose of this section assume that a feasibility study (section 11.4.4) has been completed and the company concerned intends to proceed with the development of the defined mineralisation.

11.8.1   Financing of mineral projects

Traditional financing

Generally finance to develop mineral projects (Institution of Mining & Metallurgy 1987) is obtained from three sources (Potts 1985).

BOX 11.4   Case history of Ok Tedi Mine.

The Ok Tedi Mine, Papua New Guinea (PNG), provides an example of the valuation of mineralisation and associated project financing. This porphyry copper deposit (Rush & Seegers 1998) was discovered in 1968 by Kennecott Copper Corporation (USA). Negotiations between this company and the host government on the development of the project broke down in 1975 and Kennecott left PNG. The PNG Government continued further exploration and development work. In 1976 Broken Hill Proprietary (Australia) (BHP) agreed to evaluate the deposit and established that more than 250 Mt of mineralisation at 0.7% copper (in the form of chalcopyrite) occurred in the porphyry and 115 Mt of material at 1.2 g t^{-1} gold in the capping gossan. A definitive feasibility study was completed in 1979. A consortium was formed which included BHP, Amoco Minerals (a subsidiary of the Standard Oil Company of Indiana, USA), and a German group of companies that largely represented copper smelting activities. This sponsoring consortium, managed by BHP, presented proposals for the development of the deposit to the PNG Government in November 1979 which were accepted in early 1980 (Pintz 1984). The proposals, based on the feasibility study, were that development would take place in four stages for the production of gold metal, and copper concentrates for export to Germany and other locations. A copper smelter and refinery was not envisaged in PNG.

Construction: 1981–84

Production **Stage 1:** 1984–86	To produce 22,500 t d^{-1} gossan at 2.9 g t^{-1} gold to produce about 22,000 kg of gold a year, depending on the grade and recovery
Production **Stage 2:** 1986–89	Gold production to continue as previously, but also to mine copper ore (average grade 0.7% copper) at a commencement rate of 7000 t d^{-1} and expanding to 60,000 t d^{-1} by the end of 1989. Stripping ratio for the overall open pit is 2.5:1
Production **Stage 3:** post 1990	Gold mining is phased out and copper production continued at 60,000 t d^{-1} ore. At an average grade of 0.7% copper, 80% recovery and 300 production days a year this is equivalent to 100,800 t yr^{-1} copper metal in shipped chalcopyrite concentrates containing 25–27% copper

In the early years low cost material with high value (the gold ore) was produced in order to repay rapidly the project finance. In later years after this gold ore had become exhausted copper ore would be mined from a large open pit. Copper sulfide concentrates only were to be produced on site and these concentrates sold to a German copper smelter group.

A separate project company was formed, Ok Tedi Mining Co. Ltd, with the following shareholders and sponsors: BHP 30% (managers), German Group 20%, AMOCO 30%, PNG Government 20%.

In 1981 the estimated total cost of Stage 1 was $US855M with a project debt to equity ratio of 70:30. Consequently the amount of equity required from the sponsors was $US256M and the remaining $599M was raised as project finance from a consortium of banks in several countries.

Construction was completed on time and production stage 1 was successfully concluded. However, the commencement of stage 2 coincided with a decrease in the price of copper and the ensuing delays in project development caused renegotiation of several aspects of the initial agreement between the project company and the host government. The project is continuing, although the lack of a tailings facility caused intense controversy in the early years of the twenty first century and BHP Billiton's 52% shareholding has been transferred into a company promoting sustainable development in PNG.

Equity

This is finance provided by the owners (i.e. shareholders) of the company developing the project through their purchase of shares in the company when it is floated on a stock exchange. Throughout the life of the mine these shares may be bought and sold and if the mine is successful they may be sold at a higher price than their original value. Another return to shareholders is dividends on each share which are paid out of profits after other financial commitments have been met. A successful mine may pay dividends from the start of production to the end of its life but for many mines fluctuating mineral prices mean that dividend amounts vary dramatically from year to year. Usually a reduction in the dividend leads to a drop in the share price (Kernet 1991).

Debt

In this case money can be supplied by sources outside the company, usually a group of banks. Debt finance places the company in a fundamentally different position than that of equity financing. Lenders may have the power to force it to cease trading (i.e. close down) if either interest charges or loans are not paid in a previously agreed manner. Thus, ultimately, control of the company is in the hands of the suppliers of finance and not the mining company itself.

Retained profits

A successful company may retain some of its profit and not distribute all of it in the form of dividends to shareholders. In this way a source of finance can be accumulated within a company that is preparing to develop a mineral property.

The traditional means of financing mineral development in the first half of this century was a combination of the issue of equity, debt finance, and the use of retained profit. This method was adequate while the capital cost of development was millions or tens of millions of dollars. During the last three to four decades the capital cost and size of major mineral projects has increased rapidly and large projects now cost several hundreds of millions of dollars, possibly a billion dollars. With these levels of expenditure very few, if any, mineral companies are able to finance new ventures using traditional methods. Thus companies have sought other methods of financing which made an optimum use of their financial strength and technical expertise but preserved their borrowing capability as much as possible. One method of achieving these ends is project finance.

Project finance

Project financing differs fundamentally from traditional financing. The organization providing the finance for the project looks either wholly or substantially to the cash flow of the project as the source from which the loans (and interest) are repaid and to its assets as security for the loan. In this way mineral projects are financed on their own merit rather than from the cash flow of the mining company that is promoting the scheme (the sponsor).

Lenders to the project like to see security attached to the revenue of the cash flow in the form of firm sales contracts for the mineral products. It is common for them to form a consortium to spread any lending risk as widely as practicable. Project finance is then a type of nonrecourse borrowing, which is not dependent upon the sponsor's credit. However, for this the sponsoring company isolates the new project from its other operations (Fig. 11.9) and usually has to provide written guarantees that it will be brought to a specified level of production and managed effectively.

Banks prefer to have a safety margin as protection against a deterioration in the project cash flow. They will therefore agree to finance only a proportion (say 60–80%) of the cost of a project with the sponsor providing the remainder. Full debt financing is rare. Two important considerations which decide this proportion of finance are the length of the payback period (see earlier) and the quality of the management who are to bring the new mine into production. If the loan is to be repaid over a relatively short period of time (say 3–5 years) a bank will be more likely to lend a larger proportion of the total capital cost. Management is perhaps the paramount factor in the evaluation of a project. Bad management can destroy a project which good management could turn into the next Rio Tinto plc!

As a final comment it should be remembered that the lending banks and the sponsor have a

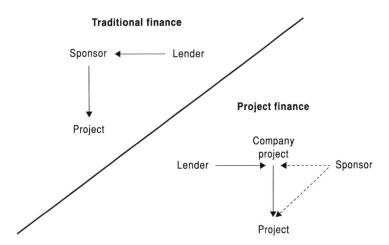

FIG. 11.9 Traditional and project
financing of capital requirements
for mine development. For
explanation see text.

difference of emphasis in project development.
A mining company will wish to achieve a positive return on its investment (i.e. the equity of
the project company) to satisfy the demands
of its shareholders. A bank, however, will wish
to ensure that its money is returned on time,
with interest and loan repayments taking priority over dividend payments to shareholders.
The payment of dividends, however, can have
considerable effect on the price of the company's shares traded on stock exchanges. Consequently, a compromise is usually reached
where the major proportion of the cash flow is
reserved for project loan repayment, a smaller
part (say 20%) is placed at the disposal of the
project company.

It is a matter of opinion as to whether cash
flows used in DCF analysis should include
drawdown and repayment of debt (i.e. project
finance), and interest payments. If not then
projects are evaluated as if they are financed
by 100% equity, although in practice this
will not happen (i.e. financing will be a combination of loan and equity). Such a procedure is said to separate investment decisions
from financing decisions (i.e. repayment of
debt, etc.). It is a fundamental principle that
a basically poor project cannot be transformed into an attractive one by the manner
in which it is financed. If an all-equity evaluation demonstrates a strong cash flow and
meets standard parameters (see earlier), then
after this it can be determined how it is best
financed.

Banks are not inherently concerned with the
rate of return of the proposed project. Their
return is more or less fixed, based upon the
interest rate at which they agree to lend money.
Banks are more concerned with protection of
their loans and they tend to examine cash
flow predicted in the feasibility study from this
perspective. A criterion is that the NPV at an
agreed rate of interest must be at least twice the
total initial loan plus interest payments. There
is the obvious implication that a bank will lend
money for a mineral project provided that its
loan and interest payments are secure, even
if actual project cash flows fall to half their
estimated level.

11.8.2 Risk in project finance

Project loans are provided on an agreed time
schedule of drawdown and repayment, based
on the technical data and cash flows in the final
feasibility report. The interest rate on loans is
usually a premium rate (say 3%) above an
agreed national base rate. As this base rate fluctuates the loan interest rate will also change,
and in times of inflation with high base rates
(as in 1990 and 1991) repayments can be onerous for a project company.

Even the best of feasibility studies may
be incorrect in actual practice and events to
which both sponsors and lenders particularly
look during the construction and operation of
a new mineral project are as follows described
below.

During construction

It is probable that the greatest risk in project financing occurs during the construction phase. This phase may legitimately take several years and during this time the loan is being used and no revenue is being generated. Mine construction is usually contracted out to major civil engineering companies and some form of completion guarantee from the managing contractor is crucial. This is usually to complete and commission the project in a specified time period and to make penalty payments to provide funds if construction is not completed on time. Complications which can delay construction include an increase in capital costs due to inflation or technological difficulties, delays in the delivery of plant and equipment, strikes, delays caused by abnormal weather, financial failure of a major contractor, and poor management.

A delay in completion and subsequent operation postpones the generation of the cash flow from which loans and interest were to be repaid. It also creates a requirement for additional money over and above that originally negotiated in order to finance this delay period. Obviously the key factors are the quality of the final project feasibility report and the reliability of the managing contractor.

During production

Adverse factors during production include: errors in the quantity and quality of ore reserve estimation, technical failure of equipment, fall in the price of the mineral product, foreign exchange fluctuations such as a currency devaluation, poor labor relations which provoke industrial strife, environmental difficulties, and government interference in the running of the project of which the final condition is expropriation. Any of these events may cause delay in the repayment of the loan. In cases of grave default the lenders have the ultimate right to take over the project and replace its management.

General

Banks, in reviewing a feasibility study, pay particular attention to the degree of confidence which is attached to estimates of mineable grade and tonnage, annual production rate, capital, and operating costs. While banks may be prepared to accept the risk of commodity prices falling below those projected, they are not prepared to accept technical risks. Of increasing importance to banks and project sponsors is the potential liability in the event of inadvertent damage to the environment. Consequently, the manner in which a feasibility study examines social and environmental issues is reviewed with great care.

11.9   SUMMARY

Cash flows in constant money are tested by sensitivity analysis in the valuation and ranking of mineralisation, mineral properties, and projects before mining companies proceed to developing a mine. A minimum acceptance rate of return (the "hurdle" rate) is usually stated, below which projects are not considered, e.g. a 20% DCF ROR before tax and a 15% DCF ROR after tax. This usually means that the initial capital investment is returned in the first 4 years or so of mineral production. Few companies consider inflation and even fewer consider a quantitative assessment of risk (section 11.6).

11.10   FURTHER READING

O'Hara (1980) and Mular (1982) provide much useful basic data on the evaluation of orebodies and the estimation of mining costs, but are now rather dated. A more relevant source is the *US Bureau of Mines Cost Estimating System Handbook* (1987), which provides a rigorous basis for cost estimating. Websites that provide current information include that of Minecost (Minecost 2004), CRU International (CRU 2004), Brook Hunt (Brook Hunt 2004), and Western Mining Engineering (Westernmine 2004). A good system for order-of-magnitude estimates is in Camm (1991) and Craig Smith (1992). Two other articles of note are the description of the Palabora copper open pit in South Africa (Crosson 1984), and the Neves Corvo copper–zinc–lead–tin project in Portugal (Bailey & Hodson 1991).

In the authors' opinion there is a scarcity of suitable textbooks and articles on the financial aspects of mineral evaluation, and Wanless (1982) remains a good choice. Aspects of the financing of mineral projects were presented in nine papers at a meeting "Finance for the mining industry" under the auspices of the Institution of Mining and Metallurgy in 1987. MINDEV 97 was an AusIMM event presenting the A–Z of "how-to" successfully develop a mineral resource project (Barnes 1997). Papers were presented by major mining companies, contractors, and equipment suppliers, covering the earliest stages of permitting, preparing feasibility studies, and assembling the develop-ment team, through to financing, procurement, construction, and commissioning. Case studies of mineral evaluation and related mine development are comparatively rare but probably the best published is that of the Ok Tedi copper-gold open pit mine (Pintz 1984) in Papua New Guinea.

For up-to-date information on environmental impact assessments, the reader is directed towards the bi-monthly publication *Environmental Impact Assessment Review* by Elsevier Science Ltd. Lord Rothschild's (1978) short paper on "Risk" is a readable and excellent introduction to this important topic.

PART II

CASE STUDIES

12

CLIFFE HILL QUARRY, LEICESTERSHIRE – DEVELOPMENT OF AGGREGATE RESERVES

MICHAEL K.G. WHATELEY AND WILLIAM L. BARRETT

12.1 INTRODUCTION

The annual production and use of construction aggregates in the UK, including both sand and gravel and hard rock, amounts to approximately 220 Mt. The materials are used for road building, construction, civil engineering, concrete, house building, chemicals, and other specialized applications. Within such a wide market rocks with particular properties are better suited to certain applications than others. The physical properties of the rocks are defined in terms of their response to different tests, most of which are covered by the relevant British Standards. Some international, American and local (but not British Standards) tests are also used.

Tarmac Quarry Products Ltd (Tarmac) is one of the leading aggregate, ready mixed concrete, and waste disposal companies in the UK; it also operates in Europe and Scandinavia, as well as the Middle and Far East. In 1999 it became a subsidiary of Anglo American plc. Tarmac operates about 100 active quarries and sandpits in the UK, spread from Ullapool in northwest Scotland to the south coast of England.

The biggest market for aggregates in the UK is the southeast of England where ironically there is no hard rock and demands have historically been met from sand and gravel sources (both land based and marine dredged) and by supplies railed in from elsewhere in England. Notwithstanding the expected efforts to expand local sand and gravel output from both land and marine sources, an aggregates supply shortfall of between 15 and 30 Mt yr^{-1} p.a. is predicted for the next decade for southeastern England. The deficiency in indigenous aggregates will have to be satisfied by increasing imports into the region from elsewhere in the UK and also from overseas. Leicestershire is well placed as a supplier and has suitable quality rock resources to benefit from this demand.

In the mid 1970s it became apparent that the profitable microdiorite (markfieldite) quarry of Cliffe Hill in Leicestershire (Fig. 12.1) belonging to Tarmac Roadstone Ltd, East Midlands was running short of recoverable reserves. Significant resources remained at the site, but these were sterilized beneath the processing plant (Fig. 12.2).

The markfieldites of the Charnwood Forest area occur as a number of relatively small igneous bodies intruded 550 Ma ago into late Proterozoic pyroclastics and metasediments. Markfieldite differs from other microdiorites in having a granophyric groundmass and it has a general uniformity and strength which puts it amongst the best and most consistent general purpose construction aggregate materials in the UK. An impressive durability means that the markfieldite can also be used for all classes of railway track ballast, as well as road stone.

Charnwood Forest contains major faults, e.g. the Thringstone Fault, which brings Carboniferous rocks against the Proterozoic sequence to the west of Stud Farm (Fig. 12.1).

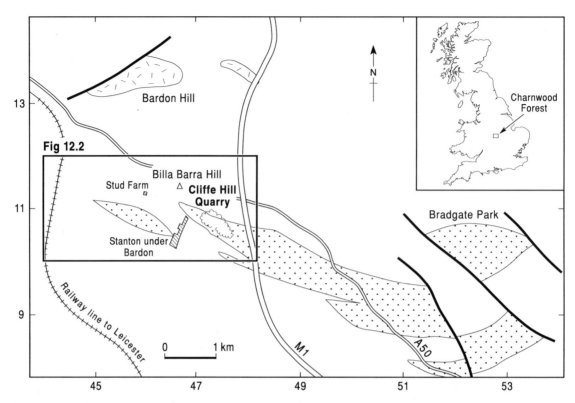

FIG. 12.1   Distribution of markfieldite (a microdiorite) in the southern part of Charnwood Forest, Leicestershire, UK. Stippled areas are known and projected areas of markfieldite as given in Evans (1968). Hatched areas are other intrusive rocks. The grid coordinates are in kilometers.

Knowing the location of faults is important in any quarry. Associated minor faulting is also important to quarrying operations. In Cliffe Hill Quarry one apparently minor shear zone was mapped. As a minor feature it could easily have been missed during core logging, however it offset the Precambrian basement by almost 60 m (Bell & Hopkins 1988).

In order to maintain its production from the English Midlands, and to maximize the exploitation of a valuable national resource, Tarmac initiated a search for alternative sources of similar quality markfieldite in the neighboring areas. This search was conducted in a number of phases with each subsequent and more costly exercise only being undertaken if clearly justified by the results of the preceding one. All costs quoted in connection with this case study relate to values current at the time of the expenditure.

This case study gives a clear example of the procedure adopted by one company in the exploration for and development of a hard rock resource. Readers wishing to expand their reading into sand and gravel and limestone resources are referred to British Geological Survey publications on the procedures for the assessment of conglomerate resources (Piper & Rogers 1980) and limestone resources (Cox et al. 1977). Additional background reading can be found in the book by Smith and Collis (2001).

12.2   PHASE 1 – ORDER-OF-MAGNITUDE STUDY

Cliffe Hill Quarry is adjacent to the M1 motorway and a railway giving it excellent access to major markets in the southeast of England. The initial desk study in 1977 concentrated on extensions to Cliffe Hill Quarry and adjacent

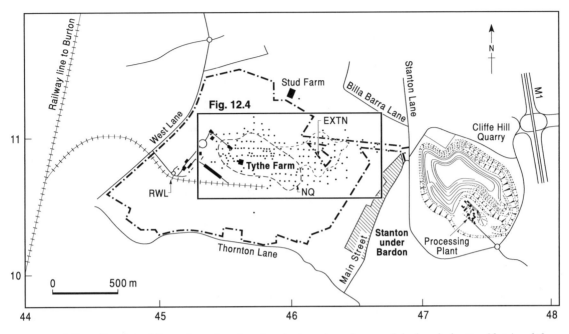

FIG. 12.2 Cliffe Hill and Stud Farm Quarries, showing the location of some of the borehole sites (dots) and the proposed new railway line (RWL). The pecked and dotted line represents the outline of the area included in the planning permission. The pecked line represents the outline of the proposed new quarry (NQ) with the possible extension shown to the east (EXTN).

land and mineral right holdings on Stud Farm. The study included a detailed literature search which indicated that both Billa Barra Hill to the northwest, and Stud Farm around 1 km to the west of the existing quarry might be of potential interest (Fig. 12.1). The 1:63,360 and 1:10,560 Geological Survey maps and the account by Evans (1968) of the Precambrian rocks of Charnwood Forest pointed to additional areas of diorite near Stud Farm. A well sunk on Tythe Farm (Fig. 12.2) at the end of the last century intersected markfieldite at a relatively shallow depth.

Subsequent ground surveys eliminated the former site as being composed of fine-grained tuff, volcanic breccia, and metasediments, but revealed scattered float boulders of markfieldite in the soils at the latter site. In view of the total absence of exposures at Stud Farm, a few pits were dug using a mechanical digger, and although these failed to reach bedrock they confirmed the existence of further pieces of markfieldite in the soils. The term bedrock in this chapter refers to the unconformity be-

tween the Triassic Mercia Mudstones and the markfieldite. The mudstones are also referred to as overburden which generally includes glacial deposits as well.

The above exercises could be regarded as separate phases or subphases, but since they took place in swift succession they have been regarded collectively as Phase 1. A cost (excluding overheads and geologist's time) of around £1500 was incurred for this work.

12.3 PHASE 2 – PRE-FEASIBILITY STUDY

The preliminary desk study had indicated areas of interest, but the Stud Farm holdings were insufficient to support a viable quarry (Bell & Hopkins 1988). Negotiations were entered into with the owner of adjacent land. An agreement was reached which allowed Tarmac access to the land for exploratory drilling. A small exploration drilling budget was approved and in mid 1978 nineteen, 150-mm continuous flight auger holes were drilled with a Dando 250 top drive, multipurpose drill rig. These holes, drilled alongside established farm tracks over the higher parts of the farm, proved that rock, harder than could be penetrated with the augers, existed at exploitable depths over an area large enough to contain around 20 Mt of material. The results from this drilling enabled the geologists to establish overburden distribution, overburden types, water-bearing zones, the depth of weathering, and the gradient of the bedrock surface (Bell & Hopkins 1988). At this stage there was no proof of the nature of the harder material since no samples were recovered. Nevertheless, it was assumed that the impenetrable rock was markfieldite and, although the drill pattern was irregular, it showed that there was only thin overburden. Geologists calculated the resources and overburden ratios for a theoretical quarry, and within the margins of error in such calculations there appeared to be sufficient resources for a viable quarry. The auger drilling in Phase 2 is estimated to have cost between £2000 and £2500.

12.4   PHASE 3 – FEASIBILITY STUDY

Phase 3 of the Cliffe Hill Quarry project included all the exploration drilling, a limited amount of geophysics, the testing of the core samples, the design of the landscaping works, the detailed specification of the processing plant, and the submission of the original planning application, and a number of subsequent amendments to it, to the relevant local government offices.

12.4.1   Drilling

The Estates and Environment Department of Tarmac acts as a contract drilling company to the operating divisions and the Stud Farm work, although welcome, caused a number of problems. With the continuing land acquisition program rig-time was becoming scarce and although it had been hoped to undertake any additional work at Stud Farm in-house, the rapid exploration success meant that the follow-up work could not be fitted into the program of the existing, company owned rigs. The company therefore took the decision to

FIG. 12.3   A top-drive, multipurpose, flight auger drilling rig used to evaluate the resources on the Stud Farm property adjacent to Cliffe Hill Quarry.

purchase a brand new six cylinder, 100-hp top-drive, multipurpose, drilling rig (Fig. 12.3) to evaluate fully the resource potential at Stud Farm. A rig of the type used, together with a back-up vehicle and the relevant in-hole equipment, would have cost around £120,000, but this figure is included as depreciation in the total drilling costs given below.

Drilling, using open-hole methods in the overburden and coring at various diameters in the bedrock, commenced in August 1979 and continued with some breaks until December 1989. During this period about 240 boreholes totalling around 17,000 m were drilled, mainly on a 50×50 m grid (Fig. 12.2). A planning application for Stud Farm was prepared in parallel with the exploration work.

The experience gained while drilling the original holes during Phase 2 indicated that certain aspects of the drilling would have to be improved if sufficient data were to be collected. The indications were that there was between

10 and 50 m of overburden. Although the flight augers could probably cope with this depth, removing the clay from the auger and clearing the hole proved slow and difficult. The clay was also expansive and, when wetted by the water flush from a diamond bit, the hole tended to close and trap the core barrel. These related problems were ultimately solved by using a drag bit and water flush. A drag bit is a bladed bit which is used when the sticky material such as clay and marl would clog up air flush bits. The drag bit drilling technique produced an open hole to bedrock almost as fast as the augers could when cleaning time was taken into account. In addition, the clay had to some extent expanded due to the water flush. By drilling with sufficient annulus size (the gap between the drill rods and the sidewall up which the drilling medium (air, water, or foam) carries the drill cuttings to the surface) and casing placement to bedrock, a good core recovery was usually possible (Bell & Hopkins 1988).

During the early drilling program the driller remarked on the apparent low penetration rate of the core barrel. At the time this had been assumed to be the result of operating in an uncased hole. Once the main program began it became apparent that, although the high rotation speed improved penetration, the productivity was lower than expected. A series of checks and experiments narrowed the problem down to polishing of the impregnated core bits. After a series of trials, conducted in conjunction with one of the leading British bit manufacturers, a bit matrix which gave the optimum balance between bit wear and penetration was developed. Once this problem had been solved the drilling proceeded at a rapid rate (Bell & Hopkins 1988). During the summer of 1981 a series of boreholes was drilled along a line joining Stud Farm with Cliffe Hill Quarry (Fig. 12.2). The aim was to investigate the route of a proposed tunnel to link the two quarries. The purpose of the boreholes was twofold: (i) to delineate the bedrock–overburden interface, in order to ensure that the tunnel which was planned to link Cliffe Hill Quarry to Stud Farm (section 12.4.5) remained in solid rock throughout its length, and (ii) to identify the engineering properties of the bedrock with particular reference to jointing and faulting. Some heavy faulting was identified in two boreholes.

Between September 1981 and August 1982 the drilling was concentrated on two areas of the site:
1 Investigation of the proposed plant site. Some 10 boreholes were drilled where access permitted.
2 Drilling of a newly acquired area to the northeast of Stud Farm.

The results confirmed the already familiar picture of highly variable overburden thicknesses, with the bedrock often "dipping away" beneath rapidly thickening Triassic Mercia Mudstone overburden.

Between April 1983 and March 1985 drilling was carried out on a 50 m grid across the area of the proposed quarry (Fig. 12.2) with the aim of:
1 confirming the geophysical interpretation;
2 identifying the volume of weathered material;
3 building up detailed knowledge of the faults, rock quality, etc., to enable detailed quarry plans to be drawn up.

The drilling results were used to produce the overburden isopach map (depth to bedrock) (Fig. 12.4). This map was initially drawn independently of the geophysical data and clearly shows the success of the previous electromagnetic and resistivity surveys in revealing the distribution of the overburden (Figs 12.5 & 12.6). Although some inaccuracy is evident in the deeper areas, with the resistivity method underestimating thickness of overburden, the two shallow ridges and steep sides were accurately delineated. Drilling indicated that these features can be attributed to faulted blocks of markfieldite alternating with late Proterozoic metasediments, as seen in Cliffe Hill Quarry.

12.4.2 Geophysics

As indicated previously, the planning application was being prepared in parallel with the exploration work and the work was coordinated by a team of geologists, engineers, estates surveyors, and landscape architects. The continual acquisition of additional borehole data during the planning of the project meant that occasionally parts of the scheme had to be redesigned. However, the boreholes revealed that the areal extent of the markfieldite was somewhat larger than originally anticipated. It became apparent that a geophysical

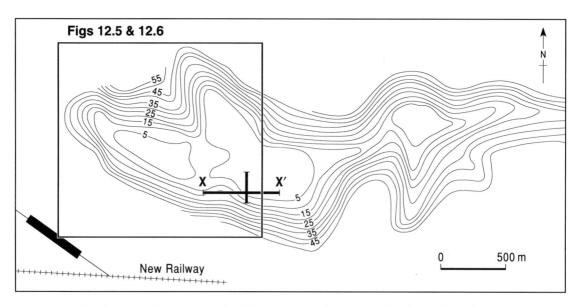

FIG. 12.4   Overburden isopach map on the Stud Farm property drawn from data derived from borehole results, conductivity measurements, and resistivity soundings.

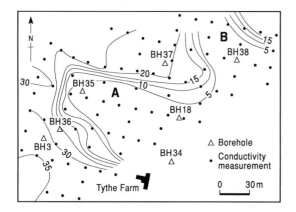

FIG. 12.5   Overburden isopach map on the Stud Farm property drawn from data derived from conductivity measurements.

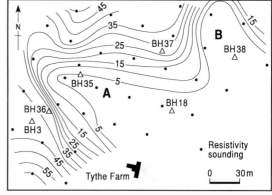

FIG. 12.6   Overburden isopach map on the Stud Farm property drawn from data derived from resistivity soundings.

investigation of the extent and form of the intrusion was essential. It was hoped that this knowledge would reduce the number of major interpretational changes that might be made in the future and also enable the drilling to be more effectively planned.

Refraction seismics

Following a trial seismic refraction survey in December 1980, a full survey was carried out across the whole of the site in February 1981. A standard 12-channel refraction seismic technique was employed, with a Nimbus seismograph recording on 12 geophones the first arrivals of the ground waves produced from the explosive source. The aim was to produce an overall picture of the bedrock topography.

The results of the survey indicated that the overburden was thin near the topographically high center of the property, and that overburden increased in all directions away from this center. Unfortunately the major geophysical surface did not correlate well with the surface of the markfieldite as identified by borehole information. It was assumed that this was because the seismic method was delineating the base of a weathered or fractured layer. In this particular deposit the depth of weathering is up to 20 m on the higher ground, but significantly less in the bedrock on the flanks of the hill where the older rocks were more deeply buried below the marls of the Triassic Marcia Mudstone. Much of the weathered rock has to be quarried conventionally and can be used for lower specification purposes. The seismic refraction method in this application had the effect of making the bedrock surface appear generally deeper than it is and also less variable at depth.

Electrical resistivity and electromagnetics

Planning permission was granted in August 1983 by the Leicestershire County Planning Department. Immediate priority was given to the detailed investigation of the site for development. An accurate overburden volume figure was required and insufficient boreholes had been drilled. Further geophysical surveys were commissioned to obtain additional detail of the overburden distribution.

Overburden thicknesses were required on a 50 m grid across the area investigated in Phase 1, and it was decided to use a combination of electrical resistivity and electromagnetic techniques to produce the detail required.

The area was traversed using the Geonics EM34 Ground Conductivity Meter (20-m and 40-m coil separations). The Mercia Mudstone, which composes most of the overburden, is considerably more conductive than the underlying markfieldite. Thus an increased thickness of the overburden produced an increased conductivity reading. The readings, reflecting the degree of conductivity of the different materials, were then used to produce a conductivity contour map. Therefore, by traversing the site it was possible to produce a qualitative picture of overburden variation. Although the assumption of a two-layer model of overburden and bedrock oversimplified the situation somewhat, e.g. an additional boulder clay layer, etc. could be present, the marked contrast in con-

ductivity meant that the method proved to be very successful in delineating the shallower (<20 m deep) bedrock features (Fig. 12.5). Although borehole information allowed some correlation of conductivity with depth, the results remained essentially qualitative and a resistivity survey was carried out to quantify the data (Fig. 12.6).

The quantification of the conductivity contours was attempted by taking a series of resistivity soundings (electrical depth probes) along sectors of equal conductivity. This technique produces a more accurate depth reading at a point, by avoiding significant lateral variations in overburden thickness. The resistivity sounding methods used employed the British Geological Survey (BGS) multicore offset sounding cable (a Werner configuration) with an Aberm Terrameter. An electrical current is passed through two electrodes and the potential difference is recorded across a further pass of electrodes to produce a resistance value in the standard array. The curves produced are computer processed to give depths to bedrock at the measurement sites.

With the highly variable nature of the overburden thickness, the electromagnetic results proved to be very important for accurately positioning the resistivity lines so that they did not cross any sharp lateral variations in bedrock topography. This greatly increased the quality of the resistivity data.

Combining the electromagnetic and resistivity results with borehole data produced an overburden isopach map (Fig. 12.4). The most important features identified from this plan are the two shallow "ridges" (marked A and B on Figs 12.5 & 12.6) away from which the overburden thickness increases rapidly. Neither of these features had been identified by the earliest drilling and, because of their narrow nature, might have remained undetected for some time. Their presence resulted in a significant reduction in calculated overburden volume.

Reflection seismics

Recent modification to seismic source, detector, recording equipment, and field techniques has improved the seismic reflection method of exploration to enable it to be used sucessfully for exploration at shallow depths. The system

100 m

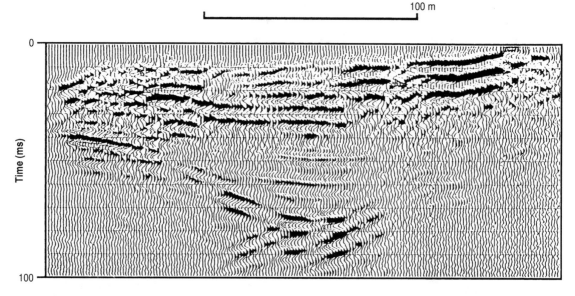

FIG. 12.7   A migrated section of part of line X–X' (Fig. 12.4), showing an image of an ancient wadi cut in markfieldite and filled with Triassic mudstone. (After Ali & Hill 1991.)

has been used recently (Ali & Hill 1991) to generate, extract and record high-resolution seismic data which eliminated noise, in particular low frequency, high amplitude, source-generated ground roll. High-frequency energy was needed to identify and distinguish them from other interfering waves.

Shallow reflection seismic methods were used on the Stud Farm property to locate and image the unconformity between the Triassic Mercia Mudstones and the markfieldite. It was assumed that the homogeneity of the overlying mudstone would result in no reflections from within their sequence, and therefore make the detection of reflections from the unconformity easier (Hill 1990). However, some reflections were obtained from within the mudstone and these are thought to represent sandy and calcareous horizons (Fig. 12.7).

Data were recorded along two lines (Fig. 12.4), usually with sixfold coverage and 1-m geophone spacing. Despite poor weather and poorly consolidated near-surface materials, some reflected waves were recognized. A probable function was chosen based on experience gained in a nearby quarry, with similar geology (Ali & Hill 1991) and stacking of the data carried out. Careful processing and correction

of the data produced a section on which the unconformity is clearly imaged as a steep sided, ancient wadi (Fig. 12.7).

12.4.3   Sample testing

In any deposit, and particularly in a new one, it is essential to determine the precise physical characteristics of the materials, both in order to establish the potential markets and to determine the type of processing plant that will be required. In some deposits a detailed chemical profile of the material may also be necessary.

The intrusive rocks of this part of Leicestershire are known to be hard, durable, and mainly consistent, but it is nevertheless essential to carry out a large number of British Standard and other special tests to quantify any variations and to establish the precise characteristics of the materials. The core samples obtained were of two sizes (35 mm and 60 mm in diameter) and all were subjected, where sample quantity permitted, to the following tests as stipulated in British Standard 812 (1975): Relative Density, Water Absorption, Aggregate Impact Value (AIV), and Aggregate Crushing Value (ACV). A few samples were also tested under the same standard for

Aggregate Abrasion Value (AAV) and Polished Stone Value (PSV).

The Leicestershire quarries have historically supplied stone to the profitable British Rail High Speed Track ballast market. In order to qualify for this market the stone must achieve a very low and consistent value in a non-British Standard wet attrition test. A larger size material is required for this test and consequently only the 60-mm cores could be used. As testing progressed, it proved possible to establish a usable correlation between the water absorption values determined on all samples and the wet attrition values.

Over 120 samples of core were tested for some, or all, of the properties mentioned above. The results provided the data necessary to determine which markets could be penetrated and what type of basic plant would be required to produce the materials to satisfy those outlets in terms of both anticipated quantity and product specification.

12.4.4 Reserve estimation

The confidence in the resources grew as the amount of data increased and several refinements of the estimation method evolved as follows:

1 After the initial phase of flight auger drilling an inferred resource estimate (USGS 1976) of some 20 Mt ± 100% was quoted.

2 By the time the first planning application was submitted, a generalized bedrock and deposit shape had been defined by widely spaced boreholes and some typical cross-sections had been drawn. A resource estimate based on the area of the cross-section, extended to the mid point between sections likely to be recovered using an average quarry configuration, gave an enhanced figure with a confidence limit of ±20%.

3 By 1989 with much of the relevant drilling completed, two sets of cross-sections (N–S and E–W) at 50-m intervals were drawn. With a provisional quarry development scheme superimposed, a more precise reserve estimate was produced. The extractable area on each cross-section was multiplied by the distance to the midpoint between the adjacent sections on each side. The N–S sections were considered best but the figures were also verified using the

E–W sections. In all cases the volumes were converted to tonnages by multiplying the volume by the average relative density of the rock in a saturated, surface dried condition. In this case the value 2.7 was used for the fresh rock. The final calculations were considered more precise and were accorded a confidence limit of ±10%.

12.4.5 Quarry planning

As the drilling and sample testing continued with a progressive clarification of the combined overburden and weathered rock thicknesses, and of the overall structure, shape, and quality of the deposit, three possible operating schemes were devised for developing the new quarry in conjunction with recovering the sizeable resources which would become available following the dismantling of the plant at the old quarry.

Bell and Hopkins (1988) described these three options as follows. A straightforward option to work out Cliffe Hill Quarry completely, while towards the end of its life, developing Stud Farm as its natural successor. This scheme had several drawbacks. The dimensions of both quarries would impose a ceiling on potential output. The extension of Cliffe Hill would require a road diversion and relocation of the existing plant and unfortunately the only feasible site for the latter would position it closer to an adjacent village. This potential plant site was also the only area suitable for overburden disposal. Even if these problems could be resolved for Cliffe Hill it would be necessary to install a second new plant when Stud Farm opened. This scheme was costed out on the basis of capital investment and the return it would yield. The main stumbling block proved to be the cost of establishing two plants, each with a relatively short life. These would have been economically viable above certain quarry outputs but neither quarry could supply the required outputs on a long-term basis.

The second option was the creation of a quarry at Stud Farm at an early date with a new processing plant and quarry. Once this was commissioned the existing Cliffe Hill plant would be demolished but crushing and screening would continue using a mobile plant. An

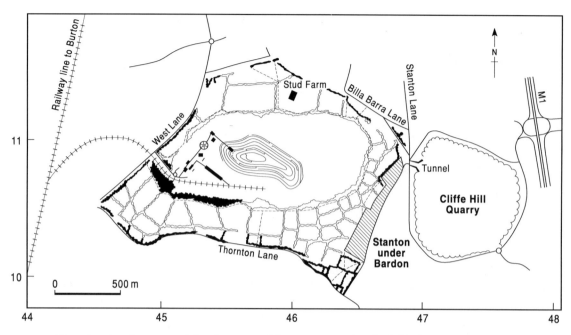

FIG. 12.8   Plan showing the proposed development of the new quarry (on the left), the current workings in Cliffe Hill Quarry, and the proposed field pattern and associated landscaping of the new quarry.

analysis of this proposed development indicated that a mobile operation of economic size at Cliffe Hill could only be located in one place and this would result in a substantial sterilization of reserves. There was also some doubt as to whether the mobile plant was capable of producing the different product types required. The financial analysis indicated that production costs using a mobile plant, which could possibly require replacement several times during the Cliffe Hill reserve extraction, combined with the additional cost of a new plant at Stud Farm, made the financial return marginal.

The third option was the most radical. This envisaged a new high capacity processing plant at Stud Farm which would be fed from both the Stud Farm and Cliffe Hill Quarries (Figs 12.2 & 12.8). The latter when extended would be linked directly to the new plant by an underground tunnel (to ensure the least disturbance to the inhabitants of Stanton under Bardon) and act as a satellite producer of primary crushed rock. Working both quarries simultaneously could supply considerably more output than a single quarry. The necessary output levels also meant that each quarry would be operating at

an optimum level from the point of view of production costs. This proposed scheme was originally received by the management with some scepticism, but when costed, it proved capable of meeting the company's financial investment and return criteria and was agreed to by all the relevant company experts before being submitted to the local planning authority in October 1980. Comprehensive but positive and wide-ranging discussions between the company and the Leicestershire county planners followed before consent was granted in August 1983.

During the discussions it became clear that the third option favored by the company would not be supported by the planners on the basis that simultaneous working on both sides of Stanton under Bardon (Fig. 12.2) would be too disruptive to the villagers. The planners did however insist that all mineable resources at both sites should be recovered at some stage. In the light of these factors, the company opted for developing a full-scale operation to work out the Stud Farm site completely, following the completion of which the remaining reserves at the old quarry would be recovered via

TABLE 12.1   Some major Phase 3 costs.

cost center	Order of magnitude of expenditure in 1987–88 £ values
1 Exploration drilling to end of 1986	280,000
2 Sample testing to the same date	25,000
3 Geological and geophysical surveys	80,000
4 Landscape design work	75,000
5 Engineering design work and research	100,000
6 Financial and administrative considerations	75,000
7 Compilation and submission of planning application	20,000
Approximate total of major costs	655,000

some suitable connection, for processing at Stud Farm.

Once planning permission had been granted, Tarmac undertook a full financial feasibility study, after which further applications were submitted in 1986, both to permit an acceleration of the development of the new quarry, and for the construction of a direct rail link from the site to the Leicester to Burton railway line (Fig. 12.2). Both proposals received planning consent in the same year. Numerous controls and legally binding agreements are associated with the planning consents, and were formulated in discussions between the planning authority and the company prior to the consents being granted.

An approximate estimate of some of the major costs for Phase 3 (as defined above) under a number of headings are tabulated in Table 12.1.

12.5   PHASE 4 – DETAILED ENGINEERING

The Phase 3 drilling, testing, and other evaluation exercises having proved, to an acceptable risk level, the existence of an economically viable quality and quantity of material, the major cost operations (collectively designated as Phase 4) commenced in 1986.

In anticipation of the commencement of major operations, extensive site investigation surveys were carried out to provide the data for the compilation of the statutory Mines & Quarries (Tips) Regulation 9 reports, without which bulk excavations and depositions on the

vast scales envisaged are prohibited in the UK. Such reports must be with the Inspectorate at least 30 days before the commencement of the operations to which they relate. The quarry development as a whole will necessitate the excavation of several million cubic meters of overburden (down to a depth of 40 m) and the disposal of this material into natural-looking landforms on the surrounding lands.

The major operations constituting Phase 4 of the development were: further drilling and testing, overburden removal, landform creation, detailed site investigation for foundation design for the plant site (Barrett 1992), plant site excavation, quarry development, blasting, road and rail access construction, the cultivation and restoration of the completed overburden disposal areas, and the erection of the plants and ancillary structures. A brief summary of the Phase 4 progress on an annual basis is as follows.

In 1986 earthworks commenced at the new site, with the clearing of hedges, the lifting and storage of topsoil from approximately half the site, and the excavation of around $1.6 \times 10^6 \, \text{m}^3$ of overburden, mainly from the proposed plant site at the western extremity. The excavated material was used to create a new landform on the southwestern perimeter of the site (Fig. 12.8). Two electricity powerlines were also re-routed during this period.

In 1987 earthworks continued with the stripping of a further $1.6 \times 10^6 \, \text{m}^3$ of overburden. This operation exposed bedrock over a workable area and completed the excavation of

Cost center	Order of magnitude of expenditure in 1988–89 £ values
1 Overburden excavation and disposal	15,000,000
2 Post Phase 3 drilling and testing	192,000
3 Landscaping design, cultivation, supervision	375,000
4 Plant purchase, erection, commissioning	30,000,000
5 Road improvements and rail link	3000,000
6 Quarry development and sundry other costs	4000,000
Approximate total of major costs	52,567,000

TABLE 12.2   Some major Phase 4 costs.

the plant area. The overburden was disposed of along the southwestern perimeter to complete this new landform. Some 8 ha of new landform were cultivated and seeded, and rock removal to form the primary crusher slot and platform commenced. Highway improvements were undertaken, and erection of the plant and associated infrastructure commenced towards the year's end.

During 1988 the erection of the processing plant, the asphalt and ready mixed concrete plants, the electrical substations, weighbridges, and offices was started and continued throughout the year. The development of quarry benches and further overburden stripping was also carried out in 1988. The overburden was deposited to create a new landform along the northern and western perimeter of the site. All completed landform sections were cultivated and seeded and tree planting commenced around the main entrance to the quarry.

In 1989 the erection of the plants and ancillary structures was essentially completed and plant commissioning commenced towards the end of the year. Further quarry development was carried out to create the approximately 200 m of 15-m-high faces that will be needed to sustain the anticipated production. The rail link was also completed and officially opened in October (Figs 12.2 & 12.8).

Cultivation and planting works continued on the completed landforms and the Central Electricity Generating Board commenced the erection of new pylons to enable the 132 kV powerline to be re-routed in 1990.

1990 saw most of the planned works completed and several million tonnes of rock produced. In the future further scheduled overburden removal and bank building phases will be required, and are incorporated with appropriate safeguards in the planning consents.

Phase 4 costs are shown in Table 12.2.

12.6   QUARRYING AND ENVIRONMENTAL IMPACT

The Stud Farm (now known as New Cliffe Hill) quarry is nearing the end of its life with annual production of 4.5 Mt of crushed aggregate, and extraction will probably be complete in 2005. As envisaged in the original plan, production will then move back to Old Cliffe Hill with processing at the New Cliffe Hill plant. In order to accomplish this, the two quarries have been linked by a 725-m-long, 9×6 m tunnel. This should enable production to continue until at least 2024, subject to a review by the planning authorities in 2007. The Cliffe Hill quarries are run by Midland Quarry Products, a joint venture of Tarmac with Hanson Quarry Products, formed in 1996 to enable more efficient working of their reserves in the area.

The Cliffe Hill quarries are a long-term operation and have had to adapt to the changing public attitudes and social makeup of the local population. During much of the twentieth century the impact of quarrying was overshadowed by that of deep level coal mining in the nearby area. Since the cessation of coal mining in the early 1990s the area has become home to

TABLE 12.3 Summary of costs by phases.

Phase	Cost in 1987–89 £ values
Phase 1 – Order-of-magnitude study	1500
Phase 2 – Pre-feasibility study (auger drilling)	2500
Phase 3 – Feasibility study (coring, testing, etc.)	650,000
Phase 4 – Detailed engineering	52,567,000
Total	53,221,000

residents commuting to nearby cities and has benefited from government aid for urban regeneration as well as the creation of a national forest in the surrounding area. Cliffe Hill has adapted to these changes by making a number of environmental initiatives and improving communication with the local population. In particular a quarry liaison committee, including representatives of local residents, local councils, and quarry management, meets every 3 months. The company also produces a 6-monthly newspaper to keep residents informed of activities at the quarries.

Although the existing operations are, at least, tolerated by local residents it would be very difficult to open a new quarry in the area. Any operation is likely to based on existing resources.

12.7 SUMMARY

The ultimate total expenditure to develop a site from greenfield into a productive and profitable quarry is high. The importance, therefore, of carrying out sufficient exploration and testing to guarantee as nearly as possible the predicted quality and quantity of material is essential. The rapidly increasing cost of each phase (Table 12.3) emphasizes the necessity for a logical approach, with each subsequent phase only being undertaken if clearly warranted by the results from the preceding one. In this particular case exploration costs accounted for approximately 1.25% of the total costs of the development of the quarry operation.

13

SOMA LIGNITE BASIN, TURKEY

MICHAEL K.G. WHATELEY

13.1 INTRODUCTION

The variables in a coal deposit (quality, thickness, etc.) are a function of the geological environment. Excellent descriptions of all aspects of coal deposits and the assessment of these deposits are given by Ward (1984), Scott (1987), and Whateley and Spears (1995). Coal or lignite is a heterogenous, organoclastic sedimentary rock mainly composed of lithified plant debris (Ward 1984), which was deposited in layers and may have vertical and lateral facies changes. These changes reflect variations in vegetation type, climate, clastic input, plant decomposition rates, structural setting, water table, etc. Coal originated as a wet spongy peat, which after burial underwent compaction and diagenesis (coalification). The coalification process proceeds at different rates in different structural settings. Coal which has been subjected to low pressure and temperature is referred to as a low rank coal, such as lignite. Coal which has been subjected to high pressure and temperature is referred to as a high rank coal, such as anthracite. Bituminous coal is a medium rank coal. Thus the properties of coal are almost entirely a reflection of the original depositional environment and diagenetic history.

Some of the properties of coal depend on the nature of the original plant material or macerals (Stach 1982, Cohen et al. 1987). These properties are measured using microscopic techniques. Large scale, subsurface changes are determined by studying the lithology of the coal and coal-bearing strata in borehole cores and down-the-hole geophysical logs. These

data form the basis for any investigation into the resource potential of a coal deposit.

The quality of a coal deposit must be assessed. This is undertaken by sampling outcrops, borehole cores (Whateley 1992), mine faces, etc. (section 10.1.3), and sending the coal to the laboratory. The coal is subjected to proximate analyses to determine the moisture, ash, volatile, and fixed carbon contents (Ward 1984). Additional tests which are often requested are sulfur content, calorific value (CV), and specific gravity (SG). The quality of coal will determine the end use to which it is put, e.g. steam coal to be burnt in a power station to generate electricity or metallurgical coal for steel making, although other uses are possible.

Since the energy crisis caused by the OPEC oil price rises which started in 1973, there has been a worldwide increase in the search for alternative fossil fuels. Lignite, bituminous coal, and anthracite are important sources of energy. The exploration for and exploitation of these fuels is important for developing countries which wish to reduce their reliance on imported oil by building coal-fired power stations at or near their own coal mines.

In 1982 Golder Associates (UK) Ltd (GA) undertook a World Bank funded feasibility study of the Soma Isiklar lignite deposit in western Turkey (Golder Associates 1983) for the state-owned coal company Turkiye Komur Isletmeleri Kurumu (TKI). The lignite from the Soma Mine was used as feedstock for the ageing 44 MW Soma A Power Station. It was the intention to increase the mine production from 1.0 to 2.5 Mt yr^{-1}, 2.0 Mt of which were to come from the expanded surface mine. The increased tonnage was needed to feed the newly constucted 660 MW Soma B Power Station. Excess lignite would be used for domestic and industrial purposes.

The GA report deals with the geological interpretation of the deposit, assessment of the reserves in terms of tonnage and quality, assessment of the geotechnical aspects of the mine area, and production of mine plans for the proposed surface and underground mines. Most of this chapter is derived from that report, although the geostatistical reserve estimations have been subsequently completed at Leicester University as student dissertations (Lebrun 1987, Zarraq 1987). The existing surface mine had suffered a major footwall failure in 1980. The steeply dipping, shaly, footwall sediments underwent noncircular failure on weak horizons over several hundred meters along strike and buried the working face. One of the intentions of the study was to establish the reasons for this failure and to incorporate safety factors into the new mine design that would reduce the risk of a repeat of this type of failure.

It is the intention in this chapter to outline the way in which the geologists who were involved on this study collected the data, presented the data for the mining and geotechnical engineers, and calculated the reserve and quality parameters. The geotechnical and mining aspects are also described. To simplify this chapter only the thick, lowermost seam in the area, designated the open pit, is described. GA also reported on the underground mining potential of the Soma Basin, which they reported might be economically mineable in the future.

13.1.1 Location

The Soma lignite deposit is in the Manisa Province of western Turkey, 10 km south of the town of Soma (Fig. 13.1). A high (1100 m) ridge separates the proposed mine site from the town. The site is on the south-facing slope of this ridge, with elevations ranging from 750 m in the north to 310 m in the south.

13.1.2 Turkish mining rights

Mining and operating rights vary in different parts of the world. Before any foreign or local company can start operations it is necessary to understand the local legal system. In Turkey all mineral and coal rights are deemed to be owned by the State and are not considered to be part of the land where they are found. The Ministry of Energy and Natural Resources administers and implements the mining laws and regulations, and grants exploration permits, exploitation permits, and leases.

13.2 EXPLORATION PROGRAMS

13.2.1 Previous work

The lignite in the Soma area has been mined since 1913, first for local domestic and industrial consumption and later as feed for the Soma A power station. TKI took over lignite production on 1979.

The database for this deposit was derived from four drilling programs (Table 13.1), field mapping, and a feasibility study conducted by TKI in 1981–82. The earliest study was conducted by Nebert (1978), which included lithological logs of 34 cored boreholes (Table 13.1) and two geological maps of the Soma area with cross-sections. Cored lignite was analyzed for moisture and ash contents, calorific value, and in some cases volatile matter and sulfur content. The holes were drilled at

TABLE 13.1 Summary of drilling exploration programs.

Drilling program	Number of holes	Total meterage	Drilling date
200	34	8790	1960
100	50	10,825	1976
300	9	1547	1981
400	29	8631	1982
Total	122	29,793	

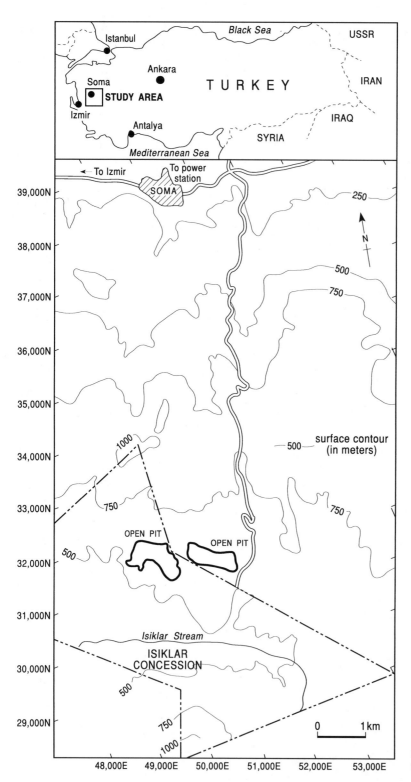

FIG. 13.1   Location diagram of
the Soma study area.

500-m centers to establish the resource potential of the basin.

The second exploration program was supervised by Otto Gold GmbH. Open-hole drilling was used with predetermined intervals cored to include lignite seams. Lignite cores were analyzed for ash and moisture content, and calorific value. The sulfur and volatile content were determined on a few samples. The program was designed to identify lateral and down-dip lignite limits, as well as to undertake infill drilling at 250 m centers within the basin.

TKI initiated the third drilling program as part of their preliminary mine feasibility study. They also used rotary drilling techniques with spot coring. Full proximate analyses and calorific value determinations were done on these cores, and occasionally sulfur analyses. These holes were drilled to obtain additional information in areas of structural complexity or areas with a paucity of data. TKI produced isopach, structure contour, isoquality, and polygonal reserve maps, as well as reserve tables, surface and underground mine plans, manpower schedules, etc.

A fourth drilling program was carried out during the feasibility study. Five of the holes were fully cored for geotechnical studies, while the remainder were rotary drilled and spot cored. Lignite cores were analyzed for proximates, calorific value, and specific gravity. Hardgrove grindability tests (section 13.6) and size analyses were also carried out on samples from the existing open pit. Although 122 boreholes were drilled, only 83 of them intersected the lowest lignite seam, or the horizon at which the equivalent of the lignite seam occurred. This was due to common drilling problems such as loss or sticking of rods in the hole, or burning the bit in at zones of serious and sudden water loss. This resulted in an overall density of 11 holes km^{-2}, equivalent to a rectangular grid roughly 330 m × 250 m, although some holes are closer and some farther apart than this. This is considered to have given sufficient density to classify the lignite in terms of measured (= proved) reserves (USGS 1976).

13.2.2 Core recovery

One of the major problems of assessing lignite or coal deposits is that of core recovery. If core is lost during drilling there is no way of determining the quality of the lost core and often the better quality, more brittle, bright sections are lost on these occasions. It is usual to have a drilling contract that requires the drillers to recover at least 95% of the core in the seam. Recoveries less than this (within reason) usually require redrilling. Techniques that help to improve core recoveries include the use of large diameter wireline (see section 10.3.2) or air flush core barrels (e.g. HQ series with a nominal hole diameter of 96.1 mm) and the use of triple tube core barrels (Cummings & Wickland 1985, Berkman 2001).

13.2.3 Geophysical logging

Down-the-hole (DTH) geophysical logging (see section 7.13) was first used in the oil industry but soon found its way into coal exploration (Ellis 1988). It is now used routinely in the evaluation of coal deposits because geophysical logs can help reduce drilling costs by enabling the use of cheaper, rotary, open-hole drilling of, say, 80% of the holes. Good comparison of open hole and cored holes is achieved with the use of DTH logs. They can also help in identifying the top and bottom of the seams, partings within the seams, lithological changes, in checking on core recovery and the depth of each hole, and ensuring that drillers do not claim for more meters than they actually drilled. Seams often have characteristic geophysical signatures which, in structurally complex areas, can help with seam correlation.

Typical DTH geophysical logs used on coal and lignite exploration programs are natural gamma, density, neutron, caliper, and resistivity, as well as sonic and slim line dip meter logs (Ellis 1988). Shale, mudstone, and marl usually have a high natural gamma response while coal has a low response, with sharp contacts often being observed. Coarsening-upward or fining-upward sequences in the clastic sections of the logs can also be inferred from the gamma logs. The density log, as its name implies, reflects the change in density of the rocks, with coal and lignite having low densities and shale and sandstone having higher densities. The neutron tool is used in estimating porosity, and the resistivity tool may be used to indicate bed boundaries. The caliper tool defines the size of the hole. The sonic log is also used in estimating the rock quality (fracture frequency), and the dip meter can be used to interpret sedimentary structures (Selley 1989).

A geophysical log showing the typical responses that the rocks in the Soma basin give, is described in section 13.3.2.

13.2.4 Sampling

It is essential to establish a procedure for sampling core and working faces so that continuity is maintained throughout an exploration project and into the mining phase. For example, at Soma the whole lignite sequence was sampled, including all parting material. The core was split and one half was retained in the field. All lithological layers in the seam greater than 30 cm thick were sampled. Thinner layers were included with adjacent layers until a minimum thickness of 30 cm was obtained. The 30 cm limit was used as it was considered by the study team that this was the minimum thickness of parting that could be mined as waste in the open pit (see also sections 13.6.1 to 13.6.3). This ensured that weighted average estimates for run-of-mine (ROM) lignite could be made.

13.2.5 Grouting

Where deep coal is likely to be mined by underground methods, all holes drilled prior to mining should be sealed using pressure grouting. This will reduce the potential water inrush hazard in underground mining.

13.3 GEOLOGY

13.3.1 Geological setting

The basement in western Turkey consists of Precambrian, Palaeozoic, and Mesozoic sedimentary and igneous rock (Campbell 1971, Brinkmann 1976) which have been subjected to various structural and metamorphic episodes. Turkey has a great variety of structures, the largest of which is the North Anatolian Fault (NAF), a major strike-slip fault system (Fig. 13.2). The NAF was formed in the late Serravallian (Sengor et al. 1985). At this time

westerly strike-slip movement of the west Anatolian Extensional Province took place. At the same time a series of NE-trending grabens, such as the Soma Graben, began to form in western Turkey. These grabens began to fill with Serravallian (Samartian) sediments, which contain thick lignite deposits.

13.3.2 Geology of the Soma Basin

The Soma Basin contains thick deposits of Miocene and Pliocene sediments (Fig. 13.3), which range in age from Serravallian (Samartian) to Pontian (Gökçen 1982), but do not contain volcanic rocks. The Tertiary sediments rest unconformably on Mesozoic basement rocks. The stratigraphy of the basin is summarized in Fig. 13.4. It contains two thick Miocene lignite seams designated the KM2 and KM3, but this case history deals only with the KM2 seam.

The Mesozoic basement forms rugged topography around the basin of limestone hills rising to 350 m above the Tertiary sediments. Boreholes which reached the basement confirmed the presence of limestone below the sediments in the basin. At Soma, the top of the basement appears to consist of debris flow deposits.

The Miocene deposits have been divided into three formations (Fig. 13.4). The basal Turgut Formation is predominantly immature sandstone and conglomerate, but there is a gradational upward fining of the sediments to the lignite horizon. Increasing amounts of sandy clay, clayey silt, and carbonaceous clay appear towards the top of the formation until these grade into the KM2 seam. The basal contact of the seam is gradational and is usually placed where the first recognizable lignite appears with less than 55% ash content. The lignite is hard, black, and bright, has numerous cleats (typical close spaced jointing of coal and lignite), and breaks with a concoidal fracture. Using the ASTM classification, the calorific value of this material would more properly result in it being termed a sub-bituminous B coal, but it is referred to as a lignite in the local terminology.

There is an extremely sharp contact between the KM2 and the overlying Sekköy Formation. This formation is a massive, hard marlstone

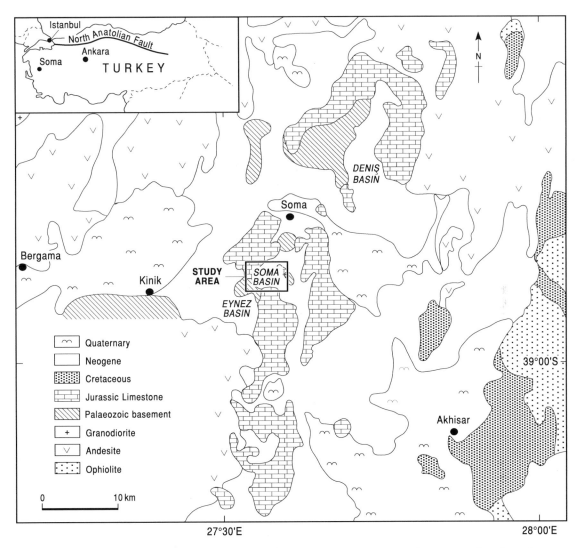

FIG. 13.2 Generalized geology of the Soma area (taken from the 1:250,000 Izmir sheet regional geological map published by MTA). The inset shows the regional structural setting for the Soma Basin in relation to the major structural feature of northern Turkey, the North Anatolian Fault (NAF).

and geophysical logs identify this contact clearly. The KM2 has a distinctly low natural gamma signature in contrast to the very high signature of the marl. The sharp change from peat (lignite) formation to the marlstone is believed to be the result of a change to an arid climate (Sengor et al. 1985).

In cores and in the open pit, the contact with the overlying freshwater limestone of the Yatagan Formation is unconformable and marked by a color change. The gamma log

shows a distinct change from the very high signature of the marl to a medium response in the limestone. The Yatagan Formation contains the KM3 Lignite which is interbedded with a series of thick calcareous clastic units, making the economic assessment of this unit more difficult. The freshwater limestone is conformably overlain by massive marlstone the basal half of which contains thin, laterally impersistent, lignite seams. Gökçen (1982) considered these marlstones to belong to the

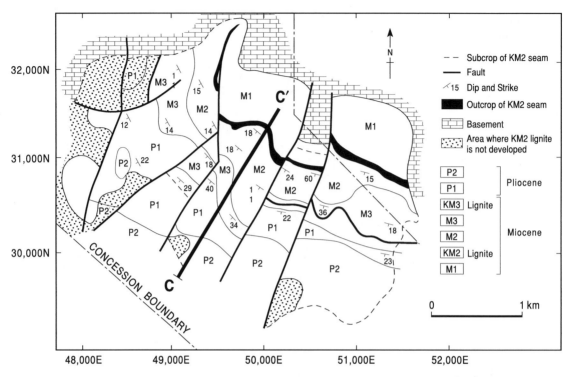

FIG. 13.3 Simplified geological map of the Soma Basin. The detail of the limit of the lignite in the south-south-west is unclear as no information was available due to the great depth of the lignite. Block numbers refer to the proposed mining blocks used in reserve and quality estimation procedures (see text).

Upper Miocene Yatagan Formation, although Golder Associates (1983) placed the Miocene–Pliocene contact at the top of the freshwater limestone.

13.3.3 Structure of the Soma Basin

The Soma Basin is a fault-controlled graben, which has been subjected to regional tilting to the southwest. The basin formed in response to stike-slip faulting along the North Anatolian Fault Zone during the mid Miocene (Sengor et al. 1985). The faulting in the Soma Basin trends NE–SW (Fig. 13.3). The major faults were defined by elevation differences in the structure contour map of the top of the KM2 seam. It is generally accepted that a disruption to the trend of a set of structural contours, that are expected to behave in a uniform way, may indicate the presence of a fault (Annels 1991, Gribble 1993, 1994). Fault positions were interpolated as lying between certain boreholes and,

as mining proceeds and more information becomes available, then the position of these faults will be defined more accurately. These faults have throws of up to 150 m and displacements of up to 300 m.

Faults of this size will affect the design of both surface and underground mines. The faults at Soma were used to define mining blocks, Blocks A, B, C, D, and E (Fig. 13.8). What is more difficult to determine is the amount of secondary faulting and fracturing formed in association with the major faulting. These minor, subparallel faults may have throws of only a few meters. It is difficult to plot these faults from drilling results, but they may have a significant effect on underground mining operations.

The sediments were deposited around the arcuate northern basin rim, and they dip to the southwest at an average of 20 degrees. These dips vary within each block, e.g. near the northern rim of the basin the dips are significantly

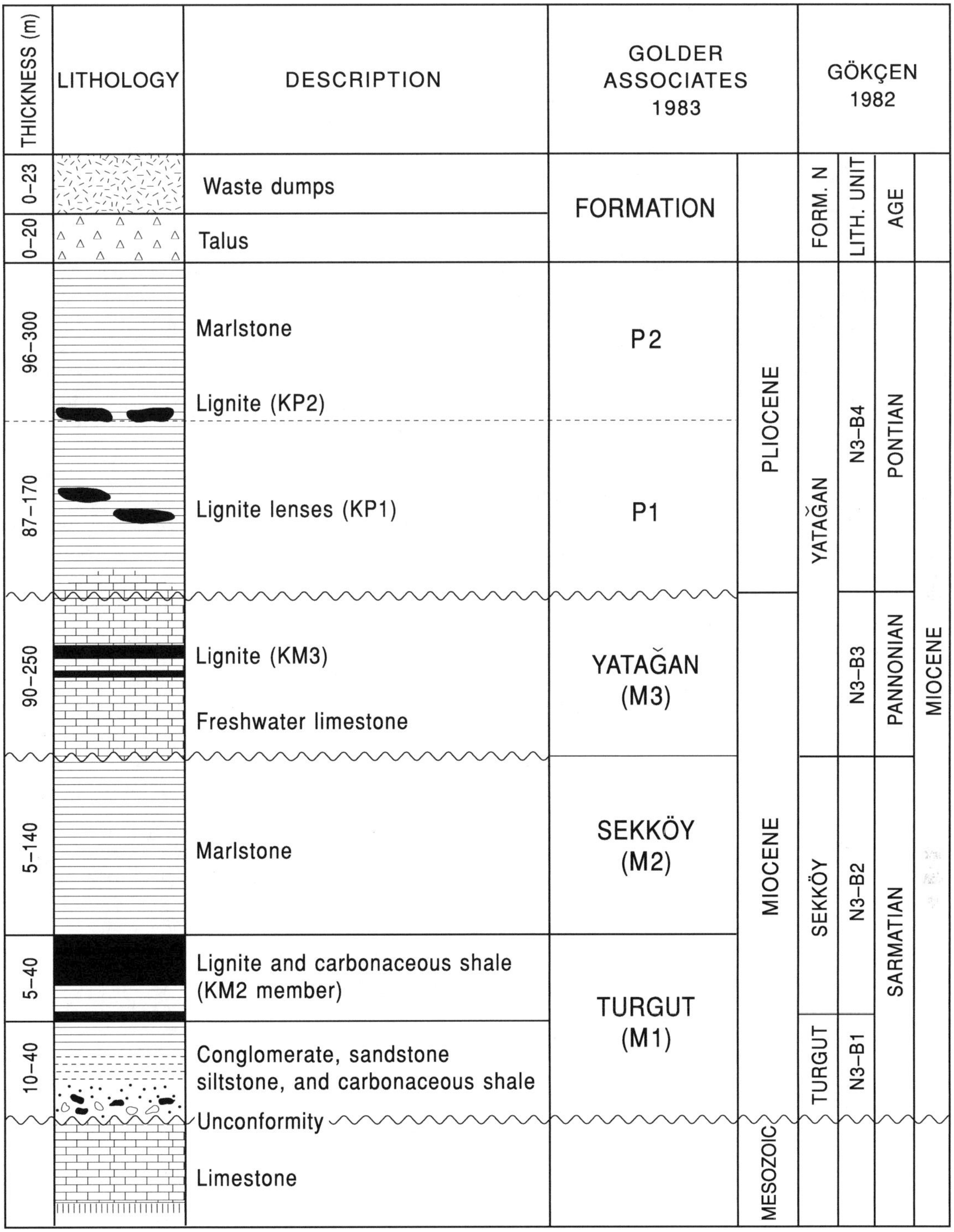

THICKNESS (m)	LITHOLOGY	DESCRIPTION	GOLDER ASSOCIATES 1983		GÖKÇEN 1982			
					FORM. N	LITH. UNIT	AGE	
0–23		Waste dumps	FORMATION					
0–20		Talus						
96–300		Marlstone	P2	PLIOCENE	YATAĞAN	N3-B4	PONTIAN	
		Lignite (KP2)						
87–170		Lignite lenses (KP1)	P1					
90–250		Lignite (KM3)	YATAĞAN (M3)	MIOCENE		N3-B3	PANNONIAN	MIOCENE
		Freshwater limestone						
5–140		Marlstone	SEKKÖY (M2)		SEKKÖY	N3-B2	SARMATIAN	
5–40		Lignite and carbonaceous shale (KM2 member)	TURGUT (M1)					
10–40		Conglomerate, sandstone siltstone, and carbonaceous shale			TURGUT	N3-B1		
		Unconformity		MESOZOIC				
		Limestone						

FIG. 13.4   A simplified stratigraphical column for the Soma Basin. The sediments have not been dated with any certainty. Comparison of traditional and recent microfossil dating (Gökçen personal communication) shows that more detailed work is still required.

steeper. A more accurate assessment of the dips was made by interpreting the structure contour map of the top of the KM2 seam (Section 13.4).

13.3.4 Depositional model for the Soma Basin

A depositional model of the Soma Basin was proposed after the isopach and ash content maps of the KM2 seam, detailed borehole logs, and stratigraphical sections were examined and interpreted in the light of ancient and modern analogs, in the manner described by McCabe (1984, 1987, 1991). The basin originated as a NE–SW fault-controlled graben in the basement rocks, which gradually infilled as the basin subsided at varying rates. Steep peripheral gradients around the north of the basin at the time of initial deposition resulted in rapid transportation of sand and gravel to the center of the basin by high energy runoff. These debris flow deposits alternate with poorly sorted, coarse-grained sandstone. Gradual lessening of the gradient, with subsequent lowering of the energy regime, resulted

in sediment load fractionation and deposition of finer sediments in the basin (sandstone and siltstone). The overall result is a fining-upward sequence from gravelly sandstone at the base to siltstone and mudstone at the top (Fig. 13.4). Temporal variations in depositional conditions are indicated by the repetition of these fining-upward sequences and rapid lateral variation. In the partially filled basin, conditions were conducive to plant growth which resulted in the formation of the thick, laterally variable KM2 lignite seam. The lignite is generally shaly at the base but improves in quality (i.e. has a lower ash content) towards the top.

Two distinct sub-basins have been recognized, the eastern Demir Basin and the western Seri Basin, separated by the Elmcik High (Fig. 13.5). The plant-forming ecosystem or mire (Moore 1987) formed around the northern rim of the basin, but carbonaceous material was transported to the more distal portions of the basin to the south and southwest. Fine-grained sediment was also transported into the basin, and deposited mainly along the northern rim of the Seri Basin, where the lignite is

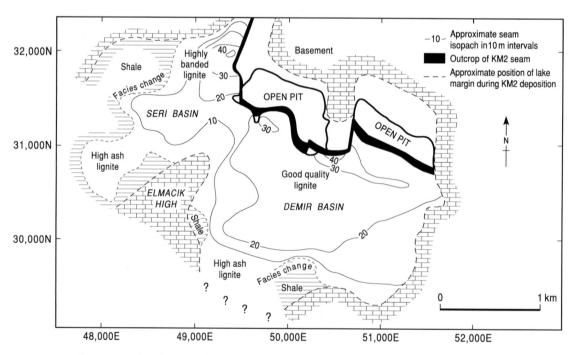

FIG. 13.5   A palaeogeographical interpretation of the Soma Basin at the time that the KM2 seam was being formed.

strongly banded having many nonlignitic partings. The overall result is a gradual facies change from low ash lignite to carbonaceous shale basinward. There is a pronounced thinning of the lignite over the Elmacik High, suggesting that this area was an active high during peat formation.

A climatic change from warm humid conditions favoring peat formation to an arid climate resulted in a sudden and basin-wide change of sedimentation to marl deposition. The fine grain size and the calcareous nature of the material suggests deposition by low energy input into a low energy water body. The resulting marlstone (Sekköy Formation) is unconformably overlain by the freshwater limestone of the Yatagan Formation. The KM3 lignite occurs at varying levels within the limestone, but is areally restricted. The KM3 has a significantly higher ash content than the KM2 lignite. Alternating marlstone and limestone continues upwards, with occasional thin, laterally impersistent seams developing.

13.4 DATA ASSESSMENT

In order to establish the reserve and quality criteria that were required for the mine design and financial analysis of the project, several maps had to be constructed, such as structure contour maps, isopach and isoquality maps, as well as stripping ratio maps and cross-sections. All the borehole logs were examined, the seams were correlated, and the top and bottom of the KM2 seam were established. Lignitic material was frequently rejected from the base of the seam because of poor quality. The top of the seam was readily identified at the sharp contact of the lignite with the overlying marlstone, but the basal contact was often drawn only after the quality data had been assessed. Once the seam thickness was established it was possible to construct the maps.

Since this study was completed several integrated computer packages have become available that calculate and plot contours quickly and accurately (Whateley 1991), e.g. Datamine, Surpac, Borsurv, PCExplore, etc.). During this study a planimeter was used to measure areas (e.g. between contours for tonnage calculations (sections 13.4.2 & 13.7)). Modern software

packages have various user-definable methods of contour construction. Two contoured surfaces (or digital terrain models (DTMs)) can be superimposed and volumes calculated, e.g. DTMs of the top and bottom of a seam, or DTMs of the surface and the top of a seam for overburden volume. The assessment of all the data up to and including the mine design and the financial analysis can now be undertaken using these computer packages. Manual methods (described in section 10.4) are still widely used by many exploration and mining companies.

13.4.1 Structure contour maps

Three structure contour maps were drawn for this study: (i) on the top of the KM2 seam; (ii) on the bottom of the KM2 seam; and (iii) on the top of the Mesozoic Basement. The KM2 seam has the greatest areal distribution, a very well defined upper contact, and the most detailed and reliable database. The elevation of the top of the KM2 seam was plotted manually at each data point and contours were constructed by linear interpolation between points. The contours were drawn at 20-m intervals (Fig. 13.6). The strike of the lignite seam at depth was assumed to be similar to that of the overlying sediments at the surface. Where the structure contours differed significantly from this, a fault was inferred. By checking against the surface geological map and cross-sections it was possible to establish the fault pattern in the basin. The structure contours were then redrawn between the faults to give the pattern shown (Fig. 13.6). It was important to determine the structure of the basin in order to assist the mining engineers with their design of the optimum open pit, by avoiding areas of unstable ground near faults and loss of lignite near these fault zones.

Once the structural pattern was established, it could be transferred to the structure maps of the base of the KM2 and the top of the basement. The structure contour map of the base of the KM2 was constructed by placing the isopach map of the KM2 on top of the structure contour map of the top of the KM2. Where the two sets of contours intersected, the seam thickness was subtracted from the top of seam elevation to give the new elevations. These

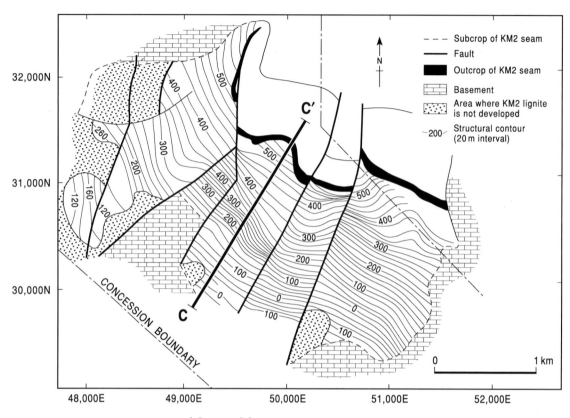

FIG. 13.6   A structure contour map of the top of the KM2 seam, Soma Basin.

contours were drawn at 20-m intervals, and were only drawn in the area designed to be the open pit, for use in mine design and to identify areas of potential slope failure.

The structure contour map of the top of the basement was also produced by linear interpolation. Only a small number of boreholes penetrated the basement, so the contours were constructed at 50-m intervals. The structure contour map of the top of the KM2 was then placed on the basement map and then the intersections of contours of the same value outlined the limits of the KM2 seam (Fig. 13.3).

13.4.2   Isopach maps

For this study, three isopach maps were drawn manually: (i) for the KM2 seam (Fig. 13.7); (ii) for the overburden material; and (iii) for the sediments between the KM2 seam and the basement (the Turgut Formation). The thickness of the KM2 was determined in each hole

between the sharp upper contact with the overlying marlstone and the base of the seam. The isopachs show the thickness of KM2 that could be extracted by surface and underground mining methods, and include waste partings within the lignite which could be separated as waste during mining (section 13.4.4). The isopach map was constructed at 5-m intervals by linear interpolation. The seam ranges in thickness from 2 to 57 m, with an average in the open pit area of some 17 m. The extreme thickness of 57 m is the result of a borehole intersecting a steeply dipping part of the seam. The seam isopach map formed the basis on which the reserve estimates were based, as well as being used to calculate the structure contours at the base of the KM2 seam.

The isopach map of the waste material above the KM2 seam (the overburden) was constructed by placing the map of the surface topography over the structure contour map of the top of the KM2 seam. Where the respective

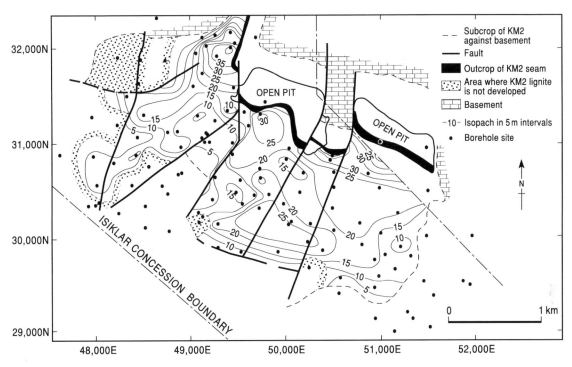

FIG. 13.7 An isopach map of the KM2 seam. Location of all the boreholes drilled in the Soma Basin are also shown.

contours intersected, the elevation of the top of the seam was subtracted from the elevation of the topographical contour and the resulting thickness value plotted. This method of constructing the overburden map provided more data points than would have been available had the borehole values alone been used, resulting in a more reliable map. The overburden above the KM2 seam varies from a few meters near the existing workings, to approximately 150 m at the deepest part of the proposed surface mine. The map was used to construct the stripping ratio map needed in the mine planning exercise.

A detailed contour map was constructed during the mine design phase of the project for use by the mining engineers to calculate the volume of overburden material which had to be removed at various stages of mining from the various mining blocks (see section 13.3.3). Contours were constructed and a planimeter was used to measure the area between adjacent overburden thickness contours which were then multiplied by the applicable vertical

overburden thickness. Products were summed to obtain the total volume for each block (Table 13.2). Pit slope volumes were calculated in a similar manner assuming a final pit slope angle of 45 degrees.

An isopach map of the Turgut Formation was constructed by subtracting the values of the structure contours at the base of the KM2 seam from the structure contours at the top of the basement. This isopach map was only drawn for areas of potential open pit development to assist the geotechnical engineers in predicting the areas of potential footwall failure in future operations (see section 13.5.1).

13.4.3 Stripping ratio maps

The stripping ratio map, or overburden ratio map shows the overburden material in m^3 as a ratio per tonne of lignite, prior to applying selective mining criteria. This map (Fig. 13.8) was constructed by placing the overburden isopach map on the isopach map of the lignite. Where the contours intersected, the thickness of the

TABLE 13.2   Estimated overburden volumes at the proposed Soma Isiklar open pit mine. For explanation of M bank m³ see section 13.8.2.

Block	Pit area			Slopes			Total volume (M bank m³)
	Area (km²)	Thickness (m)	Volume (M bank m³)	Area (km²)	Thickness (m)	Volume (M bank m³)	
A	0.146	78.9	11.52	0.148	62.7	9.28	20.80
B	0.412	90.1	37.12	0.189	54.6	10.32	47.44
D	0.072	87.6	6.31	0.096	60.3	5.79	12.10
E	0.882	84.8	74.79	0.357	70.1	25.24	100.03
Totals (mean)	1.512	(85.8)	129.74	0.790	(64.1)	50.63	180.37

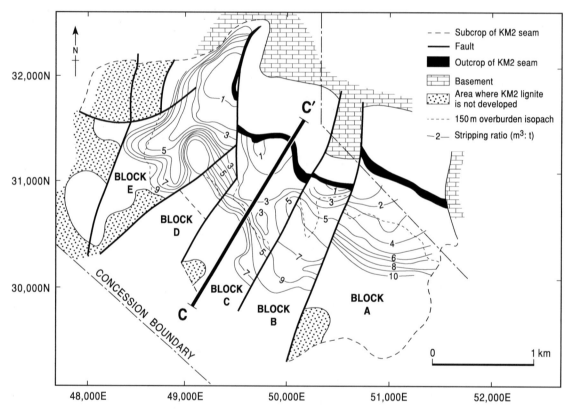

FIG. 13.8   A stripping ratio map showing the 150 m overburden isopach line used to determine the down-dip limit of the open pit mine.

lignite was multiplied by a specific gravity of 1.73 to obtain tonnes. The specific gravity was obtained by averaging the values derived from laboratory analyses of lignite from the open pit area of the mine. The product was divided into the overburden thickness to calculate the stripping ratio. Contours were then constructed at unit intervals between 1 and 10.

The stripping ratio map provides a guide to areas where mining would be preferentially started, i.e. where the ratios are lowest. The financial study suggested that open pit mining could be carried out economically where the stripping ratio is less than 7:1. As a general guide, the 7:1 stripping ratio in coal or lignite mining provides a limit to surface mining,

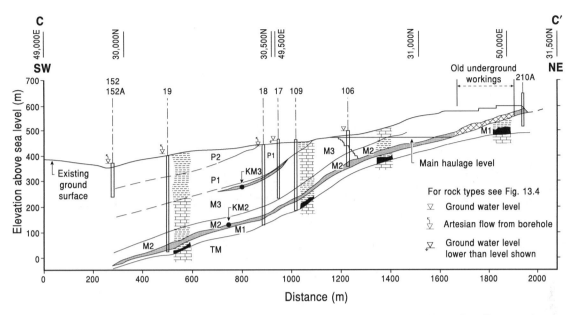

FIG. 13.9   An example of a down-dip section used to help in correlating the stratigraphy and to illustrate the hydrogeological information derived from boreholes.

although different companies operate different criteria and this figure is variable.

13.4.4   Cross-sections

East–west, north–south and down-dip sections (Fig. 13.9) were drawn at regular intervals across the basin. They were constructed by plotting the data from the structure contour and isopach maps. Boreholes which fell on or close to the section lines were also plotted. The east–west and north–south sections were cross-checked by comparing and correlating the intersection points. The sections were used to verify the structural information and the seam correlations. They were not used for reserve estimation (Reedman 1979), although this is commonly done by some coal mining companies.

13.5   GEOTECHNICAL INVESTIGATION

13.5.1   Investigation program

It is always advisable to initiate geotechnical investigations as soon as possible in an exploration program in order to avoid expensive duplicate drilling of boreholes. In this study the following program was carried out:

1 *Geotechnical logging* of five fully cored boreholes. In addition to the geological descriptions, the logs included the rock quality designation (RQD) (section 9.6) and evaluation of the point load strength (Brown 1981), which can be used to estimate the uniaxial compressive strength. Where a point load strength test was not available, the field description was used as a basis for estimating the strength class (see Table 10.8). All the cores were fully labeled and photographed (Fig. 13.10). This is particularly important where cores are prone to deterioration.

2 *Geotechnical testing* of borehole cores and samples collected from the existing open pit.

3 *Structural mapping* of overburden exposures in the existing open pit.

4 *Measurement of water levels* in exploration boreholes and measurement of flows from artesian boreholes.

5 *Sensitivity analyses* of known failures altering the variables such as position of the piezometric surface or material strength to establish typical shear strength values for design from back analysis of existing footwall failures in the open pit.

FIG. 13.10   A photograph of a box of core drilled during this study. All core recovered during this study was photographed before the core had been tested and sampled by the geotechnical engineers and before the lignite had been sampled. This ensured that a full record of the *in situ* core was available. All core box lids had a detailed descriptive label attached which gave the borehole number, box number, depths, date drilled, and a black and white and color chart to ensure that each film was processed to the same standard.

These data were used in geotechnical analyses to provide preliminary design guidelines for the proposed open pit.

13.5.2   Geotechnical conditions

Overburden

The overburden consists predominantly of limestone and marlstone with lesser amounts of mudstone and siltstone. The limestone and marlstone are moderately strong, with mean uniaxial compressive strengths of the order of 80 MPa. The siltstone and mudstone are weaker rocks with compressive strengths generally less than 30 MPa. The overburden rocks are bedded and jointed and the bedding is generally parallel to that in the lignite seam

and dips range between 15 and 45 degrees to the southwest. Jointing is mainly vertical.

Lignite

Point load strengths of selected lumps of lignite are about 1 MPa. However, the lignite is highly cleated and friable and the mass strength would be considerably lower.

Footwall

The footwall of the KM2 seam consists of variable thicknesses of mudstone, siltstone and immature sandstone overlying basement limestone. The immediate footwall has a high clay content and is generally weak, with a high slaking potential. Typical residual shear strengths for this material are c = 0–30 kPa, and Π = 14–17 degrees, where c = shear strength and Π = angle of shear resistance. A series of back analyses was carried out on exposed footwall material in the existing open pit. These analyses confirmed that the above shear strength values were realistic for design purposes, and that footwall stability is highly dependent on groundwater conditions and the local dip of the footwall.

13.5.3 Hydrogeological conditions

Standing water levels in exploration boreholes were measured using piezometers. The levels were generally close to the surface and, in the topographically lower areas, flowing artesian boreholes occurred. In some cases, artesian flows were high ($7\ \text{L s}^{-1}$) and flows could be maintained for long periods. One of the main aquifers in the strata appeared to lie in the vicinity of the KM2 seam.

13.5.4 Implications for open pit mining

The overburden rocks are moderately strong, requiring drilling and blasting prior to excavation. Controlled blasting procedures would be required on all final walls for good slope stability. Practical mining considerations would limit the maximum wall slope to about 55 degrees between haulroads and an overall slope of 45 degrees for the final highwall of the pit, including provision for 20-m-wide internal haul roads. The overall slope angle of the advancing face is determined by the equipment clearance and access requirements but should not exceed 40 degrees for stability and safety. This could be steepened to 45 degrees at the final limit when controlled blasting procedures are used. The footwall dip means that footwall failures could occur in the new pit. In order to minimize the risk, a strike advance mining method was recommended (section 13.8.1).

Dewatering the new open pit by means of boreholes would probably be required, especially prior to the initial box-cut excavation. A continuous program of borehole dewatering will probably be required, to reduce water pressure in the basement thus avoiding the possibility of footwall heave. An excellent description of the processes involved in the

dewatering of a large open pit mine is given by Cameron and Middlemis (1994).

13.6 LIGNITE QUALITY

Coal seams consist of multiple layers of carbonaceous material and noncombustible rocks (waste), such as clay, shale, marlstone, sandstone, etc. These interbedded sediments are not continuous and at Soma can be regarded as lenses within the range of the borehole spacing. This inevitably means that difficulties arise when seam quality and reserve quantities are being considered. Two alternatives are available when considering the approach to be adopted when estimating lignite quality (and reserves). The first is to assume that the zone within the lignite seam containing most of the lignite layers is mined in total without any attempt at selectively mining waste partings. This gives the *in situ* lignite quality data. The second is to consider mining the lignite layers selectively, aiming to produce a run-of-mine (ROM) product which is of acceptable quality. This gives the mineable lignite quality data.

The nonselective mining approach will result in a ROM product which would not generally meet the power station specification and would be highly variable in composition. This could be homogenized in a blending stockpile and upgraded in a washing plant, which is costly and results in losses. Selective mining in the open pit will be more expensive than bulk mining since time will be lost in moving machinery and there will also be unavoidable losses and dilution of lignite associated with this method. At Soma it was necessary to adopt the selective mining approach in the lignite quality and reserve estimations, because the lignite quality is low and further quality losses caused by bulk mining would be unacceptable.

Lignite quality was assessed by evaluating the data provided with the borehole logs obtained during the first three drilling programs (Table 13.1). Additional quality data came from samples of core submitted to the MTA laboratory in Ankara during the feasibility study. The core from the earlier programs was analyzed for ash, moisture (on an as-received basis), and calorific value. In some cases volatile matter and sulfur content were also determined. In the final drilling program the lignite core was analyzed for proximate analyses on an as-received basis as well as specific gravity and calorific value. As the power station stock-piles the lignite in the open, the as-received analyses approximate more closely to the quality of the lignite that is actually burnt. Lignite samples which do not have the surface moisture removed by air drying are said to have been analyzed on an as-received basis. Samples which are air dried until the mass of the samples remains constant (all surface moisture is assumed to have evaporated) before being analyzed are reported on an air-dried basis (section 13.6.4).

These data were placed on a computer database to facilitate the assessment of the lignite quality, using down-hole-weighted averaging techniques. The samples were weighted by sample length and specific gravity in order to calculate the weighted average quality for each composite. To ensure that a true weighted average was obtained, quality values had to be assigned to the partings within the seam. Minor partings are often not sampled during the exploration phase of a program, but during mining these partings will often be incorporated with the lignite, thus reducing the quality of the product. Partings were sampled in the last drilling program and these values were assumed to apply to similar rock types that had not been sampled earlier. Similarly, where the specific gravity of the lignite had not been determined, a value was assumed (Table 13.3). Holes where the core recovery was less than 75% were omitted as not being representative.

The possible expansion of the lignite mine was investigated for the purpose of increasing the supply of lignite to the adjacent thermal power stations. The lignite is crushed and fed directly into the furnace as a powdered fuel. The solid fuel specifications for steam-raising plants usually includes the Hardgrove Grindability Index. This test gives an indication of the ease with which a material will be crushed (Ward 1984). A low value indicates a hard rock while a high number indicates a relatively soft rock. One would expect a high number (>63.2) for a lignite, but the values obtained in tests ranged between 33 and 57. This probably reflects the amount of parting material that is included with the lignite. Sieve analyses were also carried out, and these showed that over 60% of the ROM production in the open pit is +30 mm. The domestic and industrial markets require a lump product, which the mine can easily supply.

13.6.1 Borehole sampling

Cores from the seam were split and one half was retained in the field. The entire thickness of the lignite seam was sampled including all parting material. All layers within the seam which were larger than 30 cm were sampled as individual samples. Thinner layers were included with adjacent layers until a minimum thickness of 30 cm was obtained. A maximum of 2 m of core was sampled at one time.

13.6.2 Selection criteria – surface mining option

In appraising the viability of using surface mining methods the following steps were taken. The weighted average quality of the mineable lignite seam was obtained by classifying any sample with less than 55% ash content as lignite and samples with more than 55% ash as waste. The "rules" that were applied to each sample to obtain the composite mineable thickness are shown in Fig. 13.11. The quality of the mineable reserves was calculated on all material within the seam except (i) waste partings greater than 50 cm and (ii) lignite piles less than 30 cm with partings on either side so that the total thickness of lignite plus partings above and below is greater than 50 cm. In addition, dilution and loss of lignite during mining was taken into consideration. At each interface

TABLE 13.3 Assumed ash and specific gravity values for the rock types in the Soma Isiklar Basin.

Rock type	Assumed ash content (%)	Assumed SG
Lignite		1.40
Clayey lignite		1.70
Lignitic limestone	75	2.00
Marlstone	75	2.30
Clay	75	2.40

(a)

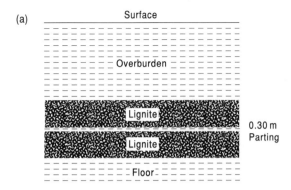

(b)

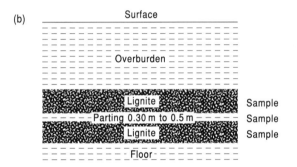

(c)

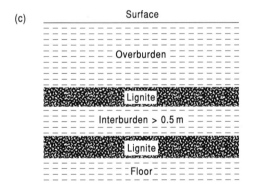

FIG. 13.11 An illustration of the sampling "rules" that were applied to all the lignite core recovered during the study in the Soma Basin. (Modified after Jagger 1977.)

but with no dilution or loss of lignite taken into consideration. Table 13.4 gives some examples of the changes in quality which arose by applying these selection criteria.

Once the areas to be mined on an annual basis had been outlined, the quality in each area was estimated using the boreholes within a 450 m radius. The mineable qualities, examples of which are seen in Table 13.4, were used to determine this quality release schedule for the open pit.

Inverse distance weighting (see section 10.4.3) was reviewed and compared to a simple weighted averaging technique. The two methods indicated little difference in the qualities. The 450-m radius of influence was chosen following a study of semi-variograms of thickness, calorific value and cumulations of thickness multiplied by calorific value. The range indicated on these semi-variograms was approximately 560 m. The distance of 450 m was selected because it is just greater than the normal two thirds to three quarters of this range which is normally selected (Annels 1991), and generally three or more boreholes fell within the radius.

13.6.3 Selection criteria – underground mining option

This section has been included to show that different mining methods require different selection criteria. During the study, alternative underground mining methods suitable for the extraction of a medium-dipping, thick seam, were examined. In-seam mining (a) and cross-seam (b) mining were considered (Fig. 13.12). As a computer database was being used, it was possible to try sensitivity analyses on the various selection criteria. These were used to assist in the selection of the best mining methods and in the calculation of the ROM tonnages and qualities. The weighted average quality of the *in situ* and mineable lignite was calculated using tabulated data examples which are shown on Table 13.5.

In-seam mining

The mineable lignite was determined once waste partings (material with >55% ash content and/or <1800 kcal kg^{-1}) greater than 1.5 m

between the lignite and a parting, a dilution by volume of 10 cm of parting was included replacing a lignite loss of 10 cm. These interfaces were taken into account at the top and bottom of the seam as well. The quality of the undiluted mineable reserves was calculated in a similar way to that of the mineable reserves,

	Borehole number	
	210	320
Lignite, vertical thickness (m)	18.40	11.80
Waste rejected from seam	0.00	3.00
Mineable vertical thickness (m)	18.40	8.80
Number of waste partings	0	3
Number of interface	2	8
Dilution (m)	0.2	0.8
In situ CV (kcal kg^{-1})	3524	2071
Undiluted CV (kcal kg^{-1})	3524	2679
Mineable CV (kcal kg^{-1})	3485	2435
In situ ash content (%)	27.6	39.9
Undiluted ash content (%)	27.6	29.7
Mineable ash content (%)	28.1	33.8
In situ SG	1.55	1.90
Undiluted SG	1.55	1.70
Mineable SG	1.56	1.78
Recovery by thickness (%)	100	75

TABLE 13.4   Examples of selective mining evaluation in the proposed open pit at Soma Isiklar, Turkey.

	Borehole number	
	218	302
Lignite vertical thickness (m)	14.70	12.10
Lignite in waste partings (m)	0.00	0.75
Waste rejected from seam (m)	0.00	2.35
Mineable vertical thickness (m)	14.70	9.00
In situ CV (kcal kg^{-1})	3662	2400
Mineable CV (kcal kg^{-1})	3595	2500
In situ ash content (%)	24.5	39.5
Mineable ash content (%)	25.2	37.6
In situ SG	1.49	1.73
Mineable SG	1.51	1.70
Recovery by thickness (%)	100	74

TABLE 13.5   Examples of selective mining evaluation in the proposed underground mine at Soma.

thick were rejected. Occasionally thin lignite beds within the thick waste partings were also rejected. The top and bottom of the *in situ* and mineable lignite are the same. It was assumed that there would be a 100% recovery of the lignite in the longwall slice, and 60% recovery of lignite and 40% dilution by waste in the caved zones above the longwall zones. The mineable quality takes into account the lignite losses and waste dilution which would occur during caving. It was considered that selective mining could be implemented above and below a waste parting greater than 1.5 m thick.

Cross-seam mining

The mineable lignite was determined once the top and bottom waste material were excluded. As this method is less selective, dilution of the lignite is inevitable. This is accounted for by expecting a 60% recovery of lignite and a 40% dilution by waste.

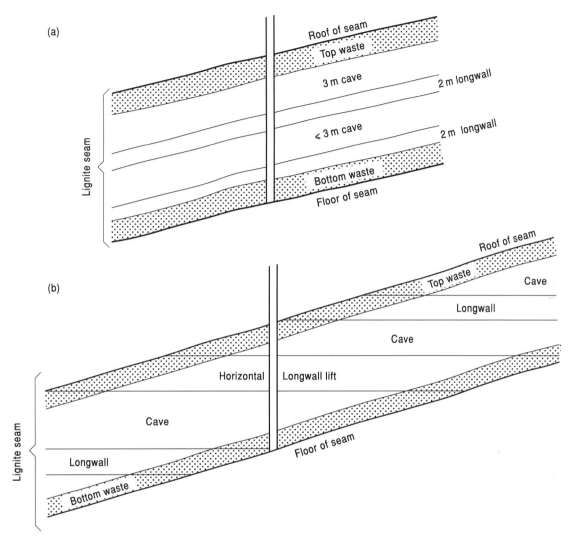

FIG. 13.12   A sketch of the two underground mining methods proposed for the deep lignite in the Soma Basin: (a) in-seam mining, in which the lignite is mined down-dip and (b) cross-seam mining, in which the lignite is mined horizontally.

13.6.4   Summary of lignite quality – surface mining option

Once the limits of the potential open pit had been established using the outcrop line and the stripping ratio limit down dip, it was possible to calculate the average quality of the *in situ* and the mineable lignite. The weighted average quality data are shown in Table 13.6. The *in situ* lignite quality includes small areas of lignite which may be included in the lignite

resource of the basin but were considered to be inaccessible to open pit mining. All the analyses were reported on an as-received basis. The air-dried moisture content was calculated for several of the lignite samples from the final series of holes. The results showed that the inherent moisture (air dried) was approximately 8% and the surface moisture was approximately 7%, totalling 15% on an as-received basis. The moisture content does not precisely represent that of the mined lignite because

	SG	CV (kcal kg^{-1})	Ash (%)	Moisture (%)
Open pit				
In situ	1.70	3069	32.7	13.5
Mineable	1.73	3103	33.0	13.5
Block A	1.73	4143	20.9	13.8
Block C	1.73	3104	32.5	14.4
Block D	1.73	2516	39.1	14.0
Block E	1.73	2816	36.6	12.9
Underground				
In situ	1.66	3337	30.0	16.2
Mineable	1.66	3467	27.5	16.2

TABLE 13.6   Average quality of the lignite in the Soma Basin.

of the presence of drilling fluid in the core samples and also the additional moisture that would be picked up during mining.

The sulfur content of the KM2 seam was analyzed in a few cases. The results indicate that the KM2 is a low sulfur lignite. The weighted average is as follows: combustible sulfur 0.53% and sulfur in the ash 0.42%, giving a total sulfur at 0.95%.

13.6.5   Summary of lignite quality – underground mining option

The areas down dip of the 7:1 stripping ratio line defined that part of the basin that could be potentially mined by underground methods. Within this area it was possible to calculate the average *in situ* lignite quality and the mineable lignite quality values (Table 13.6).

13.7   LIGNITE RESERVE ESTIMATES

The lignite reserves were calculated by assessing the data provided from borehole logs (Whateley 1992) obtained by drilling (Table 13.1), within the limits of the open pit. These limits were established between the outcrop line and the 7:1 stripping ratio line. On the northwest and the east sides of the pit, the limit is determined at the depositional edge of the seam. This limit was determined by using the basement and basal KM2 structure contour maps (section 13.4.1). The pit outline was placed on the isopach map of the total vertical thickness of the KM2 seam, and the area between each isopach was measured with the aid of a planimeter (cf. section 13.4). The areas

were multiplied by the average vertical thickness between the isopach lines (the thickness value at the midpoint) to obtain the *in situ* volume. The volume was multiplied by the average specific gravity to obtain the *in situ* tonnage.

The mineable lignite tonnage was calculated from the *in situ* tonnage by applying a recovery factor, examples of which are given in the final row of Table 13.4. The weighted average recovery for the open pit is 91%, but this varies from as low as 45% in borehole 208 to 100% in many of the remaining holes. The recoveries of mineable lignite from the total lignite will change during mining depending on the local geology. Development drilling immediately in advance of production will determine these recoveries. The reserve figures were calculated as follows: *in situ* reserves 49.4 Mt, mineable reserves 40.6 Mt, and kriged estimate of *in situ* reserves 46.3 Mt (Lebrun 1987).

13.7.1   Comparison of estimation methods

Zarraq (1987) undertook a comparison of reserve estimation methods using the Soma data. He calculated the *in situ* reserves for the whole basin using polygons, manual contours, and kriging. His reserves estimates are:

Method	Tonnage
Polygons	117.8 Mt
Manual contouring	102.5 Mt
Kriging (150 × 150 m blocks)	109.4 Mt
Statistical mean (area × average thickness)	117.5 Mt

The polygonal method has probably overestimated the reserve slightly because there are some large polygons which have thick lignite associated with them, e.g. the polygon centered on hole 208 has an area of 187,630 m² with 17.95 m of lignite, while the average area of all the polygons is only 70,000 m². Polygons also create artificial boundaries which do not exist in nature. The only way to reduce the estimation variance is to increase the sample density, which costs time and money. However, this is a quick method of providing a global estimation.

Manual contouring smooths the data by virtue of the linear interpolation used to construct the map. It appears to weight the large area of low values excessively, which may explain the lower reserve figure. Although the kriged estimate of reserves does take the spatial relationships into account, the estimation variances were generally high. Kriging gives the best estimate for each block with the lowest estimation variance, so the advantage of kriging is that one knows how reliable each block estimate is (section 9.5.1).

13.7.2 Confidence in the reserves

The confidence in the mineable reserve estimate for the open pit area was calculated using the mean and standard deviation for the thickness and specific gravity derived from the boreholes drilled in the area designated to be the surface mine. The formula

$$\frac{tS}{n}$$

was applied where S = standard deviation, n = number of boreholes, and t = t value for $n-1$ degrees of freedom at the 90% confidence level.

The results of this calculation were expressed as a percentage of the mean. The percentages of the mean results were then used in the following formula:

$$G = A^2 + B^2$$

where A = percentage of the mean for thickness, B = percentage of the mean for SG, and G = global confidence in the average expressed as a percentage.

The confidences as a percentage of the mean for the open pit area are as follows: thickness (A) 15.9%, SG (B) 5.9%, and global confidence (G) 17.0%. These results indicate that the reserves calculated from thickness and SG are sufficiently well known that they can be classified as proved within the confidence limits.

13.8 SURFACE MINE EVALUATION

This evaluation project called for a production rate of 2.13 Mt ROM lignite each year. After rejection of some waste dilution by rotary breakers, the delivered output is expected to be 2 Mt per year. This rate of production was considered appropriate to exploit the mineable reserves over a mine life of 21 years. Reserves are contained in four fault-bounded blocks (A, C, D, and E), but it was not considered advisable to include block B, because the lignite has been partly extracted using now abandoned underground workings. The presence of the faults and the varying dip angle of the footwall dictated the mining method and box-cut locations.

13.8.1 Selection of mining method

The hard massive marlstone which forms the overburden will require blasting before removal can take place and this rules out the use of a dragline or a bucket wheel excavator. Rear-dump trucks and face shovels will have to be used. Three alternative mining configurations may be considered using this equipment: (i) advance down the dip; (ii) advance up the dip; or (iii) advance along the strike. The third method, also known as terrace mining, was considered the most suitable with the particular geological and geotechnical conditions which exist at Soma (section 13.5.4).

Following this method a box-cut would be excavated down the full dip of the deposit from the outcrop until an economic stripping limit, or practical mining depth limit, is reached. This study concluded that a maximum mining depth of 150 m was feasible, although constraints were generally related to the geological configuration and consideration of maintaining footwall stability, rather than to stripping ratio economics. Advance could then be in one

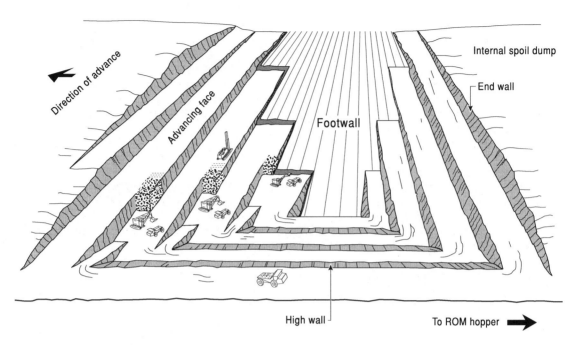

Internal spoil dump

End wall

Footwall

Direction of advance

Advancing face

High wall

To ROM hopper

FIG. 13.13   A sketch of a truck-and-shovel, terrace-mining operation proposed for the surface mine in the Soma Basin.

or both directions along the strike line, depending on the box-cut location within the proposed mining area. Waste disposal within the excavation is practical (and usually desirable) as the box-cut excavation is enlarged. Horizontal benches would be formed in the overburden along the advancing face and along the highwall formed on the deep side of the excavation. As much spoil as possible from the advancing face would be removed along the highwall benches for disposal to form an internal spoil dump within the excavation and behind the advancing face; the remainder having to be dumped outside the pit. The front of the spoil dump and the advancing face then advance in unison as mining continues along the strike of the deposit (Fig. 13.13). The final highwall is expected to have a maximum slope angle of 45 degrees.

The application of this mining method will bring several advantages: (i) a minimum area of footwall clay will be uncovered at any time, thus reducing the risk of footwall failure; (ii) internal dumping of waste reduces transport costs, helps stabilize the footwall, and begins reclamation at an early stage of mining; and (iii)

the stripping ratio is more or less constant over the life of the mine, and hence the mining costs are stabilized.

13.8.2   Mine design

Box-cut locations

Two box-cuts were proposed to take account of the adverse geological factors (Fig. 13.14). The first was sited in blocks C and D, close to the center of the open pit reserve. This location was dictated by consideration of local stripping ratios, seam dip, footwall clay thickness, and the presence of faulting. The central site means that mining will eventually advance both to the east and to the west. It was proposed that the second box-cut should be excavated at the western end of block A. Advance would be towards the west.

Slopes and access

A bench height of 12 m was proposed with excavation from horizontal benches in the overburden and lignite, with connecting ramps

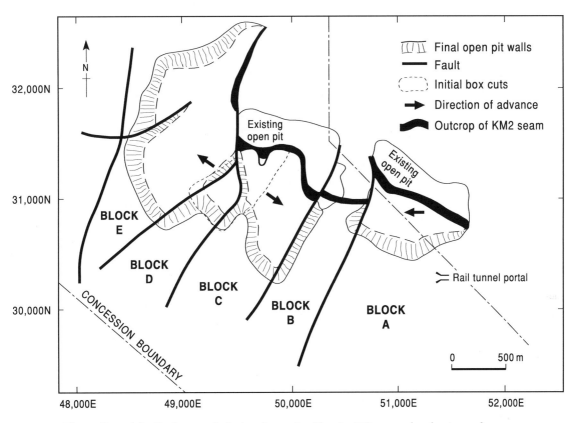

FIG. 13.14   The outline of the final open pit design determined by the 150-m overburden isopach.

at up to 8% grade located in the advancing faces (Fig. 13.13). The highwall should have an overall slope angle of 45 degrees, formed by the 12-m bench, and by berms and haulroads inside the pit for internal waste dumping (Fig. 13.15). Controlled blasting will be required for stable slopes.

Advancing faces will require an overall slope angle of 10–20 degrees, depending on the current mining activity. A working width of 65 m was proposed for each bench for independent drilling and blasting, loading and haulage activities. Lignite excavation would be from sub-benches 4 m high, subdividing the 12-m overburden benches for ease of excavation and loading (Fig. 13.15).

Haul roads

A haul road width of 20 m will be needed to provide room for trucks to pass with adequate side and center clearances (Fig. 13.15). Spoil disposed of within the pit will need be transported on haul roads at 48-m vertical intervals formed in the highwall. Ramps between haul roads, and to the rim of the pit would be formed in the advancing faces. Spoil disposal outside the pit will be via three ramps to main haulage routes on the pit rim. Lignite will be transported to the ROM lignite plant by a main haul road from the pit to the south portal of the main rail tunnel (Fig. 13.14) through the mountain and then by overland conveyer belt to the power stations near Soma (Fig. 13.1) to the north.

Spoil disposal

One-third of the 180M bank m³ of overburden (Table 13.2) could be backfilled into the mining excavation, with the remainder being disposed of in two dumps outside the pit. Bank m³ refers to the volume of rock *in situ*. Once mined the rock "swells" to occupy a greater volume

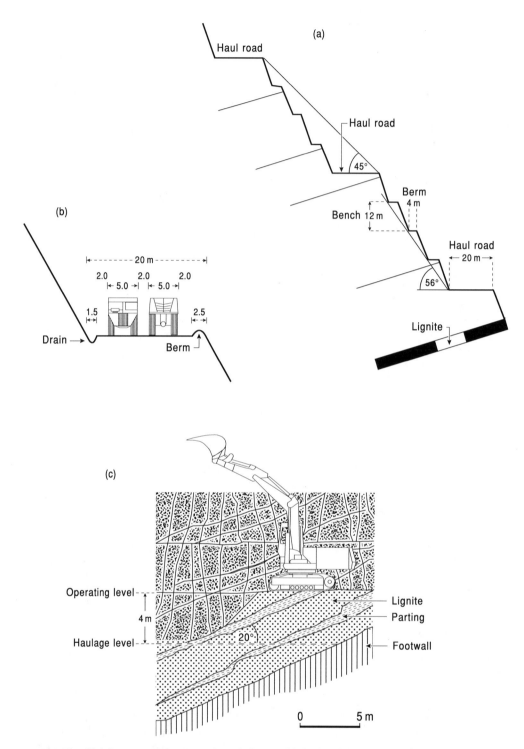

FIG. 13.15   Details of (a) the proposed open pit bench design, (b) the haul road, and (c) the lignite-mining operation.

although the increase in the void spaces results in a mass reduction m^{-3}.

13.8.3 Mining equipment

Large items of open pit equipment were not manufactured in Turkey at the time of this study. Calculations of equipment utilization, mechanical availability, and productivity were made to decide what equipment would have to be imported. The overburden equipment requirements were assessed to be three rotary blasthole rigs (250-mm-diameter holes), and electrically powered rope shovels for overburden stripping, assisted by front-end loaders. Hydraulic face shovels were recommended for lignite loading. Rear-dump trucks were selected for both overburden and lignite transport. Supporting equipment included tracked and wheeled bulldozers and a conventional fleet of pit and haul road maintenance vehicles.

13.9 SUMMARY

Available data on the reserve tonnage and quality of the Soma deposit were adequate for a feasibility evaluation, which suggested that the reserves would support a projected life of 21 years at the proposed rate of production and specified quality. Additional data were required on the groundwater regime before forecasts could be made of the effects of groundwater pressure and flows on mining. This highlights the need for careful planning of any exploration drilling program to maximize the data that can be gained from any drilling.

13.10   FURTHER READING

Ward (1984) covers a wide range of topics and his book is intended for senior students and professional geologists. He discusses coal exploitation, mining, processing, and utilization in a clear manner. Other papers on these subjects can be found in the *Bulletin de la Société géologique de France, 162 (1991)*. The Special Publications of the Geological Society edited by Annels (1991) and Whateley and Spears (1995) give some excellent case histories of coal deposit evaluation. *Stach's Textbook of Coal Petrology* (Stach 1982) is well referenced and gives detailed information on the microscopic properties of coal. The sedimentology of coal and coal-bearing strata is covered in a series of papers in a Geological Society of London Special Publication edited by Scott (1987). Further papers on the sedimentology of coal-bearing strata are to be found in the IAS Special Publication edited by Rahmani and Flores (1984), and in the *International Journal of Coal Geology*.

14

WITWATERSRAND CONGLOMERATE GOLD – WEST RAND

CHARLES J. MOON AND MICHAEL K.G. WHATELEY

14.1 INTRODUCTION

Alluvial concentrations of heavy minerals (placer deposits) are very important sources of gold, such as the rich deposits in the Yukon, the Mother Lode area of the forty-niners in northern California, and the modern-day rush in the Sierra Pelada area of the Amazonian Brazil. Even today much Russian production comes from the Lena and Magadan alluvial deposits in northeast Siberia. However modern alluvial gold output is dwarfed by production from deposits in quartz pebble conglomerates that appear to be fossilized placers. These deposits are restricted in geological time to the late Archaean and early Proterozoic and dominated by one basin, the Witwatersrand Basin in South Africa. Other gold-producing quartz pebble conglomerates occur in the Tarkwa area of Ghana, Jacobina and Moeda in the São Francisco Craton of Brazil, while the Elliot Lake area in Ontario has been a significant producer of uranium from similar rocks and carries minor gold.

This case history concentrates on the Witwatersrand (Afrikaans for ridge of white waters and often abbreviated to Rand or Wits) Basin that has been responsible for the production of 37% of all gold produced in modern times as well as significant uranium. This basin was the source of 399 t of gold metal in 2002, about 16% of annual world gold production. Mining in the Witwatersrand Basin is from large tabular orebodies that form the basis of some of the largest metalliferous underground mines in the world and which reach more than 3500 m below surface. This combination of depth and scale has been a great challenge to mining

engineers and it has produced a long and interesting exploration history involving large expenditures on deep drilling and geophysics.

14.2 GEOLOGY

Mineralized conglomerates are found in a variety of settings within the Kaapvaal Craton (Fig. 14.1) but the vast majority of production has come from the upper part of a 7000-m thick sedimentary sequence termed the Witwatersrand Supergroup. These sediments were deposited within a 350×200 km basin that formed relatively soon after the consolidation of the Kaapvaal Craton from a series of greenstone belts and granitic intrusives. The Witwatersrand sedimentation was part of a much longer depositional history which began with sedimentation in a probable rift setting of bimodal volcanics and limited sediments, termed the Dominion Group (Stanistreet & McCarthy 1991). Recent age determinations date volcanics within the succession at 3074 ± 6 and underlying granites at 3120 ± 6 Ma (Armstrong et al. in De Beer & Eglinton 1991). The basal sediments are conglomeratic and supported minor gold production and substantial uranium operations. These Dominion Group sediments form structural remnants overlain by Witwatersrand Supergroup sediments.

14.2.1 Stratigraphy of the Witwatersrand Supergroup

The Witwatersrand Supergroup is divided into two, the lower West Rand Group and the upper

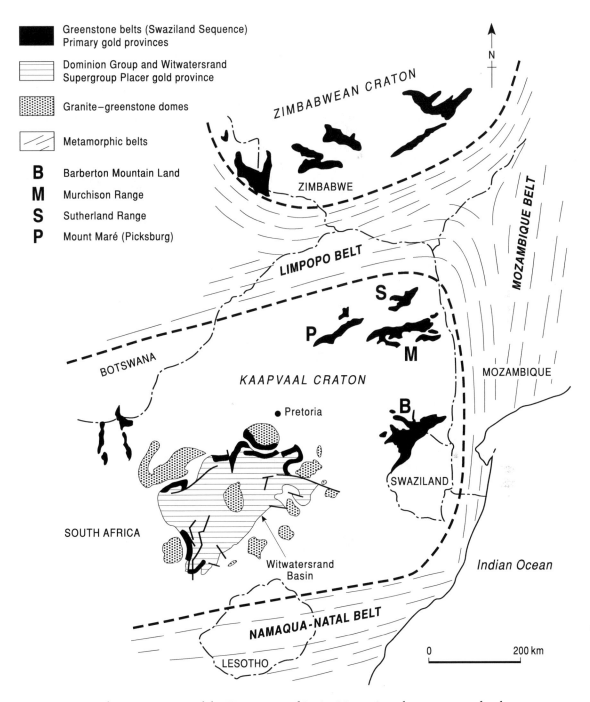

FIG. 14.1   Regional tectonic setting of the Witwatersrand Basin. Mesozoic and younger cover has been removed. (After Saager 1981.)

Central Rand Group, which contains most of the gold deposits.

The West Rand Group consists of about equal proportions of shale and sandstone and 250 m of volcanics, the Crown lavas (Tankard et al. 1982). A number of iron-rich shales are present in the lower part of the group and these give a magnetic response used in exploration. The depositional environment was littoral or subtidal on a stable shelf as indicated by the general upward coarsening nature of the macrocycles. However there are a number of upward fining cycles with conglomerate at their base, which represent more fluvial conditions and contain mineralisation. Despite their alluring names, the Bonanza, Coronation, and Promise Reefs (reef is used as a synonym for mineralized conglomerate throughout the Witwatersrand) have produced only 35 t of gold, although their mineralogy is similar to reefs in the Central Rand Group (Meyer et al. 1990). Age dates on detrital zircons give a lower age limit for deposition of 2990 ± 20 Ma and an upper limit of 2914 ± 8 Ma on the Crown Lava (De Beer & Eglington 1991).

The later stages of the deposition of the Witwatersrand Supergroup reflect major changes in the style of sedimentation together with changes in the shape and size of the basin of deposition (Stanistreet & McCarthy 1991). The Central Rand Group consists predominantly of coarse-grained subgreywacke with less than 10% conglomerate, silt, and minor lava (Tankard et al. 1982). The overall depositional setting is of fan delta complexes prograding into a closed basin, perhaps containing a lake or inland sea. The change in depositional style reflects the increasing importance of tectonism, certainly of folding, probably of faulting. Burke et al. (1986) suggested that the overall tectonic setting is that of a foreland basin related to collisional tectonics in the area of the Limpopo River to the north. Dating of detrital zircons shows that the Eldorado Formation, at the top of the group, is younger than 2910 ± 5 Ma and older than overlying Ventersdorp lavas dated at 2714 ± 8 Ma (Robb et al. 1990).

The fan delta complexes mainly have distinct entry points into the basin as indicated by palaeocurrent data, pebble size, and composition and isopach maps; these correspond with the location of the major goldfields. At present,

eight major goldfields are recognized, from south to northeast, the Orange Free State (OFS or Welkom), Klerksdorp, West Wits Line (or Far West Rand), West Rand, Central Rand, East Rand, South Rand, and Evander (Fig. 14.2). The areas between the goldfields are known as gaps, notably the Bothaville and Potchefstroom Gaps. Recent exploration has shown that there is significant mineralisation in these gaps, particularly the Bothaville Gap, where the Target mine is currently being developed. However the south side of the basin seems to be essentially barren.

Regional correlation between the different reefs is difficult as they are developed on unconformities and some reefs are developed in only one goldfield, as might be anticipated from the discrete sediment entry points. However broad correlations are possible as shown in Fig. 14.3. The reefs can be divided into two types, sheet conglomerates and channelized conglomerates (Mullins 1986). Sheet reefs have good continuity and are gravel lags overlain by mature sand, whereas the channelized reefs have well-defined channels filled with lenticular bodies of sand and gravel. Examples of channelized conglomerates are the Kimberley Reef and the Ventersdorp Contact Reef (VCR), whereas the Vaal Reef, Basal Reef, and the Carbon Leader are sheet-like. Where both sheet-like and channelized conglomerates are developed on the same unconformity, as in the case of the Steyn Reef in the OFS Goldfield, the channelized conglomerates are more proximal to the source area and change into sheet-like conglomerates over a distance of 20 km. The stratigraphical location of the major reefs is shown in Fig. 14.3.

14.2.2 Overlying lithologies

Little of the known extent of the Witwatersrand Basin sediments is exposed and most is covered by later rocks, both Precambrian and Phanerozoic: lavas and sediments of the Ventersdorp Supergroup, and sediments of the Transvaal and Karoo Sequences.

The uppermost major reef in the Witwatersrand Basin, the Ventersdorp Contact Reef, is preserved by a cover of Ventersdorp lavas and appears to have formed at a time of volcanism and block faulting. It is economically

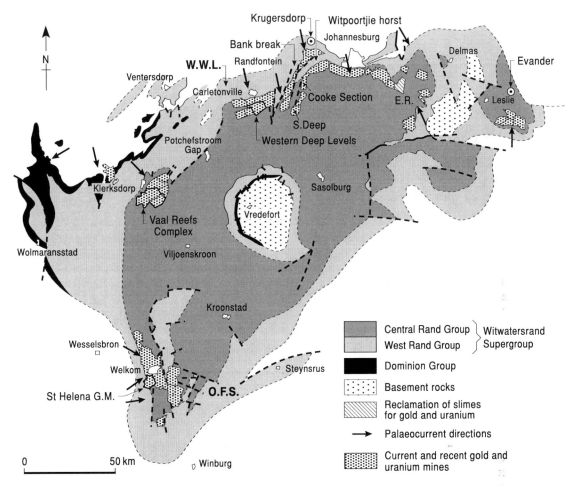

FIG. 14.2 Witwatersrand Basin with later cover removed, E.R. indicates location of East Rand Basin (Fig. 3.7). Note that the mine names are those referred to in the text and many of the current mines have new names. The new target mine is ~40 km north of Welkom. (Modified from Camisani-Calzolari et al. 1985.)

important in the West Wits Line and Klerksdorp areas, where it is heavily channelized. Although other placers have developed higher in the Ventersdorp succession, and under similar conditions, they are of little economic importance. The remainder of the Ventersdorp sequence consists largely of tholeiitic and high magnesia basalts with lesser alluvial fan and playa lake sediments.

The Transvaal Sequence was deposited from around 2600 to 2100 Ma over a wide area of the Kaapvaal Craton. The important parts of the sequence as far as exploration and mining are concerned are the basal conglomeratic unit, known as the Black Reef, and the overlying dolomites of the Malmani Subgroup. The Black Reef has been mined in a number of areas, notably on the East Rand and in the Klerksdorp areas. Mineable areas correlate with areas where the Black Reef cuts Central Rand Group mineralisation, and it seems certain that the Black Reef mineralisation is the result of reworking or remobilization of gold from below (Papenfus 1964). The dolomites overlie most of the mining area on the West Rand and Klerksdorp areas and cause considerable problems as they contain large amounts of water and are prone to the formation of sinkholes. One sinkhole swallowed the primary crusher at West Driefontein in 1962.

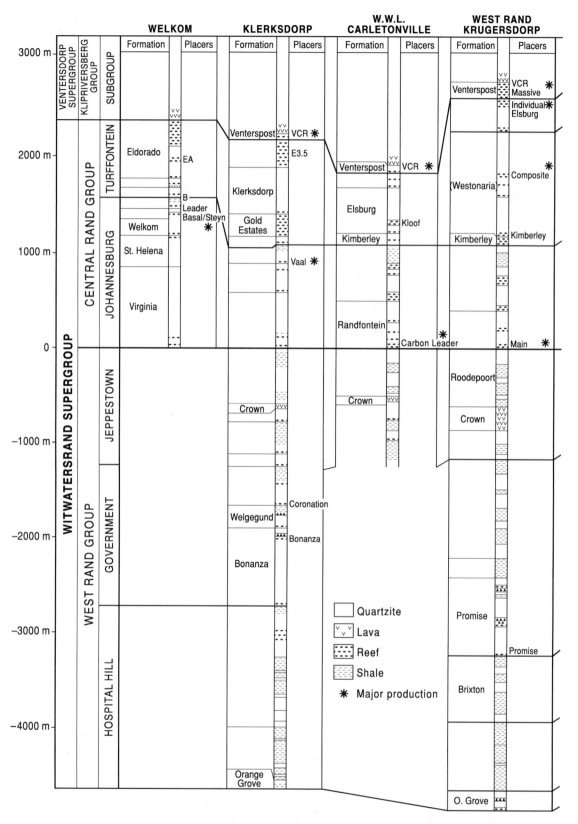

FIG. 14.3 Stratigraphical columns of the Witwatersrand Supergroup in the main gold-producing areas. Note the general correlations between goldfields and major producing conglomerates. (After Tankard et al. 1982.)

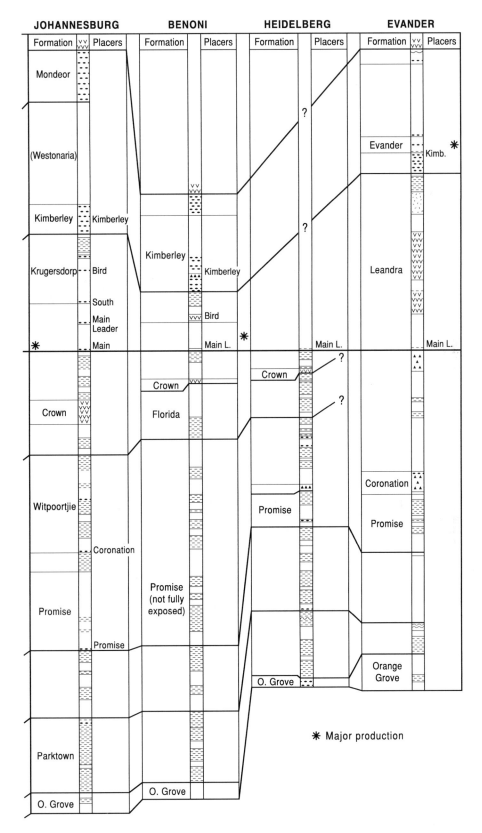

FIG. 14.3 (CONTINUED)

The youngest major sedimentary sequence found on the Witwatersrand consists of sediments of the Karoo Supergroup deposited from Carboniferous to Triassic times in glacial, marginal marine to fluvial conditions. These are important in two respects. The Ecca Group sediments are coal-bearing and besides providing a cheap source of energy have probably generated the methane that is found in a number of mines and which poses a threat of underground explosions. The Karoo sediments also have a high thermal gradient and give rise to a blanketing effect which raises the overall thermal gradient and reduces the ultimate working depths in both the OFS and Evander Goldfields.

Witwatersrand sediments have been intruded by a variety of dykes, which cause considerable disruption to mining. The oldest intrusives are dykes of Ventersdorp age and the youngest lamprophyres of post-Karoo age but the most important are dykes of "Pilanesberg" age, up to 30 m wide, on the West Rand which divide the dolomites into separate compartments for pumping purposes (Tucker & Viljoen 1986).

14.2.3   Structure and metamorphism

Although the Witwatersrand sequence has clearly been deformed and metamorphosed, there has been little emphasis on making a synthesis until the last few years. An exception is faulting of the reefs which has been examined in detail as this is a major control on the day-to-day working of the mines. The type of deformation varies through the basin but the most important for mining is block faulting which forms the boundaries of mining blocks. The main activation of these faults was in mid Ventersdorp times when there was also thrust faulting at the margin of the OFS Goldfield (Minter et al. 1986). The major faults were, however, reactivated in post-Transvaal but pre-Karoo times, although the exact timing is unclear. In the east of the basin the sediments are affected by the intrusion of the Bushveld Complex which has tilted them. Recent dating suggests that a major basinwide fracturing event, including the development of pseudo-tachylite in the Ventersdorp Contact Reef, is associated with the Vredefort event at ~2023 Ma that is probably the result of a meteorite impact (Frimmel & Minter 2002).

The occurrence of a pyrophyllite–chloritoid–chlorite–muscovite–quartz–pyrite assemblage within the pelitic sediments indicates a regional greenschist metamorphism at temperatures of $350 \pm 50°C$ that has been dated at ~2100–2000 Ma, approximately the same age as the intrusion of the Bushveld Complex (Phillips et al. 1987, Phillips & Law 2000, Frimmel & Minter 2002).

14.2.4   Distribution of mineralisation and relation to sedimentology

Gold and uranium grades appear to be controlled by sedimentological factors on both a small and a large scale. The diagram in Fig. 14.4 shows a typical relationship between grade and sedimentology for a channelized reef, with higher grade areas restricted to the main channels. The delineation of these payshoots may be the only way to mine certain conglomerates economically (Viljoen 1990). The distribution of facies can also be mapped in the sheet-like placers as shown for the Carbon Leader in Fig. 14.4b. High grade gold and uranium deposits are associated with the carbon seams and thin conglomerates, particularly in the West Driefontein area. Mapping of these sedimentary structures is also important as they form homogenous geostatistical domains for the purpose of grade calculations. In general the oligomictic placers have higher grades than the polymictic conglomerates, which contain shale and chert as well as vein quartz clasts, presumably as a result of greater reworking.

On a smaller scale, variations in sedimentary structures control grade distribution. This is exemplified by Fig. 14.5 which shows the distribution of gold through the channelized A reef (Witpan Placer) in the OFS Goldfield. In this case gold is concentrated at the top and base of the conglomerate unit. Gold is also associated with heavy mineral bands within the unit. In the sheet conglomerates gold is closely associated with carbon, if present, and the basal few centimeters may contain most of the gold in the reef. The remainder of the reef, although mostly waste, must be mined to expose the working face using a minimum stoping height of 80 cm. Jolley et al. (2004) provided a contrasting explanation for the detailed distribution of gold in the conglomerate, which

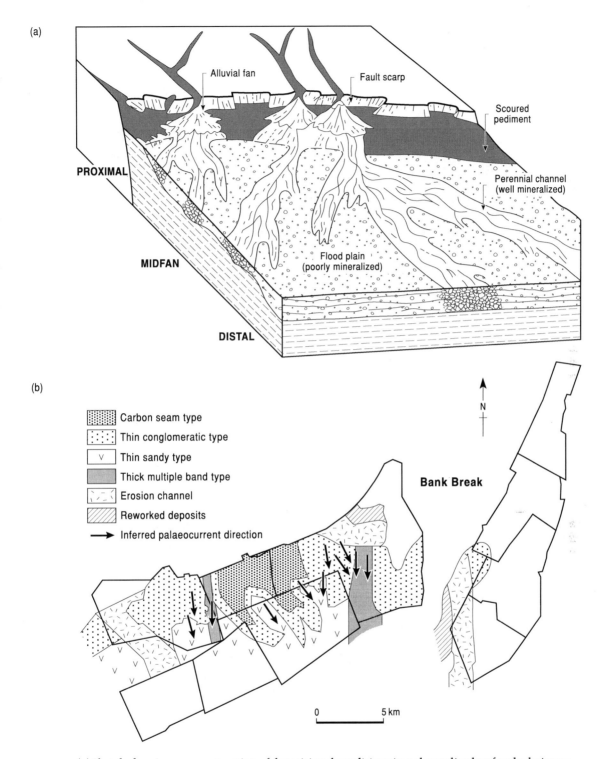

FIG. 14.4 (a) Sketch showing a reconstruction of depositional conditions in a channelized reef and relation to grade. Generalized from the Composite Reef. (From Tucker & Viljoen 1986.) (b) Facies of the Carbon Leader on the West Wits Line. Driefontein is the mine immediately west of the Bank Break. (From Viljoen 1990.)

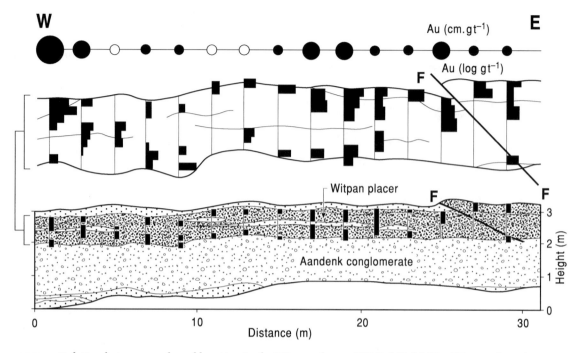

FIG. 14.5 Relation between grade and location in the Witpan placer, OFS Goldfield. The Witpan placer is stratigraphically above the B Reef of Fig. 14.3. The exact grades were omitted for commercial reasons. (After Jordaan & Austin 1986.)

they advocated were controlled by small scale fractures related to thrust fracture networks.

14.2.5 Mineralogy

The conglomerates mined (Fig. 14.6) are predominantly oligomictic with clasts of vein quartz. The matrix consists largely of secondary quartz and phyllosilicates, including sericite, chlorite, pyrophyllite, muscovite, and chloritoid. The opaque minerals have been the object of intense study and more than 70 are known; reviews are given by Feather and Koen (1975) and Phillips and Law (2000).

FIG. 14.6 Underground photograph of a typical conglomerate. The UE1a Upper Elsburg at Cooke Section, Randfontein Estates Gold Mine. Courtesy of the Sedimentological Section of the Geological Society of South Africa.

Pyrite is the predominant opaque mineral with important, but lesser amounts, of gold, uraninite, carbon, brannerite, arsenopyrite, cobaltite, galena, pyrrhotite, gersdorffite, chromite, and zircon.

Pyrite forms 3% of a typical reef and is present largely as rounded grains. The larger of these are known as buckshot pyrite, being similar to the shot used in cartridges. Pyrite occurs as grains with recognizable partings and as porous grains, often filled with chalcopyrite or pyrrhotite. Some concretionary grains are also present. At least some of the pyrite was formed by pyritization of other minerals and rocks, such as ilmenite or ferruginous pebbles. Pyrrhotite probably formed largely as a result of the metamorphism of pyrite and is most frequent near dykes. Uraninite is present mainly as rounded grains of apparently detrital origin. Gold occurs largely in the fine-grained matrix between pebbles as plate-like, euhedral, or spongy grains. The majority of gold is in the range 0.075–0.15 mm and is not often visible. Very fine-grained gold is found in the carbon seams. The majority of gold shows signs of remobilization and replacement relations with the fine-grained matrix. It also is commonly associated with secondary sulfides and sulfarsenides. Typically gold grains contain about 10% silver.

Uraninite is the dominant uranium mineral although the titanium-rich uranium mineral, brannerite, predominates in the Vaal Reef, presumably forming as a result of a reaction between ilmenite and uranium in solution. Carbon, or more correctly bitumen, is common in a number of the sheet conglomerates and is closely associated with high uranium and gold grades. Hallbauer (1975) suggests that it formed from lichen and fungi, although more recent work shows that it is a mature oil and some authors (e.g. Barnicoat et al. 1997) have argued for a source from outside the basin.

Chromite is widely distributed in the reefs and shows a sympathetic relationship with zircon. Both appear to be detrital. Two rare detrital minerals are diamond and osmium-iridium alloys. Authigenic sulfides are patchily distributed and include cobaltite, gersdorffite, sphalerite, and chalcopyrite. This assemblage is associated with remobilized gold.

14.2.6 Theories of genesis

The origin of the mineralisation has been a matter of bitter debate between those who believe it to be essentially sedimentary, developed either by normal placer-forming mechanisms, the *palaeoplacer theory*, or by precipitation of gold in suitable environments, the *synsedimentary theory*; and the hydrothermalists who believe that the conglomerates were merely the porous sediments within which gold and sulfides were deposited from hydrothermal solutions derived either from a magmatic source or from dehydration of the West Rand Group shales. The full background to the debate can be found in Pretorius (1991), Phillips and Law (2000), and Frimmel and Minter (2002).

The key arguments in favor of a placer origin are the strong spatial correlation between gold, uraninite, and unequivocally detrital zircon, and the intimate relationship between the heavy minerals and the sedimentary structures and environment. Placer advocates also point to the equal intensity of mineralisation in porous conglomerates and less porous pyritic quartzites. The problems for the placerists are mainly mineralogical: the very small particle size of the gold, its hackly shape, and low fineness (i.e. high % Ag), which are more typical of hydrothermal deposits, the presence of pyrite, uraninite, and the absence of black sands typical of modern placers. The placerists invoke metamorphism to explain the shapes of the gold grains and explain the lack of magnetite as the result of intense reworking (Reimer & Mossman 1990) or transformation into pyrite after burial. The problem of detrital pyrite and uraninite is somewhat more difficult to explain as these minerals are unstable in most modern environments, although they are known from glacially fed rivers, such as the Indus in Pakistan (for discussion see Maynard et al. 1991). However it seems very probable that the Archaean atmosphere contained less oxygen than at present and that both uraninite and pyrite would be stable under these conditions. Recent direct dating of gold and pyrite from the Basal Reef using the Re/Os technique has given ages of 3016 ± 110, indicating a detrital origin for the gold, as these ages are similar to the zircon ages and older than the ages of deposition of the sediments (Kirk et al. 2002).

By contrast, gold from the Ventersdorp Contact reef appeared to have ages reset by metamorphism. Osmium concentrations are compatible with derivation from a mantle source.

Another major problem is the source of the large quantity of gold and uranium in the basin, far more than is likely to have been derived from any Archaean greenstone belt, certainly more than from the most productive South African greenstone belt, Barberton. While it is certain that greenstone belts were part of the source area, as evidenced by the occurrence of platinum group minerals in the Evander area, it seems likely that the gold and uranium could also have been derived from altered granites. Studies by Klemd and Hallbauer (1986) and Robb and Meyer (1990) have demonstrated the presence of hydrothermally altered granites with gold and uranium mineralisation in the immediate hinterland of the Witwatersrand Basin. A neat twist to the placer theory is the suggestion by Hutchinson and Viljoen (1988) that the gold could have been derived from reworking of low grade exhalative gold deposits in the West Rand Group.

The principal arguments for a hydrothermal origin are that the gold is crystalline, sometimes replaces pebbles, may be contained within and replace pyrite that has replaced pebbles, and is associated with a suite of hydrothermal ore minerals and alteration products (sericite and chlorite) typical of hydrothermal orebodies the world over. Previous important advocates of the hydrothermal origin were Graton (1930), Bateman (1950), and Davidson (1965). The hydrothermal theory has been recently given a new breath of life by Phillips et al. (1987) and Barnicoat et al. (1997), who pointed out the lack of metamorphic studies and demonstrated that the shales show evidence of metamorphism to greenschist facies similar in many ways to the conditions of formation of greenstone belt gold deposits. They also suggested that in some areas faulting may have controlled gold distribution. Regional stratabound alteration zones have also been recognized which correlate spatially with major gold reefs (Phillips & Law 2000).

The modified placer theory, which emphasizes the control on the occurrence of ore minerals by placer-forming mechanisms but accepts some modification by metamorphism, has been widely used in exploration. If you are lucky enough to make an underground visit you should try and discuss the merits of the different theories.

14.3   ECONOMIC BACKGROUND

14.3.1   Profitability

The economics of mining in the Witwatersrand have been largely governed by the price of gold. Unlike most other commodities the gold price is not controlled by supply and demand for industrial use but by the perception of its worth. As gold is transportable and scarce the metal has been used as a way of moving and storing wealth. Its price was regulated for long periods of time by intergovernmental agreement, enforceable by the large reserves held by reserve banks such as the US hoard at Fort Knox. These large stocks were initially held when currencies were directly convertible into gold. This convertibility, the gold standard, was abandoned by the UK in 1931 and by the USA in 1933. From 1935 to 1968 the gold price was fixed at $US35 t oz^{-1} but in 1968 President Johnson allowed the gold price to float for private buyers and in 1971 President Nixon removed the fixed link between the dollar and gold and left market demand to determine the daily price. The 1970s saw extreme fluctuations in the value of gold as it was used as a hedge against inflation with sharp increases in price (see Fig. 1.8) following the increases in the oil price in 1973 and 1980. The gold price remained in the $US300–400 range in the rest of the 1980s, steadily depreciating against an inflating dollar as production has increased. The sharp increase in the gold price in 1931 set off the major exploration programs that led to the discovery of two new goldfields in South Africa. That of the 1980s led to the prospecting boom of 1985–89.

Long-term plots (Fig. 14.7) of the overall profitability of the Witwatersrand mines reflect these changes in the gold price, changes in the South African rand–US dollar exchange rate, and increases in working costs. The overall picture has been relatively healthy with spectacular profits in the early 1980s. Since then working costs have escalated sharply as

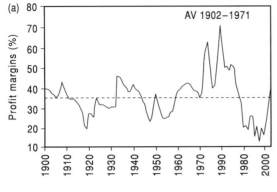

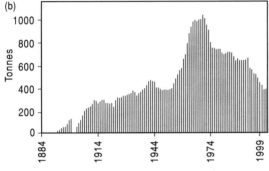

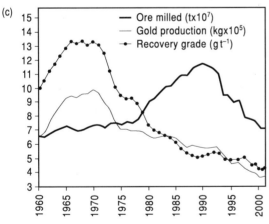

FIG. 14.7   Plots of (a) profit margins, (b) gold production, and (c) gold production, tonnes milled, and grade for the Witwatersrand Basin. (From Willis 1992, updated from Handley 2004.)

the overall recovered grade has declined, largely as a result of the exhaustion of higher grade deposits (Fig. 14.7), and South Africa has suffered sustained high inflation rates. In particular, working costs have increased due to higher wages and the increasing depth

and complexity of mining. The major political changes culminating in the 1994 elections meant that the gold mines have not been willing or able to maintain wage differentials based on skin color or migrant labor and overall labor costs have increased sharply. This led to a sharp decline in the number of operating mines and a 48% decline in the number of employees between 1988 and 1998. Reductions in unit costs will only come through increases in productivity, in particular through mechanization of mining. The taxation arrangements under which the mines operate are complex but favored the treatment of large tonnages of low grade ore. In 1999 the formula was $y = 43 - 215/x$ where x is the ratio of taxable income from gold to gross gold revenue and y the tax rate.

14.3.2 Organization of mining

Mining on the Witwatersrand requires large amounts of capital and has been dominated by large companies since its inception. The major operators in 2004 were Anglogold Ashanti, Gold Fields, Harmony Gold Mining, and South Deep. These companies have evolved from the major mining houses that dominated the South African mining industry until 1994: Anglo American Corporation, Johannesburg Consolidated Investments (JCI), Gold Fields of South Africa (GFSA), Rand Mines, Gencor, and Anglovaal. After 1994 Anglo American moved its headquarters to London and created a gold holding company, which has progressively increased its share of non-South African operations. A number of loss-making or short-life operations were sold, although Anglogold still produces 106 t from South Africa. Gold Fields, created by the merger of Gold Fields of South Africa's and Gencor's gold interests, is pursuing a similar path, although it operates the major mines at Kloof, Driefontein, and Joel. Harmony is a relatively new company created from the remnants of Rand Mines and short-life mines sold by other major producers, as well the major developing Target mine. South Deep, discussed in detail below, is a joint venture between Western Areas, a successor company to JCI, and Placer Dome of Canada. To add to the confusion since 1994, many mines have been split and renamed, e.g.

Western Deep Levels is now split into Savuka, Mponeng, and Tau Tona, and Vaal Reefs into Great Noligwa, Tau Lekoa, Kopanang, and Moab Khotsong.

14.3.3 Mineral rights

Mineral rights on the Witwatersrand were largely privately held, either by gold mining companies or in unmined areas by farmers or descendants of the original landowners. The value of these assets is usually appreciable and purchase values may be as high as R10,000 ha^{-1}; options to purchase are negotiated, usually lasting for 3 or 5 years (Scott-Russell et al. 1990). The competitive nature of exploration means that options were often not easy to obtain and option costs might rocket as soon as news leaks out of a company's interest. In many cases the mineral rights were divided into various shares and agreement was required by all parties. However the system is, at the time of writing (2004), undergoing radical change. South Africa is moving to a system in which the state will control exploration rights in a similar fashion to most of those in North America and Australia. Under the provisions of the 2002 Mineral and Petroleum Resources Development Act new exploration areas will be granted for 5 years. Existing mineral rights, either of mining or exploration, can be converted into the new system but only if they are actively being used.

14.3.4 Mining methods

The generally narrow mining widths, availability of cheap labor, moderate dip of the reefs, and great depths of most mines have led to the development of a very specific mining method, known as breast stoping. In this method (Fig. 14.8), development is kept in the footwall of the reef and a raise is cut between levels. From this raise gullies are cut at a 45-degree angle to the raise and the stope developed in a Christmas Tree manner. The stopes advance by jackleg drilling and blasting of a 2-m length of stope on each shift. Ore is scraped out from the advancing face along the gully using mechanical scrapers and the stope is cleaned carefully, often using high pressure jets, as much gold is contained in dust in the cracks in the floor of the stope. The scraper transfers the ore to footwall boxholes whence it is trammed to the main ore pass and the primary underground crusher. If the stope takes in more than one level then it is known as a longwall. The hanging wall of the advancing stope is supported by hydraulic props, later reinforced by timber packs.

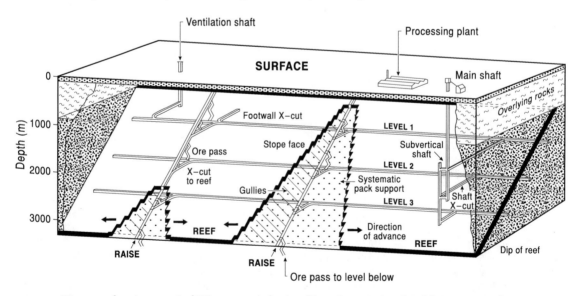

FIG. 14.8 Diagram showing a typical Witwatersrand mine. (From International Gold Mining Newletter April 1990.)

TABLE 14.1   Equipment substitution in trackless mining.

Operation	Conventional mining	Trackless mining
Drilling	Pneumatic and hydraulic hand-held drills	Hydraulic drill rigs
Stope cleaning	Scraper winches	LHDs
Gully cleaning	Gully winches	Trucks or LHDs
Tramming	Locomotives and railcars	Trucks or LHDs
Light haulage and railcars	Locomotives	Utility vehicles

LHDs, Load, haul, dump vehicles.

Mechanized mining of narrow stopes is not easy and mechanization has been most successful in wide stopes, such as those in the Elsburg reefs on the West Rand, which are often 2 m or thicker. Trackless mining was first introduced in 1984, and is seen as a way of reducing the working costs, using specialized machinery instead of the labor-intensive conventional methods. Trackless mining involves the substitution of tracks, locomotives, and railcars by free-steering, rubber-tyred (or caterpillar tracked) equipment for drilling, loading, and tramming of the ore, waste, men, and materials. The substitutions involved are shown in Table 14.1.

The potential benefits of such conversions are lower operating costs, higher productivity, greater flexibility, and improved grade control and safety (particularly in stopes, on haulages and declines, and at grizzlies). There is a concomitant change in the workforce, with the large semi- or unskilled workforce replaced by a much smaller, highly skilled workforce.

There are a number of variations in the way trackless mining is carried out, but the general use of room (bord) and pillar mining using Tamrock and Atlas Copco drilling rigs and LHD (load, haul, dump) vehicles is common to most. There are two stages of production, namely primary development and stoping followed by secondary extraction of pillars. During primary development and stoping, access ramps are driven on reef between two levels approximately 150 m apart. Stoping commences with the development of rooms approximately 17 m apart off the access ramp. Pillars 10 × 7 m are left between rooms. Drilling of blastholes is by means of boom drill rigs. The reef is blasted out into the heading making the broken ore accessible to the LHD vehicles.

During this stage of mining 71% of the ore is extracted, the remaining 29% is left in the pillars. Secondary partial extraction of the pillars takes place once primary stoping is complete. The pillars are reduced to a final size of 5 × 5 m. The area is supported by roof bolts, and the broken ore is again removed by LHD vehicle.

A major consideration at great depths is rock stability. The quartzites have considerable strength and deform in a brittle manner, causing rockbursts (explosive disintegration of rock). These events cause disruption to production and are a major cause of fatal accidents. Rockburst frequency can be reduced by back filling of stopes and careful monitoring of stress build ups.

14.4   EXPLORATION MODELS

Exploration has been carried out on a variety of scales ranging from delineating extensions of known pay shoots within a mine to the search for a new goldfield. As most of the shallow areas have been explored one of the key decisions to make before optioning land is to decide the depth limit of economic mining and exploration. The deepest mining planned in detail at present is about 4000 m but options have been taken over targets at up to 6000 m and holes drilled to 5500 m. According to Viljoen (1990) good intersections have been made at 5000 m and although there is of course a stratigraphical limit, mineralisation is present at great depths. The major constraints are the stability of mine openings, rock temperature (at 6000 m it is likely to be >85°C) and the costs of mining at these depths. At present, temperature problems are overcome by circulating refrigerated air and chilled water and amounts of these have to be increased substantially at 5000–6000 m. Currently the deepest single lift shaft is less than 3000 m (South Deep) and mining to 5000–6000 m would require two further shafts underground, termed secondary or subvertical shafts and tertiary shafts. Therefore every tonne of ore and materials would have to be transhipped at least twice. At 3000 m depth it can take a miner one and a half hours to reach the mining face and consequently production time is reduced. Mining at 6000 m may require a revision of shift patterns.

Most exploration has taken place within the known extent of the Witwatersrand Basin although there have been a number of searches for similar rocks outside it. The latter search has particularly concentrated on areas where lithologies similar to those in the basin are known either in outcrop or at depth. The possibility of West Rand Group correlatives is usually based on magnetic anomalies and the exploration model is that of a basin within mineable depth and preferably closer to postulated source areas. Examples of Witwatersrand lithologies at depth are known from deep drillholes and from surface outcrops to the Northwestern and Northern Natal-Kwazulu.

Within the basin, exploration is aimed at detecting the continuation of known reefs or new reefs within known goldfields or of detecting new goldfields in the intervening (gap) areas. The 1980s saw a great deal of exploration in the gap areas based on the premise that previous deep drilling has missed potential goldfields due to lack of understanding of the structure or controls on sedimentation. The fundamental information required is an understanding of the subsurface stratigraphy and the potential depths of the target. Upthrown blocks of the Central Rand Group are also a target.

The detection of extensions to known goldfields, either down dip or laterally, is a less difficult task as considerable information is available from the other mines. In this case it is possible to collate the grade distribution within an individual reef and project payshoots across structures as shown in Fig. 14.9. In this example sedimentation is controlled by synsedimentary structures with the highs forming areas of nondeposition or of low grade. The target areas in the structural lows can be

clearly established by structural contours and isocons of grade distribution.

In 1990, JCI were spending 5% of their exploration budget on the search for new basins, 7% on exploration for new goldfields, 12% on basin edge deposits, and 26% on down-dip extensions. The remaining 50% was spent in the search for extensions to known mines (Scott-Russell et al. 1990). In 2003 spending was very limited with Anglogold spending only US$2.4 million on exploration.

Environmental constraints on exploration have, in the past, been minimal as the Witwatersrand Basin mainly underlies flat farmland, largely used for maize and cattle production. In spite of the obvious problems generated by the extensive use of cyanide, there has been little concern about pollution. Forstner and Wittman (1979) however demonstrate that there is major contamination of surface waters associated with oxidation of pyrite and that high Zn, Pb, Co, Ni, and Mn are present as a result of this oxidation and the cyanidation process. Cyanide itself is not a problem as it breaks down under the influence of ultraviolet light and levels are within those specified in the South African Mines and Works Act. Also of immediate concern has been the dust generated by the erosion of the slimes dams. This can be countered by revegetation (Friend 1990). The rehabilitation of surface workings has now been given legal force in the Water Act and the Minerals Act of 1991.

14.5 EXPLORATION METHODS

The original discovery of gold-bearing conglomerate was made in 1886 on the farm Langlaagte by two itinerant prospectors, George Walker and George Harrison, who were resting on their way to the Barberton greenstone goldfield (Werdmuller 1986). While engaged in building a house for the owner they went for a walk and Harrison, who had been a miner in Australia, recognized the potential of the iron-rich conglomerate. He crushed and panned a sample that gave a long tail of gold. They were no doubt encouraged to prospect by the previous activities of Fred Struben, who was granted a lease to prospect veins in 1885 (Robb & Robb 1995).

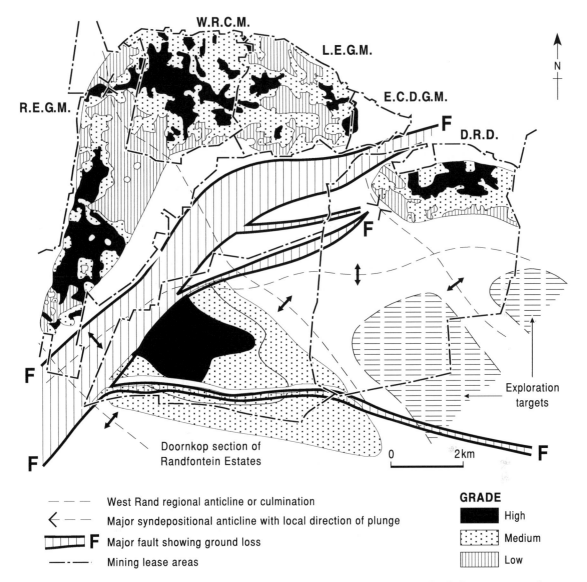

FIG. 14.9   Integration of regional grade distribution and structure to generate areas for drilling, West Rand. (After Viljoen 1990.)

14.5.1   Surface mapping and down-dip projection

After the original discovery the surface exposures were quickly evaluated by surface prospecting and the outcrop of the conglomerates mapped. In contrast to some expectations the conglomerates were not a surface enrichment but continued down dip. However this pres-

ented a major problem as the surface iron oxides turned to pyrite at around 35 m and, as all prospectors knew, this meant poor recoveries of gold by the recovery process then used, amalgamation of the gold with mercury. Fortunately the cyanide process of extraction had been developed by three Scots chemists in 1887 and was immediately demonstrated to be effective. In contrast to ores from greenstone belts,

Witwatersrand ore gives good recoveries due to the presence of gold as small free grains with very little or no gold contained in arsenopyrite, pyrite, or tellurides. The first major trial of cyanidation on tailings from the amalgamation process from the Robinson Mine yielded 6000 oz (190 kg) of gold from 10,000 tons. The success of cyanidation prompted new interest in the possible depth continuation of the reefs. This depth continuation was tested by drilling and the first drillhole in April 1890 intersected the South and Main reefs at 517 ft (157 m) and 581 ft (177 m) with grades of 9 oz 12 pennyweights $(dwt) t^{-1}$ $(328 g t^{-1})$ and $11 dwt t^{-1}$ $(18.8 g t^{-1})$. Further drillholes, notably the Rand Victoria borehole (Fig. 14.10), suggested in 1895 that the reef decreased in dip from the 60–80 degrees at outcrop to 15 degrees at 2350 ft (715 m) and that grades persisted at depth. The large potential of the area down dip was recognized by the mining houses with capital to finance mining at depth, and much of the area down dip was pegged by Goldfields and Corner House companies (Fig. 14.10). This down-dip area provided the basis for a number of major operations of which the largest, Crown Mines, eventually produced 163 Mt of ore and 1412 t of gold, an average grade of $8.7 g t^{-1}$ Au.

For the 40 years after the initial discovery exploration and mining concentrated on the areas down dip from the outcrop of conglomerates on the Central, West, and East Rand. Exploration involved drilling and development down dip from outcrop; by 1930 the Village Main Mine had reached 710 m below collar. In addition, although the average reef was payable, it was noticed that there were considerable variations in grade. Pioneering studies by Pirow (1920) and Reinecke (1927) demonstrated that these higher grade areas formed payshoots and could be correlated with larger pebbles and thicker reefs. Reinecke proposed that the control on the payshoots was sedimentary and paralleled the direction of cross-bedding.

14.5.2 Magnetics and the West Wits Line

The next stage in the exploration history was the search for blind deposits under later Transvaal and Karoo cover to the west of the West Rand Fault which forms the western limit of the West Rand Goldfield. Although a number of geologists were convinced of the continuation of the Main Reef to the west of the known outcrop and it had been tested by drilling in the period 1899–1904, results were unconvincing. A shaft sunk in 1910 to test these reefs had to be stopped at 30 m due to

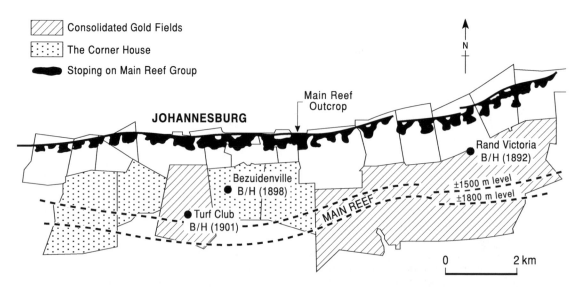

FIG. 14.10   Exploration and mining on the Main Reef, about 1900. (After Viljoen 1990.)

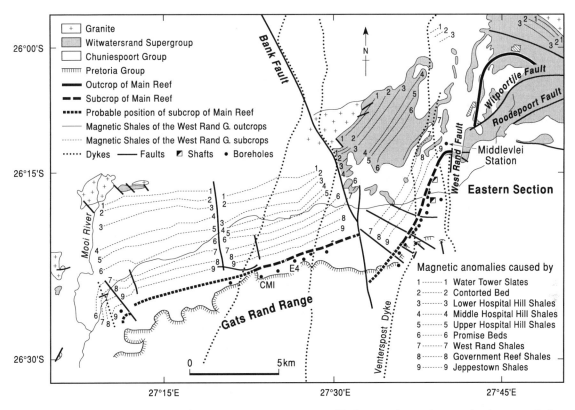

FIG. 14.11   Exploration on the West Witwatersrand line 1936. Drillhole E4 is the Carbon Leader discovery hole on the farm Driefontein, now (West) Driefontein Mine, and CM1 the discovery hole on the farm Blyvooruitzicht. (After Englebrecht 1986.)

an uncontrollable inflow of water from the Transvaal dolomites.

Wildcat drilling was prohibitively expensive and the search for the subcrop of the Main Reef was an unsolved puzzle for the period 1910–30. The solution to this problem was found by a German geophysicist, Rudolph Krahmann. One day in 1930 at a motor rally he noticed that the rocks of the West Rand Group contained magnetite and he thought that it would be possible to trace these shales under cover using newly developed magnetometers. After checking the theory over exposed West Rand Group sediments five major and four minor anomalies were found within the sequence (for fuller details see Englebrecht 1986 or the very readable account of Cartwright 1967). The depth penetration of the method was soon tested with backing from Gold Fields of South Africa who had also acquired the rights to the

cementation process that allowed shaft sinking through water-bearing formations. Finance also became more easily available with the increase in the gold price from £4.20 to £6 per ounce. The initial 11 drillholes (Fig. 14.11) down dip from the Middlevlei inlier not only intersected what was believed to be the equivalent of the Main Reef but also a further conglomerate at the base of the Ventersdorp lavas, termed the Ventersdorp Contact Reef. These holes gave rise to the Venterspost and Libanon Mines, but the real excitement came with results further to the west between the major Bank Fault, subparallel and west of the West Rand Fault, and Mooi River (Fig. 14.11). Drilling on the farm Driefontein began with three holes designed to cut the Main Reef. The first two holes were to the north of the subcrop of the Main Reef against the dolomites while the third cut poor values within the Main Reef but intersected a carbon seam with visible gold some 50–60 m below the Main Reef. The significance of this was not appreciated at first as core recovery in the very fragile carbon seams was incomplete due to grinding of core. This carbon seam later became known as the Carbon Leader Reef. In the original calculations of viability a grade of 13.6 g t^{-1} over 1.1 m was assumed, based on previous working on the Main Reef. When the first shaft was sunk by Rand Mines on the farm Blyvooruitzicht, a small piece of ground that Gold Fields did not own, the nature of the discovery soon became clear. Underground sampling around the shaft area in 1941 returned a grade of 69 g t^{-1} Au over 70 cm compared with a drillhole grade of 23.1 g t^{-1} over 55 cm. The mine management at first refused to believe the results and had the shaft resampled only to yield the same results. Similar grades were obtained when the GFSA mine at West Driefontein opened in 1952. Exploration to the west of the Bank break, along an area known as the West Wits Line, largely by GFSA, led to the development of 10 mines. Although magnetics led to the establishment of the subcrop of the Main Reef, the structural complexity of the area was not appreciated until the 1960s when the angular unconformity between the Main Reef and the Ventersdorp lavas was described. In the north and east of the area, the upper parts of the Central Rand Group have been removed by pre-Ventersdorp erosion. In addition, parts of the Carbon Leader have been removed by erosion and distinct barren areas can be recognized (Englebrecht et al. 1986).

When the importance of the finds on the immediate subcrop of the West Wits Line became apparent companies were quick to appreciate the down-dip potential. In particular Anglo American Corp. made a deliberate decision in 1943 to explore for deposits at 3000–4000 m and obtained the mineral rights to the area known as Western Deep Levels (Fig. 14.2). This is currently the deepest mine in the world and was based on 12 drillholes with an average intersection in the Carbon Leader of 3780 cm-g t^{-1} and additional mineralisation in the VCR. Production in 2003 was ~41 t of gold.

Although the magnetic method was outstandingly successful on the West Wits Line, detailed studies in other areas of the basin have encountered problems in interpretation. In particular the lateral and vertical facies variations of the beds are not understood. Magnetic anomalies can also be due to beds, other than Witwatersrand sediments, such as magnetite-rich layers in basement (Corner & Wilsher 1989).

14.5.3 Gravity and the discovery of the Orange Free State Goldfield

The increase in the gold price in the early 1930s encouraged companies to explore areas even more remote than the West Wits Line. In particular Anglo American made a discovery (now the giant Vaal Reefs Mine complex) in the Klerksdorp area, immediately north of the Vaal River, which forms the border between the Transvaal and Orange Free State provinces (Fig. 14.2). This discovery encouraged other companies, particularly Union Corporation, to explore in this area and also to the south of the Vaal River, where it was theorized that the Central Rand Group would extend under cover. In the Klerksdorp area Union Corporation had demonstrated for the first time that gravity methods could be used to predict the depth of the Central Rand Group as their density was significantly lower at 2.65 kg m^{-3} than the Ventersdorp lavas at 2.85 kg m^{-3}. This technique was selected to improve interpretation south of the Vaal River.

The area to the south of the Vaal River, particularly that around the village of Odendaalrust (north of Wesselbron, Fig. 14.2), had attracted the attention of a number of prospectors as it had been demonstrated as early as 1910 that gold could be panned from conglomerates that are stratigraphically part of the Ventersdorp Supergroup. A small venture took up the submittal from the original prospector and drilled a hole on the farm Aandenk to test the conglomerates. Although it was obvious from the beginning that the conglomerates were stratigraphically part of the Ventersdorp Supergroup, the venture persisted and intersected the top of the Central Rand Group quartzites in 1934 at a depth of 2726 ft (831 m). The hole eventually bottomed at 4046 ft (1234 m) with a best intersection of 120 in-dwt t^{-1} (521 cm-g t^{-1}) when they ran out of funds.

On the basis of the results from the Aandenk borehole and the success of the gravity method a company, Western Holdings, was floated by South African Townships to prospect in the Southern Free State in collaboration with Union Corporation. Intensive geophysical surveys showed a significant gravity anomaly over the farm St Helena flanked to the west by magnetic anomalies (Fig. 14.12). The first hole SH1, drilled in 1938, indeed cut Central Rand group quartzites at 302 m and was drilled for another 1850 m cutting three reefs at 200–300 in-dwt t^{-1} (860–1300 cm-g t^{-1}). Further drilling to the west confirmed the location of the magnetic shales. By this stage a number of other companies were drilling in the area but the stratigraphical control of the mineralisation was unknown. After routine assaying of a conglomerate one of these companies obtained an assay of 6.9 g t^{-1} over 1.05 m in what is now known to be the major gold-bearing conglomerate. Within a few months Western Holdings intersected payable values of the same reef in hole SH7 at 59 g t^{-1} over 1.5 m at a depth of 348 m. These were the first intersections of the Basal Reef (Fig. 14.3), which has been the main producing placer in the Orange Free State Goldfield. Proving the exact positions of the reefs was however difficult as the goldfield is cut by a number of major faults and interpretation of gravity data was not as simple as first thought. Gravity anomalies could reflect the depth of the Central Rand Group quartzites but they also could be due to faulting, folding, and changes in the lithology of the Ventersdorp Volcanics. However experience showed that the Central Rand Group anomalies could be recognized by their sharp gradients, whereas changes due to the Ventersdorp were much more gradual. Further exploration and exploitation of the goldfield was delayed by World War II but resumption of exploration after the war confirmed the potential of the area including a phenomenal intersection on the farm Geduld of 3377.5 dwts t^{-1} (5775 g t^{-1}) over 6 inches (15 cm). Although not the active partner in the original exploration venture, Anglo American Corp. became the dominant operator in the OFS goldfield by taking over Western Holdings and several other companies immediately after 1945.

Similar techniques were successful in the discovery by Union Corporation of the Evander Sub-basin, about 100 km east-south-east of Johannesburg (Fig. 14.2). Although the area had been known to contain concealed Central Rand Group rocks, it was only the advent of airborne magnetic and ground gravity surveys that unravelled the structure. Persistence with a drill program eventually proved that the Kimberley Reef was the only payable reef and led to the establishment of four mines (Tweedie 1986).

14.5.4   Reflection seismics, gaps, and upthrown blocks

One of the major advances of the 1980s was the use of seismic reflection surveys in detailing the structure and location of the Central Rand Group rocks at depth. Although use of the method was mooted by Roux in 1969, it was the impact of improved technology and lack of work in the oil industry that prompted a renewed trial by oilfield service companies in collaboration with the major mining houses. Initial logging of boreholes showed that there was a contrast in seismic velocities as confirmed by initial surface surveys (Pretorius et al. 1989, 1997). Although it is not possible in reconnaissance surveys to define the individual reefs, the surveys are able to define major contacts, particularly the base and top of the Central Rand Group and the contact between the Transvaal dolomites and the Pretoria shales overlying them. By 1989 Anglo American had covered the basin based on more than 10,000 line-km of 2D survey and further coverage concentrated on the detailed follow-up of target areas. Other mining houses use seismic surveys extensively and JCI (Campbell & Crotty 1990) have used 3D surveys in planning mining layouts and the location of individual longwall faces at their South Deep prospect (section 14.7.2). The cost of 3D surveys limits their use; Campbell and Crotty report that the survey at the South Deep prospect cost R2.2M, the cost of one and a half undeflected drillholes.

Pretorius et al. (1989) defined three different types of seismic survey used by Anglo American during the 1980s:

1 Reconnaissance 2D surveys of stratigraphy and structure on public roads concentrating on the gap areas where the target beds are relatively shallow.

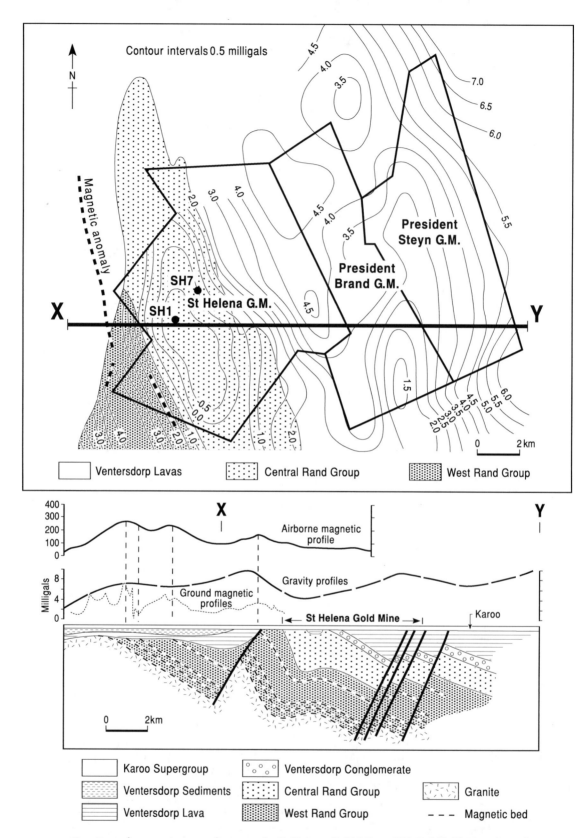

FIG. 14.12 Gravity and magnetic anomalies over the St Helena Gold Mine, OFS Goldfield. Note how the discovery hole (SH7) is on the flank of the main anomaly. (From Roux 1969.)

2 Reconnaissance 2D of deeper areas within the basin where targets might be upthrown to mineable depths by faulting.

3 Detailed 2D surveys of extensions to reserves on existing mines.

Few examples of seismic sections have been published, but the paper by Pretorius et al. (1989) provides examples of the use of the technique in reconnaissance exploration. The area shown in Fig. 14.13a is on the western edge of the basin and the seismic survey clearly delineates the upper and lower contacts of the Central Rand Group and faulting within the Transvaal sediments. The projected depth of the VCR at the upper contact is over 5000 m and drill targets can be assigned on this basis. The contrast between the seismic section and the previous interpretation of the gravity data is striking (Fig. 14.13a) In the original gravity interpretation the 3 mgal anomaly superimposed on the strong regional gradient was attributed to a horst of Central Rand Group sediments at a depth of around 2 km whereas the seismic section showed that there was no major faulting, that the Central Rand target was at 5 km, and the gravity low was the result of changes in lithology of the Pretoria Group sediments. Thus seismic and gravity data can be successfully integrated.

Figure 14.13b shows the result of a reconnaissance seismic reflection survey in the middle of the basin that detected an upthrown block of Central Rand Group sediments that are at mineable depth (<3000 m); this was confirmed by later drilling.

The third type of survey is exemplified by the seismic survey which defined the struc-ture at the margins of an established mine (Fig. 14.13c). From this section an additional area of Central Rand Group sediments at mineable depths was established.

14.5.5 Drilling

One of the major financial constraints on exploration is drilling. A 6000-m-deep hole drilled from surface may take 18 months to complete. A typical 15-hole program on a lease of 7000 ha at 3–4000 m target depth would take 20 years to complete if the holes were drilled sequentially. This posed considerable problems, not only of capital, but also of decision making, as mineral rights options are only 3 or 5 years in length and a decision to exercise the option must be taken at the end of the period. Clearly programs must have several rigs drilling simultaneously or drilling must be accelerated. Each drillhole will be deflected to provide a number of reef intersections and improve confidence in grade estimation (Fig. 14.14, Table 14.2).

Perhaps the key decision to be made is approximately how many intersections need to be made before a confident estimate can be made of a lease's value or lack of prospectivity. A quantitative approach to the problem has been given by Magri (1987), whose results are summarized in Fig. 14.14. Using an approach similar to Krige's t estimator to predict the mean of log-normally distributed data, he has produced confidence limits of estimates of the mean grade for reefs of immediate interest to his company (Anglovaal) based on knowledge of the grade distribution in mined blocks. These clearly show that making further drill

TABLE 14.2 Costs of drillholes and deflections, in 000 rand 1987 value (1R = US$0.3). (From Magri 1987.)

Depth (m)	Number of deflections						
	1	2	3	4	5	7	9
1000	93	118	131	144	158	188	220
1500	150	176	190	205	220	253	288
2000	220	249	267	286	306	348	393
2500	303	337	359	382	407	457	511
3000	408	446	473	501	530	592	658
3500	545	590	624	660	698	777	862
4000	714	768	812	858	905	1004	1111
4500	907	962	1007	1053	1102	1204	1315

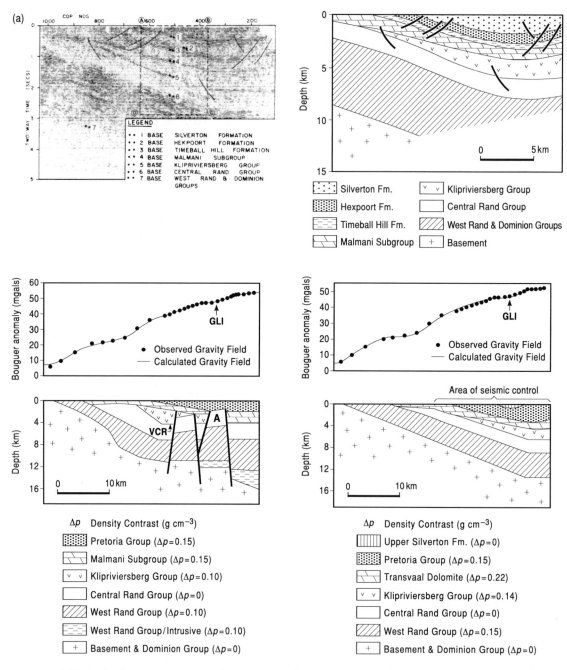

FIG. 14.13 (a) Use of reflection seismics in discriminating between two possible gravity models for targets on the western margin of the Witwatersrand Basin. The left-hand model used a block of upthrown Central Rand Group (A) to explain the gravity low while the right-hand model, confirmed by seismics, invoked near-surface Transvaal sediments.

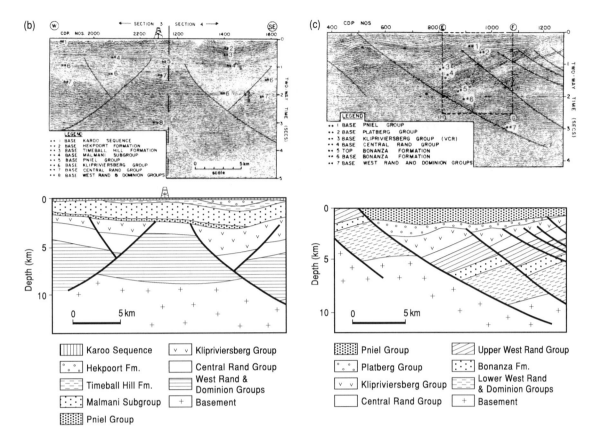

FIG. 14.13   (b), (c) Typical uses of seismic reflection data: (b) detection of upthrown blocks; (c) detection of unsuspected upthrown blocks adjacent to a lease area. (From Pretorius et al. 1989 with permission)

intersections after a certain cut-off, which varies for each reef, provides little extra information. In the example shown this is 12 or 13. In the same way it is possible to optimize the number of deflections from each drillhole – usually four are made. Magri suggested that the optimum approach to drilling an area is to decide on a minimum number of holes necessary to give a picture of an area, say 10 for a 7000-ha lease and review the data when drilling is complete. If the grade of the mean less confidence limit for the prospect is greater than the pay cut-off, then the program should be continued. If not, it should be abandoned.

The application of drilling rigs used for oil and gas exploration has helped to speed exploration and to increase the accuracy of deflections from the mother hole.

14.5.6   Underground exploration

Considerable underground exploration is necessary before a mining layout can be designed that will minimize losses of production from faulting and dykes. Figure 14.15 shows the layout of a typical underground drilling program in a faulted area.

14.5.7   Re-treating old mine dumps

One of the most significant innovations on the Witwatersrand has been the major projects to re-treat old slimes dams and sand tips. The reasons for the occurrence of the gold in waste are many; gold may be enclosed in pyrite or cyanide-resistant minerals that were not crushed or the operation of the plant was not optimal for the mineralogy treated. The

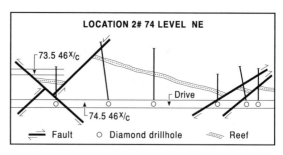

FIG. 14.15   Section showing typical underground drilling to determine reef location and fault throws, Vaal Reefs Mine. The length of the section is approximately 70 m. (After Thompson 1982.)

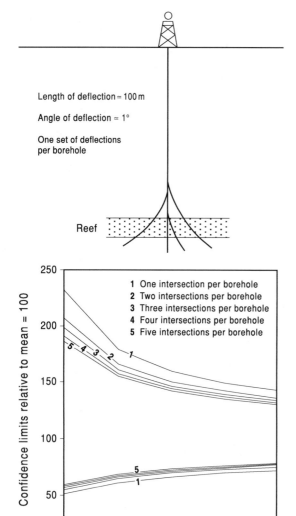

FIG. 14.14   Schematic diagram of deflections and plot of confidence limits for increasing number of drillholes and deflections per hole. There is considerable variation between placers, the case shown is the Vaal Reef. (From Magri 1987.)

average gold content in the waste is probably about 0.3 ppm but waste from the older mines on the East and Central Rand contains approximately 0.7 ppm. Advances in metallurgical techniques and increases in the gold price have

allowed profitable reworking of this waste, and by 1986 about 11 t of gold per year were produced from this source.

14.6   RESERVE ESTIMATION

Reserve estimation on a gold mine occurs in two phases, namely those resources calculated during exploration and those reserves estimated during exploitation. Ore reserves calculated during exploitation refer to those blocks of ore that have been developed sufficiently to be made available for mining. Usually, a raise would have been driven and sampled. The samples selected for use in the reserve calculation are those which are above the cut-off assigned according to the mine's valuation method (Janisch 1986, Lane 1988). The valuation of a developed reserve block has been traditionally carried out using the average grade of all the samples taken along the raises and the drives. This is steadily giving way to kriging methods (Krige 1978, Janisch 1986).

In exploration, resource estimation is not so precise. It is essential to identify the blocks which could be mined. These blocks would most probably be bounded by faults. Hence the importance of establishing the structure of an area from the borehole information, using structural contour maps and a series of cross-sections. More recently this information has been supplemented with information from 2D and 3D seismic surveys (Campbell & Crotty 1990, Pretorius et al. 1997).

The average thickness and grade for each block must be calculated. Each borehole has an average of four deflections which intersect the reef, often in random directions (Fig. 14.14), and this gives an unbiased sampling of the reef within a radius of about 5 m of the original hole. When an inclined hole intersects a dipping reef, only an apparent dip of the reef can be seen in the core, so corrections must be made, using the following formula, to calculate the true thickness, to be used in the tonnage calculations:

$$W = \cos b \, \{C(\sin d - [\cos(90 - e)\tan b]\}$$

where W = true reef thickness, b = reef dip, C = apparent reef thickness, d = borehole inclination, and e = azimuth of reef dip.

A Sichel's t average grade value (Sichel 1966) is calculated for each cluster of acceptable sample values. Each value thus calculated is either assigned to the block into which the borehole falls (modified polygonal method; Chapter 9), or the averages of all the holes, also calculated using the Sichel's t estimator, is used as the average grade of the reef for the entire area being assessed.

Usually a mine has a minimum mineable reef thickness, often of 1 m. Where reef intersections are less than 1 m the grade is recalculated to reflect the additional waste rock that would have to mined to achieve the minimum mineable stoping height and this height (in this example, 1 m) is used in tonnage calculations. Often, to establish whether the minimum mining height is going to be economic to mine, an accumulation or grade thickness product is calculated, such that $a_x = W_x \times g_x$. This accumulation is the product of the true thickness W in the sample at x and the average grade, g_x. In the gold mines, thickness is quoted in centimeters, the grade in g t^{-1}, and the product in cm g t^{-1}. Where a sample is shown to have 35 g t^{-1} over 10 cm, the accumulation is 350 cm g t^{-1}. To recalculate the grade over the minimum mining width of 100 cm (1 m) the accumulation is divided by the mining width and in this case the grade over 100 cm is 3.5 g t^{-1}. If the cut-off grade for this particular mine happened to be 3 g t^{-1}, then this block would be considered payable and the block included in the reserve calculations using the 100-cm mining width.

The tonnage is calculated for a block which has a borehole at the centre using the following formula:

$$M = W \times A \times r$$

where M = mass in tonnes, W = average true thickness of the reef (with a minimum of 1 m), A = area of the block, and r = specific gravity.

The tonnage for a particular area is then calculated by summing the tonnage for each block.

14.7 EXPLORATION AND DEVELOPMENT IN THE SOUTHERN PART OF THE WEST RAND GOLDFIELD

As discussed in section 14.4, the easily discovered, near-surface deposits have been known, and often exploited, for many years. The object of more recent exploration is to find the deeper, more stratigraphically, sedimentologically and structurally complex deposits. In areas with active mining, careful analysis and reinterpretation of geological data, including data supplied in company annual reports, can lead to new discoveries being made in areas previously thought to be barren. The area discussed in this case history is the southern extension of the West Rand Goldfield mined immediately down dip from the outcrop of the Central Rand Group. The original mines are Randfontein Estates, West Rand Consolidated, Luipardsvlei, and East Champ d'Or (Fig. 14.16). Four deposits have been discovered to the south of these mines and to the east of a major structural break known as the Witpoortjie Horst. They are the Cooke and Doornfontein sections of Randfontein Estates, Western Areas, and South Deep. The exploration of Cooke Section will be discussed in detail as it is one of the few detailed, published, exploration case histories for the Witwatersrand (Stewart 1981) and that of South Deep as it shows the application of methods developed in the 1980s.

Development of Cooke Section only started in the mid 1970s after almost a decade of exploratory drilling, but nearly 90 years after

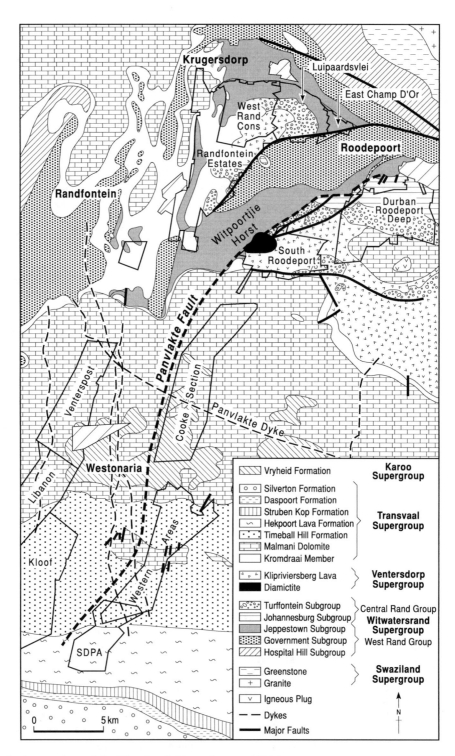

FIG. 14.16   Surface geology of the West Rand. (From Stewart 1981.)

the discovery of gold in the region. Explanations for this late discovery are that these gold-bearing rocks are concealed beneath younger sediments, the structure is more complex than in other areas, the gold-bearing conglomerates are highly variable, and these conglomerates are in a part of the sequence that had not before yielded economic quantities of gold. This latter point, in conjunction with the fact that the conglomerates which normally produced gold in other parts of the basin were not here considered to be economic, meant that other areas were explored in preference to the Cooke Section.

14.7.1   Cooke Section – previous exploration

The area had been geologically mapped by Mellor (1917). This map formed the basis of exploration until 1962 when Cousins extended this mapping southward. Mapping (Fig. 14.16) showed that Witwatersrand Supergroup rocks crop out with an approximate E–W strike from the East Rand through Johannesburg to Krugersdorp in the west (Fig. 14.2). In the west the strike of the Witwatersrand rocks starts to swing to the southwest, but the outcrop is covered by younger sediments of the Transvaal Supergroup and the Witwatersrand rocks are upthrown on to a horst, the Witpoortjie Horst, by major faulting (Fig. 14.17).

Attempts were made to trace extensions of the West Rand by drilling from as early as the beginning of this century and resulted in the recognition of the West Wits Line (section 14.5.2). Further drilling was carried out in the area to the south of Randfontein Estates between 1915 and 1950 (Fig. 14.18). This work showed that the Witpoortjie Horst had a nearer N–S strike than previously thought. This drilling also provided more detailed information about the variations in the generalized stratigraphical column (Fig. 14.19). For example, the Main Reef zone was expected to be highly productive, but nothing significant was intersected in the Cooke Section. Unfortunately, the Middle Elsburg Reefs, which were subsequently shown to be the main gold-bearing reefs of this area, were only poorly developed in these holes. One hole was drilled in what is now known to be a high grade part of Cooke Section but did not penetrate the reef as it was cut out by a dyke.

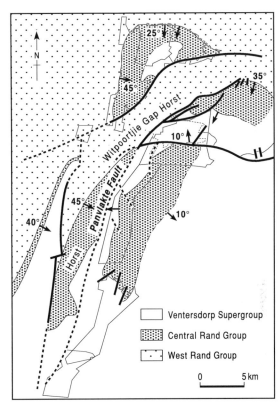

FIG. 14.17   Sub-Transvaal geology of the West Rand. Note the absence of Central Rand Group sediments on the Witpoortjie Horst (From Lednor 1986.)

Ten holes were drilled in the early 1950s (Fig. 14.18) in Cooke Section. Some were intended to obtain information about the potential of the Main Reef zone on the horst while two of the holes were drilled to test the potential of the reefs on the eastern, downthrown side of the Witpoortjie Horst. Holes were also drilled farther south, and as they immediately intersected good gold values in the Elsburg Reefs, on what is now Western Areas Gold Mine, further drilling on the Cooke Section was deferred until 1961.

Before the 1950s, borehole cores were only assayed for gold content. With the advent of interest in uranium in the 1950s, the possibility of extracting the uranium in addition to the gold was considered. Sampling and assaying of conglomerates for uranium showed that the increase in uranium price had made the uranium an attractive proposition which could

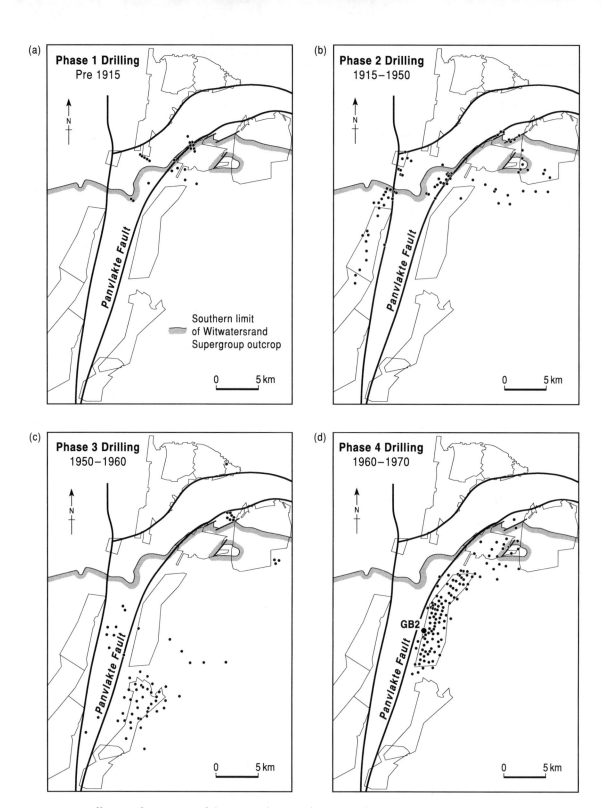

FIG. 14.18 Drilling on the West Rand. (From Tucker & Viljoen 1986.)

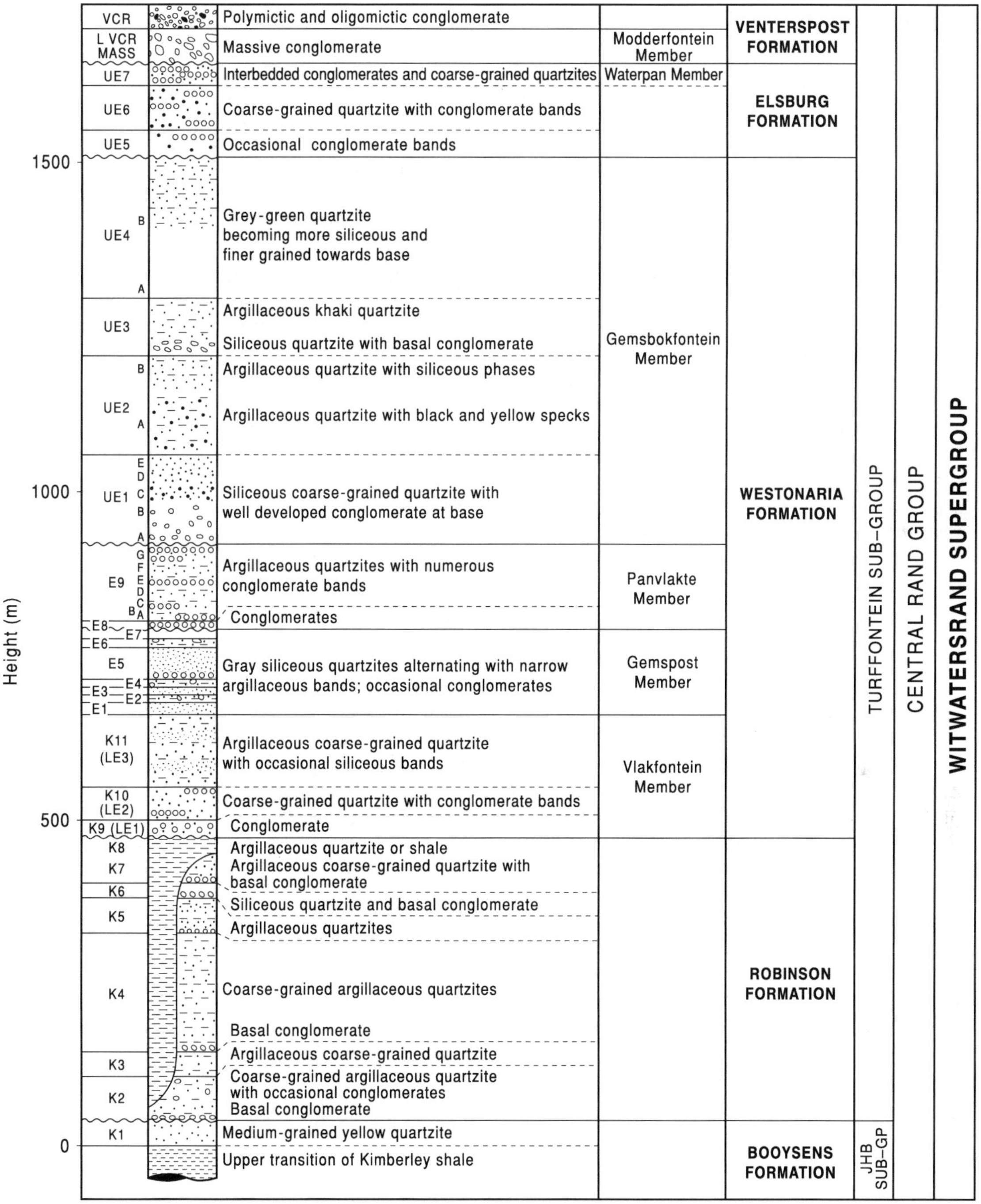

Unit	Description	Member	Formation	Sub-group	Group	Supergroup
VCR	Polymictic and oligomictic conglomerate		VENTERSPOST FORMATION			
L VCR MASS	Massive conglomerate	Modderfontein Member				
UE7	Interbedded conglomerates and coarse-grained quartzites	Waterpan Member	ELSBURG FORMATION			
UE6	Coarse-grained quartzite with conglomerate bands					
UE5	Occasional conglomerate bands					
UE4 (B, A)	Grey-green quartzite becoming more siliceous and finer grained towards base	Gemsbokfontein Member	WESTONARIA FORMATION	TURFFONTEIN SUB-GROUP	CENTRAL RAND GROUP	WITWATERSRAND SUPERGROUP
UE3	Argillaceous khaki quartzite / Siliceous quartzite with basal conglomerate					
UE2 (B, A)	Argillaceous quartzite with siliceous phases / Argillaceous quartzite with black and yellow specks					
UE1 (E D C B A)	Siliceous coarse-grained quartzite with well developed conglomerate at base					
E9 (G F E D C B A)	Argillaceous quartzites with numerous conglomerate bands	Panvlakte Member				
E8, E7, E6	Conglomerates					
E5 (E4)	Gray siliceous quartzites alternating with narrow argillaceous bands; occasional conglomerates	Gemspost Member				
E3, E2, E1						
K11 (LE3)	Argillaceous coarse-grained quartzite with occasional siliceous bands	Vlakfontein Member				
K10 (LE2)	Coarse-grained quartzite with conglomerate bands					
K9 (LE1)	Conglomerate					
K8	Argillaceous quartzite or shale		ROBINSON FORMATION			
K7	Argillaceous coarse-grained quartzite with basal conglomerate					
K6	Siliceous quartzite and basal conglomerate					
K5	Argillaceous quartzites					
K4	Coarse-grained argillaceous quartzites					
	Basal conglomerate					
K3	Argillaceous coarse-grained quartzite					
K2	Coarse-grained argillaceous quartzite with occasional conglomerates / Basal conglomerate					
K1	Medium-grained yellow quartzite / Upper transition of Kimberley shale		BOOYSENS FORMATION	JHB SUB-GP		

Height (m): 1500, 1000, 500, 0

FIG. 14.19 Detailed stratigraphical column for the Central Rand Group in the Randfontein Estates area. (After Randfontein Estates Gold Mine personal communication.)

make a deposit with marginal gold values payable. The existing data were reviewed, and the results of one of the holes (GB2) drilled in the downthrown block showed gold and uranium values which were considered worth following up (Fig. 14.18).

Geophysics

Gravity and airborne and ground magnetic surveys were carried out across the Cooke Section to obtain data that would assist in the interpretation of the structure. Despite the success of these methods elsewhere in the Witwatersrand Basin (section 14.6), neither method proved directly successful in this area and drilling was used to further evaluate the Cooke Section. The density difference between the Witwatersrand quartzites and the overlying Ventersdorp lavas was masked by the thick sequence of younger Transvaal Supergroup dolomites, which made the gravity method unsuccessful. Magnetics were used to try to establish the presence of the magnetic shale in the West Rand Group. Unfortunately the combined thickness of the overlying Central Rand Group and Transvaal Supergroup rocks was sufficient to effectively mask the magnetic signal from these shales.

Drilling

One hole drilled near the hole GB2 had revealed promising gold grades. This new hole intersected 1.32 m of Elsburg Reef at 968 m which assayed 20 g t^{-1} Au and 690 ppm U$_3$O$_8$. This success led to the planning of a full drilling program in which 88 holes were completed between 1960 and 1970 (Fig. 14.18). The holes, drilled on a grid, were spaced approximately 800 m in a N–S (strike) direction and as close as 300 m in an E–W (dip) direction.

At that time it was standard practice to drill these holes using narrow diameter (NX or BX) diamond bits. The unconsolidated overburden was drilled using larger diameter, percussion rigs (also known as drilling a pilot hole). Pilot holes are now drilled to significant depths, with much time and cost saving (Scott-Russell et al. 1990). The pilot hole would have steel casing inserted to prevent collapse of the unconsolidated material into the hole. Once coring commenced in the hard rock, this continued until the final depth was reached and a complete section of core would be available for the geologist to log. To save time and expense, four deflections were usually drilled from the original hole to obtain additional reef intersections. On average each hole took 4–5 months to drill.

All the core was taken to a central location where it was logged. After logging, most of the Transvaal Supergroup dolomites and the Ventersdorp Supergroup lavas were discarded, but the Witwatersrand Supergroup rocks were stored, in case they were needed at some future date for further study. The entire succession of Witwatersrand Supergroup rocks was then sampled, at 30-cm intervals where gold and uranium grades were expected, and at 60-cm intervals where no assay values were expected. Fire assaying was used to determine the gold content and radiometric methods were used to determine the U$_3$O$_8$ content, with suitable check assays undertaken both within one laboratory and between laboratories.

Interpretation of results

The unusual sequences intersected by the first few boreholes made it difficult to interpret the stratigraphy and the structure. Work commenced on compiling a detailed stratigraphical column relevant to this area. It was found that the initial logging had not been done in sufficient detail to compile a specific stratigraphical column. Cores were retrieved from the store on their trays and two holes were laid out side by side for re-examination. In this way detailed comparisons could be made and this was repeated many times. Marker bands, mainly fine-grained quartzites, were identified which could be used for correlation, and, after many months and many attempts, a stratigraphical column was produced which was tested against new holes. The difficulties can be gauged from Fig. 14.20 which shows the correlation between three holes (note the numerous unconformities). The process was further complicated by the presence of faults and dykes. Eventually a reliable stratigraphical column was generated and each hole could then be divided into reef zones. At this stage it was possible to study the grade and geometry of

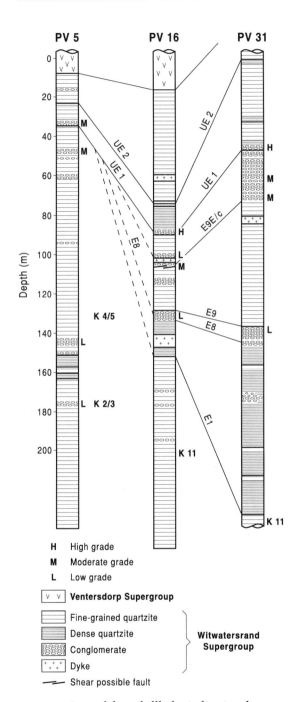

H   High grade
M   Moderate grade
L   Low grade

| v v | **Ventersdorp Supergroup** |

Fine-grained quartzite
Dense quartzite   **Witwatersrand**
Conglomerate   **Supergroup**
Dyke

Shear possible fault

FIG. 14.20   Logs of three drillholes indicating the difficulties of correlating between holes. For the location of holes see Fig. 14.21. (After Stewart 1981.)

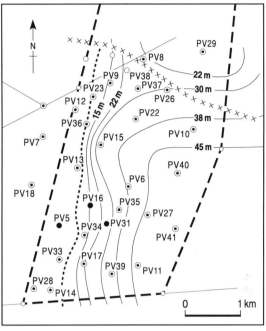

- - - - - Limit of E9E/c Conglomerate

FIG. 14.21   Typical isopach map used to interpret drilling, in this case the E9c to UE1a unit. (After Stewart 1981.)

individual reef zones, using grade distribution and isopach maps, and to determine the structure of the area, using structure contour maps. The sedimentological controls on the gold and uranium distribution mean that additional data can be gained by comparing a reef isopach map with the overlying inter-reef interval (Fig. 14.21) to identify possible depositional trends, or by investigating pebble size, and conglomerate to quartzite or gold to uranium ratios.

The drilling showed that the cover of Transvaal Supergroup dolomites varied in thickness from 100 m at the northern boundary of Cooke Section to approximately 600 m in the south. Below the dolomites older rocks were faulted. The dominant feature is the N–S Panvlakte Fault, which forms the eastern boundary of the Witpoortjie Horst. This horst was active during the later part of Witwatersrand sedimentation. Most of the Witwatersrand rocks had been eroded from the horst during the pre-Transvaal peneplanation. To the east of this fault, an

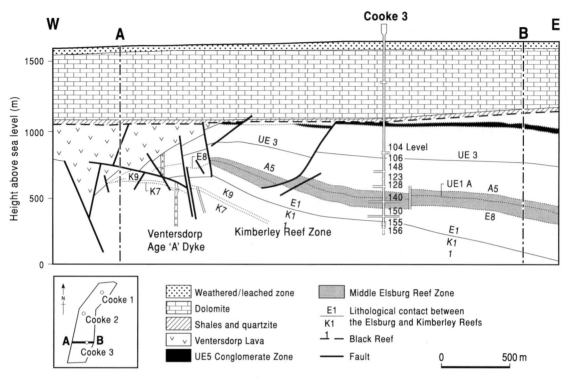

FIG. 14.22   Section through Cooke Section Mine. (After Randfontein Estates Gold Mine personal communication 1991.)

almost entire succession of Witwatersrand rocks, and some Ventersdorp Supergroup lavas, were preserved on Cooke Section (Fig. 14.22). A gentle, N–S trending anticline exists, which causes the sediments to dip from just a few degrees up to 45 degrees to the east.

Of all the recognized reefs on Cooke Section (Figs 14.20 & 14.22) only three composite units, or packages, are currently considered economically important, the UE1a–E9Gd, the E8, and the K7–K9. The former is the most important reef package that is currently being mined with only minor stoping on the E8 and E9 reefs.

Development of Cooke Section

In early 1973, shaft sinking was completed on the Cooke Section No. 1 Main Shaft at a final depth of 867 m. Towards the end of the year stoping began in the area adjacent to the shaft. By the end of 1973 some 15,544 t of ore had been mined and hoisted to the surface. This material had an average grade of 17.7 g t^{-1} Au. Cooke Section No. 2 Main Shaft was commissioned in 1977, and in 1980 the Cooke Section No. 3 Shaft was started. It was finished, at a final depth of 1373 m, in 1983.

Cooke Section has evolved into a large relatively low grade producer. By 1992 around 5 Mt of ore were being mined at an average grade of 8 g t^{-1} of gold, although this had dropped to 2.87 Mt at 4.9 g t^{-1} Au in 2002. The income from uranium has become of very minor importance with the decline in the price of that metal. Trackless mining now accounts for 70% of production and the mine is one of the most efficient on the Witwatersrand. Cooke Section has been a major success due to the nature of the reefs and its inception at a time of the sharp increase in the price of gold. Trackless mining has, however, not been successful on the Doornkop Section, immediately north of Cooke Section, due to excessive dilution and the mine was mothballed in 1999. It is however being reactivated in 2004.

14.7.2 South Deep

This prospect is one of the major undeveloped deposits in the Witwatersrand Basin with resources of more than 1000 t of contained gold. *In situ* reserves are 216 Mt at 8.4 g t^{-1} Au in a resource of 295 Mt of 10.1 g t^{-1} Au based on 5 g t^{-1} Au cut-off (South Deep 2001). The potential of the area was first recognized during the drilling of Western Areas in the mid 1960s. However the depth (>2500 m) and the exploration of other prospects in the area delayed systematic exploration until the late 1970s. It was quickly recognized that the deposit was unusually thick and provided opportunities for mechanized wide orebody mining as well as conventional stoping. The reef packages of interest are the Massive and Individual Elsburg Reefs as well as the Ventersdorp Contact Reef (Figs 14.3 & 14.19). The thicker Massive and Individual Elsburg Reefs lie to the east of a NNE trending feature known as the upper Elsburg shoreline. The Individual Elsburg Reefs, that are up to nine in number, merge within 750 m of the shoreline to form a composite reef which can be up to 70 m thick. To the west of the shoreline only the Ventersdorp Contact Reef is developed. The initial exploration was by conventional drilling from the surface (Fig. 14.23) and about 200 intersections were used in the feasibility study.

In order to make evaluation more accurate a 3D seismic survey was shot between August 1987 and September 1988. A nominal grid of 25 × 50 m reflectors was used to generate a 1:10,000 Ventersdorp Contact Reef (VCR) structure contour map. All faults over 250 m in strike length and 25 m throw are shown; these features can be extrapolated from the VCR to the Elsburgs and form the basis of mine layouts and ore reserve determinations that are more accurate than could be estimated from drilling (Campbell & Crotty 1990).

Wide orebody mining

In the South Deep Project Area, to the west of the shoreline, near the palaeoshoreline, the reefs are narrow and will be exploited by narrow orebody stoping techniques, as described above. To the east of the shoreline, the reefs increase in thickness from 2 m to over 100 m

(Fig. 14.24), and it is in these areas that the mechanized, wide orebody mining systems will be used. Room and pillar mining can be used to mine reefs up to 8 m thick, after mining a conventional stope to distress the rock but for greater thicknesses, variations on longhole stoping such as vertical crater mining will be employed (Tregoning & Barton 1990, Maptek 2004, South Deep 2004). Vertical crater mining can be developed initially in the same way as room and pillar mining. Two stopes will be established at the top and bottom of the proposed mining block (Fig. 14.25). Each reef block is planned to be 150 m wide in a strike direction and 250 m wide in a dip direction. The upper excavation is called the drilling drive and the lower, the cleaning drive. Retreat mining by means of slots up to 40 m high (between the two levels), 50 m long, and up to 7 m wide will be practised. The stopes are to be cleared using remote-controlled LHDs, and a vacuum or hydraulic cleaning system used for sweeping before backfilling. The cemented, crushed waste rock, backfill will be placed in the mined-out slots as a permanent support.

Developments 1990–2004

South Deep has an advantage over other new prospects in that the target reefs are a continuation of those mined on the south section of Western Areas and test mining of the area has been possible without the cost of sinking a new shaft. A twin haulage way has been driven from Western Areas' south shaft to the site of South Deep's main shaft, a distance of about 2000 m at a cost of around R70M, and was completed in 1991 (Haslett 1994). Underground drilling from the haulage (Fig. 14.26) and development has shown that grades obtained from surface drilling err on the low side but has confirmed the structure deduced from the seismic survey. Mining is underway to extract the reef in the area around the main shaft. Besides producing revenue from around 24,000 t per month (2.2 t of gold in 2002), the development will destress the shaft pillar. Development was also undertaken to enable underground testing of the wide orebody by drilling holes at 30-m intervals. At the time of writing (2004) both conventional (Fig 14.27) and trackless mining were underway outside the destressed area and

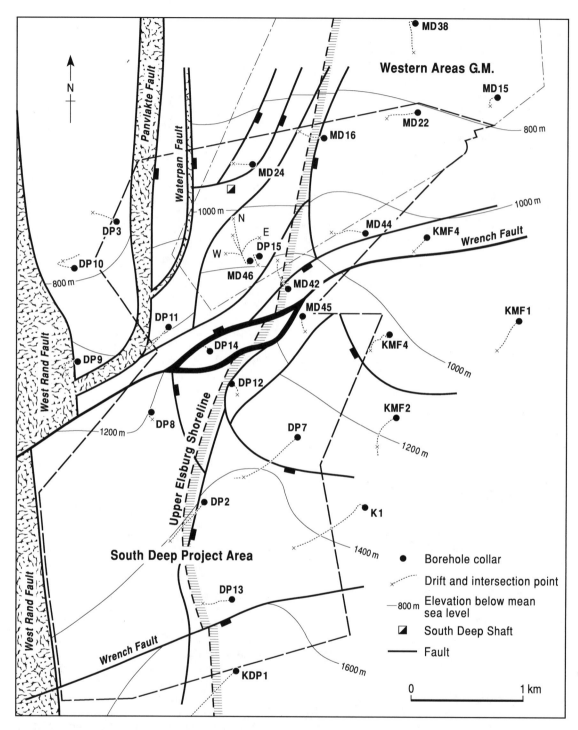

FIG. 14.23   Drill plan and simplified structure contours on VCR of the South Deep Prospect Area. (After South Deep 1990.)

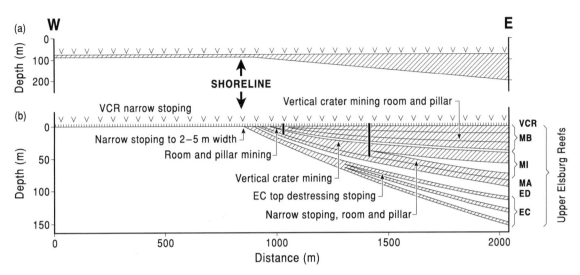

FIG. 14.24   Generalized section showing the relation of proposed mining methods to reef thickness. (After Tregonning & Barton 1990.)

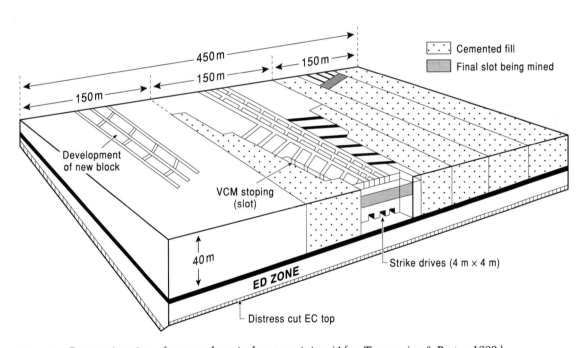

FIG. 14.25   Perspective view of proposed vertical crater mining. (After Tregonning & Barton 1990.)

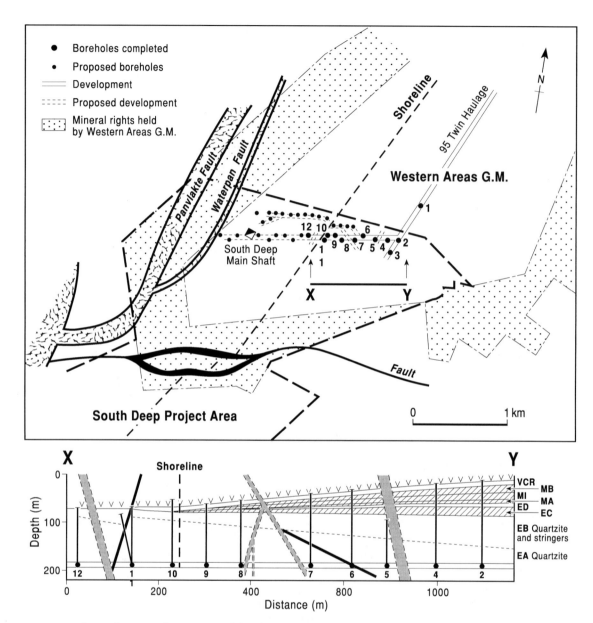

FIG. 14.26   Plan and section showing initial development at South Deep from Western Areas gold mine. Note the drilling to confirm grade and reef location. (After South Deep 1992.)

production was 1.37 Mt at 8.4 g t^{-1} Au (Western Areas 2003). The transition from conventional to trackless mining has taken considerable time and has been accompanied by the retrenchment in 1999 of 2560 staff.

Development of the deposit has had a complex history since 1990. The initial devel-

opment was financed by floating a company (South Deep Exploration Limited) on the Johannesburg Stock Exchange. This company was then merged with the Western Areas company in 1995 and subsequently the host mining house (JCI) was broken into constituent parts by commodity. In 1999 50% of the

FIG. 14.27 Underground view of workings on the Massive Elsburgs at South Deep, 2000.

project was bought for $US235 million by Placer Dome who then assumed management control. Shaft sinking began in 1995 and by 2004 the twin shaft complex had reached bottom (2994 m) and was being equipped. The overall cost of the project is estimated at R6.52 billion (~$US970M) and is anticipated for completion in 2006.

14.8 CONCLUSIONS

The Witwatersrand Basin with its large tabular orebodies is a unique gold occurrence. Exploration has evolved from the testing of down-dip extensions from outcrop through the recognition of blind districts, such as the West Wits Line, to the definition of deep targets that can support a large underground mine.

Geophysical methods have proved very successful in delineating the stratigraphical setting of the blind deposits. Magnetic and gravity methods made a major contribution in the 1930s to 1960s, as have reflection seismics in the 1980s.

Definition of mineable areas depends on an understanding of the sedimentological and structural controls on ore distribution as revealed by the very limited information available from deep drilling. Evaluation of this limited information led to the development of geostatistical methods that provide reliable estimates of grade and tonnage before a decision was made to commit large expenditures to shaft sinking. The discovery of the South Deep deposit shows that large deposits remain to be discovered, although they are likely to be very few in number and difficult to develop.

15

A VOLCANIC-ASSOCIATED MASSIVE SULFIDE DEPOSIT – KIDD CREEK, ONTARIO

ANTHONY M. EVANS AND CHARLES J. MOON

15.1 INTRODUCTION

Before reviewing the exploration history and geology of Kidd Creek it is necessary to describe and discuss the general features and different types of massive sulfide deposits, as a good knowledge of these is essential in designing an exploration program to find more deposits of this type. The term volcanic-associated massive sulfide deposits is something of a mouthful and in this chapter will often be abbreviated VMS deposits.

15.2 VOLCANIC-ASSOCIATED MASSIVE SULFIDE DEPOSITS

15.2.1 Morphology

These deposits are generally stratiform, lenticular to sheetlike bodies (see Fig. 3.8) developed at the interfaces between volcanic units or at volcanic–sedimentary units. Though often originally roughly circular or oval in plan, deformation may modify them to blade-like, rod-shaped or amoeba-like orebodies, which may even be wrapped up and stood on end, and then be mistaken for replacement pipes such as the Horne Orebody of Noranda, Quebec. The massive ore is usually underlain by a stockwork that may itself be ore grade and which appears to have been the feeder channel up which mineralizing fluids penetrated to form the overlying massive sulfide deposit. It must be remembered that slumping may upset the simple model shape mentioned above and indeed the whole of the massive ore can slide

off the stockwork and the two become completely separated. The massive ore is often capped by an exhalative, sulfide-bearing, ferruginous chert which may have a significantly greater areal extent than the orebody and hence be of considerable exploration value.

15.2.2 Classification

These ores show a progression of types. There have been a number of attempts to classify them based on the metals produced and the host rocks. Perhaps the most comprehensive is that of Barrie and Hannington (1999), who classify the deposits into five types based on host rock composition. From most primitive to most evolved these are: mafic, bimodal–mafic, mafic–siliciclastic, bimodal–felsic, and bimodal–siliciclastic. The sizes and grades of these deposits are summarized in Table 15.1.

To some extent corresponding with the above environmental classification a geochemical division into iron (pyrite deposits used for sulfuric acid production), iron-copper (Cyprus-type), iron-copper-zinc (Besshi-type), and iron-copper-zinc-lead (Kuroko- and Primitive-type) is often employed (Table 15.2, Fig. 15.1). Using a different approach from the simple chemical one Lydon (1989) has shown that, if each point on a ternary diagram of the grades of the deposits is weighted in terms of tonnes of ore metal within the deposit, then clearly there are just two major groups, Cu-Zn and Zn-Pb-Cu (Fig. 15.2). Indeed, as Lydon stresses, there are few so-called copper deposits without some zinc.

Yes, reader, the situation is confusing! – we may indeed be dealing with a continuous

TABLE 15.1 Total and average grade and tonnage for VMS types, excluding China and ex-Soviet Block countries. Figures are for combined mined and mineable reserves and resources. (From Barrie & Hannington 1999.)

(a)

Type	n	Total tonnage in billion tonnes	Total Cu in million tonnes	Total Pb in million tonnes	Total Zn in tonnes	Total Au in million tonnes $\times 10^2$	Total Ag in tonnes $\times 10^3$
Mafic	62	0.18	3.7	0.04	1.3	2.31	2.6
Bimodal-mafic	284	1.45	24.3	2.0	44.3	12.91	38.2
Mafic-siliciclastic	113	1.24	16.2	0.6	9.7	4.03	9.2
Bimodal-felsic	255	1.29	7.1	13.2	54.2	14.18	120.0
Bimodal-siliciclastic	97	2.50	21.5	24.0	55.1	4.11	60.0
Total	811	6.66					

(b)

	Average size in million tonnes	Average Cu grade in wt	Average Pb grade in wt%	Average Zn grade in wt%	Average Au grade in g t^{-1}	Average Ag grade in g t^{-1}
Mafic	2.8	2.04	0.10	1.82	2.56	20.0
Bimodal-mafic	5.1	1.88	0.75	4.22	1.52	36.5
Mafic-siliciclastic	11.0	1.74	1.83	2.43	0.84	19.8
Bimodal-felsic	5.2	1.44	1.64	5.63	2.06	92.8
Bimodal-siliciclastic	23.7	1.10	1.84	4.16	1.13	84.4

(c)

	Number of deposits >100 Mt	Number of deposits 50–100 Mt	Number of deposits 20–50 Mt	Number of deposits 10–20 Mt	Number of deposits 5–10 Mt
Mafic	0	0	3	1	7
Bimodal-mafic	1	6	9	16	20
Mafic-siliciclastic	3	1	10	7	10
Bimodal-felsic	0	3	12	19	29
Bimodal-siliciclastic	9	4	5	6	11

TABLE 15.2   Volcanic-associated massive sulfide deposit types – "conventional" terminology. (Modified from Hutchinson 1980.)

Type	Volcanic rocks	Clastic sedimentary rocks	Depositional environment	General conditions	Plate tectonic setting	Known age range
Besshi (= Kieslager) Cu–Zn±Au±Ag	Within plate (intraplate) basalts	Continent-derived greywackes and other turbidites	Deep marine sedimentation with basaltic volcanism	Rifting	Epicontinental or back-arc	Early Proterozoic Palaeozoic
Cyprus Cu(+Zn)±Au	Ophiolitic suites, tholeiitic basalts	Minor or absent	Deep marine with tholeiitic volcanism	Tensional, minor subsidence	Oceanic rifting at accreting margin	Phanerozoic
Kuroko Cu–Zn–Pb±Au±Ag	Bimodal suites, tholeiitic basalts, calc-alkaline lavas and pyroclastics	Shallow to medium depth clastics, few carbonates	Explosive volcanism, shallow marine to continental sedimentation	Rifting and regional subsidence, caldera formation	Back-arc rifting	Early Proterozoic Phanerozoic
Primitive Cu–Zn±Au±Ag	Fully differentiated suites, basaltic to rhyolitic lavas and pyroclastics	Immature greywackes, shales, mudstones	Marine <1 km depth. Mainly developed in greenstone belts	Major subsidence	Much debated: fault-bounded trough, back-arc basin?	Archaean–early Proterozoic

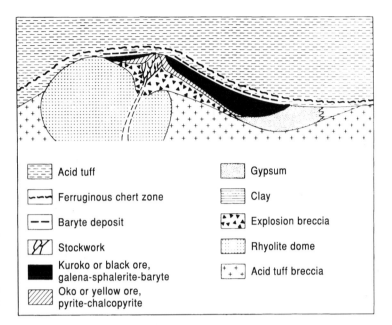

FIG. 15.1 Stratigraphical distribution of volcanic-associated massive sulfide deposits in the Noranda area, Quebec. The names on the diagram are those of various volcanic formations each of which may include both flows and pyroclastic deposits. The relative sizes of the orebodies are indicated. The largest, at the Horne Mine (H), is over 60 Mt. (After Lydon 1989.)

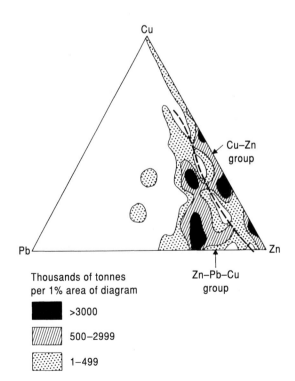

FIG. 15.2 Schematic section through a Kuroko deposit. (Modified from Sato 1977.)

spectrum that we have not yet fully recognized, as Lydon's work suggests (Fig. 15.1), but meanwhile mining geologists will continue to use the terms listed in Table 15.2 because in the present state of our knowledge, they imply useful summary descriptions of the deposits, helpful in formulating exploration models.

It is important to note that precious metals are also produced from some of these deposits; indeed in some Canadian examples of the Primitive-type they are the prime product.

15.2.3 Size, grade, mineralogy, and textures

Data from the five types defined by Barrie and Hannington (1999) are given in Table 15.1. The majority of world deposits are small and about 80% of all known deposits fall in the size range 0.1–10 Mt. Of these about a half contain less than 1 Mt (Sangster 1980). Average figures of this type tend to hide the fact that this is a deposit type that can be very big or rich or both and then very profitable to exploit. Examples of such deposits are given in Table 15.3. Large polymetallic orebodies of this type form the high quality deposits much sought after by many small to medium sized mining compan-

TABLE 15.3   Tonnages and grades for some large or high grade VMS deposits.

| | Mined ore + reserves | | | | | | | |
	(Mt)	Cu%	Zn%	Pb%	Sn%	Cd%	Ag g t^{-1}	Au g t^{-1}
Kidd Creek, Ontario	155.4	2.46	6.0	0.2	r	r	63	–
Horne, Quebec	61.3	2.18	–	–	–	–	–	4.6
Rosebery, Tasmania	19.0	0.8	15.7	4.9	–	–	132	3.0
Hercules, Tasmania	2.3	0.4	17.8	5.7	–	–	179	2.9
Rio Tinto, Spain*	500	1.6	2.0	1.0	–	–	r	r
Aznalcollar, Spain	45	0.44	3.33	1.77	–	–	67	1.0
Neves–Corvo, Portugal†	30.3	7.81	1.33	–	–	–	–	–
	2.8	13.42	1.35	–	2.57	–	–	–
	32.6	0.46	5.72	1.13	–	–	–	–

* Rio Tinto, originally a single stratiform sheet, was folded into an anticline whose crest cropped out and vast volumes were gossanized. The base metal values are for the 12 Mt San Antonio section.

† The three lines of data represent three ore types and not particular orebodies of which there are four. Ag is present in some sections. r, indicates this metal is or has been recovered, but the average grade is not available.

ies (see "Metal and mineral prices" in section in 1.2.3). The lateral continuity and shape of VMS deposits usually makes them attractive mining propositions either in open pit or as underground operations using trackless equipment. Mechanization is particularly important in countries with high labor costs, such as Canada, Sweden, or Australia.

The mineralogy of these deposits is fairly simple and often consists of over 90% iron sulfide, usually as pyrite, although pyrrhotite is well developed in some. Chalcopyrite, sphalerite, and galena may be major constituents, depending on the deposit class, bornite and chalcocite are occasionally important, arsenopyrite, magnetite, and tetrahedrite–tennantite may be present in minor amounts. With increasing magnetite content these ores grade to massive oxide ores (Solomon 1976). The gangue is principally quartz, but occasionally carbonate is developed and chlorite and sericite may be locally important. Their mineralogy results in these deposits having a high density and some, e.g. Aljustrel and Neves-Corvo in Portugal, give marked gravity anomalies, a point of great exploration significance (see section 7.4).

The vast majority of massive sulfide deposits are zoned. Galena and sphalerite are more abundant in the upper half of the orebodies whereas chalcopyrite increases towards the footwall and grades downward into chalcopyrite stockwork ore (see Figs 3.8 & 15.7). This zoning pattern is only well developed in the polymetallic deposits.

15.2.4   Wall rock alteration

Wall rock alteration is usually confined to the footwall rocks. Chloritization and sericitization are the two commonest forms. The alteration zone is pipe-shaped and contains within it and towards the center the chalcopyrite-bearing stockwork. The diameter of the alteration pipe increases upward until it is often coincident with that of the massive ore. Metamorphosed deposits commonly show alteration effects in the hanging wall. This may be due to the introduction of sulfur released by the breakdown of pyrite in the orebody. Visible chloritic or sericitic alteration is almost entirely confined to the zone around the stockwork and is only exposed at the surface in cases of near-surface, steeply dipping deposits. The alteration halo can be used as a proximity indicator in drill holes that miss the orebody as can the chemical content of the halo (see below).

In several mining districts laterally widespread zones of alteration have been reported as occurring stratigraphically below the favorable horizon. Such zones extend for up to 8 km and vertically over several hundred meters. In a successful attempt to expand the exploration target by mapping a larger part of the causative hydrothermal system, Cathles (1993) carried out a regional whole rock oxygen isotope survey to reveal the extent of the oxygen isotope alteration in the Noranda region.

15.2.5 Some important field characteristics

Association with volcanic domes

This frequent association is stressed in the literature and the Kuroko deposits of the Kosaka district, Japan, are a good example (Fig. 15.4). Many examples are cited from elsewhere, e.g. the Noranda area of Quebec, and some authors infer a genetic connexion. Certain Japanese workers, however, consider that the Kuroko deposits all formed in depressions and that the domes are late and have uplifted many of the massive sulfide deposits. In other areas, e.g. the Ambler District, North Alaska, no close association with rhyolite domes has been found (Hitzman et al. 1986).

Cluster development

Although the Japanese Kuroko deposits occur over a strike length of 800 km with more than 100 known occurrences, these are clustered into eight or nine districts. Between these districts lithologically similar rocks contain only a few isolated deposits and this tends to be the case, with a few notable exceptions, for massive sulfide occurrences of all ages. Sangster (1980) calculated the average area of a cluster to be 850 km^2 with 12 deposits containing in total 94 Mt of ore. The Noranda area (Fig. 15.1) is about 170 km^2 and contains about 20 deposits aggregating about 110 Mt. The four Neves-Corvo deposits lie in an area of 6 km^2 and have 250 Mt of possible ore; more mineralisation occurs around Algare just 2.5 km to the north (Leca 1990).

In some districts VMS deposits and associated domes appear to lie along linear fractures which may have controlled their distribution. Synvolcanic basinal areas are important orebody locations in the Iberian Pyrite Belt.

Favorable horizons

The deposits of each cluster often occur within a limited stratigraphical interval. For Primitive and Kuroko types, this is usually at the top of the felsic stage of cyclical, bimodal, calc-alkaline volcanism related to high level magma

FIG. 15.3 Data of representative VMS districts plotted on a Cu-Pb-Zn ternary diagram weighted in tonnes of ore metal contained in the deposit and then contoured. (Modified from Lydon 1989.)

chambers rather than to a particular felsic rock type (Leat et al. 1986, Rickard 1987). For example, the deposits of the Noranda area, Quebec (Fig. 15.4), lie in a narrow stratigraphical interval within a vast volcanic edifice at least 6000 m thick. A very good account of the stratigraphical relationships is given in Kerr and Gibson (1993).

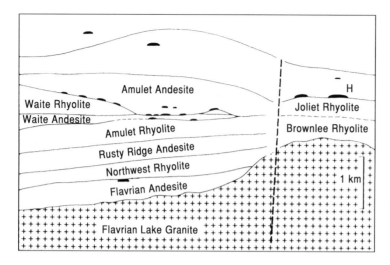

FIG. 15.4 Distribution of dacite lava domes and Kuroko deposits, Kosaka District, Japan. (Modified from Horikoshi & Sato 1970.)

Structural controls

Many massive sulfide deposits occur within recognized cauldron subsidence areas localized along synvolcanic faults and extracauldron deposits may lie within graben flanked by vents (Kerr & Gibson 1993). Detailed mapping and expertise in the interpretation of volcanic phenomena are obviously essential in the search for deposits in this terrane.

Exploration gelogists working in a favorable terrane should study all the literature on its deposits to refine their exploration model. This will help them to decide on finer points such as whether any newly found deposit is more likely to be flat lying or steeply dipping, little or badly deformed, little or much metamorphosed and to adjust their geochemical, geophysical, and drilling programs accordingly.

15.2.6 Genesis

For many decades VMS deposits were considered to be epigenetic hydrothermal replacement orebodies (Bateman 1950). In the 1950s, however, they were recognized as being syngenetic, submarine-exhalative, sedimentary orebodies, and deposits of this type have been observed in the process of formation from hydrothermal vents (black smokers) at a large number of places along sea floor spreading centers (Rona 1988).

There is a voluminous literature on the subject and the reader is advised to read the summary in Robb (2004) as well as the review volume of Barrie and Hannington (1999).

15.2.7 Volcanic facies

The association of massive sulfide with particular parts of the volcanic assemblage has long been recognized. Brecciated rhyolitic volcanics are known as "mill rock" in some mining camps because of their proximity to processing plants. The breccias are attributed to one of three processes: (i) they formed close to structural magmatic vents which also vented hot springs; (ii) they were the result of hydrothermal explosive events; and (iii) the breccias formed as a result of caldera formation with which the ore-forming process was associated (Hodgson 1989). Whichever of these processes is correct it is clear that an understanding of the development of the volcanic pile is important. This knowledge can be obtained by detailed mapping, core logging, and comparization with modern volcanic models. The most useful summary of these models is provided in the book by Cas and Wright (1987), and the reader is referred to this text as well as the paper of Gibson et al. (1999).

15.2.8 Exploration geochemistry

Exploration geochemistry has been widely and successfully used in the search for VMS deposits. In Canada the prospective areas have been glaciated and have poor surface response, but rock geochemistry is widely used to delineate drill targets at depth. In the southern hemisphere surface geochemistry has been successful even in areas of intense weathering where EM surveys have been ineffective due to conductive overburden.

Rock geochemistry

The rock geochemical signature of VMS deposits has been investigated by a number of workers but is well summarized by Govett (1989) and Franklin (1997). In a review of a large number of case histories he documents the very consistent response, regardless of age and location. Aureoles are usually extensive, more than 1 km laterally and 500 m vertically with footwall anomalies more extensive than those in the hanging wall, especially in proximal deposits. In Archaean deposits significant hanging wall anomalies appear to be absent. As discussed above Fe and Mg are enriched and Na and Ca depleted in nearly all deposits (Fig. 15.5). Copper tends to be enriched in the footwall and is depleted in the hanging wall; Mn shows the reverse behavior. Both K and Mn may be relatively depleted near the deposit and enriched further away. The ore elements, such as Zn, are commonly enriched. Copper may be depleted especially in the immediate wall rocks. The behavior of other trace elements, in particular Au, Cl, Br, As, Sb, may be distinctive but data from case histories are incomplete. Some authors have regressed trace elements against SiO_2 to compensate for fractionation. This is not always effective as some deposits, such as those in the Lac Dufault area, Noranda, have undergone addition of SiO_2 during alteration.

Cherty exhalites have been widely used as indicators of hydrothermal activity but their geochemical signatures are varied. Kalogeropoulos and Scott (1983) have used a major element ratio, $R = (K_2O + MgO) \times 100/K_2O + Na_2O + CaO + Mg$ to indicate proximity to Kuroko mineralisation in Japan. Trace and major elements, in particular zinc and manganese but also volatiles such as arsenic, are enriched sporadically in these horizons but application has had varying degrees of success. A key requirement of any successful application is that the stratigraphical position of the samples must be well understood. It is of little use comparing ratios of samples from different exhalative horizons.

Geochemistry has also been used on a regional scale to attempt to discriminate mineralized sequences of volcanics. However the variation associated with mineralisation is usually less than that caused by fractionation and trace element data must be corrected for fractionation. Studies of immobile elements (Ti, Zr, REE), that have not been markedly affected by hydrothermal alteration, may indicate the original chemistry of the volcanics. Campbell et al. (1984) have suggested that mineralized volcanics have flatter rare earth element (REE) patterns than barren volcanics which tend to be enriched in light REE and have negative Eu anomalies. Few regional rock geochemical studies have been published, but Govett (1989) suggests that many VMS deposits are associated with regional copper depletions. Both Russian and Canadian workers (Allen & Nichol 1984) suggest that analysis of sulfides, separated either by gravity or selective chemical extraction, within volcanics, may give regional anomalies of up to 10 km along strike from VMS deposits.

Surficial geochemistry

The extreme chemical contrast of the sulfides with volcanics gives rise to high contrast transition metal anomalies in soil and stream sediments. A large number of case histories are available.

15.2.9 Geophysical signatures

One of the key elements in the exploration model is the geophysical signature of the deposit. The majority of VMS discoveries have resulted from geophysics, notably in the glaciated areas of the Canadian Shield. VMS deposits respond to a number of geophysical techniques, as they contain much sulfide that often shows electrical connection, permitting the use of EM techniques, and the bodies are dense, allowing the use of gravity methods.

Mapping methods

Airborne magnetic and electromagnetic (EM) systems are widely used for mapping, both of

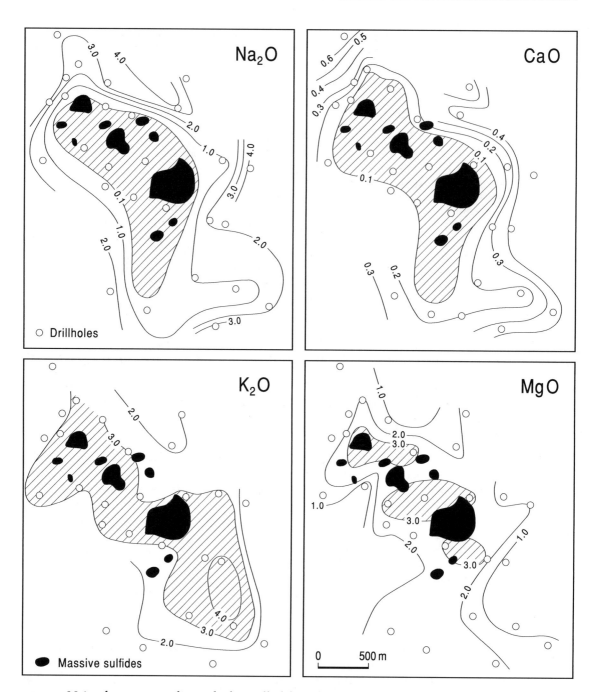

FIG. 15.5   Major element anomalies in the footwall of the Fukasawa Mine, Japan. (From Govett 1983, data from Ishikawa et al. 1976.)

general geology and of favorable lithologies. The high resolution of modern aeromagnetic systems resulted in the obsolescence of ground magnetics for all but the smallest projects. In addition it is possible to merge aeromagnetic data with remotely sensed information. An excellent example of the use of magnetics in the detailed mapping of Archaean terranes is provided by Isles et al. (1989) for the Yilgarn block in Western Australia. They highlight the use of multi-client surveys as a cost-effective method of mapping major lithologies and structural changes. Magnetic units such as banded iron formations are apparent as magnetic highs in contrast to the magnetically less disturbed areas of acid volcanics and sediments. Sometimes secondary maghemite can form in these deeply weathered environments, such as parts of the Yilgarn, and confuse the magnetic signature, although this can be helpful in detecting paleochannels as shown in Fig. 15.6. In Canada both the provincial and federal governments have flown large parts of the prospective areas and the data is available at low cost. Much of sub-Saharan Africa has also been flown although availability of data is varied (Reeves 1989).

Direct detection of sulfides

Electrical methods have been very successful, particularly in Canada. Up to the 1970s this involved the use of airborne EM systems. Their limited depth penetration (originally <150 m) and the saturation coverage of prospective areas has resulted in the delineation and drilling of virtually all near-surface conductors. More recent geophysical exploration is aimed at finding deposits at depth and has encouraged a switch to down hole methods.

Pemberton (1989) has provided a summary of EM responses from a number of Canadian VMS deposits. Figure 15.7 shows the detection of a VMS deposit in the Noranda camp using downhole pulse EM from a hole drilled some 100 m from the deposit at 700 m depth.

Airborne geophysics has been ineffective in intensely and deeply weathered areas, such as most of the Yilgarn Shield of Western Australia. The conductive nature of the overburden is responsible for the generation of near-surface responses which mask signatures from the VMS. Options that have been investigated include increasing transmitter power, shortening integration times, and increasing the number of time channels used (Smith and Pridmore 1989).

A more general problem is the excellent EM response of graphitic sediments. It is extremely difficult, if not impossible, to differentiate graphite from massive sulfides and, as graphitic sediments are often found associated with breaks in the volcanic sequence, this has resulted in many thousands of meters of wasted drilling. However, some VMS deposits, such as the Trout Lake deposit in Manitoba, are completely enclosed in graphite and would be missed if graphitic sediments were excluded from consideration.

15.2.10 Integration of exploration techniques

Perhaps the most important part is the incorporation of the factors mentioned above into an integrated exploration program. A recent example of the successful use of integration is the discovery of the Louvicourt deposit near Val d'Or, Quebec (Fig. 15.8). The program was aimed at discovering large deep extensions to the near-surface Louvem Mine which had produced for about 10 years. Drilling of surface holes to 700 m, costing $C50,000 each, was undertaken on 300 m centers, to test prospective horizons at depth (Northern Miner 1990). The spacing of 300 m was chosen as only large deposits were thought capable of supporting a mine at 600 m depth. Hole 30 intersected a stringer zone containing 0.86% Cu over

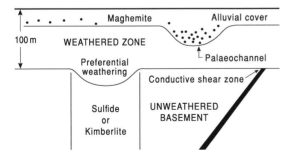

FIG. 15.6 Schematic diagram showing the problems of using magnetics and EM in deeply weathered areas, such as the Yilgarn in Western Australia. (After Smith & Pridmore 1989.)

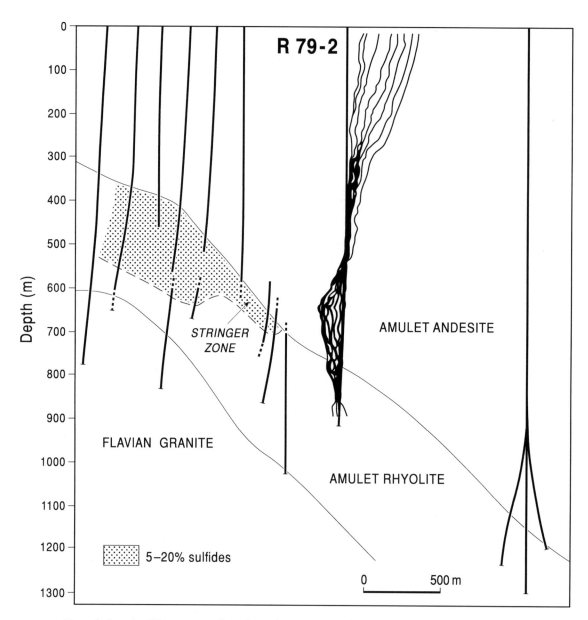

FIG. 15.7   Downhole pulse EM survey at the Ribago deposit, Noranda, Quebec. The survey clearly detects the sulfides 130 m from the hole. (After Pemberton 1989.)

118 m but more importantly downhole EM surveys showed a significant off-hole conductor and rock geochemistry showed a major element anomaly. This suggested a major zone of hydrothermal alteration associated with a different horizon (the Aur horizon) than that of the Louvem Mine. A further phase of drilling intersected further stringer zone mineralisation, leading to massive sulfides grading 1 1.8% Cu over 36 m in hole 42 and 2.8% Cu and 5.7% Zn over 43 m in hole 44. Resources are estimated at 28 Mt of 4.3% Cu, 2.1% Zn, 27.8 g t^{-1} Ag and 1.03 g t Au. The total cost of the discovery is put at \$C2.5M.

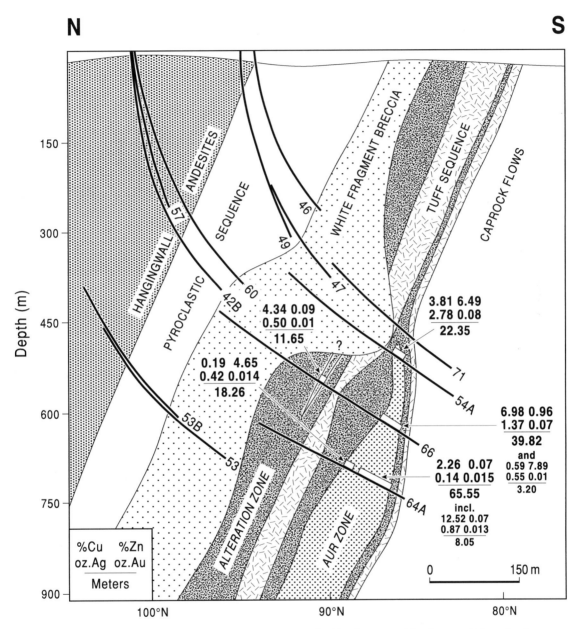

FIG. 15.8 Section of the Louvicourt deposit, Val d'Or. Note the blind nature of the deposit. (After Northern Miner Magazine 1990.)

15.2.11 Environmental impact

The highly sulfidic nature of the deposits means that they are likely to give rise to environmental problems (Taylor et al. 1995). In particular the iron and base metal sulfides are unstable when brought to the surface and, if unremediated, give rise to acid mine drainage. Many deposits also contain significant concentrations of toxic metals such as arsenic, cadmium, and mercury. The containment of tailings from flotation separation of sulfides is also expensive. These environmental problems and their associated costs have downgraded the appeal of VMS deposits for some mining companies.

15.3   THE KIDD CREEK MINE

15.3.1   Regional setting

Kidd Creek lies within the Abitibi Belt of the Superior Province of the Canadian Shield (Figs 15.9 & 15.10). With an exposed area of 1.6 million km^2 the Superior Province is the largest Archaean Province in this shield. The rocks are similar to those found at Phanerozoic subduction zones with differences due to higher mean mantle temperature and possibly greater crustal (slab?) melting at mantle depth (Hoffmann 1989). The province is composed of subparallel ENE trending belts of four contrasting types: (i) volcanic–plutonic terranes, which resemble island arcs; (ii) metasedimentary belts which resemble accretionary prisms (Percival & Williams 1989); (iii) plutonic complexes; and (iv) high grade gneiss complexes. In the southern part of the Superior Province linear, ENE trending, greenstone–granite belts alternate, on a 100–200 km scale, with metasedimentary dominated terranes.

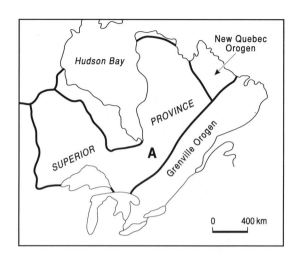

FIG. 15.9   Location of the Superior Province of the Canadian Shield. A, position of the Abitibi Belt.

Archaean greenstone belts the world over appear to be divisible into a lower dominantly volcanic section and an upper sedimentary section. The lower section may be divided into a lower part, of primarily ultramafic volcanics, and an upper volcanic part in which calcalkaline or tholeiitic, mafic to felsic rocks predominate. The ultramafic (komatiitic) part also

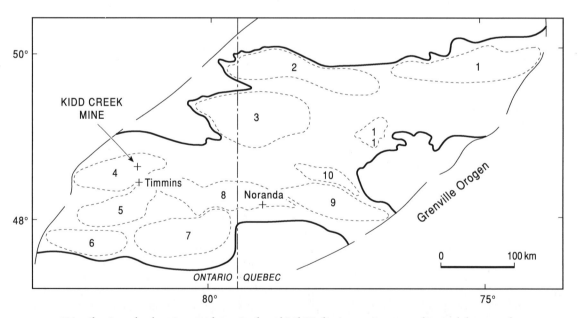

FIG. 15.10   Distribution of volcanic complexes in the Abitibi Belt. Approximate outlines of the complexes including cogenetic intrusions and sediments are shown by the heavy pecked lines. The complexes with significant developments of VMS deposits are 2, 4, 6, 8, and 9. (Modified from Sangster & Scott 1976.)

contains some mafic and felsic rocks and is thus bimodal. The tholeiitic or calc-alkaline part is dominated by basaltic, andesitic, and rhyolitic volcanics. With increase in stratigraphical height in greenstone belts there is, *inter alia*, (i) a decrease in the amount of komatiite and (ii) an increase in the relative amount of andesitic and felsic volcanics. The associated sediments are mainly exhalative chert and iron formation with some thin slates. The upper sedimentary section is dominantly clastic.

There are regional differences in the volcanic character of greenstone belts that appear to be related at least in part to their age and erosion level. Thus the oldest belts of southern Africa and Australia are typically bimodal and enriched in ultramafics and mafics (which make them prime areas for nickel exploration), whereas the younger Canadian belts have a higher proportion of calc-alkaline and tholeiitic rocks and have so far been the best ground for finding VMS deposits. Large areas of the Superior Province formed at 2.7–2.6 Ga.

The Abitibi Belt is 800 km long and 200 km wide making it the world's largest, single, continuous Archaean greenstone belt. It contains a relatively high proportion of supracrustal rocks of low metamorphic grade resulting from an unusually shallow erosion level. This belt has been divided into 11 volcanic complexes (Sangster & Scott 1976) (Fig. 15.10) which are commonly many kilometers thick and contain one or more mafic to felsic cycles. VMS ores occur in just five of these complexes and are exclusively associated with calc-alkaline volcanic rocks, especially the dacitic and rhyolitic members which form 5.5–10% of the total volume of volcanic rocks in this belt and which belong to three different affinities whose importance have been stressed by Barrie et al. (1993). Production and reserves of VMS ores at 1990 totalled 424 Mt averaging 4.4% Zn, 2.1% Cu, 0.1% Pb, 46 g t^{-1} Ag and >>1.3 g t^{-1} Au (Spooner & Barrie 1993).

15.3.2 Exploration history

The discovery of Kidd Creek is extremely well documented and was a triumph for airborne geophysics and geological understanding (Bleeker & Hester 1999). It is worth quoting the eloquent description of the discovery from a series of papers (The Ecstall Story in the *CIMM*

Bulletin May 1974 (Matulich et al. 1974)) describing the discovery and establishment of the Kidd Creek or Ecstall Mine, as it was originally known.

"The Ecstall orebody does not owe its discovery to a fantastic stroke of luck but to the foresight, dedication, tenacity and hard work of a Texasgulf [then Texas Gulf Sulfur] exploration team. The orebody was found as a result of years of patient reconnaissance with the latest exploration equipment and using the most advanced technology. It was located only 15 miles north of a gold mining center that had been operating seventeen major mines over a period of fifty years. For most of those years, prospectors had combed every square foot of the area surrounding Timmins looking for gold. In fact, a prospector had his cabin on the rock outcrop that is now occupied by the mine surface crushing plant, just 100 yards from the orebody's southern limit. As most of the gold mines had been found by surface outcrops and simple geological interpretation, the prospector's search for the rich bonanza he felt was nearby was limited to these methods. The layers of muskeg and clay over the Kidd Creek orebody protected it from these primitive efforts." (Clarke 1974)

The geological thinking behind the exploration program stemmed from the successful application of a syngenetic model. At the time of the development of the model (1954) most geologists believed that VMS deposits were replacement deposits and related to granites, although papers from both Germany and Australia had proposed an exhalative origin for these deposits (for discussion see Hodgson 1989, Stanton 1991). Walter Holyk recognized the close association of deposits with the contact between sericite-schists (rhyolites) and intervolcanic sediment and applied the model successfully in conjunction with airborne EM surveys for Texas Gulf Sulfur in the Bathurst area of New Brunswick (Hanula 1982).

Holyk then persuaded his management to apply the same techniques and model to the Canadian Shield. He selected the Timmins area based on the known occurrence of the Kam Kotia deposit at the contact between rhyolite and sediments and the lack of known EM coverage. The first stage in the exploration program was for the exploration geologist (Leo Miller) to make a compilation of the geology based on outcrop and airborne magnetics. After starting in Noranda, Miller moved to the Timmins area and investigated a number of prospects based partly on the previous work of the Ontario Department of Mines (Bleeker & Hester 1999). Mapping in 1941 had identified two outcrops that implied a contact between a fragmental rhyolite and basaltic pillow lavas (Berry 1941). Along the edge of the rhyolite outcrop, a narrow zone of sulfide-bearing siliceous rock was mapped (Fig. 15.11). During a visit to the site on July 24 1958, Miller noted the presence of a small trench dug by an unknown prospector. He also noticed traces of chalcopyrite in chert bands, although this was later shown not to be connected with the deposit.

The most promising areas were then flown by an EM system that had been mounted in a helicopter (Donohoo et al. 1970). Although the Kidd Creek anomaly was detected on the first day of the production flying program (3 March 1959), little further work was done as the exact location was not known due to lack of up-to-date photo mosaic maps and the general area of the anomaly was at the boundary of four privately owned claims, originally granted to veterans of the Boer War (Fig. 15.12). One of the claim owners was approached but would not option the claim and the area was left for the time being. Other anomalies in the area were investigated and 59 holes were drilled over the next 4 years with largely negative results. The Kidd Creek anomaly, known as Kidd 55 from its location on the fifth lot pair and fifth concession (Figs 15.11 & 15.13), was further recorded in routine flying in 1959 and 1960; this time accurate location of the anomaly was made in 1961 at the second attempt. Optioning the ground from the Hendrie estate began in January 1961 but was not completed until June 1963, by which time Miller had been transferred to North Carolina and the prospect passed to the supervision of Ken Darke. Lacking a budget to continue, Darke used money from his exploration program on Baffin Island with the permission of exploration manger Dick Mollinson. Texas Gulf Sulfur undertook ground EM surveys to confirm the location of the conductor and to delineate drill targets. Three conductors were found and drillholes sited on the central anomaly (Fig. 15.14).

The first hole cut 8.37% Zn, 1.24% Cu, and 3.9 oz t^{-1} Ag over 177 m in November 1963 as shown in Table 15.4. It was obvious that a major deposit had been discovered and the rig was moved to the east of the property to confuse scouts from other companies who might get wind of the discovery. In addition the drill crew were kept in the bush until Christmas while lawyers optioned claims adjacent to the discovery site to protect the possible extension of the deposit. When land had been optioned in April 1964 further drilling began and the initial phase lasted until October 1965, totalling 157 holes and 33,500 m and this delineated the deposit to the 335 m (1100 ft) level. The thickness of the deposit and its subcrop rendered it suitable for open pit mining and the initial drilling was aimed at determining the ore reserves for this operation and the upper part of the underground mine. Definition of the continuation of the deposit has been from underground and continues to the present.

The primary phase of exploration drilling was designed to define the ore outlines by drilling on levels 120 m (400 ft) apart. This was followed by secondary drilling to define grade and ore outlines on which the mine layout was based. An example (Fig. 15.15) shows the secondary drilling in the shallow part of the underground mine. The aim was to have an intersection every 55 m (180 ft) vertically; this corresponds to the main levels and every other sublevel. Drill spacing between sections was normally 15 m (50 ft). All the core was logged and zones of mineralisation assayed (Matulich et al. 1974).

Drilling continues on the deposit at depth. In 1995 more than 50,000 m were drilled. Of the 23,000 m drilled in 1990, underground exploration drilling accounted for 7000 m, delineation drilling 5500 m, and surface exploration drilling 10,500 m (Fenwick et al. 1991).

Drilling around the mine proved the existence of a further blind massive sulfide lens, the Southwest Orebody, in 1977. It has subsequently been mined from 785 to 975 m and is considered to be a distal equivalent of the main north and south orebodies (Brisbin et al. 1991). The area around Kidd Creek has been much prospected without the discovery of other deposits and it is still an open question as to whether Kidd Creek is an isolated deposit or whether it belongs to a cluster, whose other members have been eroded away, or remain to be found by deep prospecting.

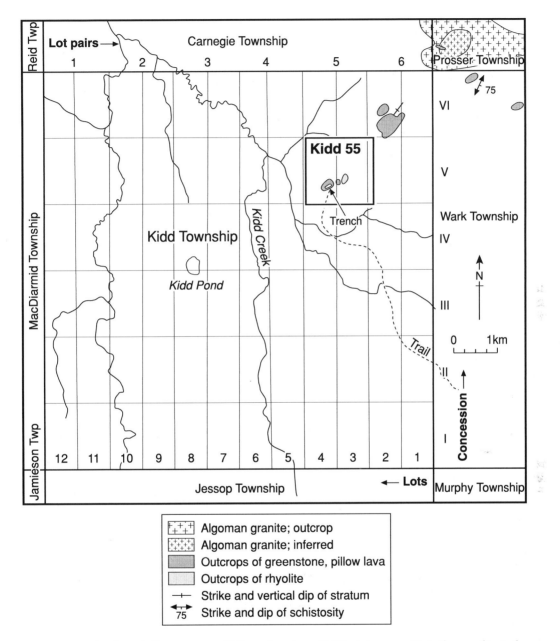

FIG. 15.11   Mapping by Ontario Department of Mines that alerted Miller's interest. Note the very limited outcrops. (After Bleeker & Hester 1999, from Berry 1941.)

15.3.3   Exploration geophysics

The discovery of the deposit prompted trials of a number of techniques to assess their response over the deposit (Donahoo et al. 1970). Besides EM with various loop configurations, induced polarization (IP), gravity, and magnetic ground surveys were also made. The ground EM surveys confirmed the three anomalies detected in the initial EM survey (Figs 15.14 & 15.16). The central anomaly also gave a very high contrast IP chargeability and resistivity anomaly as well

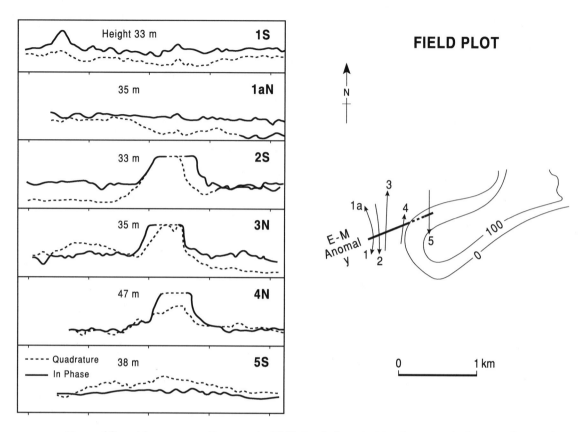

FIG. 15.12   Trace of first airborne anomalies over the Kidd Creek deposit, March 3, 1959. (After Donohoo et al. 1970.)

as a 1.6 milligal residual gravity anomaly. Only magnetics failed to define the deposit and only succeeded in outlining the peridotite.

15.3.4   Exploration geochemistry

Although geochemistry was not used during the discovery of Kidd Creek a substantial research program was undertaken immediately after the discovery to understand if surficial geochemistry would have been of use. Details can be found in Fortescue and Hornbrook (1969) and there is an excellent summary in more accessible form in Hornbrook (1975).

Kidd Creek is typical of the problems of surficial geochemical exploration in the Abitibi Belt, an area known by soils scientists as the "Clay Belt." The area was covered, before mining and logging, by dense forest with swampy peats, known as muskeg. Underlying this is a thick (up to 65 m) sequence of glacial material: consisting of from top to base of clay till,

varved clay, and lower till. These glacial deposits have a blanketing effect on the geochemical signature and vertical dispersion from the subcrop of the deposit is limited to the lower till (Fig. 15.17) as the varved clays have a lacustrine origin and are not of local derivation. If the lower till is sampled by drilling a distinct dispersion fan can be mapped down ice from the deposit subcrop (Fig. 15.18). The more mobile zinc disperses further than copper.

The blanketing effect of the glacial overburden limits the surface response to zinc anomalies in near-surface organic soils. These zinc anomalies are thought to form as a result of the decay of deciduous trembling aspen trees which tap the anomaly at relatively shallow depths (Fig. 15.19).

Recognition of the lack of surface geochemical expression of Kidd Creek and other deposits in the Abitibi Clay Belt led to the widespread use of deep overburden sampling in geochemical exploration. The basal, locally derived till

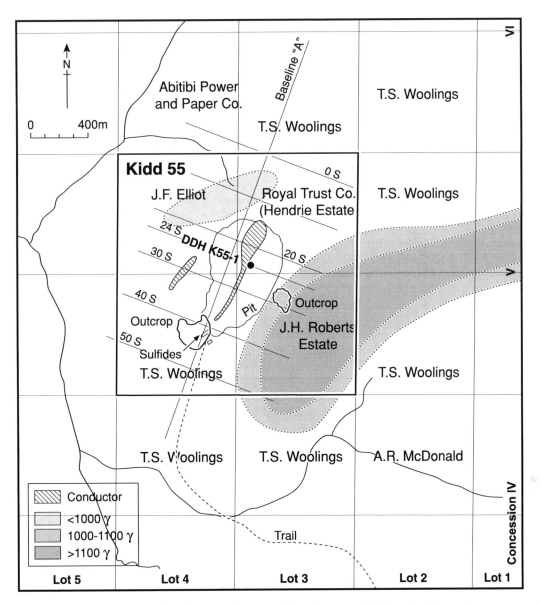

FIG. 15.13   Map of the Kidd 55 locality showing the main EM conductor and the land holdings, the prospector's trench, and initial drillhole. (From Bleeker & Hester 1999 after Miller's sketch of 1959.)

is sampled by drilling and any dispersion fans can be mapped.

15.3.5   Mine geology

The geology of the mine is discussed in detail in *Economic Geology Monograph 10* (Hannington & Barrie 1999) and there is a summary in Barrie et al. (1999). Earlier accounts include Matulich

et al. (1974), Walker et al. (1975), Coad (1984, 1985), and Brisbin et al. (1991).

Host rocks

VMS deposits in the Kidd Creek area exemplify the classic case of exhalative deposits developed at, or close to, the top of an Archaean volcanic cycle where the rhyolitic–dacitic top

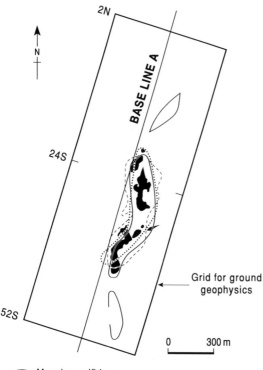

2N

N

BASE LINE A

24S

52S

Grid for ground
geophysics

0   300 m

● Massive sulfides

— Ground EM anomaly

---- Residual gravity anomaly (0.2 milligals)

······ IP and resistivity anomaly

FIG. 15.14 Composite geophysical map of initial surveys over the Kidd Creek deposit. (After Donohoo et al. 1970.)

of the lower cycle is succeeded by a dramatic change to ultramafic and mafic volcanics at the base of the succeeding cycle. Kidd Creek is no exception. This deposit occurs in an overturned volcano–sedimentary sequence (Figs 15.20 & 15.21).

The base of the sequence consists of ultramafic flows that structurally overlie greywackes of the Porcupine Group. The top of the sequence underlying the deposit consists of a number of massive rhyolite sills intruded into a felsic series of pyroclastics containing intercalated rhyolite flows and tuffs. The uppermost massive rhyolites, which have a U-Pb zircon age of 2717 ± 4 Ma (Nunes & Pyke 1981), have pervasive crackle brecciation interpreted as resulting from hydraulic fracturing. A stockwork chalcopyrite zone is present in this crackle-brecciated rhyolite with the massive ore lying immediately above. The orebody is stratigraphically overlain by a minor development of rhyolitic rocks, dated at 2711 ± 1.2 Ma by Bleeker and Parrish (1996), which are succeeded by pillowed metabasalts. The rocks of the mine area are intruded by three large masses of high Fe tholeiite, gabbroic sills, referred to as andesite–diorite in mine terminology.

The rhyolites in the immmediate area around the deposit reach 300 m in thickness, in contrast to 100 m at a distance from the

Footage	(ft)	Copper (wt%)	Zinc (wt%)	Silver (g t⁻¹)	Lead (wt%)
0–26		Clay overburden			
26–50	24	1.05	Trace	10	–
50–132	82	7.10	9.7	82	–
132–152	20	0.19	11.1	10	–
152–196	44	0.11	4.7	17	–
196–232	36	0.79	13.0	360	–
232–248	16	0.18	3.81	82	–
248–348	100	0.33	14.3	144	0.80
348–490	142	0.1	18.0	247	–
490–530	40	0.24	2.8	113	–
530–566	36	0.23	6.1	55	–
566–576	10	0.17	3.0	34	–
576–628	52	0.20	8.3	62	–
628–649	21	1.18	8.1	130	–
649–655	6	–	–	–	–

TABLE 15.4 Assays for sections from discovery hole K55-1. (From Bleker & Hester 1999. After a press release in *The Northern Miner*, April 16, 1964, with minor modifications.)

Average: 623 ft grading 1.21% Cu, 8.5% Zn, 138 g t⁻¹ Ag, minor Pb; some Cd, about 0.1% indicated in character samples Texas Gulf Sulphur discovery hole K55-1 was drilled at –60 degrees to the west, in the northern half of lot 3, concession 5, Kidd Township.

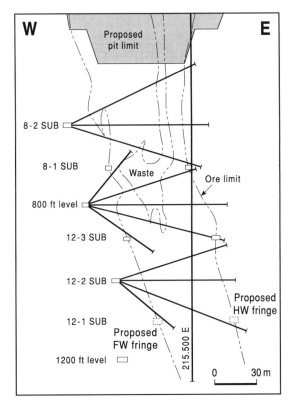

FIG. 15.15 Underground drill patterns used to define reserves. (After Matulich et al. 1974.)

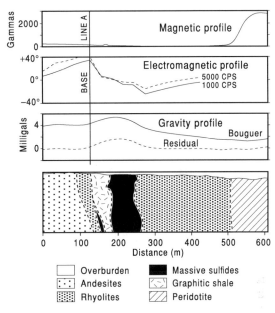

FIG. 15.16 Geophysical section on line 24S of Fig. 15.14. (After Donohoo et al. 1970.)

deposit, suggesting control of mineralisation by sub-volcanic rifting as do the geometry of the alteration and the shape of the VMS deposit (Fyon 1991).

Mineralisation

The orebody, which consists of three en echelon lenses, has a maximum thickness of 168 m, a strike length of 670 m and is known to extend to a depth of 2990 m. From 1966 to 1995 101.5 Mt of ore carrying 2.31% Cu, 0.24% Pb, 6.5% Zn, and 94 g t^{-1} Ag. were mined. At the beginning of 1996 reserves were calculated to be 32.2 Mt grading 2.48% Cu, 0.21% Pb, 6.34% Zn, and 71 g t^{-1} Ag. For a number of years tin, which may locally reach 3%, was recovered. There are three major ore types. In the stockwork there is stringer ore in which chalcopyrite stringers ramify crackle-brecciated or fragmental rhyolite. Pyrite or pyrrhotite accompanies the chalcopyrite and this ore averages 2.5% Cu. The massive ore is composed primarily of pyrite, sphalerite, chalcopyrite, galena, and pyrrhotite. Silver occurs principally as native grains with accessory acanthite, tetrahedrite–tennantite, stromeyerite, stephanite, pyrargyrite, and pearceite. Tin occurs as cassiterite with only a trace of stannite. Digenite, chalcocite, and other sulfides occur only in trace amounts.

The massive sulfides are in part associated with a graphitic–carbonaceous stratum containing carbonaceous argillite, slate, and pyroclastics. Associated with this horizon are fragmental or breccia ores that are considered to represent debris flows.

The deposit is divided into the North and Central plus South Orebodies by the Middle Shear. The orebodies can be divided into a number of zones and, as might be expected, copper-rich areas occur in the base of the massive ore, including a bornite-rich zone at the base of the South orebody. In these very high silver grades occur (up to 4457 g t^{-1}) with minor (0.35 ppm Au) associated gold values (section 15.2.2).

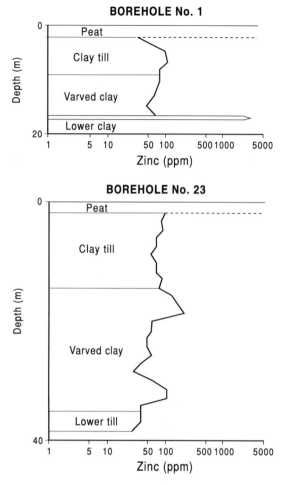

FIG. 15.17 Section through glacial overburden over the deposit and to one side of it; locations are in Fig. 15.20. Note the response is limited to lowest till and in the near surface in borehole 1. (After Hornbrook 1975.)

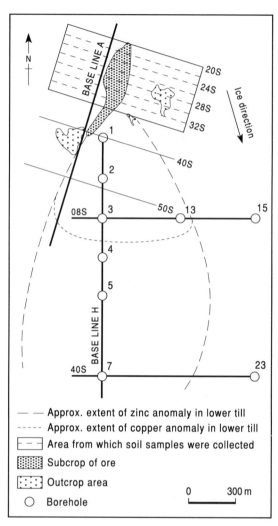

FIG. 15.18 Fan-shaped anomalies in the lowermost till at Kidd Creek. (After Hornbrook 1975.)

Wall rock alteration

At both Kidd Creek and the Hemingway Property (a drilled area 1.3 km north of the Kidd Creek orebody containing minor base metal mineralisation) early serpentinization affected the ultramafic flows and silicification the rhyolites. Later carbonate alteration was super-posed on both by pervasive CO_2 metasomatism (Schandl & Wicks 1993). Strong chloritization, sericitisation and silicification occur in the footwall stockwork zone along the length of the orebody (Barrie et al. 1999).

Genesis

The Kidd Creek Orebody is interpreted as being of exhalative origin and, apart from its large size, is similar to many other deposits of Primitive-type. Beaty et al. (1988) have shown that the hydrothermally altered rhyolites are all markedly enriched in ^{18}O compared with most other VMS deposits. Altered rocks associated with the large Iberian Pyrite Belt deposits and Crandon, Wisconsin (62 Mt of ore) show similar ^{18}O enrichment (Munha et al. 1986) and for Kidd Creek Beaty et al. have suggested that

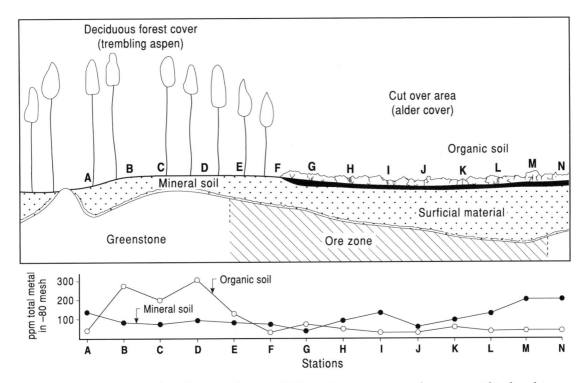

FIG. 15.19 Surface anomaly along line 24S of Fig. 15.18. The surface response in the organic soil is thought to be the result of the tapping of the deep anomaly by trembling aspen trees. (After Hornbrook 1975.)

evaporation and enrichment of normal sea water produced a high salinity fluid enriched in ^{18}O that was capable of transporting large amounts of metals, so accounting for the large size of the deposit. Taylor and Huston (1998) argued that regional scale low temperature fluid flow may be related to regional mafic-ultramafic heat sources such as the komatiite flows.

Strauss (1989) has shown that the sulfur isotopic values are similar to those of many other Archaean VMS deposits. Maas et al. (1986) using Nd isotopic data suggested that some of the base metals in the ore were leached from the rhyolitic host rocks.

15.3.6 Mining operations

The original operators were Ecstall Mining Ltd, a wholly owned subsidiary of Texas Gulf Sulfur Inc., of the USA. Following a 1981 takeover offer by Elf Aquitaine (SNEA) Kidd Creek Mines Ltd became a wholly owned subsidiary

of the government-owned Canada Development Corp. In 1986 Falconbridge Ltd of Canada, which has subsequently become part of Noranda Inc., acquired Kidd Creek.

When the decision was made in March 1965 to build the open pit–railroad–concentrator complex two of the major problems were soils and transportation. The whole vicinity of the orebody was covered with thick deposits of weak materials: muskeg, till, and clay. Only two small outcrops occurred by the pit site and one had to be reserved for the primary crusher. There was no good ground near the pit large enough for the concentrator and no roads or railroads to the pit area at that time. The concentrator was therefore built 27 km east of Timmins near a highway, a railroad, an electricity transmission line, and a gas pipeline. It was linked to the mine with a new railroad.

The open pit was designed to have an overall waste-to-ore ratio of 2.58:1 but started with a 4:1 ratio (see section 11.2.1). It was 789 × 549 m at the surface, ore was available from the pit on

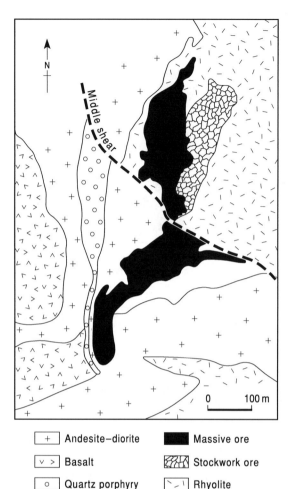

FIG. 15.20   Bedrock geological map, Kidd Creek. (After Beaty et al. 1988.)

+ Andesite–diorite

v > Basalt

o Quartz porphyry

■ Massive ore

Stockwork ore

Rhyolite

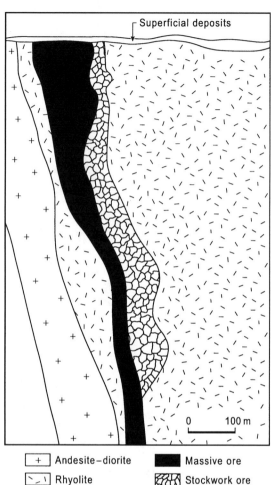

FIG. 15.21   West–east section through the North Orebody. (After Beaty et al. 1988.)

+ Andesite–diorite

Rhyolite

■ Massive ore

Stockwork ore

bench 1 by October 1965, and large scale production to supply the concentrator with 8000–9000 t day^{-1} was achieved by October 1967.

The pit had a projected production life of about 10 years, however it was decided that there were advantages in developing an underground mine fairly quickly. These advantages included: the availability of underground ore for grade control, which would become more difficult as the pit deepened; the chance to develop new systems, equipment, and skills, which was possible as a result of revolutionary changes in underground mining technology in the 1960s; and the acquisition of knowledge of local rock behavior to aid in mine planning.

Shaft sinking commenced in March 1970 and what became known as the Upper Mine with a 930-m-deep shaft was developed. The 1550-m-deep Lower Mine shaft, 100 m SW of the No. 1 shaft was completed in 1978 (Fig. 15.22). The open pit was abandoned in 1979. Blasthole and sublevel caving stopes followed by pillar recovery using blasthole methods have been employed in the Upper Mine. In the Lower Mine a modified blasthole technique with backfill of a weak concrete mix is used so that no pillars are left. Waste is mixed with cement to form backfill at the 790-m level and then trucked to the stopes. The ore crusher station is on the 1400-m level. A subvertical shaft, the

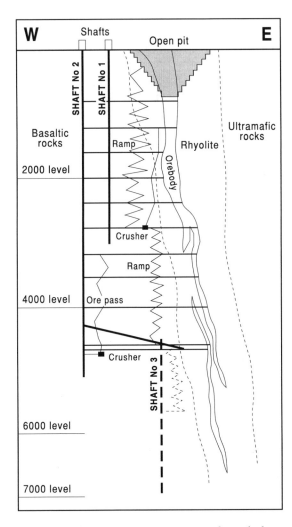

FIG. 15.22 Schematic east–west section through the Kidd Creek Mine showing location of shafts and ramp. (After Brisbin et al. 1991.)

No. 3 shaft, is being sunk to access the lower part of the deposit to a depth of 3100 m. Overall production was approximately 2 Mt from the No. 1 and No. 2 mines in 2001.

15.3.7 Rock mechanics

Considerable use has been made of rock mechanics instrumentation such as exten-someters, strain cells, and blast vibration monitors to detect and predict ground instability leading to safer operations, the design of more stable underground workings, and thus greater productivity.

15.3.8 The concentrator and smelter

The highly automated concentrator (see section 10.2.2) has undergone a number of developments and improvements and by 1981 its capacity was up to 12,250 t day^{-1} (about 4.5 Mt yr^{-1}) and it was producing seven different concentrates: copper concentrates with high and low silver contents, two similar zinc concentrates, lead, tin and pyrite concentrates. After about 9 years of planning and the testing of various smelting processes, a smelter and refinery complex was built near Timmins and went into production in June 1981. This complex treats two thirds of the copper concentrates and produces 90,000 t of cathode copper. Most of the rest of the copper concentrate is custom smelted and refined by Noranda Inc. Much of the zinc concentrate is treated in the electrolytic zinc plant which can produce 134,000 t zinc metal per year. The remaining zinc concentrates and all the lead concentrates are sold to custom smelters.

15.4 CONCLUSIONS

VMS deposits can constitute very rewarding targets as they can be large, high grade, polymetallic orebodies like Kidd Creek. They should be sought for using an integrated combination of geological (sections 15.2 & 15.3.1), geochemical (sections 15.2.9 & 15.3.4), geophysical (sections 7.4, 7.9, 7.10, 15.2.10 & 15.3.3), and drilling (section 15.2.11) methods. In the immediate World War II years, when geochemical prospecting was in its infancy, the use of airborne EM led to many successful discoveries in the Canadian Shield. Experience and research since then has shown that sophisticated geochemical procedures (section 15.3.4) and other geophysical methods can play an important part in the search for more VMS orebodies.

16

DISSEMINATED PRECIOUS METALS – TRINITY MINE, NEVADA

MICHAEL K.G. WHATELEY, TIMOTHY BELL AND CHARLES J. MOON

Disseminated precious metal deposits, particularly of gold, have been the most popular target for metallic exploration through the 1980s and 1990s. Their popularity was based on the high price of gold during this period (see section 1.2.3), the low cost of mining near-surface deposits in open pits, and improvements in gold recovery techniques. The key metallurgical innovation was the development of the cheap heap leach method to recover precious metals from low grade oxidized ores. This case history deals with the evaluation of the Trinity silver mine.

16.1 BACKGROUND

16.1.1 Overview of deposit types

A variety of deposit types have been mined in this way but the most important in Nevada (discussed in detail by Bonham 1989) are:
1 Sediment hosted deposits of the Carlin type.
2 Epithermal deposits, such as Round Mountain.
3 Porphyry-related systems rich in gold.

Carlin-type deposits

These deposits are responsible for the majority of gold production in Nevada, 261 t in 2001. Typically the gold occurs as micronmeter-sized grains which are invisible to the naked eye ("noseeum" gold), within impure limestones or calcareous silstones. The host rocks appear to be little different from the surrounding rocks and the margins of the deposit are only distinguishable by assay. An excellent summary is given in Bagby and Berger (1985), updated in Berger and Bagby (1991) and Hofstra and Cline (2000).

Individual deposits occur mainly within major linear trends, up to 34 km in length, the most important of which is the Carlin Trend in Northeast Nevada. This trend and type of deposit are named after the Carlin Mine, which was the first significant deposit of this type to be recognized, although others had previously been mined. The deposits vary from broadly tabular to highly irregular within favorable beds and adjacent to structures such as faults that have acted as fluid pathways. The faulting and fluid flow has often resulted in brecciation and several deposits are hosted in breccias.

Silicification is the commonest form of alteration and the occurrence of jasperoid, which replaces carbonate, is widespread in many deposits. Gold is thought to occur predominantly as native metal although its grain size is very small, usually less than 1 μm, and the metal's exact mineralogical form remains unknown in some mines. Gold forms films on sulfide and amorphous carbon and is particularly associated with arsenian pyrite. Pyrite is the most common sulfide and may be accompanied by marcasite and arsenic, antimony and mercury sulfides. The arsenic sulfides realgar and orpiment, which are readily distinguishable by their bright color, occur in many deposits as

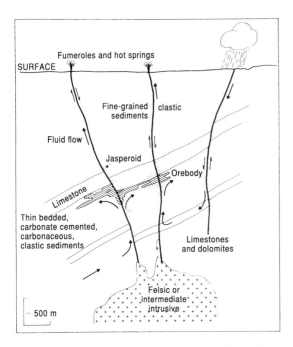

FIG. 16.1 Diagram for the formation of Carlin style deposits. (After Sawkins 1984.)

does stibnite and cinnabar. The general geochemical association is of As, Sb, Tl, Ba, W and Hg, in addition to Au and Ag.

The deposits are spatially associated with granites and many authors favor intrusion driven hydrothermal systems (Fig. 16.1). Berger and Bagby (1991) favor a mixing model in which magmatic fluids have been mixed with meteoric water causing precipitation in favorable host rocks.

Epithermal systems

Much of the bonanza-style precious metal mineralisation in the western USA is hosted in volcanic rocks, e.g. the Comstock Lode which produced 5890 t Ag and 256 t Au at an average grade of about 300 g t^{-1} Ag and 13 g t^{-1} Au in the late nineteenth century (Vikre 1989). More recent exploration has resulted in definition of bulk mineable targets such as Round Mountain with reserves of 196 Mt of 1.2 g t^{-1} Au. Volcanic hosted deposits can be divided into three general types based on mineralogy and wall rock alteration: acid sulfate (or alunite-kaolinite), adularia sericite, and alkalic (Heald

et al. 1987, Bonham 1989, Henley 1991, Hedenquist et al. 2000).

The high sulfidation or acid sulfate type generally contains more sulfide and is characterized by the occurrence of the copper sulfide, enargite. Other base metal sulfides are frequently present as is pyrite. Advanced argillic alteration is common as is alunite $(K_2Al_6(OH)_{12}(SO_4)_4)$ and kaolinite. The low sulfidation or adularia sercite type generally has less sulfide and sercitic to intermediate argillic alteration. Alunite may be present but it is only of supergene origin. The structural setting of both types may be similar (Fig. 16.2) but the differences between the two types probably reflect their distance from the heat source (Heald et al. 1987). There are obvious parallels between the formation of these deposits and modern geothermal systems. Evidence from modern day geothermal systems, such as the deposition of gold on pipes used to tap the systems for geothermal power, suggests that the gold may have been deposited rapidly. Several major gold deposits occur in alkalic volcanics that may be spatially associated with alkalic porphyries. The deposits are characterized by the occurrence of the gold as tellurides with quartz–carbonate–fluorite–roscoelite (vanadium mica)–adularia alteration and regional propylitic alteration (Bonham 1989).

The Hot Spring deposits of Bonham (1989) appear to be a variant of the epithermal type and are characterized by the occurrence of a recognizable paleosurface that is usually associated with a siliceous sinter zone interbedded with hydrothermal breccia. The sinter passes downward into a zone of silicification and stockwork with a zone of acid leaching. The precious metals are generally restricted to within 300 m of the paleosurface and probably reflect rapid deposition. There is an overall spatial association with major centers of andesitic to rhyolitic volcanism

Gold-rich porphyries

Gold is of major economic importance in several porphyry deposits that produce copper as their main product, such as Ok Tedi (see p. 273). More recent research (Sillitoe 1991) suggested that there may be related copper-deficient systems that contain economic gold

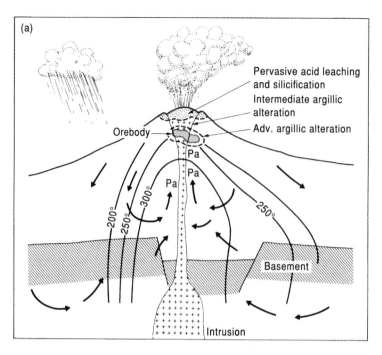

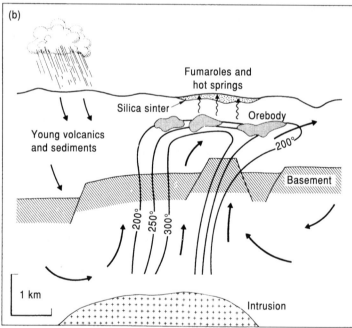

FIG. 16.2   Diagrams showing the formation of two types of epithermal precious metal deposit in volcanic terranes. (a) Acid sulfate type. Pa, propylitic alteration. Note that the mineralisation occurs within the heat source. (b) Adularia–sericite type. The upwelling plume of hydrothermal fluid is outlined by the 200°C isotherm. The mushroom-shaped top reflects fluid flow in the plane of major fracture systems, a much narrower thermal anomaly would be present perpendicular to such structures. The heat source responsible for the buoyancy of the plume is shown as an intrusion several kilometers below the mineralised zone. In both diagrams the arrows indicate circulation of meteoric water. (a) is drawn to the same scale as (b). (Based on Heald et al. 1987 with modifications from Henley 1991.)

contents. The deposits are associated with granitic intrusives that vary in composition but with quartz monzonite and diorite as the most important hosts. Alteration shows a characteristic concentric zoning from inner potassic through phyllic to propylitic zones. Gold generally correlates with copper grade and is associated with K alteration (Sillitoe

1991). In some cases high sulfidation gold systems are mined at higher levels than the porphyry, although a direct connection is often hard to prove. For further discussion of porphyry copper deposits see Evans (1993) and Robb (2004).

16.1.2 Geochemistry

One of the major features in the discovery of disseminated gold deposits has been the further developments in gold geochemistry. The recognition of "noseeum" gold deposits depends on the ability to detect fine-grained gold. As this is usually in the micron size class it does not often form grains large enough to be panned. Much gold remains encased in fine-grained silica and therefore sampling and analytical techniques must be devised that accurately determine all the gold present. This analytical problem has largely been overcome, firstly by the improvement of graphite furnace AAS and more latterly by the application of neutron activation analysis and ICP–MS (see section 8.2.3). Graphite furnace AAS methods have relied on the separation of gold from the rock or soil matrix using an organic solvent. This allows analysis with detection limits of lower than 5 ppb.

As gold is present as discrete grains within the sample either encased in the matrix or as free gold, sampling methods have to be devised that will allow the analytical aliquot to represent accurately the original sample. Investigations by Clifton et al. (1969), well summarized by Nichol et al. (1989), have shown that it is necessary to have 20 gold grains in each sample to achieve adequate precision. In real terms this means taking stream sediment samples of around 10 kg and carefully subsampling. The geologist who takes a 1-kg sample can not expect to make an accurate interpretation of analytical data. An alternative strategy, which has been used where rocks have been oxidized, is to take a large sample and leach it with cyanide; this technique is known as Bulk Leach Extractable Gold (BLEG). This technique has found wide acceptance in Australia and the Pacific Rim countries.

Most but not all gold deposits have concentrations of pathfinder elements associated with the gold. In Nevada the most widely used elements are As, Sb, Hg, Tl, and W for sedimentary hosted deposits, with the addition of base metals for most deposits. Although pathfinder elements can be extremely informative as to the nature of the gold anomaly, they should not be used without gold as pathfinder elements may be displaced from gold. In addition some gold deposits, particularly adularia–sericite-type deposits, have no pathfinder signature and the exploration geologist must rely largely on gold geochemistry. An example is the Ovacik deposit in Turkey (see section 4.2.1) that was discovered by the follow-up of BLEG geochemistry.

In Nevada large stream sediment samples and extensive soil sampling have proved effective where overburden is residual. However in some areas the overburden has been transported and surface sampling is ineffective. Chip sampling of apparently mineralized rocks, particularly jasperoids, has been successful although jasperoidal anomalies are often slightly displaced from the associated deposits.

There are a large number of case histories on gold exploration and the reader is advised to read Zeegers and Leduc (1991) and case histories on Nevada in Lovering and McCarthy (1977).

16.1.3 Geophysics

Disseminated gold deposits are difficult geophysical targets. Electrical and EM methods have been used to map structure and identify high grade veins, e.g. at Hishikari, Japan (Johnson & Fujita 1985), but are not usually effective in detecting mineralisation except indirectly. On occasion unoxidized sulfides are present allowing the use of induced polarization (IP). One of the more successful uses of geophysics in Nevada has been the detection of silicified bedrock under Tertiary cover (Fig. 16.3). The section shows the results of a Controlled Source Audio Magneto Telluric Survey that clearly detected a zone of silicification under 50 m of cover. Further case studies can be found in Paterson and Hallof (1991).

16.1.4 Mining and metallurgy

The recognition that disseminated precious metal deposits could be successfully bulk

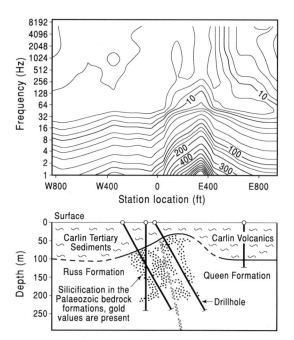

FIG. 16.3 Section showing the application of a CSAMT survey in detecting areas of silification under Tertiary cover. Top: contoured data; bottom: drill section. (After Paterson & Halof 1991.)

mined and processed, sometimes with spectacular returns, has been responsible for the allocation of the large exploration budgets spent in Nevada. Oxidized deposits are generally mined in open pits, usually with low stripping ratios. One of the keys to a successful operation is effective grade control (section 16.5) as different grade ores are treated differently in most operations. The higher grade material is usually milled and extracted with cyanide in a conventional manner whereas the lower grade ore is crudely crushed and placed on leach pads. These leach pads are built on plastic liners, sprinkled with cyanide, using a system similar to a garden sprinkler, and the cyanide allowed to percolate through the crushed ore. The cyanide is then pumped to a central treatment plant where the gold is extracted by passing the cyanide through activated carbon. Marginal ore and waste are stockpiled separately to allow for the later treatment of marginal material if the gold price increases.

One of the interesting developments in Nevada is the recognition that some of the low

grade material has previously unsuspected high grade material with different metallurgical characteristics beneath it. For example, at the Post deposit on the Carlin trend high grade material in which gold is difficult to extract, or refractory, underlies the main surface deposit. The grade of this material is so much higher that it is worthwhile investing in underground mining and pressure leach extraction to recover the gold.

16.2   TRINITY MINE, NEVADA

The Trinity Silver Mine was an open pit, heap leach, silver mining operation extracting rhyolite-hosted, disseminated, hydrothermal, silver oxide mineralisation. The mine was brought into production on September 3, 1987 and mining ceased on August 29, 1988. The heap leach operation continued for just over one more year. Our choice of a silver mine rather than a gold producer has been governed entirely by the availability of data. Most of our discussion is of course applicable to any disseminated precious metal deposit. US measures are used throughout this chapter as the mining industry in the USA has yet to metricate. Thus oz t^{-1} are troy ounces per short ton of 2000 avoirdupois pounds (lb).

This case study focuses upon the statistical assessment of blasthole assay data, for the purpose of determining the distribution and quality of mineralisation. It is essential for reliable ore reserve calculations to derive the best grade estimate for an orebody. The actual mined reserves and estimated reserves calculated by US Borax and Chemical Corporation (US Borax) from exploration borehole data are also described. The Surpac Mining System Software was used for data management and statistical evaluation. This work was undertaken as an MSc dissertation at Leicester University (Bell 1989) under the auspices of US Borax.

The Trinity Mine is in Pershing County, Northwest Nevada (Fig. 16.4), on the northwest flanks of the Trinity Range, 25 km northnorthwest of Lovelock. The area is covered by the souhtwest part of the USGS Natchez Spring 7.5′ topographic map and Pershing County Geological Map. Elevations vary from 1200 m to 2100 m (3900–7000 ft). The mineralisation

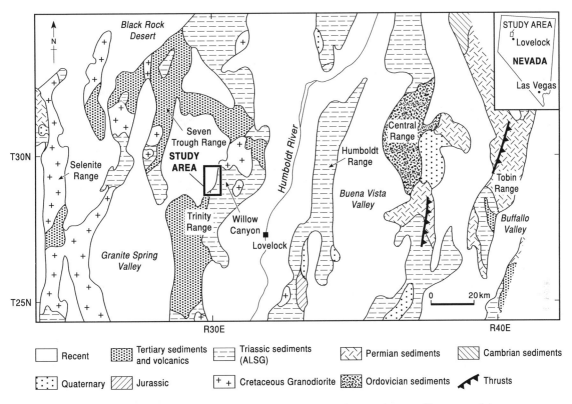

FIG. 16.4 Sketch map of Pershing County, Nevada showing the outline geology and location of the Trinity silver deposit. ALSG, Auld Lang Syne Group. (After Ashleman 1988.)

at Trinity principally lies between 1615 and 1675 m (5300 and 5500 ft).

16.3 EXPLORATION

16.3.1 Previous work

Mineralisation was first discovered in the Trinity Range by G. Lovelock in 1859 and limited, unrecorded production from Pb-Ag and Ag-Au-Cu-Zn veins occurred between 1864 and 1942 (Ashleman 1983). Evidence of minor prospecting during the 1950s was found within Willow Canyon (Fig. 16.5) (Ashleman 1988). Geophysical exploration and trenching during the 1960s was carried out by Phelps Dodge within Triassic sediments to the north of Willow Canyon. Later a geochemical exploration program aimed at locating gold mineralisation was undertaken by Knox–Kaufman Inc. on behalf of US Borax, but they were unable to locate previously reported gold anomalies. However in 1982 a significant silver show was found within altered rhyolites. As a result, agreements were reached between five existing claim holders, Southern Pacific Land Company (SPLC), holders of the surface rights and US Borax, after which the Seka claims were staked. The Trinity joint venture between Sante Fe Pacific Mines Inc., the land holder and a subsidiary of SPLC, and Pacific Coast Mines Inc., a subsidiary of US Borax, was operated by US Borax and mining was contracted to Lost Dutchman Construction.

16.3.2 US Borax exploration

Various stages of exploration were completed by US Borax during the period 1982–86. In 1982 and 1983 mapping, geochemical and geophysical techniques were used to located payable

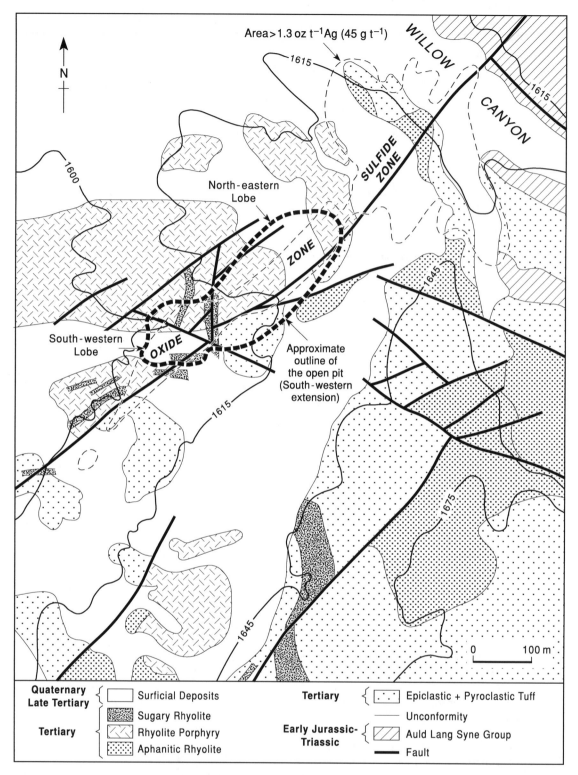

FIG. 16.5 Sketch map of the geology in the vicinity of the Trinity silver deposit, Nevada. (After Ashleman 1988.)

sulfide mineralisation. In 1983 drilling commenced within an area known to host sulfide mineralisation. Exploration continued from 1985 to 1986 in an attempt to find additional reserves. Although assessment of the sulfide ore continued with metallurgical testing and feasibility studies, the possibility of a high grade oxidized zone to the southwest of the main target area became the focus of attention. Efforts were thereafter concentrated upon the evaluation of this southwestern extension (Fig. 16.5).

16.3.3 Mapping

Initial mapping of the claim area at 1:500 was followed by detailed mapping of the sulfide zone at a scale of 1:100. Mapping of structure and lithology helped define the surface extent of mineralisation and alteration, and the controls of the distribution of mineralisation. When the potential of the silver mineralisation within the oxidized zone was established in mid 1986, detailed mapping of the southwestern extension commenced. Mapping showed the host rock, of both oxide and sulfide mineralisation, to be fractured rhyolite porphyry and defined an area of low grade surface mineralisation intruded by barren rhyolitic dykes (Fig. 16.5). These barren dykes divide the southwest extension into northeast and southwest lobes. The rhyolite dykes were seen to be intruded along east and north-northeast faults and to disrupt the stratigraphical sequence within the southwest lobe.

16.3.4 Geochemistry and geophysics

A geochemical soil survey was completed over the oxide and sulfide zones and samples were analyzed for Pb, Zn, and Ag. The mobility of Ag and Zn in the soil horizons placed constraints upon the significance of anomalies reflected by these elements. Lead is more stable and was therefore used as a pathfinder element for Ag mineralisation. Significant Pb anomalies were located over the sulfide zone. Anomalies of greater than 100 ppm defined potential mineralized areas and higher levels (>1000 ppm) coincided with the surface intersection of Ag mineralisation. Surface rock samples were taken as part of a reconnaissance survey, and

contour maps of the results helped define the extent of surface mineralisation.

Between 1985 and 1986 rock geochemistry was applied to the oxide zone in the southwest extension. Rock chips from drilling, trenching and surface exposure were analyzed for Pb, Zn, Ag, and Au. High grade samples containing 30–50 ppm Ag were reported and localized samples containing 40–90 ppm Ag were recorded as the zone of oxidization was traced southward along strike. Anomalies from the rock and soil geochemical surveys were used in the planning and layout of the drilling program within the sulfide zone.

Geophysical surveys, including gradient array IP, magnetic and gamma ray spectrometry, were carried out concurrently over the sulfide mineralisation. The gradient induced polarization (IP) survey required three arrangements of transmitting electrodes with a pole spacing of 12,000 ft (3650 m). Readings were taken every 500 ft (150 m) along lines spaced 1000 ft (300 m) apart (Ashleman 1983). A time domain chargeability anomaly coincided with the area of sulfide mineralisation. A NNW trending belt of high resistivity was located adjacent to the eastern margin of an area of known subsurface mineralisation. The magnetic survey proved to be of little value due to the poor contrast in magnetic values. Gamma ray spectrometry using K, Th, and U channels was tested, but only subtle differences were seen between altered and unaltered rocks and the suvey was discontinued (Ashleman 1983).

16.3.5 Drilling

Successive stages of drilling on a 100 ft (33 m) grid orientated 11 degrees west of true north (Fig. 16.6) totalled 29,880 m (98,031 ft) drilled (Table 16.1). Most of the drilling was within areas of principal sulfide mineralisation, focusing on the surface exposure of silver mineralisation, lead, and IP anomalies and areas of favorable geology. Drilling defined a body of silver mineralisation dipping to the west.

The drilling grid was extended to explore for additional reserves and high grade silver oxide mineralisation was intercepted to the southwest (Fig. 16.5). In 1985 drilling was concentrated within the southwest extension.

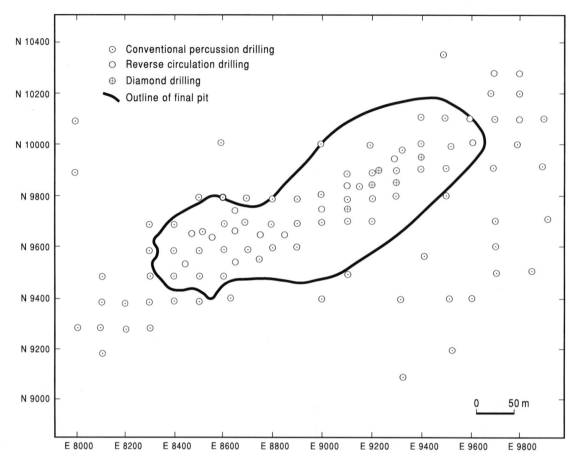

FIG. 16.6   Map showing the location of the exploration boreholes in relation to the final outline of the Trinity mine open pit from which the oxide mineralisation was extracted.

TABLE 16.1   Summary of drilling exploration programs.

Drilling methods	Number of holes	Total meterage
Percussion	199	22,857 (74,990 ft)
Reverse circulation	39	6257 (20,530 ft)
Cored	10	765 (2511 ft)
Total	258	29,879 (98,031 ft)

Ninety-two of the holes were drilled to define the oxide zone. Although mineralisation was located south of the southwest extension along strike and at depth, it was noted to be discontinous.

Diamond drill holes were sampled every 1–3 ft (0.3–0.9 m) for metallurgical testing, while percussion holes were sampled as 5 ft (1.5 m) composites (Ashleman 1986). Samples were sent to the laboratory and analyzed for Ag, Au, Pb, Zn, and As. Analyses on the samples from the exploration holes were carried out using atomic absorption spectrometry (AAS), which gave the total silver content. Samples recording

Ag in excess of 1000 ppm were reanalyzed using fire assay.

16.4 THE GEOLOGY OF PERSHING COUNTY AND THE TRINITY DISTRICT

16.4.1 Regional setting – Pershing County

The Trinity Mining District of Pershing County is part of the Pacific Rim Metallogenic Belt. The mountain ranges of Pershing County reflect the surface expression of major NE–SW structural lineaments sequentially repeated (basin and range) across Nevada. Much of the present geological setting is the response to regional compression subsequently modified by extension. The regional geology is outlined in Fig. 16.4.

Cambrian to Permian shallow marine carbonates and clastics are exposed to the northeast and east of Pershing County. The Harmony Formation (U. Cambrian) is composed of quartzite, sandstone, conglomerate, and limestone which pass laterally into clastic and detrital sediments of the Valmy Formation (Ordovician). Carboniferous to Permian sediments (limestone, sandstone, and chert) and volcanics (andestic flows and pyroclastics) form most of the Tobin Range to the east. During the Mesozoic fault-bounded, deep water, sedimentary basins dominated the area, with associated widespread volcanic and silicic intrusive activity.

Triassic cover is widespread in central Pershing County. The lower units to the east consist of interbedded andesites, rhyolites, and sediments, locally intruded by leucogranites and rhyolite porphyries. To the southeast the Middle Trias is exposed as calcareous detrital and clastic sediments, limestone, and dolomite. The upper unit which forms the major sedimentary cover within the Trinity, South Humbolt, and Severn Troughs Range, defines cyclic units of sandstone and mudstone with limestone and dolomite. Adjacent to Cretaceous granodiorite intrusions, the Triassic to Jurassic sediments are locally metamorphosed to slates, phyllites, hornfels, and quartzite.

The Auld Lang Syne Group (ALSG) forms the majority of the Triassic sedimentary sequence and reflects the shallow marine conditions of a westerly prograding delta (Johnson 1977). Tertiary basaltic and andestic volcanics pass into a thick, laterally variable, silicic, volcanoclastic sequence consisting of a complex pile of rhyolite domes, plugs, flows, and tuffs. The Tertiary rhyolitic volcanics are overlain and interdigitate with lacustrine and shallow lake sediments, later intruded by silicic dykes, sills, and stocks. The Tertiary sequence is covered by Quaternary to Recent gravel and alluvial deposits. The Miocene to Pliocene deposits form a thick stratified, well-bedded sequence of tuffs, shales, sandstone, clay and gravel, locally interbedded with basaltic and andestic flows.

Various structural phases are apparent within Pershing County (Johnson 1977). Pre-Cenozoic large scale compressional folding and thrusting during the Sonoma Orogeny (mid Jurassic) resulted in the south and southeast movement of basinal sediments and volcanics over shelf sediments. These folded, imbricate, thrust slices were later offset by dextral strike–slip faulting that has northwest arcuate and discontinuous trends (Stewart 1980). Such thrusting is apparent to the east and southeast within the Central and Tobin Ranges (Fig. 16.4). Crustal extension during the Cenozoic resulted in northerly trending basin and range faulting. Associated with the basin and range faulting are effusive phases of bimodal volcanism. Stewart (1980) records both pre- and post-basin and range N–E and E–W fault sets, and implies possible strike–slip components to the normal faults.

Tertiary silver mineralisation is reviewed by Smith (1988b), who classifies the silver-bearing deposits into three categories; disseminated type, vein type, and carbonate hosted. The former two types are outlined and compared with the Trinity deposit in Table 16.2. A further classification into four subtypes (Bonham 1988) includes volcanic-hosted epithermal, silver-base metal limestone replacement deposits, sedimentary-hosted stockworks, and volcanic-hosted stockworks. Pershing County is dominated by Pb-Zn-Ag-Au veins hosted within and adjacent to Cretaceous granodiorites, with minor occurrences in Triassic sediments. Mineralisation tends to be concentrated in shear zones and along hornfels dykes forming pinch and swell structures along oblique joint sets (Bonham 1988).

TABLE 16.2   Outline of silver deposits of Nevada, including a comparison with the Trinity deposit. (Adapted from Smith 1988.)

	Disseminated	Vein	Trinity
Host rock	Tertiary felsic volcanics and clastic sediments, Devonian limestone	Tertiary andesites and rhyolites	Teritary porphyry rhyolite, minor occurrence in tuff, and argillite
Type of mineralisation	Discrete grains and vein stockworks	Banding in quartz veins	Disseminations, microfractures, veinlets, and breccia infill (vein stockwork)
Morphology	Tabular, podiform bodies, defined by assay cut-off of <300 g t^{-1}	Veins defined by fault contacts with assay cut-off between 200 and 300 g t^{-1}. Change in cut-off has little effect on tonnage	Lenticular and stacked tabular, sharp assay boundaries of <30 g t^{-1}
Mean grade	270 g t^{-1}	500–800 g t^{-1}	160–250 g t^{-1}
Geological control	Structural, rock permeability and proximity to conduit	Structural and fracture density, intersection of fault and fractures	Structural, rock permeability, fault intersections, percentage fracture density
Zoning	Increased mineral content with depth, high Pb–Zn–Mn, low Au, Ag:Au>100	Sharp vertical zoning, sulfides increase with depth, Ag:Au>50	Sulfide-oxide zoning, high Ag:Au
Mineralogy	Acanthite, tetrahedrite, native Ag, pyrite and Pb–Zn base metals	Acanthite, low pyrite and sulfide content	Sulfides: friebergite–pyragyrite, chalcopyrite, pyrite, stannite. Oxides
Alteration	Silica–sericite (carbonate, clay, chlorite, potassic feldspar)	Extensive and variable, hematite–magnetite, propylitic, adularia and illite	Silicification, adularia–quartz–sericite, minor propylitic, illite and kaolinite
Process of formation	Hypogene	Hypogene, hydrofracturing, and brecciation	Epithermal and hydrotectonic fracturing
Hydrothermal solutions	165–200°C, 5 wt% NaCl, low salinity groundwater mixing with oilfield brines and wall rock alteration	200–300°C, meteoric solutions, precipitation induced by mixing, boiling	Not available

16.4.2   The geology of the Trinity Mine

Stratigraphy

The local stratigraphy of the Trinity Mine and surrounding area is outlined in Fig. 16.5 and Table 16.3. The ALSG is described by Johnson (1977) as a fine-grained clastic shelf and basin facies with interbedded turbidites. The local silicic and calcareous units of the ALSG (Ashleman 1988) have been subjected to both low grade regional metamorphism and contact metamorphism. The latter is associated with the intrusion of Cretaceous granodiorite dykes and stocks located to the northeast of the mine (Fig. 16.4). Outcrops of the ALSG to the south are dominated by phyllites and slates which grade into coarse-grained facies (siltstones, sandstones, and quartzites) northwards. To the east the unit is represented by carbonaceous

TABLE 16.3   Stratigraphy of the Trinity area. (From Bell 1989.)

Age	Unit name	Lithologies	Intrusions
Quaternary	Surficial deposits	Gravel, alluvium, and colluvium	
Pleistocene	Volcanics, fluvial deposits, Upper Tuff	Basalt, fanglomerates and channel sands, phreatic-clastic welded and unwelded, tuff	Latitic and rhyolitic dykes
Tertiary (early Pliocene)	Rhyolite porphyry	Rhyolite flows, agglomerates, and breccias	Rhyolitic, exogenous domes, porphyritic to aphanitic
	Lower tuff	Epiclastic and pyroclastic airfall, reworked tuff	
Cretaceous	Argillite breccia	Subangular to angular argillite clasts (1–10 cm) in fine-grained matrix. Breccia is gradational, tectonic in origin, and fault associated	Granodiorite, medium-grained, hypidiomorphic (90 Ma)
Triassic–early Jurassic	Auld Lang Syne Group	Shelf and basin facies. Shale, siltstone, slate, phyllites, and argillites. Siliceous and carbonaceous sandstones and limestones. Quartzites	

siltstones and impure limestones (Ashleman 1988). The transition between the Tertiary rhyolite volcanics and the ALSG is marked by an argillite breccia. The breccia is fine-grained and matrix supported, it contains semiangular to angular argillite clasts and is closely associated with faulting.

Tertiary volcanism resulted in a complex sequence of rhyolitic flows superseded by epiclastic and pyroclastic deposits. The Tertiary rhyolites dominate the geology of the Trinity area, with extensive exposure to the south and flanking the Trinity Range to the north. Latitic and "sugary" rhyolitic dykes disrupt the sequence, with effusive phases and breccias spatially associated with rhyolite domes. The volcanics are capped by Pliocene to Pleistocene fluvial sediments and locally by basaltic flows. Quaternary alluvium infills valleys and flanks the Trinity Range forming channel fill and alluvial fan deposits.

Structural geology

Polyphase deformation involving compressional tectonics and granodiorite intrusions resulted in two phases of metamorphism of the Mesozoic sediments: (i) low grade regional and (ii) contact metamorphism. Pre-Tertiary folding and faulting of the ALSG resulted in isoclinal folds (Nevadian Orogeny). Tertiary deformation produced large scale, north-trending open folds and low to high angle faulting.

Mid Miocene tectonic activity marked the onset of extensional tectonics, resulting in NNE and NE trending, high angle, basin, and range faulting. This was preceded by N, NNW, and WNW fault sets and localized thrust zones (Ashleman 1988). Faulting appears to postdate the Tertiary rhyolites and predate alteration and mineralisation; however age relationships may not be so easily defined. Fault displacement and reactivation in conjunction with localized NE thrusts and oblique ENE shear zones (Johnson 1977, Stewart 1980) confuses the structural history of the area. The structural history of the Trinity area appears to be one of initial NE trending block faults and associated thrusts and shear zones, later offset by NW trending normal faults.

Alteration

The ALSG was not generally receptive to Tertiary hydrothermal alteration, except locally along faults and breccia zones. Tertiary volcanics were receptive to hydrothermal fluids which resulted in varying degrees of alteration and hydrofracturing. Rhyolitic tuffs and porphyritic flows have been extensively altered, defining an alteration halo that extends 2.5 km beyond the main mineralized zone (Fig. 16.7). Sericitization, silicification and quartz–adularia–sericite (QAS) alteration tends to be most intense along faults, within permeable lithologies and breccia zones. Minor propylitic alteration and kaolinite–illite clays are also reported (Ashleman 1988). QAS altera-

tion is most extensive within the rhyolitic porphyry and tuffs. Silicification is restricted to discrete W dipping lenticular zones within the upper zone of the deposit, with minor extensions along dykes and veins. Silicification is concentrated within the NE lobe extending from and parallel to NE and ENE faults. Patchy Fe staining and limonite is spatially associated with high grade Ag mineralisation.

Mineralisation

Silver mineralisation in Pershing County is reported to be hosted within breccias peripheral to rhyolite domes with mineralisation concentrated in microfractures (Johnson 1977).

FIG. 16.7   Aerial photograph of Trinity Mine. The light areas in the center represent waste and low grade ore stockpiles. The medium tone areas in the foreground show the extent of the alteration in the area. The rectangular area is the heap on which the ore is placed for cyanide leaching. The ponds in which the pregnant solutions collect are seen next to the mine buildings. (Photograph supplied by US Borax.)

However, the Trinity silver orebody is described as a hydrothermal, volcanic hosted, silver-base metal (Cu-Pb-Zn-As) deposit (Ashleman 1988). Bonham (1988) describes it as occurring in stockwork breccias formed in and adjacent to a rhyolite dome of Miocene age. Both sulfide and oxide ore is present, but high grade zones amenable to cyanide leaching are restricted to the oxide zone. The principal sulfide phases identified in pan concentrates from rotary drill cuttings consist of pyrite–marcasite–arsenopyrite–sphalerite–galena with lesser amounts associated with chalcopyrite–pyrrhotite–stannite. They were deposited with silica along stockwork fractures. The silver-bearing sulfide phases consist of freibergite (silver-rich tetrahedrite) and pyrargyrite. Acanthite and native silver are also present in trace amounts. Silver in this zone is fine-grained (0.01–0.5 mm), locked within sulfide phases as inclusions and intergrowths, imposing significant metallurgical problems (section 16.6).

Only the southwest oxide zone, on which this chapter concentrates, has been exploited. The Ag mineralogy is essentially bromine-rich cerargyrite (Ag(Cl,Br)) with minor amounts of acanthite, often closely associated with limonite and Fe staining. There are some chloride and sulfide minerals present too. Payable mineralisation is mainly restricted to porphyritic rhyolite flow units (75% of total) as disseminations, fracture infills, veinlets, and replacements of potassic feldspars. Weak mineralisation occurs within rhyolitic tuffs, breccia, and argillites (25% of total) associated with QAS alteration. High grade zones (>10 oz t^{-1} (>343 g t^{-1})) exhibit a spatial relationship with areas of strong jointing, sericitic alteration, and limonite (Ashleman 1988).

Base metals (Pb-Zn-Cu-As) exhibit a spatial association with the distribution of silver mineralisation, i.e. a gradual increase to the south. Lead shows the strongest association with silver in the oxide zone while zinc values tend to be low.

A sharp redox boundary within the oxide zone marks the transition from upper oxide ore into lower sulfide ore. The oxide orebody is subdivided into a continuous NE lobe and fragmented SW lobe, the former containing the highest grades. The NE lobe trends NE to ENE defined by sharp but irregular boundaries, and dips steeply (55–70 degrees) to the W and NW. The SW lobe is disrupted by postmineralisation faulting and dykes. The oxide orebody defines a linear feature 75 × 610 m (250 × 2000 ft) extending to depths of 90–135 m (300–450 ft) forming lenticular to stacked tabular bodies.

Mineralisation appears to be confined to the northwest side of a NE trending fault zone with the southern boundary defined by a steep, possibly listric, normal fault which brought tuff into contact with the rhyolitic porphyry. NW trending offshoots cross the footwall contact into the rhyolitic tuffs and parallel NW faults. The hanging wall, or northern contact, is not marked by any significant lithological change other than a brecciated form of the rhyolitic porphyry at a high stratigraphical level. This boundary has been inferred to reflect a change in host rock permeability or a high angle reverse fault consistent with basin and range faulting.

NE trending basin and range faulting offset by a cogenetic or later, minor, NW fault system probably acted as feeders for the invasion of early Pliocene hydrothermal fluids. Hydro-fracturing and brittle failure resulted in the formation of a conjugate fracture zone. Hydrothermal fluids exploited the fracture zone and, in conjunction with alteration, made the host rock amenable to mineralisation. Mineralisation is concentrated at fault intersections (northwest offshoots) and in a zone parallel but displaced from the fault planes. The sharp footwall contact is either the function of mineralisation being dissipated by a highly porous tuffaceous rhyolite or ponded by an impermeable fault gouge. Alternatively, both the sharp north and south boundaries may reflect the effects of fault reactivation, i.e. dextral strike slip displacing mineralisation along strike and producing further uplift and erosion. This however is complicated by NW–SE offshoots that show no displacement.

16.5 DEVELOPMENT AND MINING

The exploration drilling program defined the ore zone on which the block model for the reserve estimation and the open pit mine plan were based. Drilling continued during mining

FIG. 16.8   Photograph showing the color-coded flags used to delineate ore from waste in the pit. The dark flags denote ore blocks and the pale flags denote waste blocks.

to confirm ore reserves. Pit modification was required during mining to improve ore control (section 16.8.4), slope stability, ore shoot excursion, and operational access. The final pit dimensions were $1400 \times 500 \times 246$ ft ($427 \times 152 \times 75$ m) (Fig. 16.6), with 15-ft (4.5-m) benches attaining slopes of 60–72 degrees. Pit slopes were 52 degrees to the west and 43 degrees to the east, and safety benches were cut every 45 ft (14 m). Production haulage ramps were 55 ft (16.5 m) wide with 4 ft (1.2 m) berms and had a maximum gradient of 10%.

Standard mining methods of drilling and blasting, front end loading, and haulage by truck were carried out by the contractor. A total of 16,500 t of ore and waste were removed daily in a 10-hour shift, 5 days a week. Drilling for blasting, using a $5^5/_8$ in (143 mm) down-the-hole hammer, was carried out at 15 ft (4.5 m) centers, and at 12 ft (3.7 m) centers in areas of blocky silicification.

Grade control was based upon the blasthole samples and pit geology. The orebody was mined principally as a vein type deposit (Perry 1989). The eastern boundary (hanging wall contact) was taken to be a steeply dipping range front fault and the western boundary was an assay boundary. Blastholes were sampled, by sample pan cut or manual cone cuts, each representing a 15 ft (4.5 m) composite. In contrast to the AAS analytical method used on the exploration borehole samples, the blasthole samples were analyzed by a cyanide

(CN) leach method, giving a more representative extractable silver content. Statistical and financial analyses had shown that the silver grades could be subdivided into four groups: (i) ore at >1.3 oz t^{-1} (45 g t^{-1}); (ii) low grade ore at $0.9–1.3$ oz t^{-1} ($31–45$ g t^{-1}); (iii) lean ore at $0.5–0.9$ oz t^{-1} ($17–31$ g t^{-1}); and (iv) waste at <0.5 oz t^{-1} (<17 g t^{-1}). All conversions of ounces per short ton to grammes per tonne have been made using a factor of 34.285 (Berkman 1989). Each bench was divided into blocks, each block centered on a blasthole. The assayed grade for each blasthole was assigned to the block. The grades were color coded and each block was marked on the bench in the pit by the surveyors with the appropriate color-coded flags (Fig. 16.8). This ensured that waste was correctly identified and sent to the waste pile, while the different grade materials were sent either to the cyanide leach heap or to one of the two low grade stockpiles.

Sulfide ore was not amenable to heap leaching and where encountered was selectively mined and hauled to low and high grade sulfide ore stockpiles (Fig. 16.8) (Perry 1989). Sulfide ore was differentiated from oxide ore on the basis of color (yellow oxide ore and grey sulfide ore) and a percentage comparison of total AAS assays (or fire assay) and cyanide leachable assays i.e.:

$$\frac{AAS - CN \times 100}{AAS}$$

a percentage difference of 6–17% defined the oxide ore and the sulfide ore was marked by a 24–40% difference.

16.6   MINERAL PROCESSING

Metallurgical testing of core samples determined that cyanide leaching was the most cost-effective method of treating the oxide ore. An excellent overview of heap leach technology is given by Dorey et al. (1988). Sulfide ore resulted in high cyanide consumption and required fine grinding to give a 78–84% recovery. The oxide ore was amenable to direct leaching resulting in a 94–97% recovery. Flotation tests on the sulfide ore liberated 90–95% of the silver and 90% of the Pb and Zn providing the pH values of the collectors, which suppressed Fe and As, were high. Oxide ore recoveries by flotation were low (50–60%). Laboratory tests indicated that it would be possible to heap leach the oxide ore, but the sulfide ore had to be stockpiled with a view to possible flotation extraction at a later date. A flow sheet of the process is shown in Fig. 16.9.

Mined ore was hauled and dumped on to an ore surge pile and fed in to a primary jaw crusher (Fig. 16.9). The $-^3/_4$ in (−19 mm) product passed over a two-deck vibrating screen on to an in-line agglomerator. Secondary crushing of the oversized ore was completed by cone crushers in parallel circuit. The crushed ore was automatically sampled and weighed.

Ten pounds (4.5 kg) of cement was added to 1 t of ore prior to agglomeration. The agglomerated ore was screened to remove the $-^3/_8$ in (−10 mm) fines, and was then spread on the leach pad with a slough stacker. The 700×1100 ft (210×335 m) leach pad was lined with 60-mm heavy duty polyurethane (HDPE) and divided into six cells. Each cell had a 183 t capacity when stacked to 41 ft (12.5 m). Primary and secondary leaching utilized drip feed emitters providing a flow rate of 0.005–0.008 gallons per minute per square foot (200–350 ml min^{-1} m^{-2}). Sprinklers were used during the final stage of leaching and rinsing.

Pregnant solutions passed into the pregnant pond and were then processed by a Merrill–Crowe Zn precipitator plant with a 1000 gallons min^{-1} (3785 L min^{-1}) capacity.

After initial filtering the precipitate was dried, analyzed, fluxed, and smelted (Fig. 16.9), to produce silver doré bullion (99% silver). It was estimated that 75% of the silver in the oxide ore was recovered by cyanide leaching.

16.7   ENVIRONMENTAL CONSIDERATIONS

Any mining project which will result in the alteration of the surface of a property, either through the addition of mine buildings or because of mining, is required to submit a Planning Application (in the UK) or a Plan of Operation (in the USA) to the authority responsible for administering the area. This application should describe the proposed mining operation and the infrastructure requirements and include a schedule of the proposed activities (Thatcher et al. 1988). Often the operator will include a plan of the procedures he proposes to implement to protect the environment during and after mining (reclamation).

In the case of a precious metal heap leach operation, additional permits and approvals are required before a heap leach operation can be commissioned. Thatcher et al. (1988) discuss the various regulatory aspects and permitting requirements for precious metal heap leach operations. These requirements include air quality, and surface and groundwater quality permits as well as cyanide neutralization requirements. In the latter case, Nevada specifies detailed heap rinsing procedures and analytical methods for cyanide detection as operating conditions in their water quality or discharge permits (Thatcher et al. 1988). In Nevada operators are expected to obtain cyanide neutralization levels of 0.2 mg per litre free CN$^-$ as a target concentration, but most operators are unable to meet these standards. Similarly, there should be only 1 ppm cyanide in the residual weak acid digest (WAD) or <10 ppm CN in the soil. More realistically, they expect the company to attempt to meet these criteria and then show that the cyanide remaining behind will not affect or put at risk the groundwater.

The Trinity operators had several alternatives to consider with regard to cyanide degradation and heap detoxification. These are described in some detail by Smith (1988a), and the various methods are outlined here. The

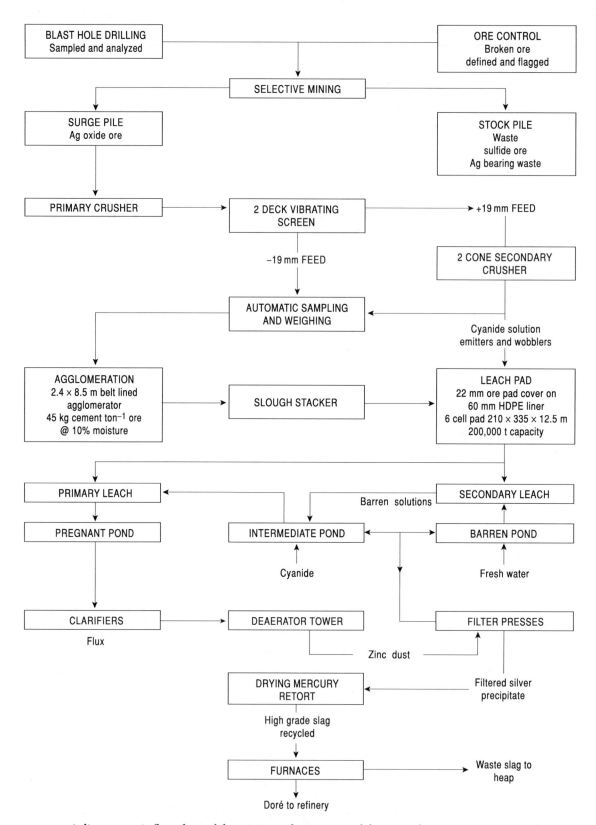

FIG. 16.9 A diagrammatic flow sheet of the mining and processing of the ore at the Trinity Mine, Nevada.

methods fall into two categories, namely natural processes or chemical treatment.

16.7.1 Natural degradation and detoxification

1 *Passive abandonment.* High UV levels and strong winds for 9 months of the year would make this a possible alternative. The natural degradation of the cyanide would be fast in the near-surface layers, but would slow rapidly with depth.

2 *Rinsing with barren solutions.* Rinsing to clean the entire heap has been tested empirically in the laboratory. A column in the laboratory was rinsed with fresh water and the cyanide levels were reduced from 700 to 50 ppm. Unfortunately this took between 12 and 21 months to achieve and extremely large quatities of water (of the order of 4500 L t^{-1}) were required.

16.7.2 Chemical treatment

Five types of chemical treatment are currently available and were considered by US Borax.

1 *Acid leach.* The leachate thus produced would be collected in the ponds, neutralized with lime, and the metals which precipitated would be removed mechanically.

2 *Sulfur dioxide and air oxidation method.* (Patented by INCO). These are added in conjunction with lime and a copper catalyst, but special equipment is required and this method is expensive.

3 *Alkaline chlorination.* Either chlorine gas or calcium hypochlorite are used in the oxidation process but these methods have very high reagent costs, and careful pH control with the second method is required to prevent the formation of toxic cyanogen chloride.

4 *Hydrogen peroxide oxidation.* This method is clean and nontoxic, but the H_2O_2 is expensive. This method is used by OK Tedi in PNG where H_2O_2 is added to the recirculation pond and the cyanide-free water is then recirculated through the heap using the original sprinkler system.

5 *Ferrous sulfate.* Chemical destruction of the cyanide is used whereby 1 mole of ferrous sulfate ties up 6 moles of cyanide to produce Prussian Blue. This reaction takes place in the pond so that the recirculated water is free from cyanide. Both the last two methods are quick acting, especially with the initial flush. Thereafter there is a longer term diffusion mechanism which completes the flushing procedure.

16.7.3 Summary

There are advantages and disadvantages with each process as outlined above. The State of Nevada discourages the use of hypochlorite and peroxide neutralization methods (Thatcher et al. 1988), but the INCO SO_2–air and the peroxide processes are thought to be more technically efficient (Smith 1988a). However the final conclusion may well be that each of the major detoxification processes in use has its application in a particular set of circumstances which each operator must establish for each property.

16.8   GRADE ESTIMATION

It is essential for reliable ore reserve calculations to derive the best Ag grade estimate for an orebody. The 5220 bench of the Trinity silver deposit was chosen to illustrate the problems and procedures of grade estimation. The 5220 level exhaustive dataset contains 1390 blasthole samples. No spatial bias or clustering effects are evident due to the regular sample pattern (15 ft (4.5 m) centers).

For ore reserve calculations a global estimate of mean silver is required. The global estimate must, however, be controlled by local estimates because of the spatial variation of silver values. Measures of variability are important in evaluating the accuracy of a grade estimate, especially within a spatial context, and have strong implications for mine planning.

In many geological environments the problem of grade estimation is intrinsically linked to the type of statistical distribution. Many problems can be solved from an assessment of the univariate statistics, histograms, and probability plots (Davis 1986), but this assessment can not be divorced from the spatial context of the data (Isaaks & Srivastava 1989). The following sections explore the problems of grade estimation and derive simple solutions for modeling skewed distributions.

16.8.1   Univariate statistics

For the purpose of grade estimation the distribution of assay values within the population is a prime requirement, especially regarding frequencies above a specified lower limit, e.g. cut-off grades. Most statistical parameters are applied to the Gaussian distribution, yet within a geological environment a normal distribution is not always immediately evident. It is common to find a large number of small values and a few large values, the lognormal distribution is then a good alternative (Isaaks & Srivastava 1989). The histogram (Fig. 16.10a) describes the character of the distribution. The clustering of points at low values and the tail extending toward the high values indicates that the assays on the 5220 bench do not conform to a normal distribution. Most of the cumulative frequencies for the lower values plot in a relatively straight line, but departures occur toward the higher values (Fig. 16.10b). This is further evident from the curved nature of the normal probability plot (Fig. 16.10c).

The departure from a straight line at lower values on the log normal probability plot (Fig 16.10d) indicates that the data also do not fit a log-normal distribution. Although it is possible to model a "best fit" distribution to the overall population, errors will occur during grade estimation. This is evident from the summary statistics (Table 16.4). It is clear from the frequency distributions that a global estimate based upon a normal distribution will be biased toward high values (overestimated). Similarly the log transformed data will be biased by low values (underestimated) (Fig. 16.10d). Due to the mathematical complications that arise from a log transformation, such a transformation is best avoided. It is a preferred alternative to try to establish Gaussian distributions within the exhaustive dataset in order to improve grade estimation.

The important features of the distribution are captured by the univariate statistics (Table 16.4) and illustrate immediately the problem of grade estimation for the 5220 bench. The summary statistics provide measures of location, spread, and shape. The mean Ag content of the bench is recorded as 1.43 oz t^{-1} (49 g t^{-1}), yet other estimates of central tendency, the median (0.33) and the trimmed mean (0.48), invoke caution in using the mean as an accurate global estimate. The high variance and standard deviation describe the strong variability of the data values. The implication is that insufficient confidence can be placed upon the mean as an accurate global estimate. A strong positive skew with a long tail of high values to the right is evident from the high positive skewness (+10.62). The degree of asymmetry is also supported by the coefficient of variance (2.58). Knudsen (1988) implies that if the coefficient of variance is greater than 1.2 a log-normal distribution could be modeled. The Sichel-T estimator can be used to estimate the mean of a log normal distribution, and is calculated to be 1.28.

The frequency histogram (Fig. 16.10a) illustrates how the data are proportioned. It is important to note that even though the data range is 0–85 oz t^{-1} Ag, only 5% is > 7.2, 2% > 12.5, and 0.5% > 20 oz t^{-1} Ag. The upper quartile indicates that 75% of the data lie between a restricted range of 0–1.1 oz t^{-1} Ag, and the mean which is higher (1.43) does not reflect the majority of the data. It is obvious therefore that the magnitude of the positive skew will have a disproportionate effect upon the mean.

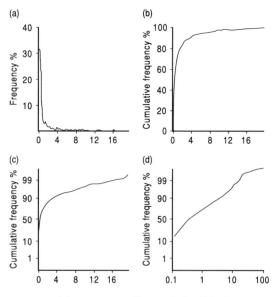

FIG. 16.10   (a) Histogram of (cyanide leach) silver values (oz t^{-1}) obtained from blasthole samples on the 5220 bench, Trinity Mine, Nevada. (b) Cumulative frequency plot. (c) Normal probability plot. (d) Log-normal probability plot.

TABLE 16.4 Univariate statistics of the silver values derived from cyanide leach analysis of the blast hole samples on the 5220 level, Trinity Mine, Nevada.

	Total population	Excluding >20 oz t⁻¹ Ag	1.3–20 oz t⁻¹ Ag
Number of assays	1390	1385	309
Mean	1.43 (49)	1.30 (45)	4.65 (159)
Median	0.33 (11)	0.33 (11)	2.30 (79)
Trimean	0.48 (16)	0.48 (16)	3.45 (118)
Sichel-T	1.28	1.22	4.52
Variance	13.59	6.93	16.36
Standard deviation	3.69	2.63	4.04
Standard error	0.10	0.07	0.36
Skewness	10.62	3.89	1.84
Coefficient of variance	2.58	2.02	0.87
Lower quartile	0.16 (5)	0.16 (5)	1.90 (65)
Upper quartile	1.11 (38)	1.09 (37)	5.84 (200)
Interquartile range	0.95 (33)	0.93 (32)	3.93 (135)

Silver values are in oz t⁻¹ (g t⁻¹).

The initial statistical evaluation illustrates a potential error in the global estimate resulting in possible overestimation and bias toward erratic high values. More detail is required given that the mean is only a robust estimate if applied to a Gaussian distribution and a log transformation is not adopted. It is important at this stage to investigate the distribution further. Having ascertained that the data do not conform to a normal or log-normal distribution, the next step is to evaluate potential modality and departures from a continuous distribution.

The probability plot and histogram are very useful in checking for multiple populations. Although breaks in the graphs do not always imply multiple populations, they represent changes in the character of the distribution over different class intervals. The initial step is to explode the lower end of the histogram by making the class interval smaller (Fig. 16.11) and searching for potential breaks in the population. Breaks in the distribution are chosen on the basis of significant changes in the frequency between class limits or repeated patterns reflecting polymodality. If subpopulations are found, explanations must be sought and may depend upon sample support, geological control, or population mixing. The positively skewed distribution of the data may be a function of the overprinting of multiple populations. Defining subgroups which may

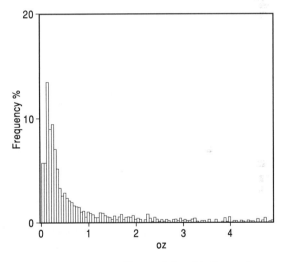

FIG. 16.11 Histogram of (cyanide leach) silver values (oz t⁻¹) obtained from blasthole samples on the 5220 bench, Trinity Mine, Nevada, with class intervals reduced to 0.05 oz t⁻¹. The higher values (>5 oz t⁻¹) are omitted for clarity.

approximate normal distributions could improve the global estimate.

Such an evaluation reveals a number of breaks within the population of silver grade on the 5220 level. The following class breaks have been interpreted: 0.4, 0.8, 1.3, 5, 10, and 20 oz t⁻¹ Ag (14, 27, 45, 171, 343, and 686 g t⁻¹). Those values greater than 20 oz t⁻¹ are

TABLE 16.5   Univariate statistics of the subpopulations of the silver values derived from cyanide leach analysis of the blasthole samples on the 5220 level, Trinity Mine, Nevada.

	Silver values in oz t^{-1} (g t^{-1})					
	0.0–0.4	0.4–0.8	0.8–1.3	1.3–5	5–10	10–20
Number of assays	762	212	1032	215	731	
Mean	0.18 (6)	0.57 (20)	1.03 (35)	2.56 (88)	7.46 (256)	14.37 (493)
Median	0.17 (6)	0.55 (19)	1.01 (35)	2.25 (77)	7.34 (252)	13.60 (466)
Trimean	0.17 (6)	0.56 (19)	1.02 (35)	2.35 (81)	7.44 (255)	13.93 (478)
Sichel-T	0.20	0.57	1.03	2.56	7.46	14.37
Variance	0.01	0.01	0.02	1.04	2.48	10.79
Standard deviation	0.10	0.11	0.15	1.02	1.57	3.28
Standard error	0.00	0.01	0.02	0.07	0.21	0.58
Skewness	0.19	0.28	0.19	0.74	0.08	0.25
Coefficient of variance	0.55	0.19	0.15	0.40	0.21	0.23
Lower quartile	0.10 (3)	0.47 (16)	0.89 (31)	1.71 (59)	5.98 (205)	11.18 (383)
Upper quartile	0.26 (9)	0.67 (23)	1.16 (40)	3.20 (110)	9.09 (312)	16.70 (573)
Interquartile range	0.16 (6)	0.20 (7)	0.27 (9)	1.49 (51)	3.11 (107)	5.52 (189)

considered outliers (5 out of 1390 samples) and are not evaluated. To assess the statistical significance of the subgroups univariate statistics are calculated for each group (Table 16.5). In all instances normal distributions are approximated with significant reductions in the variance, coefficient of variance, and skewness. Estimates of central tendency lie within statistically acceptable limits indicating improved confidence in the mean of each group. The 1.3–5 oz t^{-1} group is slightly skewed and a further split possible. However, improved detail may not result in enhanced accuracy or confidence in the mean.

The exhaustive dataset can be split into subgroups according to grade classes and the grade estimate for each group is improved. The statistical and financial evaluation undertaken by US Borax resulted in a similar grouping (section 16.5). US Borax applied a 1.3 oz t^{-1} cut-off to the data such that everything above 1.3 oz t^{-1} was mined as ore. It is important to evaluate the effect of cut-off on the global estimate and the implications in relation to the subgroups.

16.8.2   Outlier and cut-off grade evaluation

Those values in excess of 20 oz t^{-1} (686 g t^{-1}) Ag have been classed as anomalously high values, and if rejected have a significant effect upon the univariate statistics and grade estimate (Table 16.4) despite only representing 0.5% of the data. The mean is reduced by 10% to 1.3 oz t^{-1} (45 g t^{-1}) Ag while the variance and skewness are reduced by up to half. However the distribution is still strongly skewed and the problems of estimation still remain. If the 1.3 oz t^{-1} cut-off is then also applied the distribution becomes flatter, the global estimate more precise, but accuracy is reduced because of increased variability (Table 16.4). The 1.3 oz t^{-1} cut-off also reflects a major break in the population and improved estimates may be a function of a separate population approximating to a normal distribution. With a reduced skewness the variability of values become more symmetrical about the mean.

A further investigation of outlier removal (Fig. 16.12) shows that as the tail effect of the distribution is reduced the mean grade and standard deviation decreases in a linear fashion. The decline is more gradual for the total population than when the 1.3 g t^{-1} cut-off is applied. It is obvious that as values are included or excluded the magnitude of the mean varies. This must be viewed in relation to the number of samples, the effect of which is evident with values above or below 20 oz t^{-1} Ag. Although the magnitude of the grade estimate can be increased by including outliers the rate of

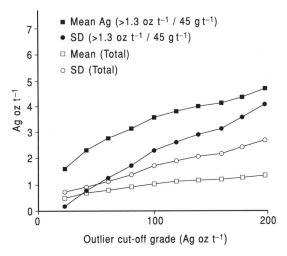

FIG. 16.12 Graph showing changes in mean grade and standard deviation when different outlying (high) grades are removed from the database of the silver values from the 5220 bench, Trinity Mine, Nevada.

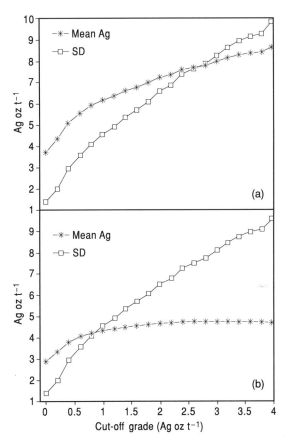

FIG. 16.13 Graphs showing the average silver grade when varying cut-offs are applied to the database of the silver values for the 5220 bench, Trinity Mine, Nevada: (a) total population; (b) with values >20 oz t⁻¹ removed.

improvement in the ore reserve estimation is reduced. A grade estimate must also evaluate how the values fluctuate around the mean, therefore the relationship of the standard deviation to the mean is important.

Within a mining environment the confidence of the grade estimate is derived from the standard deviation. The value of the standard deviation, however, is subject to the group of values from which the estimate was obtained and the spatial context. The standard deviation expresses confidence in symmetrical intervals, therefore when the standard deviation exceeds the mean the magnitude of underestimates is significantly different from the magnitude of overestimates. This is illustrated in Fig. 16.12 for the total population such that the global estimate would exceed the local estimate in the majority of cases. The degree of variability and error increase in magnitude as the tail effect of the distribution is more prominent, therefore the above effect becomes more pronounced.

The relationship of the standard deviation to the mean changes when the same evaluation is applied to ore grade material. The graph (Fig. 16.12) indicates that the estimate for the ore grade material is more reliable and subject to less fluctuation, although as the distribution

expands, variability increases. The grade cut-off curves reinforce the relationship of the mean to the standard deviation for the total population (Fig. 16.13a) and the effect of removing the 20 oz t⁻¹ outliers (Fig. 16.13b). As extreme values of the distribution are altered the degree of variability declines. A point is achieved where the number of local underestimates and overestimates as compared with the global estimate balance each other. As low or high values are included or removed the mean changes accordingly and when an approximate normal distribution is modeled the variability reaches a constant (Fig. 16.13b). At this point there is no improvement in the confidence of the grade estimate and, in terms of reserve

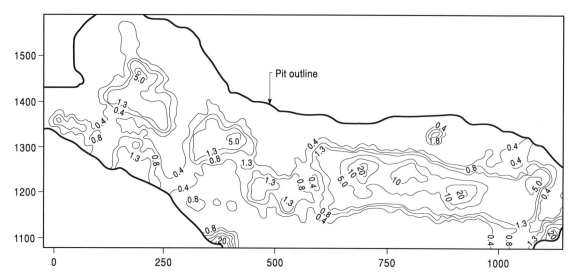

FIG. 16.14   A contour map of the silver grades (oz t^{-1}) on the 5220 bench, Trinity Mine. Contours are coded according to grade classes.

estimation, it is the spatial distribution of silver that becomes important.

16.8.3   Spatial distribution

Although it is possible to improve grade estimates by evaluating the statistical distribution, the spatial features of the dataset are important. A spatial evaluation will focus upon the location of high values, zoning, trends, and continuity. A contour map of the 5220 level (Fig. 16.14) shows an E–W trend to the data which fragments and changes to a NW–SE trend to the west. Zoning occurs and discrete pods of ore grade material are evident. Low grade values less than 0.8 oz t^{-1} (27 g t^{-1}) cover a greater area. The closeness of the contours especially at the 1.3 oz t^{-1} Ag (45 g t^{-1}) contour indicates a steep gradient from high to low values. The abrupt change from waste to ore grade material marked by a steep gradient will be reflected by a high local variance.

It is obvious that the subgroups previously defined are not randomly distributed but conform to a zonal arrangement. The indicator maps (Fig. 16.15) support the spatial integrity of the subgroups, illustrating the concentric zoning and isolated groupings of the data. The statistics of each subgroup can then be applied to specific areas of the 5220 level and local

estimates can be applied to the contoured zones. A further assessment of local estimates can be achieved by using moving windows (Isaaks & Srivastava 1989, Hatton 1994a). Statistics were calculated for 100 ft (30 m) square windows overlapping by 50 ft (15 m) (Fig. 16.16). The contour plots of the mean and standard deviation for each window (Fig. 16.17) show that the average silver values and variability change locally across the area. Zones of erratic ore grades can be located and flagged for the purpose of mine planning and grade control.

In general the change in variability reflects the change in the mean, although at the edge of the high grade zone to the east, the variability increases at a greater rate. High variability is a function of the mixing of two populations or the transition from one population or zone to another. This is seen in the profile taken along N1300 (Fig. 16.18a). A profile passing through the high grade zone (N1200, Fig. 16.18b) shows a reduced change in variability compared with a pronounced increase in the magnitude of the mean. The strong relationship between the local mean and variability (Fig. 16.18) is referred to as a proportional effect which implies that the variability is predictable although the data do not reflect a normal distribution. The contour plots (Fig. 16.17) show that in certain

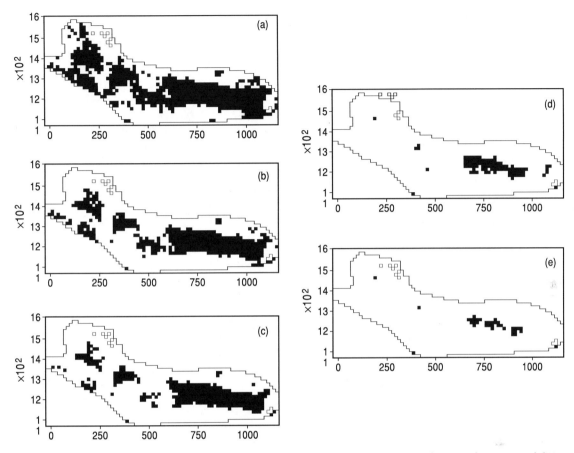

FIG. 16.15 Indicator maps of the Ag grades (oz t⁻¹) of the 5220 bench demonstrating the spatial integrity of the data. The grade boundaries are at (a) 0.4, (b) 0.8, (c) 1.3, (d) 5.0 and (e) 10.0 oz t⁻¹ (14, 27, 45, 171, and 343 g t⁻¹).

locations the change in local variance mirrors or is greater than the change in the mean (the proportional effect). However elsewhere the variability remains roughly constant whilst the local mean fluctuates.

16.8.4 Semi-variograms

Semi-variograms are used to quantify the spatial continuity of the data (see Chapter 9). The range (a) of the semi-variogram defines a radius around which the local estimate has the least variability. The omnidirectional semi-variogram of the 5220 bench (Fig. 16.19a) shows a range of 120 ft (35 m). When samples larger than 20 oz t⁻¹ (686 g t⁻¹) were excluded (Fig. 16.19b) the sill was reduced by a third from

13 to 8 oz Ag². This again demonstrates the effect that the few high values have on the variability.

Directional semi-variograms show different ranges in different directions (Fig. 16.19a–d). Continuity was greater in the E–W direction (300 ft) (Fig. 16.19c) compared with all other directions. The minimum continuity was 90 ft in a N–S direction (Fig. 16.19a). An E–W elipse measuring 300 × 90 ft (Fig. 16.19f) would be the optimal moving window shape and size to give the lowest variability for a local estimate.

The NW–SE directional semi-variogram (Fig. 16.19d) shows a "hole effect" (Journel & Huijbregts 1978, Clark 1979) which results from alternate comparison of high and low zones in the NW half of the deposit.

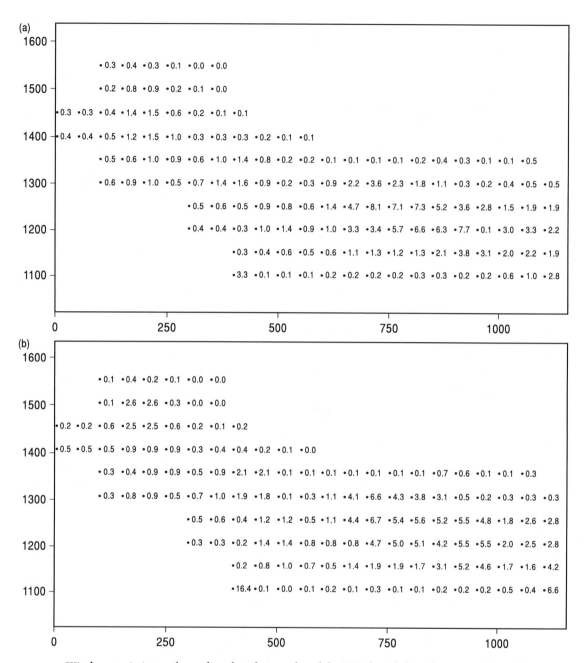

FIG. 16.16   Window statistics performed on the silver grades of the 5220 bench, based on a transformed grid oriented E–W with 30 m square, moving window overlapping by 15 m: (a) mean Ag oz t^{-1}; (b) standard deviation.

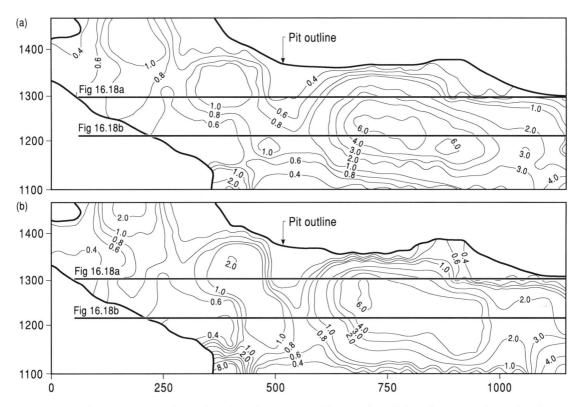

FIG. 16.17 Window statistics performed on the silver grades of the 5220 bench, based on a transformed grid oriented E–W with 30 m square, moving window overlapping by 15 m: (a) contours of mean Ag oz t^{-1}, (b) contours of standard deviation. (From Bell & Whateley 1994.)

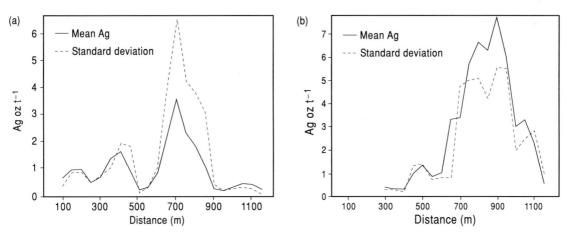

FIG. 16.18 Profiles 1 and 2 (a and b respectively) of the window mean and standard deviation showing the proportional effect and highlighting areas of expected high error. (From Bell & Whateley 1994.)

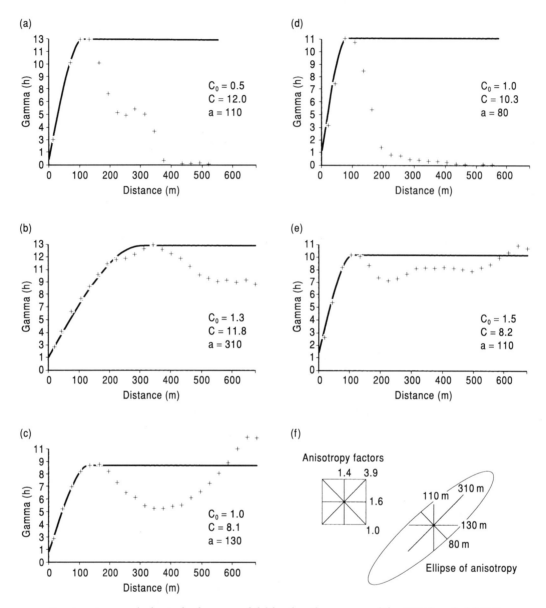

FIG. 16.19 Semi-variograms (spherical scheme models) for the silver assays of the 5220 bench: (a) N–S direction; (b) NE–SW direction; (c) E–W direction; (d) SE–NW direction; (e) omni-directional; (f) ellipse of anisotropy and anisotropy factors. (From Bell & Whateley 1994.)

16.8.5 Conclusions

Attempting a global estimate of the grade for the entire bench is irrelevant in a mining situation because geologists and mining engineers are only interested in ore grade material. The skewed population means that ore grade material will be underestimated. It is essential to understand what constitutes the population, in order that sensible subdivisions into subpopulations can be made. It is incorrect to evaluate an exhaustive dataset if the data are derived from different episodes of mineralisation, when their physical and chemical characteristics differ. If the global estimate is used to make local estimates of grade, then the grade will be overestimated, because no account is made of the spatial relationship of the samples or of the high variability.

To improve the grade estimation it is necessary to try to establish normal distributions within the whole skewed population, using geology, structure, mineralisation, alteration, etc., rather than transforming the whole population. The variability within each subpopulation is thus reduced, which increases the confidence in the subpopulation mean.

The high values cause extreme variability which influences the mean. By excluding them the mean and variance are reduced and the estimation becomes more reliable.

By applying these techniques to the data above cut-off grade (omitting outliers), a population which approximates a normal distribution is achieved. This improves the confidence in the estimation of the mean. By subdividing the data above cut-off into subpopulations, variability is further reduced.

Indicator maps show that subgroups, as defined by statistical parameters, are spatially arranged. High local variance is a function of boundaries between these populations. Application of the moving window statistics should therefore be viewed in terms of the spatially distributed groups, trying to avoid crossing subpopulation boundaries.

From the univariate statistics and spatial distribution of the data the conclusion is that a global estimate must be derived from local estimates rather than the whole population.

In hindsight, the best ore reserve estimation could be derived using the average of the estimates from discrete zones or the average of the estimates derived from the areas between contours of the grades. Bell and Whateley (1994) undertook global estimation of the grade by applying a number of estimation techniques to the original exploration borehole data of the 5220 level. They compared the results with the blasthole dataset and concluded that linear interpolation provided grade estimates with the lowest variance.

16.9 ORE RESERVE ESTIMATION

During the life of the project, from the exploration stages to the feasibility and mine design studies, different ore reserve calculations were carried out, each with a different confidence level. The confidence in the reserve estimate grew as the amount of borehole data increased.

16.9.1 Evaluation of initial exploration data

Immediately after exploration was initiated in early 1982, it was realized that a significant tonnage of disseminated silver existed. During the 1982–83 exploration program the oxide mineralisation had not been fully evaluated, and the initial reserve estimates only took the sulfide mineralisation into account.

16.9.2 Evaluation of additional exploration data

Additional exploration drilling took place in 1986 with the intention of identifying additional sulfide mineralisation, and to collect samples for metallurgical and engineering studies. In mid 1986 it became apparent that a small area of high grade silver oxide mineralisation could possibly support a heap leach operation. The reserve potential of the oxide zone was calculated by several members of the joint venture, using different methods and cut-off grades. Table 16.6 summarizes the results.

The tonnage calculations vary from 0.932 to 1.304 Mt where a Ag cut-off grade of 1.5 oz t^{-1} (50 g t^{-1}) was used. Where the cut-off grade was raised to 2 and 3 oz t^{-1} (69 and 103 g t^{-1}), as one would expect, the tonnage was reduced but the average grade increased. These differences in tonnage and average grade were accounted for when thin, lower grade intersections used in earlier calculations were omitted, and modified polygons were used. No reference was made to the geology of the deposit in any of the above calculations.

This demonstrates the difficulty in calculating reserve estimates in the exploration phases of a program (Bell & Whateley 1994). Different methods, by different people using differing cut-off grades, produces a variety of estimates. It is not possible to quantify the error of estimation using these 2D manual methods of reserve estimation. It is not even practical to compare the early reserve estimates with the final tonnage and grade extracted, because in the final estimate a cut-off grade of 1.3 oz t^{-1} (45 g t^{-1}) was used.

TABLE 16.6   Summary of the ore reserve calculation for the silver oxide zone, Trinity Mine, Nevada.

Method	Tonnage factor ft³ t⁻¹ (SG)	Composite length ft (m)	Ag cut-off grade oz t⁻¹ (g t⁻¹)	Tonnage × 10⁶	Average Ag grade oz t⁻¹ (g t⁻¹)	Total Ag oz (g) × 10⁶
Polygons–USB	13.3 (2.41)	20 (6.1)	1.5 (50)	0.967	6.95 (238)	6.22 (213)
N–S cross-sections, USB	13.3 (2.41)	10 (3.0)	1.5 (50)	1.304	6.16 (211)	8.03 (275)
E–W cross-sections, USB	13.3 (2.41)	10 (3.0)	1.5 (50)	1.293	5.90 (200)	7.63 (262)
N–S cross-sections, USB	13.3 (2.41)	10 (3.0)	1.5 (50)	0.932	7.69 (264)	7.17 (246)
Polygons, Santa Fe	13.3 (2.41)	20 (6.1)	3.0 (100)	0.669	9.10 (310)	6.09 (209)
N–S cross sections, Santa Fe and USB	13.3 (2.41)	10 (3.0)	2.0 (69)	0.870	8.00 (274)	6.96 (240)

16.9.3   Reserve estimate for mine planning

In 1987 a computerized 3D block modeling technique was used to estimate reserves and the data thus generated were used for mine planning. A conventional 2D inverse distance squared ($1/D^2$, see section 9.5.1) technique was used to interpolate grades from borehole information to the block centers, using a 150 ft (46 m) search radius (Baele 1987). The blocks were given a 25×25 ft (7.6×7.6 m) area and were 15 ft (4.6 m) deep. Borehole assays were collected at 5 ft (1.5 m) intervals, but were composited by a weighted average technique over 15 ft (4.6 m) to conform to the block model. Geology and alteration were incorporated into the block model by assigning a rock and alteration code to each block. Tonnage and grade was estimated at various cut-off grades (Table 16.7).

Despite the geological control, this method of reserve estimation appears to have overestimated the reserve potential (in comparison with the early ore reserve estimates). This may be partially a result of an edge effect where the majority of any given block at the edge of the deposit may lie outside the mineralized area, but because of the shape of the boundary they have been included in the reserve estimate. An additional factor is the use of 13.7 cu ft t⁻¹ (2.34 t m⁻³) as the ore and waste specific gravity, to calculate *in situ* tonnage compared with only 13.3 cu ft t⁻¹ (2.41 t⁻¹ m⁻³) used in the exploration evaluation.

16.9.4   Updated tonnage and grade estimate

The grade and tonnage of the ore mined from the first three benches was significantly different from that predicted by the $1/D^2$ block model. An updated reserve estimate was hand-calculated using cross-sections to evaluate the discrepancy. The geology and structure from the mined-out area were plotted on the cross-sections and used to project the orebody

TABLE 16.7   Reserve estimates on the silver oxide zone, using the $1/D^2$ block model, Trinity Mine, Nevada.

Tonnage × 10⁶	Tonnage factor ft³ t⁻¹ (SG)	Composite length ft (m)	Ag cut-off grade oz t⁻¹ (g t⁻¹)	Average Ag grade oz t⁻¹ (g t⁻¹)
2.65	13.7 (2.34)	15 (4.5)	0.5 (17)	3.05 (105)
2.05	13.7 (2.34)	15 (4.5)	1.0 (34)	3.76 (129)
1.65	13.7 (2.34)	15 (4.5)	1.5 (50)	4.46 (153)
1.26	13.7 (2.34)	15 (4.5)	2.0 (69)	5.19 (178)
0.88	13.7 (2.34)	15 (4.5)	3.0 (100)	6.33 (217)
0.63	13.7 (2.34)	15 (4.5)	4.0 (137)	7.52 (258)

TABLE 16.8 Updated reserve estimate on the silver oxide zone, Trinity Mine, Nevada.

Method	Tonnage factor ft³ t⁻¹ (SG)	Composite length ft (m)	Ag cut-off grade oz t⁻¹ (g t⁻¹)	Tonnage × 10⁶	Average Ag grade oz t⁻¹ (g t⁻¹)	Total Ag oz (g) × 10⁶
N–S cross-sections	13.7 (2.34)	15 (4.5)	1.5 (50)	0.899	6.94 (238)	6.24 (214)
Bench plans*	13.7 (2.34)	15 (4.5)	1.6 (55)	0.859	7.33 (251)	6.29 (216)
Blocks 15 × 15 × 15 ft (4.5 × 4.5 × 4.5 m)	13.7 (2.54)	15 (4.5)	1.6 (55)	0.963	6.55 (225)	6.31 (217)

* These estimates exclude the first three benches which had already been mined out.

geometry downwards (Reim 1988). Reim felt that the unsatisfactory grade control given by the $1/D^2$ method could be accounted for by the inadequate structural control in the $1/D^2$ model, variability in grade, and paucity of sampling data within the southwest portion of the oxide ore body. Additional drilling increased the sample density in poorly represented areas and this helped define the orebody geometry. The reserve estimate calculated using the updated cross-sections is shown in Table 16.8.

For mine planning purposes a mineable reserve was calculated on a bench by bench basis. Mineralized zones were transferred from the cross-sections on to 15 ft (4.5 m) bench plans. Polygons were constructed around each borehole and a reserve calculated (Table 16.8).

16.9.5 Grade control and ore reserve estimation using blastholes

Mine grade control was exercised by sampling and assaying the blasthole cuttings. Mined blocks were $15 \times 15 \times 15$ ft ($4.6 \times 4.6 \times 4.6$ m) and only blocks >1.3 oz t⁻¹ (45 g t⁻¹) were mined as ore. A final reserve was calculated using these blocks and a slightly higher reserve with a lower grade was estimated (Table 16.8).

16.9.6 Factors affecting ore reserve calculations

A number of factors influenced the validity of ore reserve estimates, namely geological control, sampling method, cut-off grade, sample mean, analytical technique, specific gravity and dilution factors:

1 *Geological control.* It is most important to establish the geological and structural control of mineralisation as early in a project as possible. In the example used here, the $1/D^2$ block model had good geological control, but the structural control was lacking. This resulted in overestimation of reserves.

2 *Sampling methods.* It is important to control the sampling method to ensure consistent and reliable sampling, e.g. down-the-variation in silver values derived from either reverse circulation or percussion drilling may show disparity.

3 *Cut-off grade.* This varied between estimation methods from 3.0 oz t⁻¹ (103 g t⁻¹) in the exploration assessment to 1.3 oz t⁻¹ (45 g t⁻¹) during mining (Tables 16.6–16.8). Comparisons of reserve estimates using different cut-off grades can be achieved if grade–tonnage graphs are constructed. These are difficult to construct during the exploration phase as data are sparse. Normally economic criteria are used to establish the cut-off grade (Lane 1988). US Borax assumed a commodity price of $6.50 per oz and expected silver recoveries between 65 and 79% (amongst other criteria) to establish the 1.3 oz t⁻¹ (45 g t⁻¹) cut-off grade used during mining.

4 *Sample mean.* The mean grade is dependent upon the method used to estimate it (section 16.8.2). The arithmetic mean probably overestimated the mean in this study but this could have been improved by establishing subpopulations with normal distibutions and averaging the estimates from these discrete zones.

5 *Analytical technique.* In order to compare samples used to estimate reserves using exploration boreholes and blastholes, a simple linear regression of cyanide-extractable silver on total recoverable silver was performed. A correlation coefficient of 0.94 indicated a strong positive relationship. Above 10 oz t^{-1} (343 g t^{-1}) Ag, the conversion of total silver to cyanide-eachable silver was less accurate. Changing analytical procedure during an exploration or feasibility exercise, without undertaking an overlapping series of analyses which can be used for correlative purposes, introduces an unnecessary risk of error.

6 *Specific gravity.* Initially, densities of 13.0 and 13.3 ft^3 t^{-1} (2.34 and 2.41 g cm^{-3}) were used to calculate tonnages. Density was determined on a number of different rock types and an average of 13.7 ft^3 t^{-1} (2.34 g t^{-1}) was calculated, which was used for the final tonnage calculations.

7 *Mining dilution.* Dilution of ore by waste during mining is inevitable but careful grade control will minimize the risk of dilution. Dilution usually occurs in ore deposits where grade is highly variable, as at the Trinity Mine. The assumption that a blasthole in the center of a $15 \times 15 \times 15$ ft block ($4.5 \times 4.5 \times 4.5$ m) is representative of that block is not necessarily true. The arbitrary boundary between a block showing a grade less than cut-off and a high grade block is drawn half way between the two holes. The low grade ore may extend farther into the high grade block than one expects, thus adding to the mining dilution problem. US

Borax initially expected a 20% dilution, but this was later modified to 7% after further assessment during mining.

16.10   SUMMARY AND CONCLUSIONS

The Trinity oxide orebody defined a small but economic silver deposit, amenable to cyanide heap leaching. The orebody was fairly continuous to the northeast but became fragmented and fault controlled to the southwest. The ore zone was defined by stacked tabular to lenticular bodies dipping steeply to the west and northwest and offset by N–S and NW–SE cross-faulting. Mineralisation tended to be controlled by faulting, alteration, and fracture density. The sharp ore grade boundaries along the footwall and hanging wall suggest geological control. These boundaries may reflect change in host rock permeability or sharp fault contacts.

The Trinity deposit illustrates the estimation problems encountered in deposits which have highly skewed data. The problem is compounded when trying to estimate grade and resource potential from a relatively low number of exploratory boreholes. Recognizing the importance that a few very high values may have upon the estimation of the global mean is important. If the variability of the data can be reduced the confidence in the mean goes up.

17

DIAMOND EXPLORATION – EKATI AND DIAVIK MINES

CHARLES J. MOON

The discovery of major diamond deposits in the barren lands of the Canadian Arctic was one of the great exploration successes of the 1990s. Targeting of the remote discovery area resulted from remarkable persistence by a junior company, although two major mining companies undertook the detailed exploration and development.

17.1 WORLD DIAMOND DEPOSITS AND EXPLORATION

Diamond is the hardest mineral known and has significant industrial uses, e.g. in the cutting of other materials and diamond drilling, but it is its use in jewellery that is most important for the exploration geologist (Fig. 17.1). Although 90% of industrial diamonds are produced synthetically, it is, at present, uneconomic to produce the much more valuable gem diamonds artificially.

Diamond deposits have many similarities to industrial mineral deposits in that the value of the raw material is only a small proportion of the value of the finished product. The world rough diamond market in 2001 was worth $US7.9 billion whereas the retail diamond jewellery market was worth $US56 billion of which $US13.5 billion was the value of polished stones (Diamond Facts 2003). Unlike other minerals, wholesaling of gem diamonds

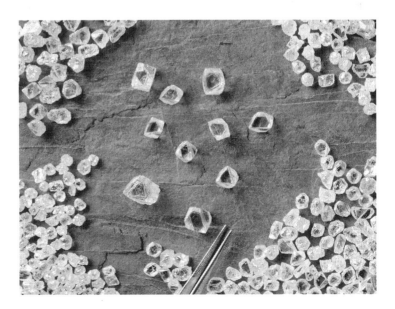

FIG. 17.1 The final product: diamonds from the Diavik Mine. (Courtesy Diavik Diamond Mines Inc.)

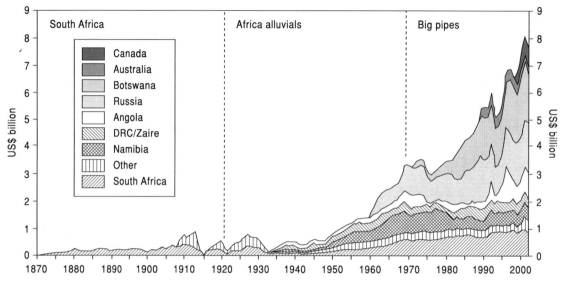

FIG. 17.2   World Production of Gem Diamonds. (De Beers 2003a.)

has been dominated by a single company, the Diamond Trading Company (formerly Central Selling Organisation) owned by the De Beers Group, which also controls many of the diamond mining companies. The De Beers Group has also recently set up a company to market retail diamonds. Unlike other industrial minerals, however, individual mineral grains have a value that depends on their particular characteristics, especially if the diamonds are large.

The value of an individual diamond is determined by the four Cs:

Carat, i.e. size. The term carat derives from the bean of the carob tree, which was used as a measure in ancient times. The measure was metricated as 0.2 g in 1914.

Color. Although diamonds are mainly thought of as colorless, colored or "fancy" stones are sought after. Colors range from pink through brown to blue. Even inclusion-rich black stones have their advocates.

Clarity. This reflects the number of fractures and inclusions in the stones and also affects the way in which diamonds are cut.

Cut. The appearance after cutting determines the return of light ("brilliancy") and color flashes ("dispersion" or "fire").

17.1.1   World diamond deposits

Diamonds are mined on a large scale in only a few countries: South Africa, Botswana, Namibia, Angola, Democratic Republic of Congo, Russia, Australia, and, recently, Canada (Fig. 17.2). Diamond production in 2001 is estimated at 117 Mct (23.4 t) of which 10% are gems, 55% near gems (stones with many inclusions but usable in cheap jewellery), and 35% industrial (Diamond Facts 2003). Of the gems and near gems approximately 73% were from bedrock deposits and 27% from alluvial workings (USGS 2003). However, the quantity is perhaps less important than the quality and hence monetary value (Table 17.1). Although Australian production is by far the largest by mass (26 Mct) the price per carat is low ($US10.60 ct^{-1}) and the value of production at $US278M is less than the $US417M produced in Namibia from 1.49 Mct at an average price of $US280 ct^{-1}.

Bedrock sources

Although diamonds have been found in a number of different bedrock sources, with the exception of minor (approximately 10% of

TABLE 17.1 World diamond production by country. (From Diamond Facts 2003.)

Country	Production (000 ct)	Average ($ ct^{-1})	Revenue ($ USM)
Angola	5175	141.0	730
Australia	26,152	10.6	278
Botswana	26,416	82.0	2160
Brazil	650	40.0	26
Central African Republic	450	146.0	65
Canada	3691	134.0	450
Congo (Dem. Rep.)	19,500	21.0	526
Ghana	870	21.0	18
Guinea	400	200.0	80
Namibia	1490	280.0	417
Russia	20,000	83.0	1665
Sierra Leone	300	230.0	69
South Africa	11,158	77.0	958
Tanzania	191	120.0	23
Venezuela	350	115.0	40
Other	530	118.0	63
Total	117,323	65.0	7568

world output) production from lithified placer deposits, two rare igneous rock types are the hosts for virtually all commercial bedrock diamond deposits. The rock types are lamproite and the predominant host, kimberlite. Age dating shows that the diamonds are much older than their host rocks in both types of igneous rock and are therefore present as xenocrysts (Richardson et al. 1984, 1990). Laboratory and theoretical investigations demonstrate that diamonds form at high temperatures and pressures, corresponding to depths of at least 150 km for a low continental geothermal gradient (Boyd & Gurney 1986). Kimberlites and lamproites are two of the few rock types generated at these depths that are capable of bringing diamonds close to the surface. It must however be noted that only a few lamproites and kimberlites carry economic diamond deposits. De Beers reported that of the 2000 kimberlites tested over a period of 20 years only approximately 2.5% (50) carried grades that might be economic (Diamond Geology 2003).

Morphology of kimberlites and lamproites

Most diamond-bearing kimberlite and lamproites occur in pipe-like diatremes which are generally grouped in clusters and result from the eruption of volatile-rich magmas (Nixon 1995). Typically pipes are small (Fig. 17.3) with sizes of 1–10 ha (120–350 m diameter) common, although the largest pipes such as the Mwadui and Orapa kimberlites and the Argyle lamproite exceed 1 km^2. At depth pipes coalesce into dykes (Fig. 17.4) which tend to be thin (often <1 m) but elongated (up to tens of kilometers), often becoming more complex in shape. Where the magmas reached the paleo-surface they generally formed maars, volcanic craters filled with lacustrine sediments, known to reach up to a depth of 300 m. Other types of eruption are also known and fissure-style eruptions may generate kimberlitic tuffs that interfinger with other sediments, such as at Fort a la Corne, Saskatchewan (Nixon 1995). In the classic pipes of Kimberley, South Africa the crater facies sediments are absent and the pipes are carrot shaped with walls dipping inwards at approximately 80 degrees. The diatreme zone at Kimberley has a depth of 500 m from the present surface to its base with probably 1400 m of overlying sediments removed by erosion (Clement et al. 1986).

In nonglacial areas, kimberlites weather and oxidize from the "blue ground" and "hardebank" (resistant kimberlite), gradually becoming "yellow ground" towards the surface. This

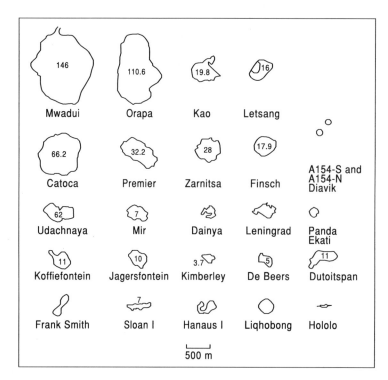

FIG. 17.3 Pipe sizes. The size of the Panda pipe is taken from Nowicki et al. (2003) and the sizes of the A154 pipes digitized from an Ikonos image. (Modified after Helmstaedt 1993.)

is colored by the weathering of serpentine to saponite and smectite and release of iron. In the upper levels of pipes kimberlite often forms an "agglomerate" of xenoliths of country rock and nodules that can be mistaken in exploration for a water-borne conglomerate (Erlich & Hausel 2002).

Kimberlite petrology

Kimberlite is ill-defined petrologically and often divided into two distinct groups. Mitchell (1991) defines Group I kimberlites as complex hybrid rocks consisting of minerals representing:

1 Fragments of upper mantle xenoliths (mainly peridotites and ecologies, some including diamond).
2 Minerals of the megacryst (large crystals) or discrete nodule suite (rounded anhedral grains of magnesian ilmenite, Cr-poor titanian pyrope garnet, olivine, Cr-poor clinopyroxene, phlogopite, enstatite, and Ti-poor chromite).
3 Primary phenocryst and groundmass minerals (olivine, phlogopite, perovskite, spinel, monticellite, apatite, calcite, primary serpentine).

Group II kimberlites have also been termed micaceous kimberlite or orangeite (Mitchell 1995). They consist mainly of rounded olivine macrocrysts in a matrix of phlogopite and diopside with rare spinel, perovskite, and calcite.

Lamproite geology

Lamproites are defined as potash- and magnesia-rich rocks consisting of one or more of the following phenocryst and/or groundmass phases: leucite, Ti-rich phlogopite, clinopyroxene, amphibole (typically Ti-rich potassic richterite) olivine, and sanidine. Accessories may include priderite, apatite, nepheline, spinel, perovskite, wadeite, and ilmenite. Xenoliths and xenocrysts including olivine, pyroxene, garnet and spinel may be present, as well as diamond.

Peridotite and eclogite xenocryst and xenolith petrology

Diamonds have been found associated with several different types of xenolith but the

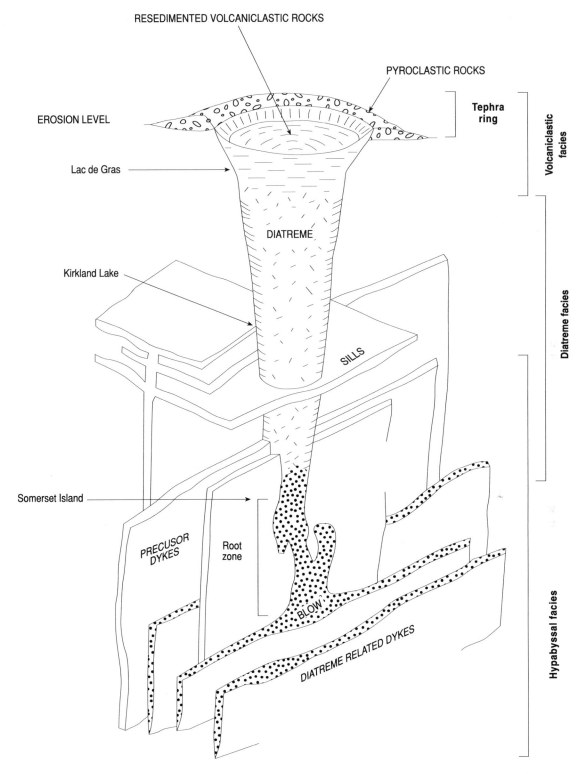

RESEDIMENTED VOLCANICLASTIC ROCKS

PYROCLASTIC ROCKS

EROSION LEVEL

Tephra ring

Volcaniclastic facies

Lac de Gras

DIATREME

Kirkland Lake

SILLS

Diatreme facies

Somerset Island

PRECUSOR DYKES

Root zone

BLOW

DIATREME RELATED DYKES

Hypabyssal facies

FIG. 17.4 Model of South African kimberlite pipe formation with Canadian deposits for comparison. The approximate vertical scale of the diatreme facies is 2000 m. (Taken from McClenaghan & Kjarsgaard 2001, after Mitchell 1986.)

two most important types are eclogite and peridotite:

1 *Eclogite*. These are the most common xenoliths in kimberlite and consist mainly of omphacite clinopyroxene and almandine–pyrope garnet. Diamond inclusions in this type of material are common and can reach the phenomenal grade of 18,000 ct t^{-1} at Orapa, Botswana (Robinson et al. 1984). This observation, coupled with the observation that diamonds in the interiors of eclogite xenoliths are pristine compared to the corroded nature of stones in kimberlite, suggests that diamonds could have been derived by disaggregation of eclogite during upward transport in kimberlite. The ultimate origin of eclogites is uncertain but they could result from melting of garnet–peridotite or as fragments of subducted crust (Helmstaedt 1993).

2 *Peridotite*. Although garnet–peridotite xenoliths are relatively rare in kimberlites they are thought to be the most common xenoliths based on the frequency of peridotite inclusions in diamond and the likelihood that it is the most common rock type in the mantle. The lack of peridotite xenoliths has been ascribed to their preferential disaggregation during transport in the kimberlite (Boyd & Nixon 1978). The peridotite xenoliths consist of olivine, clinopyroxene, orthopyroxene, and garnet. They are also enriched in diamond, although much less strongly than eclogite, with grades of up to 650 ct t^{-1} at the Finsch mine (summary of Helmstaedt 1993).

Dating of the diamond with eclogitic inclusions has given isotopic ages of 2700–990 Ma whereas diamonds with peridotitic inclusions gave dates of approximately 3300 Ma (Richardson et al. 1984). This could mean that the two different types of diamond could have formed by different processes although both are associated with the roots beneath Archean cratonic areas (Fig. 17.5). The host kimberlites and lamproite intrusives vary greatly in age,

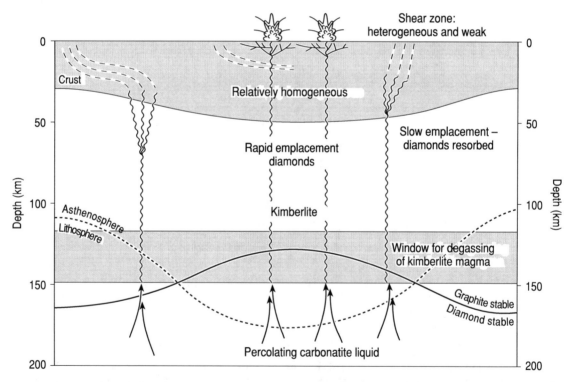

FIG. 17.5   Section through the crust beneath southern Africa. Note the deep keel beneath the craton that is associated with diamond generation. (From Vearncombe & Vearncombe 2002.)

e.g. pipes in southern Africa range from 60 to 1200 Ma, although many economic pipes are approximately 80–90 Ma in age.

Structural control of lamproites and kimberlites

Kimberlites and lamproites are restricted to continental intraplate settings with most kimberlites intruding Precambrian rocks (Fig. 17.6). Clifford (1966) was one of the first workers to recognize that economically viable diamond deposits in kimberlite are restricted to Precambrian cratons, particularly those of Archean age (Fig. 17.6). This appears in the literature as Clifford's rule and has guided much area selection. The exact time control has been a matter of much debate and some authors have divided Precambrian areas into archons and protons, with archons being older than 2.5 Ga (Levinson 1998, Hart 2002) and protons consisting of Proterozoic mobile belts adjacent

to archons (Janse & Sheahan 1995). However the diamond favorability criteria are by no means set in stone and many recent discoveries in southern Africa have been in the Limpopo Mobile Belt, which experienced a 2.0-Ga deformation event (Barton et al. 2003). Readers should also note that not all archons host economic kimberlites, e.g. South American cratons, are almost barren in spite of much exploration.

The relationship between pipe emplacement, particularly kimberlite emplacement, and structure has been much debated but no clear consensus has emerged. Vearncombe and Vearncombe (2002) presented a review of the distribution of kimberlites in southern Africa (Fig. 17.7). They conclude that kimberlite emplacement is related to corridors parallel to, but not within prominent shear zones and crustal faults (Fig. 17.7). Jaques and Milligan (2003) report that similar deep structures can be recognized in Australia based

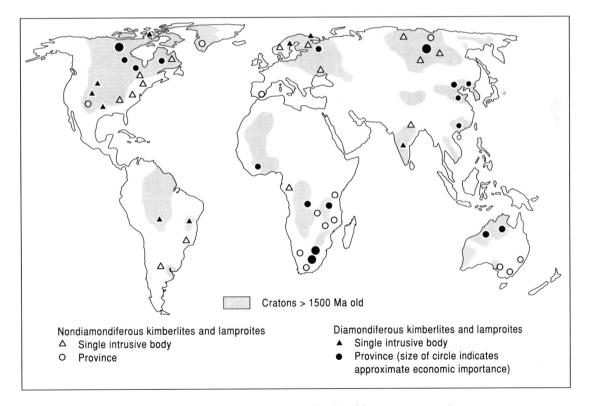

Cratons > 1500 Ma old

Nondiamondiferous kimberlites and lamproites
△ Single intrusive body
○ Province

Diamondiferous kimberlites and lamproites
▲ Single intrusive body
● Province (size of circle indicates approximate economic importance)

FIG. 17.6 World distribution of kimberlites and lamproite. (Updated from Evans 1993.)

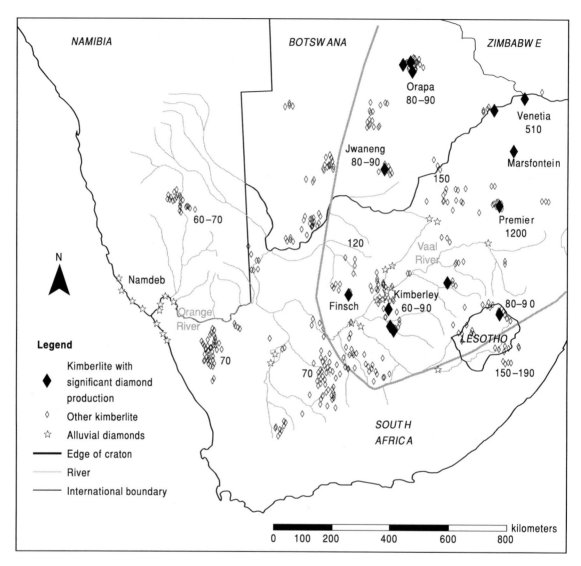

FIG. 17.7   Kimberlite and alluvial diamond distribution in southern Africa, numbers are ages of kimberlites in Ma (De Beers 2003b). (The kimberlite locations are modified from Vearncombe and Vearncombe 2002 and Lynn et al. 1998, as well as LANDSAT images.)

on geophysical data sets and suggest that the intrusions may be at the edge of different lithospheric blocks.

Other primary sources

Lithified placer deposits are the major other source of current diamond production. For example, Ghanaian alluvial production, which has been substantial although of poor quality, has a source in Precambrian sediments as has some alluvial production in Ivory Coast (Janse & Sheahan 1995). Although diamonds are known from high grade metamorphic massifs in Kazakhstan, Norway, and China the diamonds appear small and of poor quality (Nixon 1995). They, like the diamonds reported from ophiolites, appear the result of rapid uplift from the diamond stability field. Diamonds are also known to occur as the result of meteorite impact, particularly at Popigay in northern Russia where stones up to 2 mm in size are found in related alluvial deposits (Erlich & Hausel 2002).

Placer diamonds

Placer diamond production has, with notable exceptions, been mainly small scale. Even this small-scale alluvial production is however important in west African countries, notably Liberia, Sierra Leone, and Guinea. Major recent placer production has been derived from the alluvial deposits in Angola as well as the large alluvial and beach placers of southern Namibia and northwestern South Africa. These deposits are apparently derived by transport of diamonds in the paleo- and present Orange and Vaal drainage systems from the interior of South Africa as shown in Fig. 17.7 (Moore & Moore 2004). Longshore currents then reworked these diamonds, particularly concentrating them at the bedrock–coarse sediment interface in a beach environment. The onshore reserves in Namibia are largely exhausted and future mining will concentrate on offshore deposits. These diamonds are of particular interest as 95% are gems, and represent the largest gem diamond resource in the world. However the mining environment is difficult and some recent private ventures have not been successful.

17.1.2 Geochemistry

Although diamond exploration usually relies on a combination of methods, geochemical techniques (in the broad sense) have been very successful in detecting diamondiferous pipes. A number of distinctive, dense minerals (Table 17.2, Fig. 17.8) are associated with kimberlites and lamproites and it is these minerals that have been used to indicate the locations of pipes, hence their name indicator (or, in translation from Russian, satellite) minerals. The chemistry of some of these minerals also indicates their sources and transport history and, by comparison with known deposits, can be used to predict whether the source is diamondiferous as well as some indication of grade.

The indicator minerals used since the late nineteenth century and shown in Fig. 17.8 are garnet (both pyrope and eclogitic), chrome diopside, chromite, and picro (Mg-rich) ilmenite (Gurney & Zweistra 1995, Muggeridge 1995).

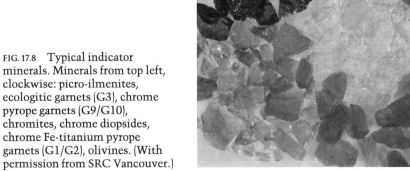

FIG. 17.8 Typical indicator minerals. Minerals from top left, clockwise: picro-ilmenites, ecologitic garnets (G3), chrome pyrope garnets (G9/G10), chromites, chrome diopsides, chrome Fe-titanium pyrope garnets (G1/G2), olivines. (With permission from SRC Vancouver.)

TABLE 17.2   Indicator mineral characteristics. (Simplified from McClenaghan and Kjarsgaard 2001.)

Mineral	Composition	Color	Magnetism	SG	Hardness	Diagnostic features	Main source rocks
Pyrope garnet	Mg Al silicate	Purple, red, orange	None to very weak	3.5	7.5	Anhedral, kelyphite rim, color	Kimberlite, lamproite, peridotite, lamprophyre
Mg-ilmenite	Mg Fe Ti oxide	Black	Medium to strong	4.5–5	5–6	Anhedral, rounded, glassy lustre, conchoidal fracture	Kimberlite
Cr-diopside	Ca Mg silicate	Emerald green	None to very weak	3.2–3.6	5–6	Anhedral, blocky, color	Kimberlite, lamproite, peridotite, other ultramafics
Cr-spinel	Mg Fe Cr Al oxide	Black, reddish brown	Weak to medium	4.3–4.6	5.5	Octahedral to irregular shape, reddish brown on edges	Lamproite, kimberlite, carbonatite, various basic and ultramafic rocks
Forsteritic olivine	Mg silicate	Pale yellow to green	None to very weak	3.2–3.3	6–7	Color, irregular crystal apices	Peridotite, lamproite, kimberlite, carbonatite
Diamond	C	Colorless, pale colors	None	3.5	10	Admantine lustre, crystal form	Kimberlite, lamproite, certain lamprophyres, metamorphic rocks

Typical abundances in the host kimberlite are hard to assess, but Erlich and Hausel (2002) estimate ilmenite as between 1 and 10 wt% of some kimberlites. At the Premier pipe, ilmenite was estimated at 0.5% of kimberlite, and thought to be the highest of any kimberlite mined by De Beers in South Africa (Bartlett 1994). The ilmenite–garnet–diopside ratio at the Premier mine was estimated at 100:1:1. Lamproites have very different indicator mineral assemblages: chromite, Nb-rich rutile, ilmenite, zircon, tourmaline, and garnet have been reported as most useful, although garnets are much less abundant than in kimberlites (Erlich & Hausel 2002).

Heavy mineral samples are generally strongly diluted with nonkimberlitic material, and finding the signal of a kimberlite on a regional basis requires detecting very small numbers of indicator minerals; sometimes a single grain may be significant. Correct sampling and processing of samples is therefore very important. Although orientation studies are required to optimize sample size, generally 10–40 kg is used, with some lamproite searches requiring larger samples (Muggeridge 1995).

The size of indicator minerals is generally in the range 2 mm to 0.3 mm and samples are screened in the field through a 2-mm screen and then bagged. Many exploration programs in the past have panned or jigged samples in the field but this approach has been largely replaced by laboratory separation of the whole sample. Laboratory processing is complex and expensive with typical processing costs of several hundred dollars per sample. The key phase in processing is generation of a heavy mineral concentrate using either a shaking table or dense media separation. The heavy mineral concentrate is then split into ferromagnetic, paramagnetic, and nonmagnetic fractions using a magnetic separator, followed by size fractionation if any of the fractions are heavier than 9 g. Each fraction is then visually examined under a binocular microscope and any indicator minerals counted and mounted on stubs for possible examination by a scanning electron microscope or electron microprobe. The binocular examination is time consuming and requires training to recognize the minerals, as well as immunity to boredom once the training is complete. The number of ilmenite or garnet grains can then be plotted to highlight areas for follow-up (see Fig. 8.16).

Indicator mineral chemistry

The study of the chemistry of indicator minerals requires the use of a scanning electron microscope or electron microprobe, typically at a cost of $US10 per analysis. The aim is to identify the diamond potential of the intrusive from which the minerals have been derived by answering three questions (Fipke et al. 1995):

1 How much diamond-bearing peridotite and eclogite are present?
2 What was the diamond grade of the source rocks?
3 How well were the diamonds preserved during transport to the surface?

The first question can be addressed by examining the chemistry of garnets and chromites to assess the contribution of disaggregated harzburgite (olivine and orthopyroxene) material. Plotting of CaO against Cr_2O_3 was used by Gurney (1984) to define a line to the left of which 85% of known Cr-pyrope diamond inclusions plot (Fig. 17.9). Garnets plotting to the left of the line are termed G10 garnets and those to the right G9 after a garnet classification of Dawson and Stephens (1975). Lower calcium, higher chromium garnets are associated with higher diamond contents. The line at 2% Cr_2O_3 was used to discriminate between garnets of peridotitic and low-Ca garnets which are probably of eclogitic or megacrystic origin. A plot of MgO against Cr_2O_3 for chromite defines a highly restricted Cr-rich field for diamond-related chromite. The diamond potential of eclogitic garnets can be estimated from a plot of Na_2O against TiO_2. In general the Na_2O content of garnets from eclogitic diamond inclusions is greater than 0.07% Na_2O.

The preservation of diamonds during transport from the mantle to the surface is estimated by attempting to assess the oxygen fugacity of the magma. It had been observed that the diamond content of kimberlite correlated with trends in picritic ilmenite, particularly Fe_2O_3 against MgO, with low Mg representing highly oxidized kimberlite (Gurney & Zweistra 1995). It should be noted that the diamond potential can only be downgraded by estimates of the diamond preservation potential.

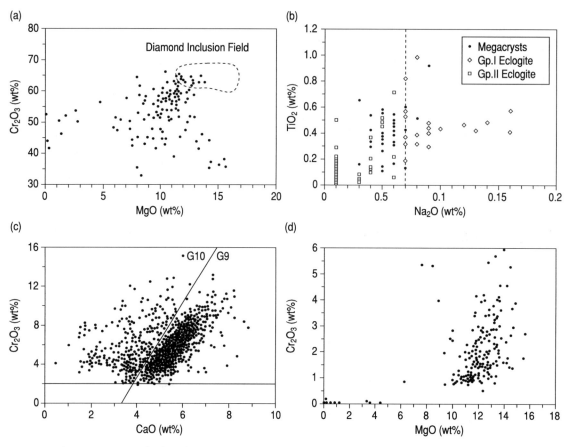

FIG. 17.9   Electron microprobe data from indicator minerals collected during the Ekati discovery. For discussion see text. (From Fipke et al. 1995.)

Conventional exploration geochemistry

Conventional geochemistry has also been used where there is a geochemical contrast between the kimberlitic material and their host rocks. Elements of use include Ba, Ni. Cr, V, Co, Mg, K, partial Sr, Nb, Ta, P, and light rare earth elements (McClenaghan & Kjarsgaard 2001).

17.1.3   Geophysics

Geophysical techniques are widely used to detect kimberlite and lamproite pipes, particularly at the reconnaissance stage. The techniques used will depend on the climatic and weathering history, the country rock into which the pipes have been intruded, and the age of the pipes. In general the magnetic, electrical (conductivity and induced polarisation), density, and seismic velocity properties of a pipe are significantly different from the country rock.

Airborne magnetic and electromagnetic (EM) surveys have been widely used for direct kimberlite detection, although responses can be highly variable with some pipes in a cluster detectable and others not (Macnae 1995, Smith & Fountain 2003). In addition, close flight line spacing is needed as the targets are relatively small. For example, in northwest Canada, regional government surveys flown at 800 m line spacing and 300 m terrain clearance detected only one pipe, whereas company surveys flown at 125 m line spacing and 30 m terrain clearance were able to detect more than 100 pipes (section 17.2) (St. Pierre 1999). Most pipes are present as discrete bull's eye anomalies although some may be associated with linear features. The formation of clay minerals during weathering of kimberlite diatremes or crater facies kimberlite often reduces the magnetic response but produces a very conductive cover to pipes, which makes them amenable to detection by EM techniques (Macnae 1995). A good example is the Point Lake pipe discussed in detail in section 17.2.1 (Smith et al. 1996). In other areas such as Western Australia kimberlite can be more resistive than conductive soils. Airborne gravity gradiometers, such as the Falcon system of BHP Billiton, are being widely used to search for pipes, especially where the pipes are covered by sand, as in parts of Botswana. This technology will detect large (>200 m diameter) pipes using a 100 m line spacing, and in test flights over the Ekati area about 55% of 136 known pipes were detected (Liu et al. 2001, Falcon 2003). In general kimberlite is less dense than country rock. Radiometrics have also been used where intrusives crop out and where there is a contrast with country rock.

Ground geophysics is also widely used to define and refine the source of the airborne anomalies. The individual phases of intrusion can also often be defined, which considerably aids in the definition of grade (Macnae 1995). In addition to the methods mentioned above, seismic surveys have been used both in Yakutia (Macnae 1995) and at Snap Lake in northern Canada (Kirkley et al. 2003), although their expense has restricted their use to definition of pipes and dyke geometry.

17.1.4 Remote sensing

The circular nature of many pipes and their cross-cutting relationship with local stratigraphy make their outcrops relatively easy to recognize on air photography and high resolution satellite imagery. The distinctive mineralogical composition of pipes can also be detected using spectral measurements in the shortwave red infrared wavelengths for minerals such as phlogopite and serpentine. However Kruse and Boardman (2000) reported great difficulty in interpreting Hymap airborne scanner data over kimberlites in Wyoming due to the intense weathering and very limited outcrop of the distinctive minerals.

17.1.5 Evaluation

The relative scarcity of diamonds in hardrock deposits and the great variability of value of individual diamonds cause many problems in evaluation and valuation when a pipe has been found. In some low grade primary, and many alluvial, deposits the recovery of large stones is crucial to viability. Rombouts (2003), who has provided some of the few published discussions of the problem, suggests that is not uncommon for a single stone to represent 30% of the entire value of a parcel of diamonds. During commercial mining only stones larger than 1 mm (macrodiamonds) are generally recovered, although some Russian mines recover down to 0.2 mm for use as diamond powder.

Evaluation begins once a pipe has been discovered and consists of a number of stages which improve confidence in the grade and value of the diamonds. These stages may take considerable time, often years. In addition as the pipes occur in clusters there are often a number of targets to test and it is usually not clear which pipe in the cluster is likely to be of most economic interest. Within pipes with a number of differing kimberlite phases, the phases are likely to have different grades that need to be identified. Thomas (2003) divided the evaluation phases into four:

1 Initial delineation drillhole.
2 Delineation drilling: determine the size and geometry of the pipe and whether there are multiple intrusive phases.
3 Mini bulk sampling: large diameter drilling (or pitting) to recover a sample that is processed through commercial dense medium plant to recover diamonds larger than 1.0 mm.
4 Underground or surface bulk sampling: recover 5–10,000 ct by dense medium separation for sorting and valuation.

Initial confirmation of the kimberlite is normally by core drilling but analysis of the core, as well as that of the delineation stage (2 above), can provide information on the indicator assemblage and possible grade. The core is dissolved either by caustic fusion or hydrofluoric acid and minerals separated at a

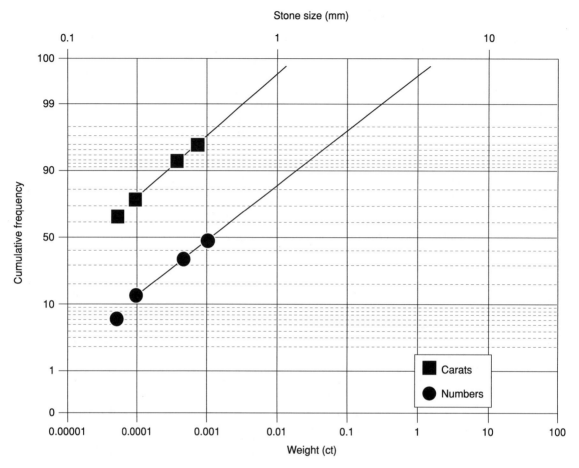

FIG. 17.10   Use of microdiamonds in predicting overall grade using a logarithmic (x axis) probability (y axis) plot. The sample of 52 kg contained 58 stones with a total mass of 0.0169 ct. Extrapolation gives a grade of 0.06 ct t^{-1} above 0.01^{-1}ct or 1 mm. (Taken from Rombouts 2003.)

cost of approximately $US350 per 10 kg of sample. The grade of the macrodiamonds can be estimated from the microdiamond (<1 mm) distribution, as the central part of the distribution is log-normal (Fig. 17.10). Poor recovery of very fine stones and the chance occurrence of large stones may cause variation from log-normality at the ends of the graph. A log–log plot of stone size against cumulative grade will provide an estimate of overall grade. Typically stage 1 will cost about $US10,000. Stage 3 will take samples of 200 t or more. In Canada this will comprise about 10 large diameter drillholes depending on the grade. The fourth stage in which representative diamonds are re-

quired for commercial valuation is much more expensive and may reach the order of $US10M.

17.1.6   History of discovery

Kimberley, South Africa and the discovery of kimberlite pipes

Diamonds have been known since at least the fourth century BC but it appears that India was virtually the world's sole source of diamonds until 1730, when diamonds were discovered in Brazil (Levinson 1998). All the Indian production was considered alluvial, although one area was an unrecognized primary deposit.

Similarly, all the diamonds produced in Brazil from 1730 to 1870 were also alluvial mainly from Minas Gerais and Bahia states. Total production from 1730 to 1891 in Brazil has been estimated at 13 Mct (Levinson 1998).

The breakthrough in recognition of a bedrock source for diamonds came in the 1870s in South Africa. Alluvial diamonds were followed from their discovery on the banks of the Orange River, northwest of Hopetown (Fig. 17.7) to a number of sources in the Kimberley area. These sources were then mined vertically, first in yellow ground and then into blue ground. Individual claims were amalgamated into the major operations at the Kimberley and De Beers diggings by Cecil Rhodes, an entrepreneur who founded the De Beers company, which eventually controlled mining of these two main pipes (Hart 2002).

The nature of the blue ground was first recognized by Emil Cohen as a volcanic rock in 1872. The term pipe was first applied to the cylindrical form of the blue ground in 1874 and kimberlite was first applied to the diamondiferous rock type in 1897. Three of the pipes at Bultfontein, Dutoitspan, and Wesselton mines are still in underground production, although output is relatively small (Table 17.3). The De Beers and Kimberley mines are now closed although the latter now forms part of a well-known museum.

Other African pipes and alluvials

The success of the Kimberley mines led to the search for other diamond deposits in South Africa. The most notable discovery was in 1902 by a former bricklayer, Thomas Cullinan, who found the Premier pipe, approximately 100 km north of Johannesburg, that was three time larger at surface than the largest Kimberley pipe. In 1905 the pipe produced the largest known diamond, the Cullinan of 3106 ct, the main part of which is now in the British crown jewels. De Beers eventually bought control of the Cullinan company in 1914 and production continues to the present in an underground mine. Although the grade is low the pipe produces a considerable number of large (>100 ct) stones.

The next major diamond development in southern Africa was not of a kimberlite pipe but of alluvial diamonds on the coastline of the Atlantic Ocean to the south and north of the Orange River. Although the stones are relatively small they are mainly gems and were initially discovered in sand dunes in, the then, German South West Africa by a railway worker. The deposits were taken over by South Africa after the World War I, the area closed off to outsiders, and sold to a company controlled by Anglo American Corporation. These concessions were then merged with the mines of De Beers in 1929 and are currently operated by Namdeb, a joint venture between De Beers and the Namibian government.

As a result of the abundant supply of diamonds there was relatively little further systematic exploration in Africa until the 1950s. However, in the early 1940s an ex De Beers employee, John Williamson, discovered a major pipe at Mwadui in what is now Tanzania. Looking for the source of previously known alluvial diamonds, he followed ilmenite indicator minerals until he eventually panned a diamond (Krajick 2001).

Starting in the mid 1950s De Beers began large scale exploration in Africa, based around indicator mineral sampling, backed up by air photography and geophysics. The most significant discoveries resulting from this program were made in central Botswana. De Beers developed techniques to sample the surface deflation layer for indicator minerals that are useful even in areas of thick sand cover. The heavy minerals are thought to be brought to the surface (at least partly) by the action of termites (bioturbation) and concentrated by wind action. De Beers geologists collected samples, usually of the –2 mm +1 mm fraction, at 10- to 15-m intervals that were then aggregated into 8-km-long sections.

Although indicator minerals and alluvial diamonds were known in Botswana in the early 1960s, no kimberlites were known (Baldock et al. 1976). A small occurrence of alluvial diamonds was traced westwards from the far northeast corner of Botswana but the dispersion trail of diamonds was cut off to the west. Working on the premise that the trail had been cut off by Pliocene upwarping, Gavin Lamont moved the sampling further west. This sampling confirmed the occurrence of indicator minerals and located a high amplitude

TABLE 17.3 Pipe sizes and grade. (Mainly after Janse & Sheahan 1995, Jennings 1995, De Beers 2001.)

Pipe	Country	Surface area (ha)	Year discovered	Total production (Mct or reserve)	Grade (cpht)	$ ct^{-1}	$ t ore^{-1}
Kimberley, Kimberley	South Africa	3.7	1871	32.7	100	200	200
De Beers, Kimberley	South Africa	5.1	1871	36.4	90	75	65
Bultfontein, Kimberley	South Africa	9.7	1869	36.2	40	75	30
Dutoitspan, Kimberley	South Africa	10.8	1869	21.3	20	75	15
Finsch	South Africa	17.9	1960	93 (to 1995)	80	40	32
Venetia	South Africa	12.7	1980	~35 (115)	136	100	136
Jwaneng	Botswana	45	1972	171 (413)	147	110	161.7
Orapa	Botswana	110.6	1967	146 (320)	68	54	34
Mwadui	Tanzania	146	1940		6	150	9
Marsfontein	South Africa		1997	1.83	169	165	278
Argyle	Australia	46	1979	541	549	9	49.4
Mir	Russia	7	1955				
Internationalya	Russia	1.7			400	120	480
Udachnaya	Russia	52	1955	100	100	100	
Lomonosov	Russia	118			75	80	60
Grib	Russia	16	1996	R 67	69	79	55
Panda Ekati	Canada	3	1993	R 12.6 (pit)	109	130	142
Koala Ekati	Canada		1993	R 14.6 (pit)	76	122	93
Fox Ekati	Canada		1993	R 17.7	40	125	50
Misery Ekati	Canada		1993	R 12.6	426	26	111
A154-S Diavik	Canada		1994	R 10.8	467	59	276
A154-N Diavik	Canada		1994	R 2.7	289	33	95
A418 Diavik	Canada		1994	R 8.3	388	53	206
A21 Diavik	Canada		1994	R 3.7	289	36	104

cpht, carats per hundred tonnes; R, pre-mining reserve.

FIG. 17.11   Aerial view of Kimberley Pipe. The view shows the city of Kimberley to the east of the "Big Hole." The pit in the background is the De Beers pipe.

indicator mineral anomaly associated with a circular structure on air photography. The first pits in 1967 returned 12 diamonds within 45 minutes of running a rotary pan test. This led to the establishment of the major mine at Orapa, accompanied by the location of the nearby smaller Leklhatkane pipes in 1971. De Beers then moved to the south under thicker (up to 100 m) sand discovering the first pipe in the Jwaneng area in 1972 and the Jwaneng pipe (DK2 of Fig. 8.16) in 1973 (Lock 1987). The Jwaneng pipe is, and has been, one of the world's most profitable mines.

South Africa has also been the target of much systematic exploration. The major discoveries include the Finsch mine, found by a prospector in the 1950s, and Venetia, discovered by De Beers in 1980. Smaller pipes and fissures, such as Marsfontein, make attractive targets for smaller companies (Davies 2000). The Marsfontein pipe paid back its investment in 1 week!

Russian pipes

Until 1991, little was publicly known of diamond exploration or mining in the USSR. It is now apparent that the major producing area in central Siberia was discovered as the result of a systematic search organized after World War II (Erlich & Hausel 2002). Although diamonds were known from the Urals placer gold workings since 1829, the discovery resulted from the use of geological mapping and heavy mineral sampling (Fig. 17.12). Follow-up of diamond occurrences in black sands using the occurrence of pyrope garnets and mapping eventually led to the discovery of the Zarnitsa pipe in 1954. Among the pipes discovered shortly after were the high grade Mir (1955), Udachnaya (1957), and Aikhal (1960) pipes (Table 17.3). In the 1970s a further cluster of pipes was discovered in the area to the east of the White Sea, near the city of Akhangl'sk. The pipes include the

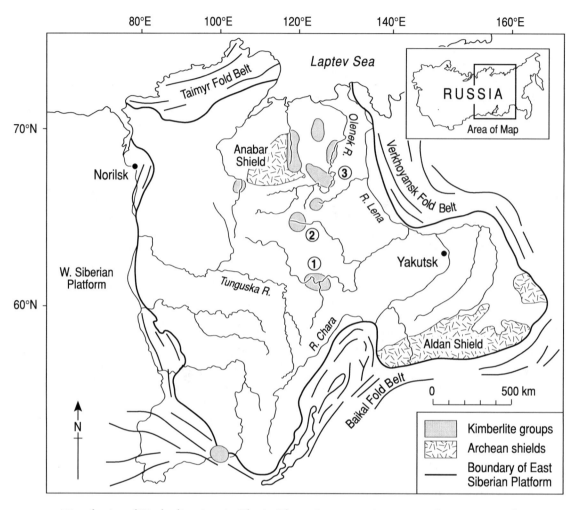

FIG. 17.12 Distribution of Kimberlite pipes in Siberia. The major economic pipes are the Mir pipe in cluster 1, Udachnaya in cluster 2. The Leningrad pipe is in cluster 3. (After Pearson et al. 1997.)

Lomonosov deposit, consisting of several pipes, and the Grib pipe, which was discovered by a non-Russian junior company in 1996.

Argyle

Australia was an obvious target to search for diamonds with its large Archean cratons, particularly the Yilgarn and Pilbara shield area in Western Australia as well as the Gawler craton in South Australia. However until 1970 most of the limited production of diamonds (200,000 ct) had been recovered from alluvial deposits in the Paleozoic fold belt area of Western Australia (Smith et al. 1990). The

north of western Australia, including the central older Kimberley block, was targeted by junior exploration companies based on the southern African model, the known occurrence of ultra-potassic intrusives, which were thought analogous to the barren alkalic rocks off craton in southern Africa, as well as a good drainage network. The rivers allowed the use of sediment sampling by helicopter for indicator minerals to define follow-up areas (Fig. 17.13). Follow-up of indicator minerals in the north Kimberley area showed barren kimberlite dykes with pyrope and picro–ilmenite indicator assemblage but in the west Kimberley area, a breccia pipe of lamproitic affinity was

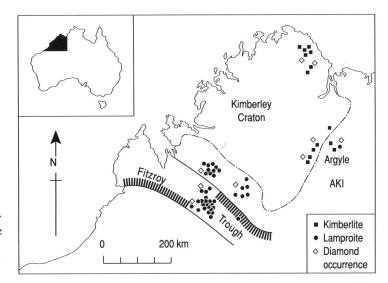

FIG. 17.13 Lamproites and Kimberlite distribution: Kimberley area of Western Australia. Note the occurrence of the Argyle mine off the craton. (From Evans 1993, after Atkinson et al. 1984.)

found to contain diamonds associated with chromite as an indicator mineral. This was the first time that diamonds had been reported from lamproite. This discovery led to the discovery of the (then) uneconomic Ellendale cluster of pipes. Further regional sampling by CRA, who had farmed into the project, based on 40-kg samples returned two diamonds in a sample. This was followed 20 km upstream to locate the Argyle AK1 pipe and some associated alluvial deposits. The grade of the deposit is very high with 75 Mt proven and probable reserves at approximately 6700 carats per hundred tonnes (cpht). However the quality of diamonds is low: only 5% gem, 40% near gem, and 55% industrial stones with an average price in 1982 of $US11 ct^{-1}.

The Argyle discovery was important in that it was the first significant discovery of a diamondiferous nonkimberlitic pipe and this pipe was in a mobile zone, if Precambrian in age.

17.2 CANADIAN DIAMOND DISCOVERIES: EKATI AND DIAVIK

Although Canada has the largest area of Archean shield in the world and is thus a favorable area, it was not until 1991 that economic diamond deposits were discovered. A summary of early work by Brummer (1978)

shows that diamonds were known from eastern Canada as early as 1920. A diamond was found together with indicator minerals in till and thin kimberlitic dykes discovered in an up-ice direction in the Kirkland Lake gold mining area of Ontario, but further sampling was discouraging. De Beers also has a long history of exploration in Canada and investigated kimberlite in Arctic Canada on Somerset Island in 1974–76. In a joint venture with Cominco, at least 19 pipes were discovered and sampled, although grades were low.

17.2.1 Ekati discovery

Although major companies had been involved in the diamond search in Canada as detailed by Brummer, a junior company, led by two geologists, Chuck Fipke and Stewart Blusson, located the remote Lac de Gras target area in 1988. The history of the Ekati discovery is well documented in the very readable book by Krajick (2001) and the career of Fipke described in a book by Frolick (1999).

Fipke set up a consulting mineralogical company after working as an exploration geologist for Kennecott, Cominco and JCI in southeast Asia, Brazil, and southern Africa for 7 years. During his stay in South Africa he had visited Kimberley and the Finsch mine and became interested in using large heavy mineral samples

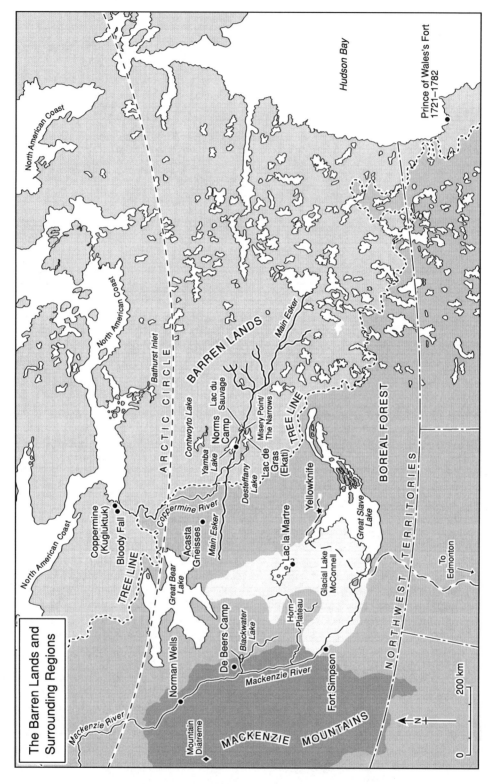

FIG. 17.14   Regional setting, Lac de Gras exploration. (After Krajick 2001.)

to target areas for follow-up. This was the technology he developed for his own business that he set up in 1977 in his home town in southern British Columbia. One of his clients was Superior Oil, which was engaged in a diamond search in the USA headed by exploration manager, Hugo Dummett, a former De Beers geologist. Blusson's career was initially more academic. He had worked for the Geological Survey of Canada but then became involved in exploration as a consultant and on his own account. He had the added advantage of being a qualified helicopter pilot.

Fipke began his Canadian search as a junior partner with Superior Oil, in the Rocky Mountains area near Kimberley British Columbia in a company called C.F. Minerals. Cominco and other companies had investigated a number of small nondiamondiferous kimberlite pipes discovered in 1976 during logging operations (Fig. 17.14). The Superior-funded search located a number of pipes, including the Jack pipe, north of Kimberley, based on indicator mineral analysis, particularly chrome diopsides. The Jack Pipe samples however contained only one diamond and few garnets. In spite of this lack of success, the joint venture moved north and examined a pipe named Mountain Diatreme that was discovered by geology summer students exploring for gold (Fig. 17.14). When they were examining this pipe they learned from a helicopter pilot that De Beers (exploring under the name Diapros in Canada) had a large exploration program 250 km to the east in the Blackwater River area. They then (1982) staked an area adjacent to Diapros's claims after finding favorable indicator minerals, particularly G10 garnets. However Diapros failed to find any kimberlites and withdrew, as the De Beers board was unwilling to pursue the dispersion trail to the more remote areas to the east. At this stage (1981) Superior withdrew from diamond exploration, due to changes in senior management, and terminated the joint venture leaving the claims to C.F. Minerals.

In order to finance further exploration, Fipke and Blusson floated a company, Dia Met Minerals, raising funds from individuals in Fipke's home town and from corporate investors in Vancouver. Convinced that the source of the indicator minerals was to the east, they confirmed the eastern extension of the glacial train

associated with a major esker by sampling around Lac de Marte. In 1984 they eventually raised enough finance for further regional-scale heavy mineral sampling to establish the source of the dispersion train, which eventually took place in 1985. By mid 1988 processing of the sampling was complete and a pattern became clear. The area east of Lac de Gras (Fig. 17.15) was essentially barren and one sample G71 from an esker beach contained a huge number of indicator minerals: 100 ilmenites, 2376 chrome diopsides, and 7205 pyrope garnet grains relative to the usual 10 or 20 of the regional dispersion train. The area to the north of Lac de Gras was obviously the area of interest (Fig. 17.15), although the complex glacial history made it difficult to tie down the location of any other pipes. This lack of definition meant that a large land position was needed so that any pipes could be located and tested. The major problem was raising the finance to do this, each claim of 2582.5 acres would cost $C1 per acre to register and have to be physically staked. Staking required putting a wooden post in the ground every 458 m (1500 ft) and marking it, an intensive operation requiring float plane or helicopter support. By the end of the summer season they had staked 385,000 acres in the name of Norm's Manufacturing Company to avoid raising suspicion at the Yellowknife Mining Recorder's office. In April 1990, as soon as conditions permitted, Fipke staked further ground to the south and recognized that a lake surrounded by steep cliffs at a place called Misery Point could be the location of a pipe. While collecting a till sample of a beach at this location, Fipke's son Mark found a pea-sized chrome diopside that suggested that a kimberlite pipe was very close, probably under the lake.

The next stage was to find a joint venture partner to undertake detailed exploration, test any pipes, and most of all maintain the expenditure requirements on the large land position ($C2 per acre per year, a total of $C770,000). They approached Dummett, by then North American exploration manager for BHP Minerals and agreed a joint venture. He was presented with data on indicator mineral abundances and results of an analysis by consultant John Gurney. They showed that the garnets and ilmenites suggested very high

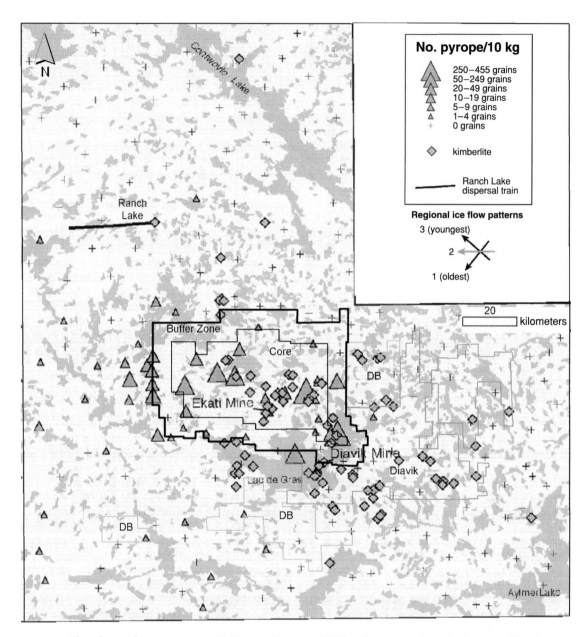

FIG. 17.15   Abundance of Cr pyrope in 0.25–0.5 mm fraction of 10 kg till samples: Lac de Gras area. The data are from the Geological Survey of Canada but are similar to that Diamet obtained from their regional sampling. The broad halo around Lac de Gras is clear. The narrow dispersion train from a solitary pipe at Ranch Lake to the north of the main esker is not detected. Some licence blocks are shown for comparison. DB, De Beers Canada Inc; Diavik, Diavik Diamond Mines and joint ventures. (Garnet data from McClenaghan & Kjarsgaard 2001.)

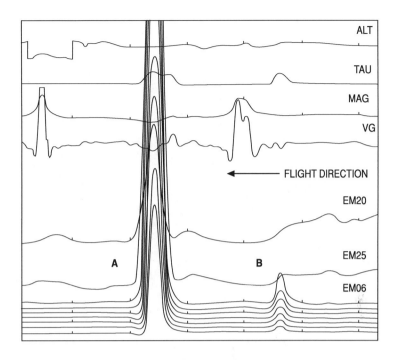

ALT

TAU

MAG

VG

◀—— FLIGHT DIRECTION

EM20

EM25

EM06

A   B

FIG. 17.16   Point Lake Geophysics. Individual flight line across the Point Lake pipe. (Courtesy Fugro.)

diamond potential (Fig. 17.9). A minimum expenditure of $US2M was established with BHP agreeing to fund exploration and development up to $C500M in exchange for 51% of any mine, Dia Met retained 29% and Fipke and Blusson each were to personally get 10% of any mine. The new joint venture engaged in staking, in an attempt to protect its position and built a buffer zone of 408,000 acres around the original claim block (Fig. 17.15). Geophysical methods were then used to locate any pipes. Initially magnetic and EM methods were used at the Point Lake prospect and confirmed the circular anomaly (Fig. 17.16). This was tested in September 1991 and the first hole cut kimberlite from 455–920 ft. Caustic dissolution of parts of the initial intersection returned 81 small diamonds from 59 kg. The stock price of Dia Met Minerals began to rise from $C0.68 to $C1, and in order to forestall any accusation of insider trading on November 7, 1991 the following press release was made: "The following information has been released to the Vancouver Stock Exchange, Canada: The BHP–Dia Met diamond exploration joint venture, in Canada's Northwest Territories, announces that corehole PL 91-1 at Point Lake intersected kimberlite from 455 feet to the

end of the hole at 920 feet. A 59-kg sample of the kimberlite yielded 81 small diamonds, all measuring less than 2 mm in diameter. Some of the diamonds are gem quality."

Although it was not clear whether there was a deposit or that it was economic, the release was enough to trigger the greatest staking rush in Canadian history. By 1994, more than 40M acres had been staked. Initial large positions were taken by De Beers (now exploring as Monopros) which took a land immediately to the west of the BHP joint venture and a syndicate part financed by Aber Resources which took a position to the south (Fig. 17.17).

The BHP–Dia Met joint venture (eventually renamed Ekati from the aboriginal name for Lac de Gras) systematically explored the claim blocks using airborne geophysics and heavy mineral sampling in an attempt to define further targets for drilling. At Point Lake the initial diamond drilling was followed up by large diameter drilling in early 1992 to take a 160-t sample for processing. The initial sample confirmed that the pipe was diamondiferous, 101 ct diamonds were recovered of which 25% were gems to give an average grade of 56 cpht. An airborne geophysical survey in the spring of 1992 located a number of pipes,

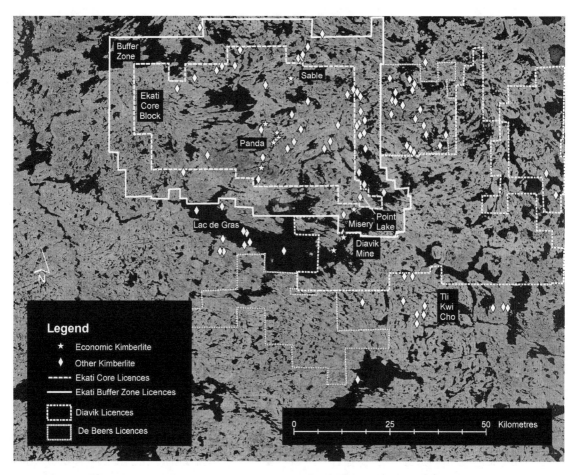

FIG. 17.17   Landsat image: Lac de Gras Area. The licence blocks and kimberlite locations are shown. (Ekati kimberlites from Nowicki et al. 2003, licence blocks and other kimberlites as at 2003.)

including the higher grade Koala pipe. This pipe, along with the Fox and Panda pipes, was bulk sampled by drilling and underground development between 1993 and 1995 (Fig. 17.18). The underground sampling collected bulk samples of 3500–8200 t. The Panda pipe returned 3244 ct from 3402 t with an average price of \$US130 ct^{-1} to give a in-ore value of \$US124 t^{-1}. Similarly the Koala pipe returned a grade of 95 cpht at a price of \$US122 ct^{-1} (\$US116 t^{-1}) from a 1550-t sample and the large Fox pipe 2199 ct from 8223 t at price of 125 ct^{-1} for \$US34 t^{-1}. By comparison with other deposits worldwide (Table 17.3) it is clear that the Ekati pipes are significant, especially if they are mined as a group.

One of the major problems that had discouraged potential explorers from the start was the pristine Arctic environment. The issue of development therefore needed to be addressed and BHP started a \$C14M environmental baseline study. Although the population of the Northwest Territories is very low, the aboriginal and Inuit inhabitants are highly dependent on the caribou and fishing. The claim block contains over 8000 lakes so water quality is important. Climatically the area is extreme; the average temperature is −11°C and winter temperatures can go down to −70°C including wind chill (Fig. 17.18). After application for a mining review in 1995, a panel was established by the federal government and made a

FIG. 17.18   Winter drilling. (Courtesy Diavik Diamond Mines Inc.)

series of recommendations that led to the granting of conditional approval of the project in August 1996. One of the major conditions was the involvement of the three different Aboriginal groups and one Inuit group. Another major recommendation was that commitments made by BHP were to be formalized and that an independent monitoring agency should be established to assess the mine performance (Independent Environmental Monitoring Agency 1998). A significant issue was the impact of mining and processing on water quality. The major problem was that tailings from the Fox pipe would have a pH of 9.5 and might impact on fish. By contrast, schist bedrock from the Misery pipe will be acid producing and will need to be neutralized. The only major water quality problem was the impact of sewage waste on the lake in which it was disposed. Another concern was the impact of mining on caribou migration although the

dumps have been tailored to allow their free movement.

Project construction was difficult as access for bulk items is restricted to shipment on a 475-km-long ice road usable for 10 weeks in winter. Any other items and personnel have to be flown in from Yellowknife. The mine has a workforce of 760 employees with accommodation for a maximum of 600 people. Commercial production began in October 1998, only 1.5 years after the beginning of construction. The overall mine plan included six pipes (Table 17.3) with a further two at the permitting stage. The initial mining was an open pit on the high value Panda pipe. Subsequently mining started at the Koala pit, the very small Koala North pit (Fig. 17.19), and the larger, but more distant, Misery open pit. Underground production is planned from the Panda, Koala North, and Koala pipes, with the small Koala North deposit as a test for underground mining. The Fox

FIG. 17.19   View of the Koala, Koala North (center), and Panda (nearest) pipes, Ekati in 2002. (Courtesy BHP Billiton Diamonds.)

and Beartooth pipes will be mined next followed by the Sable and Pigeon deposits. Total resources at the end of the feasibility study were estimated at 85 Mt with a grade of 1.09 ct t^{-1} and a price of $US84 ct^{-1}. In 2001 BHP (by then BHP Billiton) bought Dia Met Minerals for $US436M and currently (2004) holds 80% of the Ekati joint venture.

By 2003 BHP had discovered 150 pipes on the Ekati property (Nowicki et al. 2003). All pipes have been core drilled, usually with a single hole, and the core samples used for indicator or micro-diamond analysis. These analyses are then used to prioritize pipes for initial bulk sampling by reverse circulation drilling. Typically this produces samples of 50–200 t that are processed on site in a dense media separation plant. Diamonds were originally marketed partly (35%) through the Diamond Trading Company but are now sold totally in the Antwerp market.

17.2.2   Diavik

The most significant other project in northern Canada is at Diavik, immediately south of the Ekati buffer zone (Fig. 17.20). Fipke and Blusson were not alone in their interest in the area (see Hart 2002). Another geologist, Chris Jennings, had been active in the Lac de Gras area for a year before the discovery announcement at Ekati. This activity was based on his diamond exploration experiences in Botswana and Canada with Superior and its then Canadian associate Falconbridge, and subsequently he had also explored the indicator train east of the Blackwater Lake area with Selco. A geologist working on his instructions had already found indicators and 0.1-ct diamonds on the shore of Lac de Gras near, but outside, the Dia Met–BHP claim blocks. However, the company for whom he was working at the time was focused on gold and not interested in diamonds.

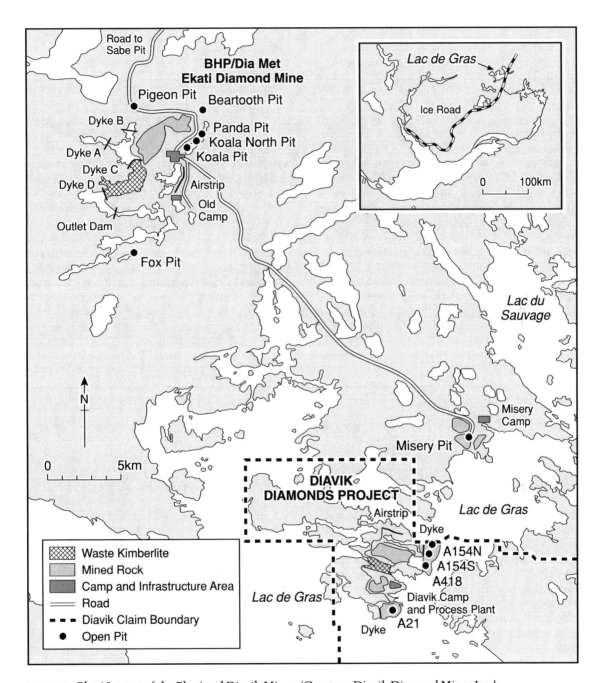

FIG. 17.20   Ekati Layout of the Ekati and Diavik Mines. (Courtesy Diavik Diamond Mines Inc.)

He was approached by a group of explorers led by longtime Yellowknife prospector, Grenville Thomas and Robert Gannicott to stake ground immediately after the Dia Met–BHP press release of November 1991. The syndicate (known as Aber Resources) staked the claims in the winter season. As Monopros had immediately staked the area adjoining Aber Resources' first choice northeast corner of the Dia Met–BHP block, the Aber syndicate started staking at its second choice area at the southeast corner. Aber Resources was, like Dia Met, dependent on larger companies to undertake the systematic expensive exploration and agreed a joint venture with Kennecott Canada, a subsidiary of Rio Tinto plc. Kennecott took an option on the property with the agreement that it would earn 60% of the property by spending $C10 million over 4 years with Jennings receiving a 1% royalty for his work on the property. The joint venture flew the area with airborne geophysics, magnetics, and EM, and began drilling the most prominent anomalies in 1992–93. Although kimberlite was intersected the micro diamond counts were not encouraging. Between April 1992 and January 1994 20,500 line kilometers of airborne geophysics and 1700 heavy mineral samples were taken. A total of 25 kimberlites were discovered based on 65 diamond drillholes totalling 6630 m.

Kennecott was also involved in other projects, notably the Tli Kwi Cho project, also a joint venture with Aber Resources but with other junior company partners. At this location airborne geophysics located a complex pipe, which yielded 39 microdiamonds and 15 macrodiamonds from 86 kg of core. Spurred on by the need to compete for exploration funds with this DHK joint venture, the Aber Resources–Kenecott joint venture drilled lower amplitude magnetic targets under Lac de Gras which had distinct indicator trails with G10 garnets. Drilling of target A21 from the Lac de Gras ice in April 1994 intersected kimberlites with many indicator minerals. Initial results showed the pipe was diamondiferous with 154 stones from a 155-kg sample taken from the property. As the ice was beginning to melt on Lac de Gras the rig was moved quickly north to test a further target A154 which appeared to be a twin pipe. The first hole intersected kimberlite including a 3-ct diamond visible in the core in May 1994. The initial core sample of 750 kg contained 402 macro diamonds and 894 micro diamonds. Of the macro diamonds seven exceeded 0.2 ct in mass. Drilling from a barge in November 1994 Kennecott intersected the A154 north pipe and the A418 pipe, approximately 800 m to the southwest, in the spring of 1995. Delineation drilling of the A154 pipes began in 1995 using inclined 15.2-cm diamond holes from the shore of East Island and other small islands in 1995. As an example the A154S pipe was delineated by 39 contacts from 6792 m in 26 holes (Thomas 2003). Mini bulk samples from large diameter core drilling provided 56.5 t of kimberlite from which 255.6 ct were recovered and valued at $US56.70 per carat or a phenomenal $US255 per tonne. Underground bulk sampling of the A154S and A418 pipes in 1996 and 1997 was next undertaken to generate large parcels of diamonds for sorting and valuing. The bulk sample of the A154S pipe returned 2585 dry tonnes of kimberlite containing 12,763 ct and that of the A418 3350 t containing 8323 ct. These stones were valued both by Rio Tinto's valuers in Perth, Australia and independently in Antwerp. The cost of underground sampling and deep delineation drilling as well as preliminary environmental studies was $C23.5M. Testing of the A418 North and A21 pipes followed in 1996 and 1997.

The very high revenues projected from the two high grade pipes led to commissioning of a pre-feasibility study, which was completed in September 1997. A project description was submitted to the Federal government in March 1998 triggering a formal environmental assessment. The feasibility study recommended mining the A154S and N pipes in one open pit to 285 m depth for 10 years. Open pits were also planned for the A418 and A21 pipes (Table 17.4). Underground mining was considered economic for the A154S and A418 pipes. One of the most important considerations was that unlike Ekati, the pipes were underneath a major albeit shallow lake. The lake would have to be partially drained and dykes constructed (Fig. 17.21). The environmental review process was similar to that undertaken for the Ekati development with a major commitment to win approval from the Aboriginal and Inuit inhabitants of the area. One hiccup was that

TABLE 17.4   Ore reserves Diavik. (From Ellis Consulting 2000.)

Pipe	A-418	A-154S	A-154N	A-21	Total
Open pit					
Tonnes (Mt)	4.2	9.3	2.7	3.7	19.9
Grade (ct t⁻¹)	3.79	4.69	2.89	2.89	3.92
Carats (Mct)	17.0	43.6	7.8	10.8	78.2
Underground					
Tonnes (Mt)	4.1	1.5			5.7
Grade (ct t⁻¹)	3.97	4.55			4.12
Carats (Mct)	17.4	7			23.3
Total reserve					
Tonnes (Mt)	8.3	10.8	2.7	3.7	25.6
Grade (ct t⁻¹)	3.88	4.67	2.89	2.89	3.96
Carats (Mct)	32.4	50.6	7.8	10.8	101.5
Valuation					
Price ($US ct⁻¹)	53	59	33	36	53
Value (US$M)	1715	2984	256	389	5344

FIG. 17.21   Aerial view of pre-stripping of the A154 pipes at Diavik in 2002. (Courtesy Diavik Diamond Mines Inc.)

TABLE 17.5   Projected cash flow, Diavik in 1998 $CM. (From Ellis Consulting 2000.)

Year	Capital expenditure	Resource income	Private royalties	Operating costs	Closure costs	Capital recovery	Resource profits	After-tax income
1995–1999	206							
2000	123							
2001	430							
2002	430							
2003	246	199	4	75	0	120	0	0
2004	0	398	8	125	0	266	0	0
2005	0	569	11	143	0	415	0	0
2006	73	569	11	143	0	415	0	0
2007	73	569	11	143	0	363	52	28
2008	15	569	11	143	0	15	400	215
2009	15	569	11	143	0	15	400	215
2010	89	551	11	143	0	89	308	165
2011	89	539	11	143	0	89	296	159
2012	15	521	10	143	0	15	353	189
2013	15	416	8	143	9	15	241	129
2014	15	401	8	143	9	15	226	121
2015	60	329	7	143	9	60	110	59
2016	15	323	6	168	9	15	124	67
2017	15	312	6	140	9	15	141	76
2018	15	422	8	152	9	15	236	127
2019	15	417	8	152	9	15	232	125
2020	15	353	7	131	9	15	191	103
2021	0	233	5	89	9	0	130	70
2022	0	197	4	75	9	0	108	58
Total	1968	8456	169	2675	94	1968	3550	1905

construction in 2000 required the use of the ice road to move materials from Yellowknife. However the agreement was only signed in March and the key land use permit issued, causing some delay in construction. The total cost of the exploration and development process, including feasibility studies and public review process, was $C206M.

As part of the review Diavik Diamond Mines Inc. has released an independent assessment of the resource revenue shown in Table 17.5 (Ellis Consulting 2000). Reserves are estimated to be a total of 25.6 Mt at 396 cpht with an average price of $US53 ct^{-1} (Table 17.4). The high grade A154-S pipe is being mined first optimizing cash flow with underground mining scheduled for 2008. The overall cost of construction was estimated at $C1286M and production began successfully in January 2003, $C50 million under budget and 6 months ahead of schedule.

Since the Ekati and Diavik discoveries, three other projects have been discovered that appear to have potential to become mines at Snap Lake, Gahcho Kue, and Jericho in the Slave Province. There have also been a number of less successful ventures reminding investors

of the difficult process of proving a viable deposit. At Tli Kwi Cho, Kennecott elected to move directly from the delineation stage to bulk sampling and spent $C15M extracting an underground bulk sample in 1994, at the same time as the Diavik pipes were discovered. Although the pipe returned reasonable grades at 35.9 cpht the quality was poor with less than 30% gems and the deposit was clearly not of economic interest.

17.3   SUMMARY

The discovery of the Ekati and Diavik was the result of perseverance by a small company aided eventually by major company investment. The technology of regional sampling for indicator minerals was well known, although the detailed chemical investigation of heavy minerals grains required refinement and investment in technology. Although the targets were small, many were easily detectable by airborne geophysics. The remoteness of the area did not in the end prove to be an obstacle as the deposits were high grade with a ready market.

INTERNET LINKS

The following are some of the more useful internet links by chapter and current at 2005.

CHAPTER 1:

Mining Companies

Index of companies with links http://www.goldsheetlinks.com

Mining Information

Infomine http://www.infomine.com
Mining Journal http://www.mining-journal.com
Mineweb http://www.mineweb.net
Northern Miner http://www.northernminer.com
Reflections http://www.reflections.com.au/miningandexploration
Breaking New Ground: Mines And Minerals Sustainable Development Project. http://www.iied.org/mmsd/finalreport/index.html

CHAPTER 2: MINERALOGY

Athena mineralogy database – includes search by elements http://un2sg4.unige.ch/athena/mineral/mineral.html
Barthelmy's Mineralogy database http://webmineral.com/
Mineralogy Database http://mindat.org
Rob Ixer's Virtual Atlas of Opaque and Ore Minerals http://www.smenet.org/opaque-ore/

CHAPTER 3: DEPOSIT MODELS (ON LINE)

British Columbia http://www.em.gov.bc.ca/Mining/Geolsurv/MetallicMinerals
MineralDepositProfiles/
USGS http://minerals.cr.usgs.gov/team/depmod.html

CHAPTER 6: REMOTE SENSING

Data: Landsat, Global Land Cover (free) http://glcf.umiacs.umd.edu/data/
Landsat Mr. Sid format coverage (free) https://zulu.ssc.nasa.gov/mrsid/
ASTER http://asterweb.jpl.nasa.gov/
Topography (90 m resolution) http://www2.jpl.nasa.gov/srtm/

CHAPTER 7: GEOPHYSICS

Fugro http://www.fugro.nl/geoscience/airborne/minexpl.asp

CHAPTER 8: GEOCHEMISTRY

Association of Applied Geochemists http://www.appliedgeochemists.org/
FOREGS Geochemistry http://www.gsf.fi/foregs/geochem/

CHAPTER 9: DATA MANAGEMENT–SOME SOFTWARE PROVIDERS

Acquire http://www.acquire.com.au/
Arcview/ArcGIS http://www.esri.com/
Datamine http://www.datamine.co.uk/
Geosoft http://www.geosoft.com/
Interdex/Surpac http://www.surpac.com/
Mapinfo http://www.mapinfo.com/
Micromine http://www.micromine.com/
Vulcan http://www.vulcan3d.com/

CHAPTER 10: SAMPLING AND EVALUATION

Boart drilling http://www.boartlongyear.com/

Web page on Mine Sampling

http://pws.prserv.net/Goodsampling.com/sampling.html

CHAPTER 12: CLIFFE HILL

Midland Quarry Products http://www.mqp.co.uk/

CHAPTER 14: WITWATERSRAND

Western Areas http://www.westernareas.co.za/
Placer Dome http://www.placerdome.com/operations/southdeep/southdeep.html

CHAPTER 15: KIDD CREEK

Kidd Creek Mine http://www.falconbridge.com/our_business/copper_kidd_creek.html
Kidd Creek Monograph http://www.nrcan.gc.ca/gsc/mrd/projects/kiddcreek monograph10_e.html

CHAPTER 17: DIAMONDS

De Beers http://www.debeersgroup.com/
Ekati/BHP Diamonds http://ekati.bhpbilliton.com/
Diavik http://www.diavik.ca/
NWT Diamond site http://www.gov.nt.ca/RWED/diamond/

REFERENCES

Abrams M.J., Conel J.E. & Lang H.R. (1984) Joint NASA/GEOSAT test case report. *Am. Assoc. Petroleum Geol.*, 3 vols.

Abrams M. & Hook S. (2002) *Aster Users Handbook, Version 2.* Jet Propulsion Laboratory, Pasadena, USA. http://asterweb.jpl.nasa.gov/documents/documents.htm.

AcQuire (2004) *Acquire Corporate Website.* http://www.acquire.com.au.

Adams S.S. (1985) Using geological information to develop exploration strategies for epithermal deposits. In Berger B.R. & Bethke P.M. (eds), *Geology and Geochemistry of Epithermal Systems*, 273–298. Economic Geology Publishing Co., Littleton, Colorado.

AGSO (Australian Geological Survey Organisation) (1998) Geology and mineral potential of major Australian mineral provinces. *AGSO J Aust. Geol. Geophys.*, **17**(3), 260; (4), 320.

Akçay M., Özkan H.M., Moon C.J. & Scott B.C. (1996) Secondary dispersion from gold deposits in West Turkey. *J. Geochem. Explor.*, **56**, 197–218.

Ali J.W. & Hill I.A. (1991) Reflection seismics for shallow geological investigations: a case study from central England. *J. Geol. Soc. London*, **148**, 219–222.

Allen M.E.T. & Nichol I. (1984) Heavy-mineral concentrates from rocks in exploration for massive sulphide deposits. *J. Geochem. Explor.*, **21**, 149–165.

Allum J.A.E. (1966) *Photogeology and Regional Mapping.* Pergamon Press, Oxford.

Andrew R.L. (1978) *Gossan Evaluation.* Short Course, Rhodes University, Grahamstown.

Annels A.E. (1991) *Mineral Deposit Evaluation: a practical approach.* Chapman and Hall, London.

Anon. (1977) Consolidated Rambler finds zinc bearing mercury. *World Mining*, **March**, 79.

Anthony M. (1988) Technical change – the impending revolution. In *Australasian Institute of Mining and Metallurgy Sydney Branch, Minerals and Exploration at the Crossroads*, 19–25, Sydney.

Arcatlas (1996) *Arcatlas CD-ROM.* Russian Academy of Sciences & ESRI, ESRI, Redlands, California, USA.

Arc-SDM (2001) *Spatial Data Modeller for Arcview.* Natural Resources Canada. http://cgpd.cgkn.net/sdm/default_e.htm.

Ashleman J.C. (1983) *Trinity Silver Project 1982–1983, Progress Summary Report.* Unpublished Report EX84–10, US Borax.

Ashleman J.C. (1986) *Trinity Venture Project, 1986 Summary Report.* Unpublished Report, EX 87–3, US Borax.

Ashleman J.C. (1988) *The Trinity Silver Deposit, Pershing County, Nevada.* Unpublished Field Guide.

Atkinson W.J., Hughes F.E. & Smith C.B. (1984) A review of the kimberlitic rocks of Western Australia. In Kornprobst J. (ed.) *Kimberlites I: kimberlites and related rocks*, 195–224. Elsevier, Amsterdam.

Australian Drilling Industry Training Committee (1997) *Drilling: the manual of methods, applications, and management.* CRC Press, Boca Raton.

Austr. Inst. Geoscientists (1987) *Meaningful Sampling in Gold Exploration*, bull. 7. Duncan and Associates, Leederville.

Austr. Inst. Geoscientists (1988) *Sample Preparation and Analyses for Gold and Platinum Group Elements*, bull. **8**, Duncan and Assoc., Leederville.

Australasian Joint Ore Reserves Committee (2003) *The JORC Code.* http://www.jorc.org/main.php. Australasian Institute of Mining and Metallurgy, Carlton, Victoria.

Baele S.M. (1987) *Silver Ore Reserve Estimation, Trinity Heap Leach Project, Pershing County,*

Nevada. Unpublished Report, MD 87–3, US Borax.

Bagby W.C. & Berger B.R. (1985) Geologic characteristics of sediment-hosted, disseminated precious-metal deposits in the western United States. In Berger B.R. & Bethke P.M. (eds), *Geology and Geochemistry of Epithermal Systems,* 169–202. *Reviews in Economic Geology,* vol. 2. Publishing Co., Littleton, Colorado.

Bailey D.C. & Hodson D.I. (1991) The Neves-Corvo Project – the successes and challenges. *Trans. Instn Min. Metall., Sect. A,* **100,** A1–A10.

Baldock J.W., Hepworth J.V. & Marengwa B.S. (1976) Gold, basemetals, and diamonds in Botswana. *Econ. Geol.,* **71,** 139–156.

Ball Aerospace & Technologies Corp. (2002) *Quickbird.* Boulder, Colorado. http://www.ball.com/aerospace/quickbird.html.

Barnes E. (1997) *MINDEV 97, The Third International Conference on Mine Project Development.* Australasian Institute of Mining and Metallurgy, Carlton, Victoria, 280 pp.

Barnes J.W. & Lisle R.J. (2003) *Basic Geological Mapping,* 4th edn. Wiley, New York.

Barnicoat A.C., Henderson I.H.C., Knipe R.J., et al. (1997) Hydrothermal gold mineralization in the Witwatersrand basin. *Nature,* **386,** 820–824.

Barrett W.L. (1987) Auger drilling. *Trans. Instn Min. Metall., Sect. B,* **96,** B165–169.

Barrett W.L. (1992) A case history of pre-extraction site investigation and quarry design, Cliffe Hill Quarry, Leicestershire. In Annels A.E. (ed.) *Case Histories and Methods in Mineral Resource Evaluation,* 69–76, spec. publ. **63.** Geological Society, London.

Barrie C.T. & Hannington M.T. (1999) Classification of volcanic-associated massive sulfide deposits based on host-rock composition In Barrie C.T. & Hannington M.T. (eds) *Volcanic Associated Massive Sulfide Deposits and Examples in Modern And Ancient Settings,* 1–12. *Reviews in Economic Geology,* vol. 8. Economic Geology Publishing Co., Littleton, Colorado.

Barrie C.T., Ludden J.N. & Green T.H. (1993) Geochemistry of volcanic rocks associated with Cu-Zn and Ni-Cu deposits in the Abitibi Subprovince. *Economic Geology,* **88,** 1341–1358.

Barrie C.T., Hannington M.T. & Bleeker W. (1999) The giant Kidd Creek volcanic-associated massive sulphide deposit, Abitibi subprovince, Canada. In Barrie C.T. & Hannington M.T. (eds) *Volcanic Associated Massive Sulfide Deposits and Examples in Modern And Ancient Settings,* 247–260. Reviews in Economic Geology, vol. 8.

Economic Geology Publishing Co., Littleton, Colorado.

Bartlett P.J. (1994) Geology of the Premier Mine. In Annhaeuser C.R. (ed.) *Proceedings of the XVth CMMI Congress,* 201–214. South African I.M.M., Johannesburg.

Barton J.M., Barnett W.P., Barton E.S., et al. (2003) The geology of the area surrounding the Venetia kimberlite pipes, Limpopo Belt, South Africa: a complex interplay of nappe tectonics and granitoid magmatism. *S. Afr. J. Geol.,* **106,** 109–128.

Barton P.B. (1986) Commodity/Geochemical Index. In Cox D.P. & Singer D. (1986) *Mineral Deposit Models.* US Geological Survey Bulletin 1693, Appendix C, 15 pp. Reston, VA.

Bateman A.M. (1950) *Economic Mineral Deposits.* Wiley, New York.

BCGS (British Columbia Geological Survey) (2004) *British Columbia Mineral Deposit Profiles Version 2.1.* http://www.em.gov.bc.ca/Mining/Geolsurv/MetallicMinerals/MineralDeposit-Profiles/default.htm.

Beaty D.W., Taylor H.P. Jr & Coad P.R. (1988) An oxygen isotope study of the Kidd Creek, Ontario, volcanogenic massive sulfide deposit: evidence for a high ^{18}O ore fluid. *Econ. Geol.,* **83,** 1–17.

Bell A.R. & Hopkins D.A. (1988) From farmland to tarmac. *Extractive Industry Geology 1985,* 19–26.

Bell T. (1989) *Ore Reconciliation and Statistical Evaluation of the Trinity Silver Mine, Pershing County, Nevada.* Unpublished MSc Dissertation, University of Leicester.

Bell T. & Whateley M.K.G. (1994) Evaluation of grade estimation techniques. In Whateley M.K.G. & Harvey P.K. (eds), *Mineral Resource Evaluation II, Methods and Case Histories,* spec. publ. **79,** Geol. Soc. London.

Bennett M.J., Beer K.E., Jones R.C., Turton K., Rollin K.E., Tombs J.M.C. & Patrick D.J. (1981) *Mineral Investigations Near Bodmin, Cornwall. Part 3 – The Mulberry and Wheal Prosper Area.* British Geological Survey Mineral Reconnaissance Report **48,** Keyworth.

Berger B.R. & Bagby W.C. (1991) The geology and origin of Carlin-type gold deposits. In Foster R.P. (ed.) *Gold Metallogeny and Exploration,* 210–48. Blackie, Glasgow.

Berkman D.A. (1989 & 2001) *Field Geologist's Manual,* 3rd and 4th Editions. Australasian Institute of Mining and Metallurgy, Parkville.

Berry L.G. (1941) Geology of the Bigwater Lake area. *Ontario Department of Mines Annual Report,* **48,**

1–9 including map 48n, Bigwater Lake area, 1 inch to 1 mile, Toronto.

Berry L.G., Mason B. & Dietrich R.V. (1983) *Mineralogy.* Freeman, San Francisco.

Beus A.A. & Grigorian S.V. (1977) *Geochemical Methods for Mineral Deposits.* Applied Publishing, Wilmette, Illinois.

BHP Billiton (2003) *BHP Billiton Website.* http://www.bhpbilliton.com.

Bieniawski Z.T. (1976) Rock mass classification in rock engineering. In Bieniawski Z.T. (ed.), *Exploration for Rock Engineering,* vol. 1, 97–106. A.A Balkema, Cape Town.

Bishop J.R. & Lewis R.J.G. (eds) (1996) DHEM Special Volume. *Explor. Geophys.,* **27**, 37–177.

Blain C.F. (2000) Fifty-year trends in minerals discovery – commodity and ore-type targets. *Explor. Mining Geol.,* **9**, 1–11.

Blanchard R. (1968) *Interpretation of Leached Outcrops.* Nevada Bureau of Mines Bull. **66**, Reno.

Blanchard R. & Boswell P.F. (1934) Additional limonite types of galena and sphalerite derivation. *Econ. Geol.,* **29**, 671–690.

Bleeker W. & Hester B. (1999) Discovery of the Kidd Creek massive sulphide orebody: a historical perspective. In Hannington M.D. & Barrie C.T. (eds) *The Giant Kidd Creek Volcanogenic Massive Sulfide Deposit, Western Abitibi Subprovince, Canada.* Economic Geology Monograph 10, 31–42. Economic Geology Publishing Co., Littleton, Colorado.

Bleeker W. & Parrish R.R. (1996) Stratigraphy and U-Pb zircon geochronology of Kidd Creek: implications for the formation of giant volcanogenic massive sulphide deposits and the tectonic history of the Abitibi greenstone belt. *Can. J. Earth Sci.,* **33**, 1213–1231.

Bloom L. (ed.) (2001) *Writing Geochemical Reports,* 2nd cdn. Association of Exploration Geochemists, spec. publ. **15**. Nepean, Ontario, Canada.

Bonham H.F. (1988) Silver deposits of Nevada. In *Silver: exploration, mining and treatment,* 25–31. Institution of Mining and Metallurgy, London.

Bonham H.F. Jr (1989) Bulk mineable gold deposits of the Western United States In Keays R.R., Ramsay W.R.H. & Groves D.I. (eds) *The Geology of Gold Deposits – the perspective in 1988,* 193–207. Economic Geology Monograph **6**. El Paso.

Bonham-Carter G.F. (1994) *Geographic Information Systems for Geoscientists: modelling with GIS.* Pergamon, New York.

Bonham-Carter C.F., Agterberg F.P. & Wright D.F. (1988) Integration of geological datasets for gold exploration in Nova Scotia. *Photogrammetric Engineering Remote Sens.,* **54**, 1585–1592.

Boyd F.R. & Gurney J.J. (1986) Diamonds and the African lithosphere. *Science,* **232**, 472–477.

Boyd F.R. & Nixon P.H. (1978) Ultramafic nodules from the Kimberley pipes, South Africa. *Geochim. Cosmochim. Acta,* **42**, 1367–1371.

Bradshaw P.M.D. (ed.) (1975) Conceptual models in exploration geochemistry – the Canadian Shield and Canadian Cordillera. *J. Geochem. Explor.,* **4**, 1–213.

Brady B.H.G. & Brown E.T. (1985) *Rock Mechanics for Underground Mining.* George Allen and Unwin, London.

Brassington R. (1988) *Field Hydrogeology.* Geological Society of London Professional Handbook Series, Wiley, Chichester.

Brinkmann R. (1976) *Geology of Turkey.* Elsevier, Amsterdam.

Brisbin D., Kelly V. & Cook R. (1991) Kidd Creek Mine. In Fyon J.A. & Green A.H. (eds), *Geology and Ore Deposits of the Timmins District, Ontario,* 66–71. Geological Survey of Canada Open File Report **2161**.

British Standard 812 (1975) *Method for Sampling and Testing of Mineral Aggregates, Sands and Fillers.* British Standards Institution, London.

Brook Hunt (2004) Brook Hunt website. www.brookhunt.com.

Brooks R.R. (1983) *Biological Methods of Prospecting for Minerals.* Wiley, New York.

Brown E.T. (ed.) (1981) Rock Characterization, Testing and Monitoring. International Society for Rock Mechanics, Pergamon Press, Oxford.

Brown W.M., Gedeon T.D., Groves D.I. & Barnes R.G. (2000) Artificial neural networks: a new method for mineral prospectivity mapping. *Australian J. Earth Sci.,* **47**, 757–770.

Brummer J.J. (1978) Diamonds in Canda. *C.I.M. Bull.,* **71**, 64–70.

Brundtland G.H. (1987) *Our Common Future.* World Commission on Environment and Development, Oxford University Press, Oxford.

Burke K., Kidd W.S.F. & Kusky T.M. (1986) Archaean foreland basin tectonics in the Witwatersrand, South Africa. *Tectonics,* **5**, 439–456.

Bustillo R.M. & Lopez J.C. (1997) *Manual de Evaluacion y Diseño de Explotaciones Mineras.* Entorno Grafico, Madrid.

Butt C.R.M. & Smith R.E. (1980) Conceptual models in exploration geochemistry – Australia. *J. Geochem. Explor.,* **12**, 89–365.

Butt C.R.M. & Zeegers H. (eds) (1992) *Regolith Exploration Geochemistry in Tropical and*

Subtropical Terrains. Handbook of Exploration Geochemistry, vol. 4. Elsevier, Amsterdam.

Cady J.W. (1980) Calculation of gravity and magnetic anomalies of finite length right polygonal prisms. *Geophysics*, **45**, 1507–1512.

Cameron E.M., Hamilton S.M., Leybourne M.I., Hall G.E.M. & McClenaghan M.B. (2004) Finding deeply buried deposits using geochemistry. *Geochem. Explor. Environ. Analysis*, **4**, 7–32.

Cameron R.I. & Middlemis H. (1994) Computer modelling of dewatering a major open pit mine: case study from Nevada, USA. In Whateley M.K.G. & Harvey P.K. (eds), *Mineral Resource Evaluation II: methods and case histories*, 205–215, spec. publ. **79**. Geological Society, London.

Camisani-Calzolari F.A.G.M., Ainslie L.C. & van der Merwe P.J. (1985) *Uranium in South Africa 1985.* South African Atomic Energy Corporation of South Africa, Pelindaba.

Camm T.W. (1991) *Simplified Cost Models for Prefeasibility Mineral Evaluations. Bureau of Mines Information Circular 9298.* United States Department of the Interior, Washington DC.

Campbell A.S. (ed.) (1971) Geology and history of Turkey. *Petroleum Exploration Society of Lybia*, 13th Annual Conference.

Campbell G. & Crotty J.H. (1990) 3-D seismic mapping for mine planning purposes at the South Deep prospect. In Ross-Watt R.A.J. & Robinson P.D.K. (eds), *Technical Challenges in Deep Level Mining*, 569–597. South African Institute of Mining and Metallurgy, Johannesburg.

Campbell I.H., Lesher C.M., Coad P., Franklin J.M., Gorton M.P. & Thurston P.C. (1984) Rare-earth mobility in alteration pipes below Cu-Zn-sulphide deposits. *Chem. Geol.*, **45**, 181–202.

Cartwright A.P. (1967) *Gold Paved the Way.* Purnell, Johannesburg.

Carver R.N., Chenoweth L.M., Mazzucchelli R.H., Oates C.J. & Robbins T.W. (1987) Lag – a geochemical; sampling medium for arid regions. *J. Geochem. Explor.*, **28**, 183–199.

Cas R.A.F. & Wright J. (1987) *Volcanic Successions: modern and ancient.* Allen & Unwin, London.

Cathles L.M. (1993) Oxygen isotope alteration in the Noranda Mining District, Abitibi Greenstone Belt, Quebec. *Econ. Geol.*, **88**, 1483–1511.

Chadwick J. 2002. Northparkes. *Mining Magazine*, **187**, 8–14.

Chaussier J-B. & Morer J. (1987) *Mineral Prospecting Manual.* Elsevier, Amsterdam.

Chen W. (1988) Mesozoic and Cenozoic sandstone-hosted copper deposits in South China. *Mineral. Deposita*, **23**, 262–267.

Chinn C.T. & Ascough G.L. (1997) Mineral potential mapping using an expert system and GIS. In Gubins A.G. (ed.) (1997) *Geophysics and Geochemistry at the Millenium: Proceedings of Exploration 97*, 89–86. Prospectors & Developers Association of Canada, Toronto.

Christison Scientific Equipment Catalogue (2002) *Particle Size Reduction.* http://www.christison.com/CAT7.html. Gateshead, UK.

Clark I. (1979) *Practical Geostatistics.* Elsevier Applied Science, Amsterdam.

Clark R.J. McH., Homeniuk L.A. & Bonnar R. (1982) Uranium geology in the Athabaska and a comparison with other Canadian Proterozoic Basins. *CIM Bull.*, **75**(April), 91–98.

Clarke P.R. (1974) The Ecstall Story: introduction. *Can. Inst. Min. Metall. Bull.*, **77**, 51–55.

Clifford J.A., Meldrum A.H., Parker, R.H.T. & Earls G. (1990) 1980–1990: a decade of gold exploration in Northern Ireland and Scotland. *Trans. Inst. Min. Metall., Sect. B*, **99**, B133–138.

Clifford J.A., Earls G., Meldrum A.H. & Moore N. (1992) Gold in the Sperrin mountains, Northern Ireland: an exploration case history In Bowden A.A., Earls G., O'Connor P.G. & Pyne J.F. (eds) *The Irish Minerals Industry 1980–1990*, 77–87. Irish Association of Economic Geology, Dublin.

Clement C.R., Harris J.W., Robinson D.N. & Hawthorne J.B. (1986) The De Beers kimberlite pipe – a historic South African diamond mine. In Anhausser C.R. & Maske S. (eds), *Mineral Deposits of South Africa*, 2193–2214. Geological Society of South Africa, Johannesburg.

Clifford T.N. (1966) Tectono-metallogenic units and metallogenic provinces of Africa. *Earth Plan. Sci Lett.*, **1**, 421–434.

Clifton H.E., Hunter R.E., Swanson F.J. & Phillips R.L. (1969) *Sample Size and Meaningful Gold Analysis.* US Geol. Surv. Prof. Paper **625-C**.

Closs L.G. & Nichol I. (1989) Design and planning of geochemical programs. In Garland G.D. (ed.) *Proceedings of Exploration '87*, 569–583, spec. vol. **3**. Ontario Geological Survey, Toronto.

Coad P.R. (1984) Kidd Creek Mine. In Gibson H.L. et al. (eds) *Surface Geology and Volcanogenic Base Metal Massive Sulphide Deposits and Gold Deposits of Noranda and Timmins.* Geological Association of Canada Field Trip Guide Book **14**.

Coad P.R. (1985) Rhyolite geology at Kidd Creek – a progress report. *Can. Inst. Min. Metall. Bull.*, **78**, 70–83.

Cohen A.D., Spackman W. & Raymond R. (1987) Interpreting the characteristics of coal seams from chemical, physical and petrographic studies of peat deposits. In Scott A.C. (ed.), *Coal and Coal-bearing Strata: recent advances*, 107–125, spec. publ. **32**. Geological Society, London.

Coker W.B., Hornbrook E.H.W. & Cameron E.M. (1979) Lake sediment geochemistry applied to mineral exploration. In Hood P.J. (ed.), *Geophysics and Geochemistry in the Search for Metallic Ores*, 435–447. Econ. Geol. Rept **31**. Geological Survey Canada.

Colman T.B. (2000) *Exploration for Metalliferous and Related Minerals: a guide*, 2nd edn. British Geological Survey, Keyworth.

Cook D.R. (1987) A crisis for economic geologists and the future of the Society. *Econ. Geol.*, **82**, 792–804.

Coope J.A. (1991) *Monitoring Laboratory Performance with Standard Reference Samples*. Abstract, 15th Geochemical Exploration Symposium, Reno.

Cooper P.C. & Sternberg M. (1988) Deep directional core drilling with Navi-Drill. *Engrg. Min. J.*, **189**(7), 48–49, 59.

Corbett J. (1990) Overview of geophysical methods applied to gold exploration in Nevada. *Leading Edge*, **9**, 17–26.

Corner B. & Wilshire W.A. (1989) Structure of the Witwatersrand Basin derived from the interpretation of aeromagnetic and gravity data. In Garland G.D. (ed.), *Proceedings of Exploration '87*, 532–545, spec. vol. **3**. Ontario Geological Survey.

Cox D.P. & Singer D. (1986) *Mineral Deposit Models*. US Geol. Surv. Bull. **1693**. Reston.

Cox F.C., Bridge D.McC. & Hull J.H. (1977) *Procedure for the Assessment of Limestone Resources*. Mineral Assessment Report **30**. Institute of Geological Sciences, London.

Craig J.R. & Vaughan D.J. (1994) *Ore Microscopy and Ore Petrography*, 2nd edn. Wiley, New York.

Craig Smith R. (1992) PREVAL: Prefeasibility Software Program for Evaluating Mineral Properties. Information Circular 9307. United States Department of Interior, Washington DC.

Crosson C.C. (1984) Evolutionary development of Palabora. *Trans. Instn Min. Metall., Sect. A*, **93**, A58–A69.

Crowson P.C.F. (1988) A perspective on worldwide exploration for minerals. In Tilton J.E., Eggert R.G. & Landsberg H.H. (eds), *World Mineral Exploration*, 21–103. Resources for the Future, Washington.

Crowson P. (1998) *Inside Mining: the economics of the supply and demand of minerals and metals*. Mining Journal Books, London.

Crowson P. (2003) *Astride Mining: issues and policies for the mining industry*. Mining Journal Books, London.

CRU (2004) Commodities Research Unit website. http://www.crugroup.com

Cummings J.D. & Wickland A.P. (1985) *Diamond Drill Handbook*. J.K. Smit and Sons, Toronto.

Curtin G.C., King H.D. & Moiser E.L. (1974) Movement of elements into the atmosphere from coniferous trees in the subalpine forests of Colorado and Idaho. *J. Geochem. Explor.*, **3**, 245–263.

Danchin P.D. (1989) Obituary of J. Dalrymple. *Geobulletin* 2nd Quarter, 21–22.

Davenport P.H., Friske P.W.B. & DiLabio R.N.W. (1997) The application of lake sediment geochemistry to mineral exploration: recent advances and examples from Canada. In Gubins, A.G. (ed.) *Geophysics and Geochemistry at The Millenium: Proceedings of Exploration 97*, 249–260. Prospectors & Developers Association of Canada, Toronto.

David M. (1988) *Handbook of Applied Advanced Geostatistical Ore Reserve Estimation*. Developments in Geomathematics, **6**. Elsevier, Amsterdam.

Davidson C.F. (1965) The mode of origin of banket orebodies. *Trans. Instn Min. Metall.*, **74**, 319–338.

Davies M.H. (1987) Prospects for copper. *Bull. Institn Min. Metall.*, **September**, 3–6.

Davies P. (2000) Discovery of Marsfontein M-1 pipe. http://www.msaprojects.co.za/discm1pipe.htm.

Davis J.C. (1986, 2003) *Statistics and Data Analysis in Geology*. Wiley, New York.

Dawson C.W. & Tokle V. (1999) Directional Core Drilling: the state of the art. In Hilton D.E. & Samuelson K. (eds) *Proceedings of the 1999 Rapid Excavation and Tunnelling Conference*, 351–362. Society for Mining, Metallurgy and Exploration, Littleton, CO.

Dawson J.B. & Stephens W.E. (1975) Statistical classification of garnets from kimberlite and associated xenoliths. *J. Geol.*, **83**, 589–607.

De Beer J.H. & Eglington B.M. (1991) Archaean sedimentation on the Kaapvaal Craton in relation to tectonism in granite-greenstone terrains: geophysical and geochronological constraints. *J. Afr. Earth Sci.*, **13**, 27–44.

De Beers (2001) Technical and Financial Report (background to offer by DBI). http://www.angloamerican.co.uk.

De Beers (2003a) *De Beers Exploration and Operations Overview*. www.angloamerican.co.uk.

De Beers (2003b) *Venetia Analysts Visit.* www. angloamerican.co.uk.

Decisioneering (2003), *Crystal Ball.* http://www. decisioneering.com/monte-carlo-simulation.html.

Dentith M.C., Frankcombe K.F., Ho S.E., Shepherd J.M., Groves D.I. & Trench A. (eds) (1995) *Geophysical Signatures of Western Australian Mineral Deposits.* Geology and Geophysics Department (Key Centre) & UWA Extension, University of Western Australia, Pubn **26**, and Australian Society of Exploration Geophysicists, spec. publ. **7**, 454 pp.

Department of Land Information (2003) Government of Western Australia, Perth. http://www.dola.wa. gov.au/.

D'Ercole C. Groves D.I. & Knox-Robinson C.M. (2000) Using fuzzy logic in a geographic information system environment to enhance conceptually based prospectivity analysis of Mississippi Valley-type mineralization. *Aust. J. Earth Sci.,* **47**, 913–927.

Deutsch C.V. (2002) *Geostatistical Reservoir Modelling.* Oxford University Press, New York.

Deutsch C.V. & Journel A.G. (1998) *GSLIB, Geostatistical Software Library and User's Guide,* 2nd edn. Oxford University Press.

Deutsch M., Wiesnet D.R. & Rango A. (eds) (1981) *Satellite Hydrogeology.* American Water Resources Association, Minneapolis, MN.

Diamond Facts (2003) *North West Territories Government Web Site.* http://www.gov.nt.ca/ RWED/diamond/.

Diamond Geology (2003) *De Beers Group Website.* http://www.debeersgroup.com/exploration.

Dickinson R.T., Jones C.M. & Wagstaff J.D. (1986) MWDatanet – acquisition and processing of downhole and surface data during drilling. *Trans. Instn Min. Metall., Sect. B,* **95**, B140–148.

DigitalGlobe (2004) *QuickBird Specifications.* http://www.digitalglobe.com/products/quickbird. shtml, Longmont, CO.

Dines H.G. (1956) *The Metalliferous Mining Region of South-west England.* HMSO, London.

Donohoo H.V., Podolsky G. & Clayton R.H. (1970) Early geophysical exploration at Kidd Creek Mine. *Mining Congress Journal,* **May,** 44–53.

Dorey R., Van Zyl D. & Kiel J. (1988) Overview of heap leaching technology. In Van Zyl D., Hutchison I. & Kiel J. (eds), *Introduction to Evaluation, Design and Operation of Precious Metal Heap Leaching Projects,* 3–22. Society of Mining Engineers, Denver, CO.

Dozy J.J., Erdman D.A., Jong W.J., Krol G.L. &. Schouten C. (1939) Geological results of the Carstenz Expedition 1936. *Leidsche Geologische Mededeelingen,* **11**, 68–131.

Drury S.A. (1986) Remote sensing of geological structure in European agricultural terrains. *Geol. Mag.,* **123**, 113–121.

Drury S.A. (2001) *Image Interpretation in Geology,* 3rd edn. Allen & Unwin, London.

Du Toit A.L. (1954) *The Geology of South Africa.* Oliver and Boyd, Edinburgh.

Dunn C.E. (2001) Biogeochemical methods in the Canadian Shield and Cordillera. In McClenaghan M.B., Bobrowsky P.T., Hall G.E.M. & Cook S.J. (eds) *Drift Exploration in Glacial Terrain,* 151–164, spec. publ. **185**. Geological Society, London.

Dunn P.R., Battey G.C., Miezitis Y. & McKay A.D. (1990a) The distribution and occurrence of uranium. In Glasson K.R. & Rattigan J.H. (eds), *Geological Aspects of the Discovery of Some Important Mineral Deposits in Australia,* 455–462. Australasian Institution of Mining and Metallurgy, Melbourne.

Dunn P.R., Battey G.C., Miezitis Y. & McKay A.D. (1990b) The uranium deposits of the Northern Territory. In Glasson K.R. & Rattigan J.H. (eds), *Geological Aspects of the Discovery of Some Important Mineral Deposits in Australia,* 463–476. Australasian Institute of Mining and Metallurgy, Melbourne.

E3 (2004) *Environmental Excellence in Exploration (Prospectors & Developers Association of Canada) Web Site.* http://www.e3mining.com.

Eaton D.W., Milkereit B. & Salisbury M. (2003) Seismic methods for deep mineral exploration: mature technologies adapted to new targets. *Leading Edge,* **22**, 580–585.

Eckstrand O.R. (ed.) (1984) *Canadian Mineral Deposit Types: a geological synopsis.* Geol. Surv. Canada Econ. Rept **36**. Ottawa.

Eckstrand O.R., Sinclair W.D. & Thorpe R.I. (eds) (1995) *Geology of Canadian Mineral Deposit Types.* Geology of Canada **8**. Geological Survey of Canada.

Economagic (2003) *Producer Price Index Data.* http://www.economagic.com/blsppi.htm.

Edwards A.C. (ed.) (2001) *Mineral Resource and Ore Reserve Estimation: the AusIMM guide to good practice.* Australasian IMM monograph series, **23**, 421–434. Australasian Institute of Mining and Metallurgy, Carlton, Victoria.

Eggert R.G. (1988) Base and precious metal exploration by major corporations. In Tilton J.E., Eggert R.G. & Landsberg H.H. (eds), *World Mineral Exploration,* 105–144. Resources for the Future, Washington DC.

Eggington H.F. (ed.) (1985) *Australian Drillers Guide.* Australian Drilling Training Committee, Macquarie Centre, New South Wales.

Eimon P.I. (1988) Exploration for and evaluation of epithermal deposits. In *Gold Update 87–88, Epithermal Gold*, 68–80. Course Manual. Department of Geology, University of Southampton, Southampton.

Eldorado Gold (2003) *Eldorado Gold Website.* http://www.eldoradogold.com.

Ellis D. (1988) *Well Logging for Earth Scientists.* Elsevier, Amsterdam.

Ellis Consulting (2000) *The Diavik Diamonds Project: the distribution of the project resource income.* http://www.diavik.ca/pdf/ellisreport.pdf.

Emerson D.W. (1980) *The Geophysics of the Elura Orebody.* Australian Society of Exploration Geophysicists, Sydney.

Engineering Group Working Party (1977) *Quart. J. Eng. Geol.*, **10**, Geol. Soc., London.

Englebrecht C.J. (1986) The West Wits Line. In Antrobus E.S.A. (ed.), *Witwatersrand Gold – 100 Years*, 199–225. Geological Society of South Africa, Johannesburg.

Englebrecht C.J., Baumbach G.W.S., Matthysen J.L. & Fletcher P. (1986) The West Wits Line. In Annhaeusser C.R. & Maske S. (eds), *Mineral Deposits of Southern Africa*, 599–648. Geological Society of South Africa, Johannesburg.

Environment Agency (2002) *Environmental Impact Assessment (EIA): a handbook for scoping projects.* Environment Agency, London.

Erlich E.I. & Hausel W.D. (2002) *Diamond Deposits: origin, exploration, and history of discovery.* Society for Mining, Metallurgy, and Exploration, Littleton, CO.

Erseçen N. (1989) *Known Ore and Mineral Resources of Turkey.* M.T.A., Ankara.

ESRI (2004) *ESRI Website.* http://www.esri.com.

European Space Agency (2004) *ERS 1 and 2.* Paris, France. http://www.esa.int/export/esaSA GGGWBR8RVDC_earth_0.html .

Evans A.M. (1968) Charnwood Forest. In Sylvester-Bradley P.C. & Ford T.D. (eds), *The Geology of the East Midlands*, 1–12. Leicester University Press, Leicester.

Evans A.M. (1993) *Ore Geology and Industrial Minerals – an introduction.* Blackwell Scientific Publications, Oxford.

Falcon (2003) *BHP Billiton Falcon Website.* http://falcon.bhpbilliton.com/.

Farmer I. (1983) *Engineering Behaviour of Rocks*, 2nd edn. Chapman & Hall, London.

Fauth H., Hindel R., Siewers U. & Zinner J. (1985) *Geochemischer Atlas Bundesrepublik Deutschland.* Bundesanstalt fur Geowissenschaften und Rohstoffe, Hannover.

Feather C.E. & Koen G.M. (1975) The mineralogy of the Witwatersrand reefs. *Miner. Sci. Engrg*, **7**, 189–224.

Fenwick K.G., Newsome J.W. & Pitts A.E. (1991) *Report of Activities 1990.* Ontario Geological Survey Misc. Paper **152**. Toronto.

Fipke C.E., Gurney J.J. & Moore R.O. (1995) *Diamond Exploration Techniques Emphasizing Indicator Mineral Chemistry and Canadian Examples.* Geological Survey of Canada Bulletin **423**.

Fletcher W.K. (1981) *Analytical Methods in Exploration Geochemistry. Handbook of Exploration Geochemistry*, vol. 1, Elsevier, Amsterdam.

Fletcher W.K. (1987) Analysis of soil samples. In Fletcher W.K., Hoffman S.J., Mehrtens M.B., Sinclair A.J. & Thomson I. (eds), *Exploration Geochemistry: design and interpretation of soil surveys*, 73–96. *Reviews in Economic Geology*, vol. 3. Society of Economic Geologists, El Paso.

Fletcher W.K. (1997) Stream sediment geochemistry in today's exploration world. In Gubins A.G. (ed.), *Geophysics and Geochemistry at the Millenium: Proceedings of Exploration 97*, 249–260. Prospectors & Developers Association of Canada, Toronto.

Fletcher W.K., Hoffman S.J., Mehrtens M.B., Sinclair A.J. & Thomson I. (1987) *Exploration Geochemistry: design and interpretation of soil surveys. Reviews in Economic Geology* vol. 3. Society of Economic Geologists, El Paso.

Fluor Mining and Metals Inc. (1978) *The Use of Geostatistics in Exploration and Development.* Unpublished report, Fluor Metals Inc.

Forstner U. & Wittman G.T.W. (1979) *Metal Pollution in the Aquatic Environment.* Springer-Verlag, Berlin.

Fortescue J.A.C. & Hornbrook D.H.W. (1969) Two quick projects: one at a massive sulphide orebody near Timmins, Ontario and the other at a copper deposit in Gaspe Park, Quebec. In *Progress Report on Biogeochemical Research at the Geological Survey of Canada*, 39–63. Geological Survey of Canada Paper **67-23**. Ottawa.

FracSIS (2002) *Fracsis Website.* http://www.fractal.csiro.au/fracsis/.

François-Bongarçon D.M. & Gy P. (2002) The most common error in applying 'Gy's Formula' in the theory of mineral sampling and the history of the liberation factor. *J.S. Afr. Inst. Min. Metall.*, **102**, 475–484.

Francké J.C. & Yelf R. (2003) Applications of GPR for surface mining. In *Proceedings of the 10th International Conference on Ground Pemetrating Radar (GPR 2004)*. Delft University of Technology, Delft, Netherlands

Franklin J.M. (1997) Lithogeochemical and mineralogical methods for base metal and gold exploration In Gubins A.G. (ed.) *Geophysics and Geochemistry at the Millenium: Proceedings Of Exploration 97*, 191–208. Prospectors & Developers Association of Canada, Toronto.

Fraser Institute (2003) *The Fraser Institute Annual Survey of Mining Companies.* http://www.fraserinstitute.ca.

Friend T. (1990) The Vegetation Unit – caring for the environment. *Mining Survey*, **1990**(1), 18–22.

Frimmel H.E. & Minter W.E.L. (2002) Recent developments concerning the geological history of the Witwatersrand gold deposits, South Africa. In Goldfarb R. & Nielsen R.L. (eds) *Global Exploration 2002– integrated methods for discovery.* Special Publ. **9**, 17–45. Society for Economic Geology, Littleton.

Frolick V. (1999) *Fire Into Ice.* Raincoast Books, Vancouver.

Fry N. (1991) *The Field Description Of Metamorphic Rocks*, 2nd edn. Wiley, New York.

Fyon A. (1991) Volcanic-associated massive base metal sulphide mineralization. In Fyon J.A. & Green (eds), *Geology and Ore Deposits of the Timmins District, Ontario (Field Trip 6).* Geol. Surv. Canada Open File Rept **216**.

Gableman J.W. & Conel J.E. (1985) Section 7, Uranium Commodity Report. In Abrams M.J., Conel J.E. & Lang H.R. (eds), *The Joint NASA/Geosat Test Case Report*, 3 vols. American Association of Petroleum Geologists, Tulsa.

Gardiner P.R. (1989) The recent Irish experience in promoting mineral exploration. In Chadwick J.R. (eds), *Mineral Exploration Programmes '89*, paper **28**. International Mining, Madrid.

Garrett R.G. (1989) The role of computers in exploration geochemistry. In Garland G.D. (ed.) *Proceedings of Exploration '87*, spec. vol. 3, 586–608. Ontario Geological Survey, Toronto.

Gatzweiler R., Scheling B. & Tan B. (1981) Exploration of the Key Lake uranium deposits, Saskatchewan, Canada. In IAEA (eds), *Uranium Exploration Case Histories*, 195–220. International Atomic Energy Agency, Vienna.

Geelhoed B. & Glass H.J. (2001) A new model for sampling of particulate materials and determination of the minimal sample size. *Geostand. Newsl. J. Geostand. Geoanalysis*, **25**, 65–81.

Gentry D.W. (1988) Minerals project evaluation – an overview. *Trans. Instn Min. Metall.*, Sect. A, **97**, A25–A35.

Geocover (2004) *Geocover (Earthsat Corp.) Website.* http://www.geocover.com.

Geosoft (2004) *Geosoft Website.* http://www.geosoft.com.

Geovariances (2001) *Isatis Software Manual*, 3rd edn. Ecole des Mines de Paris, Fontainebleau.

Gibson H.L., Morton R.L. & Hudak G.J. (1999) Submarine volcanic processes, deposits and environments favourable for the location of volcanic-associated massive sulphide deposits In Barrie C.T. & Hannington M.T. (eds) *Volcanic Associated Massive Sulfide Deposits and Examples in Modern And Ancient Settings*, 13–52. *Reviews in Economic Geology*, **8**. Economic Geology Publishing Co., Littleton, CO.

Gilluly J., Waters A.C. & Woodford A.O. (1959) *Principles of Geology.* Freeman, San Francisco.

Glasson K.R. & Rattigan J.H. (eds) (1990) *Geological Aspects of the Discovery of Some Important Mineral Deposits in Australia.* Australasian. Institute of Mining and Metallurgy, Melbourne.

GLCF (Global Land Cover Facility) (2004) *Global Land Cover Facility.* http//esip.umiacs.umd.edu.

Globe (Globe Team) (2004) *Globe Website.* http://www.ngdc.noaa.gov/seg/topo/globe.shtml.

Gocht W.R., Zantop H. & Eggert R.G. (1988) *International Mineral Economics.* Springer-Verlag, Berlin.

Goetz A.F.H., Rock B.N. & Rowan L.C. (1983) Remote sensing for exploration: an overview. *Econ. Geol.*, **78**, 573–590.

Goetz A.F.H. & Rowan L.C. (1981) Geologic remote sensing. *Science*, **211**, 781–791.

Gökçen N. (1982) The ostracod biostratigraphy of the Denizli–Mugla Neogene Sequence. *Bull. Inst. Earth Sci.* **9**, 111–131.

Golder Associates (1983) *Feasibility Report on the Soma-Isiklar Lignite Deposit, Manisa Province, Western Turkey.* Unpublished Report for Turkiye Komur Isletmeleri Kurumu, 8 vols.

Goldfarb R.J. & Neilsen R.L. (eds) (2002) *Integrated Methods for Discovery: global exploration in the twenty-first century.* Society of Economic Geologists, spec. publ. **9**. Society of Economic Geologists, Littleton, CO.

Goldspear (1987) *Gold Panning. Practical and useful instructions.* Goldspear (UK) Ltd, Beaconsfield.

Goldstein J.I., Newbury D.E., Echlin P., Joy D.C., Fiori C.E. & Lifshin E. (1981) *Scanning Electron*

Microscopy and X-ray Microanalysis. Plenum, New York.

Goode J.R., Davie M.J., Smith L.D. & Lattanzi C.R. (1991) Back to basics: the feasibility study. *Can. Inst. Min. Metall. Bull.*, **84**, 53–61.

Goodman R.E. (1976) *Methods in Geological Engineering in Discontinuous Rocks*. West Publishing Co., St Paul.

Goodman R.E. (1989) *Introduction to Rock Mechanics*, 2nd edn. Wiley, New York.

Goold D. & Willis A. (1998) *The Bre-X Fraud*. McClelland & Stewart, Toronto.

Gorman P.A. (1994) A review and evaluation of the costs of exploration, acquisition and development of copper related projects in Chile. In Whateley M.K.G. & Harvey P.K. (eds), *Mineral Resource Evaluation II: Methods and Case Histories*, 123–128, spec. publ. **79**. Geological Society, London.

Govett G.J.S. (1983) *Rock Geochemistry in Mineral Exploration. Handbook of Exploration Geochemistry*, vol. 3. Elsevier, Amsterdam.

Govett G.J.S. (1989) Bedrock geochemistry in mineral exploration. In Garland G.D. (ed.) *Proceedings of Exploration '87*, 273–299, spec. vol. 3. Ontario Geological Survey, Toronto.

Grant F.S. & West G.F. (1965) *Interpretation Theory in Applied Geophysics*. McGraw-Hill, New York.

Graton L.C. (1930) Hydrothermal origin of the Rand gold deposits, Part I, testimony of the conglomerates. *Econ. Geol.* 25 (suppl. to no. 3) 185pp.

Grayson R.L. (2001) Hazard identification, risk assessment and hazard control. In Karmis M. (ed.) *Mine Health and Safety Management*, 275–289. Society for Mining, Metallurgy and Exploration, Littleton, CO.

Gribble P. (1993) *Fault Interpretation from Coal Exploration Borehole Data using Surpac Software*, abstract volume. Mineral Resource Evaluation 1993 Conference, Leicester University.

Gribble P. (1994) Fault interpretation from coal exploration borehole data using Surpac software. In Whateley M.K.G. & Harvey P.K. (eds), *Mineral Resource Evaluation*, 29–35, spec. publ. **79**. Geological Society, London.

GSC (1995) *Generalized Geological Map of the World, CD-ROM*. Geological Survey of Canada Open File **2915d**.

Gu Y. (2002) Automated scanning electron microscope based mineral liberation analysis. *J. Min. Mater. Character. Engrg*, **2**, 33–41.

Gubins A.G. (ed.) (1997) *Geophysics and Geochemistry at the Millenium: Proceedings of Exploration 97*. Prospectors & Developers Association of Canada, Toronto.

Gulson B.L. (1986) *Lead Isotopes in Mineral Exploration*. Elsevier, Amsterdam.

Gunn A.G. (1989) Drainage and overburden geochemistry in exploration for platinum-group element mineralisation in the Unst Ophiolite, Shetland, U.K. *J. Geochem. Explor.*, **31**, 209–236.

Gurney J.J. (1984) A correlation between garnets and diamonds in kimberlites. In Glover J.E. & Harris P.G. (eds) *Kimberlite Occurrence and Origin: a basis for conceptual models in exploration*, **8**, 143–166. University of Western Australia, Perth, Western Australia.

Gurney J.J. & Zweistra P. (1995) The interpretation of the major element compositions of mantle minerals in diamond exploration. *J. Geochem. Explor.*, **53**, 293–310.

Gy M. (1992) *Sampling of Heterogeneous and Dynamic Material Systems*. Elsevier, New York.

Hale M. & Plant J.A. (1994) *Drainage Geochemistry. Handbook of Exploration Geochemistry*. Elsevier, Amsterdam.

Hallbauer D.K. (1975) The plant origin of Witwatersrand carbon. *Miner. Sci. Eng.*, **7**, 111–131.

Hallberg J.A. (1984) A geochemical aid to igneous rock identification in deeply weathered terrain. *J. Geochem. Explor.*, **20**, 1–8.

Handley G.A. & Carrey R. (1990) Big Bell gold deposit. In Hughes F. (ed.) *Geology of the Mineral Deposits of Australia and Papua New Guinea*, 217–220. Australas. Inst. Min. Metall. Monograph **14**. Melbourne.

Handley J.R.F. (2004) *Historic Overview of the Wiwatersrand Goldfields*. Handley, Howick, Natal, South Africa.

Hannington M.D. & Barrie C.T. (1999) *The Giant Kidd Creek Volcanogenic Massive Sulfide Deposit, Western Abitibi Subprovince, Canada*. Economic Geology Monograph **10**. Economic Geology Publishing Co., Littleton, CO.

Hanula M.R. (ed.) (1982) *The Discovers*. Pitt Publishing, Toronto.

Harben P.W. & Kuzvart M. (1997) *Industrial Minerals: a global geology*, 3rd edn. International Minerals Information, London.

Hart M. (2002) *Diamond, the History of A Cold-Blooded Love Affair*. Fourth Estate, London.

Haslett M.J. (1994) The South Deep project – geology and planning for the future. In Anhaeusser CR (ed.) *Proceedings XVth CMMI Congress, Vol. 3 – Geology*, 71–83. S. Afr. I.M.M. Symposium Series **14**.

Hatton W. (1994a) INTMOV. A program for the interactive analysis of spatial data. In Whateley M.K.G. & Harvey P.K. (eds), *Mineral Resource Evaluation II: methods and case histories*, 37–43. Spec. Publ. **79**. Geological Society, London.

Hatton W. (1994b) *Exploration Risk Assessment in the UK Coal Measures.* Unpublished PhD, University of Leicester.

Hawkes H.E. (1976) The downstream dilution of stream sediment anomalies. *J. Geochem. Explor.,* **6**, 345–358.

Hawkins M.A. (1991) Large fine fraction stream samples – a practical method to maximize catchment size, minimize nugget effect and allow detection of coarse or fine gold. In *15th International Geoechemical Symposium, Reno*, abstract.

Heald P., Foley N.K. & Hayba D.O. (1987) Comparative anatomy of volcanic-hosted epithermal deposits: acid sulphate and adularia-sercite types. *Econ. Geol.,* **82**, 1–26.

Hedenquist J.W., Arribas R.A. & Gonzalez-Urien E. (2000) Exploration for epithermal gold deposits. In Hagemann S.G. & Brown P.E. (eds) *Gold in 2000,* 245–279. *Reviews in Economic Geology,* **13**. Economic Geology Publishing Co., Littleton, CO.

Hefferman V. (1998) *Worldwide Mineral Exploration.* Financial Times Energy, London.

Heitt D.G., Dunbar W.W., Thompson T.B., Jackson R.G. (2003) Geology and geochemistry of the Deep Star gold deposit, Carlin Trend, Nevada. *Econ. Geol.,* **98**, 1107–1135.

Helmstaedt H. (1993) Primary diamond deposits – what controls their size grade and location? In Whiting B.H., Mason R. & Hodgson C.J. (eds) *Giant Ore Deposits.* Society of Economic Geologists Special Publication **2**, 13–80. Society of Economic Geologists, Littleton, Colorado.

Henley R.W. (1991) Epithermal gold deposits in volcanic terranes In Foster R.P. (ed.) *Gold Metallogeny and Exploration,* 133–164. Blackie, Glasgow.

Henley S. (1981) *Nonparametric Geostatistics.* Applied Science Publishers, London.

Heyl A.V. (1972) The 38th Parallel Lineament and its relationship to ore deposits. *Econ. Geol.,* **67**, 879–894.

Hill I.A. (1990) *Shallow Reflection Seismic Methods in the Extractive and Engineering Industries.* Unpublished Workshop Notes, University of Leicester.

Hitzman M.W., Proffett J.M. Jr, Schmidt J.M. & Smith T.E. (1986) Geology and mineralization of the Ambler District, Northwestern Alaska. *Econ. Geol.,* **81**, 1592–1618.

Hodgson C.J. (1989) Use (and abuses) of ore deposit models in mineral exploration. In Garland G.D. (ed.), *Proceedings of Exploration '87,* 31–45, spec. vol. **3**. Ontario Geological Survey, Toronto.

Hoek E. & Bray J. (1977) *Rock Slope Engineering.* Institution of Mining and Metallurgy, London.

Hoek E. & Brown E.T. (1980) *Underground Excavation in Rock.* Institution of Mining and Metallurgy, London.

Hoeve J. (1984) Host rock alteration and its application as an ore guide at the Midwest Lake Uranium deposit, northern Saskatchewan. *Can. Inst. Min. Metall. Bull.,* **77**(August), 63–72.

Hoffman S.J. (1987) Soil sampling. In Fletcher W.K., Hoffman S.J., Mehrtens M.B., Sinclair A.J. & Thomson I. (eds), *Exploration Geochemistry: design and interpretation of soil surveys,* 39–77. *Reviews in Economic Geology,* vol. 3. Society of Economic Geologists, El Paso.

Hoffmann P.F. (1989) Precambrian geology and tectonic history of North America. In Bally A.W. & Palmer A.R. (eds), *The Geology of North America: an overview,* 447–512. Geological Society of America, Boulder, CO.

Hofstra A.H. & Cline J.S. (2000) Characteristics and models for Carlin-type deposits. In Hagemann S.G. & Brown P.E. *Gold in 2000.* Reviews in Economic Geology, vol. 13, 163–220. Economic Geology Publishing Co., Littleton, CO.

Hoover D.B., Heran W.D. & Hill P.L. (eds) (1992) *The Geophysical Expression of Selected Mineral Deposit Models.* United States Department of the Interior, Geological Survey Open File Report **92–557**.

Horikoshi E. & Sato T. (1970) Volcanic activity and ore deposition in the Kosaka Mine. In Tatsumi T. (ed.), *Volcanism and Ore Genesis,* 181–195. University of Tokio Press, Tokio.

Hornbrook D.H.W. (1975) Kidd Creek Cu-Zn-Ag deposit Ontario (case history) *J. Geochem. Explor.,* **4**, 165–168.

Hosking K.F.G. (1971) Problems associated with the application of geochemical methods of exploration in Cornwall, England. In Boyle R.W. (ed.), *Geochemical Exploration,* spec. vol. 11, 176–189. Canadian Institution of Mining and Metallurgy, Toronto.

Howarth R.J. (ed.) (1982) *Statistics and Data Analysis in Exploration Geochemistry,* vol. 2. *Handbook of Exploration Geochemistry.* Elsevier, Amsterdam.

Howarth R.J. (1984) Statistical applications in geochemical prospecting: a survey of recent developments. *J. Geochem. Explor.,* **21**, 41–61.

Humphreys D. (2003) *The Relocation of Global Industrial Production: what it means for mining.* http://www.riotinto.com.

Huston D.L. & Large R.R. (1989) A chemical model for the concentration of gold in volcanogenic massive sulphide deposits. *Ore Geol. Rev.*, **4**, 171–200.

Hustrulid W.A. & Bullock R.L. (Eds) (2001) *Underground Mining Methods: engineering fundamentals and international case studies.* Society for Mining, Metallurgy and Exploration, Littleton, CO.

Hutchinson R.W. (1980) Massive base metal sulphide deposits as guides to tectonic evolution. In Strangeway D.W. (ed.), *The Continental Crust and Its Mineral Deposits*, 659–684. Geol. Assoc. Canada, Spec. Pap. **20**.

Huchinson R.W. & Grauch R.I. (eds) (1991) *Historical Perspectives of Genetic Concepts and Case Histories of Famous Discoveries.* Monograph **8**, Econ. Geol.

Hutchinson R.W. & Viljoen R.P. (1988) Re-evaluation of gold sources in Witwatersrand ores. *S. Afr. J. Geol.*, **91**, 157–173.

Hutchison C.S. (1974) *Laboratory Handbook of Petrographic Techniques.* Wiley, New York.

Hutchison C.S. (1983) *Economic Deposits and Their Tectonic Setting.* Macmillan, London.

IG (1989) *Code of Practice for Geological Visits to Quarries, Mines and Caves.* Institute of Geologists, London.

IMM Working Group (2001) *Code For Reporting Of Mineral Exploration Results, Mineral Resources and Mineral Reserves (The Reporting Code)* http://www.imm.org.uk/rescode/reportingcode.doc.

Independent Environmental Monitoring Agency (1998) *Independent Environmental Monitoring Agency Report for 1998.* http://www.monitoringagency.net/.

Ineson P.R. (1989) *Introduction to Practical Ore Microscopy.* Longman Scientific & Technical, Harlow.

Infomine (2004) *Infomine Web Site, Vancouver.* www.infomine.com.

Infoterra (2004) *Ikonos.* http://www.infoterra-global.com/ikonos.htm.

Institution of Mining and Metallurgy (1987) Finance for the Mining Industry. *Trans. Instn Min. Metall.*, *Sect. A*, **96**, A7–A36.

Isaaks E.H. & Srivastava R.M. (1989) *An Introduction to Applied Geostatistics.* Oxford University Press, Oxford.

Ishikawa Y., Sawaguchi T., Iwaya S. & Horiuchi M. (1976) Delineation of prospecting targets for Kuroko deposits based on modes of volcanism of the underlying dacite and alteration halos. *Min. Geol.*, **26**, 105–117 (in Japanese).

Isles D.J., Harman P.G. & Cunneen J.P. (1989) The contribution of high resolution aeromagnetics to Archaean gold exploration in the Kalgoorlie region, Western Australia. In Keays R.R., Ramsay W.R.H. & Groves D.I. (eds) *The Geology of Gold Deposits – the perspective in 1988*, 389–397. Econ. Geol. Monograph **6**. El Paso.

Ixer R.A. & Duller P.R. (1998) *Virtual Atlas of Opaque and Ore Minerals in their Associations.* http://www.smenet.org/opaque-ore/.

Jackson D.G. & Andrew R.L. (1990) Kintyre uranium deposit. In Hughes F. (ed.), *Geology of the Mineral Deposits of Australia and Papua New Guinea*, 653–658. Australas. Inst. Min. Metall. Monograph **14**. Melbourne.

Jacobsen J.B.E. & McCarthy T.S. (1976) The copper-bearing breccia pipes of the Messina District, South Africa. *Mineral. Deposita*, **11**, 33–45.

Jagger F. (1977) Bore core evaluation for coal mine design. In Whitmore R.L. (ed.), *Coal Borehole Evaluation*, 72–81. Symposium, Australasian Institution of Mining and Metallurgy, Parkville.

Janisch P.R. (1986) Gold in South Africa *J.S. Afr. Inst. Min. Metall.*, **86**, 273–316.

Janse A.J.A. & Sheahan P.A. (1995) Catalog of world wide diamond and kimberlite occurrences – a selective and annotative approach. *J. Geochem. Explor.*, **53**, 73–111.

Jaques A.L. & Milligan P.R. (2003) Patterns and controls on the distribution of diamond pipes in Australia. In *8th International Kimberlite Conference, Victoria*, abstracts.

Jeffery R.G., Sampsom D.B., Seymour K.M. & Walker I.W. (1991) A review of current exploration and evaluation practices in the search for gold in the Archacan of Western Australia. In *World Gold '91*, 313–322. Australasian Institution of Mining and Metallurgy, Cairns.

Jenner J.W. (1986) Drilling and the year 2000. *Trans. Instn Min. Metall., Sect. B*, **95**, B77–144.

Johnson I.M. & Fujita M. (1985) The Hishikari gold deposit: an airborne EM discovery. *CIM Bull.*, **78**, 61–66.

Johnson M.G. (1977) *Geology and Mineral Deposits of Pershing County, Nevada*, bull. **80**. Nevada Bureau of Mines and Geology.

Jolley S.J., Freeman S.R., Barnicoat A.C., et al. (2004) Structural controls on Witwatersrand gold mineralization. *J. Struct. Geol.*, **26**, 1067–1086.

Jones M.P. (1987) *Applied Mineralogy.* Graham & Trotman, London.

Jordaan M. & Austin M. (1986) *Gold-placer Sedimentology. Excursion guide 1A Geocongress '86.* Geological Society of South Africa, Johannesburg.

Journel A.G. & Huijbregts C-H.J. (1978) *Mining Geostatistics.* Academic Press, London.

Kalogeropolous S.I. & Scott S.D. (1983) Mineralogy and geochemistry of tuffaceous exhalites (Tetsuekiei) of the Fukazwa Mine, Hokuroku District, Japan. In Ohmoto H. & Skinner B.J. (eds), *Kuroko and Related Volcanogenic Massive Sulfide Deposits.* Econ. Geol. Monogr., **5**, 412–432.

Kauranne K., Salimen R. & Eriksson K. (1992) *Regolith Exploration Geochemistry in Arctic and Temperate Terrains. Handbook of Exploration Geochemistry,* vol. 5. Elsevier, Amsterdam.

Kearey P., Brooks M. & Hill I. (2002) *An Introduction to Geophysical Exploration,* 3rd edn. Blackwell Science Ltd, Oxford.

Kelly T., Buckingham D., DiFrancesco C., et al. (2001) *Historical Statistics for Mineral and Material Commodities in the United States.* US Geological Survey Open-File Report 01-006. //minerals.usgs.gov/minerals/pubs/of 01-006/.

Kernet C. (1991) *Mining Equities: evaluation and trading.* Woodhead Publishing Ltd, Cambridge.

Kerr D.J. & Gibson H.L. (1993) A comparison of the Horne volcanogenic massive sulfide deposit and intracauldron deposits of the Mine Sequence, Noranda, Quebec. *Econ. Geol.,* **88**, 1419–1442.

Kesler S.E. (1994) *Mineral Resources, Economics, and the Environment.* Macmillan, Oxford.

Kilburn L.C. (1990) Valuation of mineral properties which do not contain exploitable reserves. *CIM Bull.,* **83**, 90–93.

King B.M. (2000) *Optimal Mine Scheduling Policies.* Unpublished PhD thesis, University of London.

Kirk J., Ruiz J., Chesley J., Walshe J. & England G. (2002) A major Archean, gold- and crust-forming event in the Kaapvaal craton, South Africa. *Science,* **297**, 1856–1858.

Kirkham R.V., Sinclair W.D., Thorpe R.I. & Duke J.M. (eds) (1993) *Mineral Deposit Modeling.* Geological Association of Canada Special Paper **40**, St John's, Newfoundland, Canada.

Kirkley M., Mogg T. & McBean D. (2003) Snap Lake field trip guide. In Kjarsgaard B.A. (ed.) *VIIIth international Kimberlite Conference Field Trip Guidebook.* Geological Survey of Canada, CD-ROM.

Klemd R. & Hallbauer D.K. (1987) Hydrothermally altered peraluminous Archaean granites as a provenance model for Witwatersrand sediments. *Mineral. Deposita,* **22**, 227–235.

Knox-Robinson C.M. (2000) Vectoral fuzzy logic: a novel technique for enhanced mineral prospectivity mapping, with reference to orogenic gold mineralisation potential of the Kalgoorlie terrane, Western Australia. *Australian J. Earthsci.,* **47**, 929–941.

Knudsen H.P. (1988) *A Short Course on Geostatistical Ore Reserve Estimation.* Unpublished Rept, Montana Tech., Butte.

Knudsen H.P. & Kim Y.C. (1978) A comparative study of the geostatistical ore reserve estimation method over the conventional methods. *Min. Eng.,* **30**(1), 54–58.

Krajick K. (2001) *Barren Lands: an epic search for diamonds in the North American Arctic.* W.H. Freeman & Co., New York.

Kreiter V.M. (1968) *Geological Prospecting and Exploration.* Mir, Moscow.

Krige D.G. (1978) *Log Normal – De Wijsian geostatistics for ore evaluation.* South African Institution of Mining and Metallurgy Monograph **1**. Johannesburg.

Kruse F.A. & Boardman J.W. (2000) Characterization and mapping of kimberlites and related diatremes using hyperspectral remote sensing. In *Proceedings, 2000 IEEE AeroSpace Conference, 18–24 March 2000, Big Sky, Montana.*

Krynine D. & Judd W.R. (1957) *Principles of Engineering Geology and Geotechnics.* MaGraw-Hill, New York.

Kunasz I.A. (1982) Foote Mineral Company – Kings Mountain Operation. In Cerny P. (ed.), *Short Course in Granitic Pegmatites in Science and Industry,* 505–512. Mineralogical Society of Canada, Winnipeg.

Lafitte P. (ed.) (1970) *Metallogenic Map of Europe,* 9 Sheets. UNESCO, Paris.

Lane K.F. (1988) *The Economic Definition of Ore – Cut-off grades in theory and practice.* Mining Journal Books, London.

Larocque A.C.L. & Cabri L.J. (1998) Ion-microprobe quantification of precious metals in sulphide minerals. In McKibben M.A., Shanks W.C. & Ridley W.I. (eds), *Applications of Microanalytical Techniques to Understanding Mineralizing Processes. Reviews in Economic Geology* **7**, 155–168.

Larson L.T. (1989) Geology and gold mineralization in west Turkey. *Min. Eng.,* **41**, 1099–1102.

Lawrence M.J. (1998) Australian project valuation lessons for Canadian developers. In *PDAC (Prospectors & Developers Association of Canada) Mineral Property Valuation and Investor Concer,* 69–96. Short Course Notes, March 1998.

Leaman D.E. (1991) Exploration significance of gravity surveys, Roseberry Mine, Tasmania. *Exploration Geophysics,* **22**, 231–234.

Leake R.C., Cameron D.G., Bland D.J., Styles M.T. & Rollin K.E. (1992) *Exploration for Gold in the South Hams District of Devon.* Mineral Reconnaissance Programme Report **121**. British Geological Survey.

Leat P.T., Jackson S.E., Thorpe R.S. & Stillman C.J. (1986) Geochemistry of bimodal basalt-subalkaline/peralkaline rhyolite provinces within the southern British Caledonides. *J. Geol. Soc. London,* **143**, 259–273.

Lebrun S. (1987) *The Development of a Menu Driven Computer Program, 'The Geostats Program', to Carry Out Geostatistical Analysis of Geological Data,* 2 vols. Unpubl. MSc Dissertation, University of Leicester.

Leca X. (1990) Discovery of a concealed massive sulphide deposit at Neves-Corvo, southern Portugal – a case history. *Trans. Instn Min. Metall. Sect B,* **99**, B139–B152.

Lednor M. (1986) The West Rand Goldfield. In Antrobus E.S.A. (ed.), *Witwatersrand Gold – 100 Years,* 49–110. Geological Society of South Africa, Johannesburg.

Levinson A.A. (1980) *Introduction to Exploration Geochemistry.* Applied Publishing, Wilmette, IL.

Levinson A.A. (1998) Diamond sources and their discovery. In Harlow G.E. (ed.) *The Nature of Diamonds,* 72–104. Cambridge University Press.

Levinson A.A., Bradshaw P.M.D. & Thomson I. (eds) (1987) *Practical Problems in Exploration Geochemistry.* Applied Publishing, Wilmette, IL.

Lillesand T.M., Kieffer R.W. & Chipman J.W. (2004) *Remote Sensing and Image Interpretation,* 5th edn. Wiley, New York.

Lindley I.D. (1987) The discovery and exploration of the Wild Dog gold-silver deposit, east New Britain, Papua New Guinea. In *Pacific Rim Congress '87,* 283–286, Australasian Institution of Mining and Metallurgy, Melbourne.

Liu G., Diorio P., Stone P., et al. (2001) Detecting kimberlite pipes at Ekati with airborne gravity gradiometry. Abstract ASEG 15th Geophysical Conference and Exhibition, August 2001, Brisbane.

LME (2003) *Data & Prices, London Metal Exchange.* http://www.lme.co.uk/data_prices/home.html.

Lo C.P. (1986) *Applied Remote Sensing.* Longman Scientific & Technical, New York.

Lock N.P. (1985) Kimberlite exploration in the Kalahari region of Southern Botswana with emphasis on the Jwaneng Kimberlite Province. In *Prospecting in Areas of Desert Terrain,* 183–190. Institution of Mining and Metallurgy, London.

Longley P.A., Goodchild M.F., Maguire D.J. & Rhind D.W. (1999) *Geographical Information Systems,* 2 vols. Wiley, New York.

Longley P.A., Goodchild M.F., Maguire D.J. & Rhind D.W. (2001) *Geographic Information Systems and Science.* Wiley, New York.

Lovell J.S. & Reid A.R. (1989) Carbon dioxide/oxygen in the exploration for sulphide mineralization. In Garland G.D. (ed.) *Proceedings of Exploration '87,* 457–472, spec. vol. 3. Ontario Geological Survey, Toronto.

Lovering T.G. & McCarthy J.H. Jr (eds) (1977) Conceptual models in exploration geochemistry – the Basin and Range Province of the western United States and Northern Mexico. *J. Geochem. Explor.,* **9**, 89–365.

Lumina (2003) *Analytica.* http://lumina.com/software/.

Lydon J.W. (1989) Volcanogenic massive sulphide deposits parts 1 & 2. In Roberts R.G. & Shehan P.A. (eds), *Ore Deposit Models,* 145–181. Geological Association of Canada, Memorial University, Newfoundland.

Lynn M.D., Wipplinger P.E. & Wilson M.G.C. (1998) Diamonds. In Anhaeusser C.G. & Wilson M.G.C. (eds) *The Mineral Resources of South Africa.* Handbook Council for Geoscience **16**, 232–258.

Maas R., McColloch M.T., Campbell I.H. & Coad P.R. (1986) Sm-Nd and Rb-Sr dating of an Archaean massive sulfide deposit: Kidd Creek, Ontario. *Geology,* **14**, 585–588.

MacDonald A. (2002) North America: an industry in transition. *Mines and Minerals Sustainable Development Project.* Earthscan, London, CD-ROM.

Machamer J.F., Tolbert G.E. & L'Esperance R.L. (1991) Discovery of Serra dos Carajas. In Huchinson R.W. & Grauch R.I. (eds), *Historical Perspectives of Genetic Concepts and Case Histories of Famous Discoveries,* 275–285. Monograph **8**. Economic Geology Publishing Co., El Paso.

Mackenzie B. & Woodall R. (1988) Economic productivity of base metal exploration in Australia and Canada. In Tilton J.E., Eggert R.G. & Landsberg H.H. (eds), *World Mineral Exploration,* 21–104. Resources for the Future, Washington.

Macnae J. (1995) Applications of geophysics for the detection and exploration of kimberlites and lamproites. *J. Geochem Expl.,* **53**, 213–244.

Magri E.J. (1987) Economic optimization of boreholes and deflections in deep gold exploration. *J. S. Afr. Inst. Min. Metall.,* **87**, 307–321.

Majoribanks R.W. (1997) *Geological Methods in Mineral Exploration and Mining.* Chapman & Hall, London.

Maptek (2004) *Maptek Web Site.* http://www.maptek.com/pdf/VCS_southdeep.pdf.

Mason A.A.C. (1953) The Vulcan tin mine. In Edwards A.B. (ed.), *Geology of Australian Ore Deposits*, 718–721. Australasian Institution of Mining and Metallurgy, Melbourne.

Mather P.H. (1987) *Computer Processing of Remotely Sensed Images.* John Wiley & Sons, Chichester.

Matulich A., Amos A.C., Walker R.R., et al. (1974) The Ecstall Story; the geology department. *Can. Inst. Min. Metall. Bull.*, **77**, 228–235.

Maynard J.B., Ritger S.D. & Sutton S.J. (1991) Chemistry of sands from the modern Indus River and the Archean Witwatersrand Basin: implications for the composition of the Archean atmosphere. *Geology*, **19**, 265–268.

Mazzucchelli R.H. (1989) Exploration geochemistry in areas of deeply weathered terrain: weathered bedrock geochemistry. In Garland G.D. (ed.) *Proceedings of Exploration '87*, 300–311, spec. vol. 3. Ontario Geological Survey, Toronto.

McCabe P.J. (1984) Depositional environments of coal and coal-bearing strata. In Rahmani R.A. & Flores R.M. (eds), *Sedimentology of Coal and Coal-bearing Strata*, 13–42. Spec. Publ. 7. International Association of Sedimentologists.

McCabe P.J. (1987) Facies studies of coal and coal-bearing strata. In Scott A.C. (ed.), *Coal and Coal-bearing Strata: recent advances*, 51–66. Spec. Publ. **32**. Geological Society, London.

McCabe P.J. (1991) Geology of coal: environments of deposition. In Gluskoter H.J., Rice D.D. & Taylor R.B. (eds), *Economic Geology, US. The Geology of North America*, vol. P-2, 469–482. Geological Society of America, Boulder, CO.

McClay K.R. (1991) *The Mapping of Geological Structures.* Wiley, New York.

McClenaghan M.B. & Kjarsgaard B.A. (2001) Indicator mineral and geochemical methods for diamond exploration in the glaciated terrain of Canada. In McClenaghan M.B., Bobrowsky P.T., Hall G.E.M. & Cook S.J. (eds), *Drift Exploration in Glacial Terrain*, 83–124, spec. publ. **185**. Geological Society of London.

McLean A.C. & Gribble C.D. (1985) *Geology for Civil Engineers*, 2nd edn. George Allen & Unwin, London.

McMullan S.R., Matthews R.B. & Robertshaw P. (1989) Exploration geophysics for Athabasca uranium deposits. In Garland G. (ed.), *Proceedings of Exploration '87*, 547–68. Ontario Geological Survey, Toronto.

McNish J. (1999) *The Big Score.* Doubleday, Canada.

McVey H. (1989) Industrial Minerals – can we live without them? *Industr. Min.*, **259**, 74–75.

Mellor E.T. (1917) *The Geology of the Witwatersrand: Map and Explanation.* Spec. Pub. **3**. Geological Survey of South Africa, Johannesburg.

Meyer F.M., Tainton S. & Saager R. (1990) The mineralogy and geochemistry of small-pebble conglomerates from the Promise Formation in the West Rand and Klerksdorp areas. *S. Afr. J. Geol.*, **93**, 103–117.

Micromine (2004) *Micromine Website.* http://www.micromine.com.au.

Milsom J. (2002) *Field Geophysics*, 3rd edn. John Wiley & Sons, Chichester, 232 pp.

Minecost (2004) *World Mine Cost Data Exchange.* www.minecost.com.

Mining Journal (weekly) *Mining Journal Website*, London. www.mining-journal.com.

Minter W.E.L., Hill W.C.N., Kidger R.J., Kingsley C.S. & Snowden P.A. (1986) The Welkom Goldfield. In Annhaeusser C.R. & Maske S. (eds), *Mineral Deposits of Southern Africa*, 497–539. Geological Society of South Africa, Johannesburg.

Mitcham T.W. (1974) Origin of breccia pipes. *Econ. Geol.*, **69**, 412–413.

Mitchell R.H. (1986) *Kimberlites.* Plenum Press, London.

Mitchell R.H. (1991) Kimberlites and lamproites: primary sources of diamond. *Geosci. Canada*, **18**, 1–16.

Mitchell R.H. (1995) *Kimberlites, Orangeites, and Related Rocks.* Plenum Press, London.

MMSD (2002) *Breaking New Ground: mines and minerals sustainable development project.* Earthscan, London. (http://www.iied.org/mmsd/finalreport/index.html).

Moeskops P.G. (1977) Yilgarn nickel gossan geochemistry – a review with new data. *J. Geochem. Explor.*, **8**, 247–258.

Moon C.J. (1973) *A Carborne Radiometric Survey of the Prince Albert-Beaufort West area, Cape Province.* Open File Report Geol. Surv. S. Africa, **G227**. Pretoria.

Moon C.J. (1999) Towards a more quantitative relationship for the downstream dilution of point source geochemical anomalies. *J. Geochem.Expl.*, **65**, 111–132.

Moon C.J. & Whateley M.K.G. (1989) A retrospective analysis of commercial exploration strategies for uranium in the Karoo Basin, South Africa. In Chadwick J.R. (ed.), *Mineral Exploration Programmes '89*. Mining International, Madrid.

Moon W.M. (1990) Integration of geophysical and geological data using evidential belief function. *IEEE Trans. Geosci. Remote Sensing*, **28**, 711–720.

Moore J.M. & Moore A.E. (2004) The roles of primary kimberlitic and secondary Dwyka glacial sources in the development of alluvial and marine diamond deposits in Southern Africa. *J. Afr. Earth Sci.*, **38**, 115–134.

Moore P.D. (1987) Ecological and hydrological aspects of peat formation. In Scott A. (ed.), *Coal and Coal-bearing Stata*, 17–24. Spec. Publ. **32**. Geological Society, London.

Moore R.L. (1998) Case study – the impact of mineral deposit models on exploration strategies of a major mining company. In Prospectors & Developers Association of Canada (ed.) *The Fundamentals of Exploration and Mining. Short Course Notes*, 95–104. Toronto.

Morgan J.D. (1989) Stockpiling. In Carr D.D. & Herz N. (eds), *Concise Encyclopedia of Mineral Resources*, 288–292. Pergamon Press, Oxford.

Morris R.O. (1986) Drilling – potential developments. *Trans. Instn Min. Metall., Sect. B*, **95**, B77–78.

Morrissey C.J. (1986) New trends in geological concepts. *Trans. Instn Min. Metall., Sect. B*, **95**, B54–57.

Moseley F. (1981) *Methods in Field Geology*. Freeman, Oxford.

Muggeridge M.T. (1995) Pathfinder sampling techniques for locating primary sources of diamond: recovery of indicator minerals, diamonds and geochemical signatures. *J. Geochem Expl.*, **53**, 125–144.

Mular A.L. (1982) *Mining and Mineral Processing Equipment Costs and Preliminary Cost Estimations*, spec. vol. **25**. Canadian Institution of Mining and Metallurgy, Montreal.

Mullins M.P. (ed.) (1986) *Gold Placer Sedimentology. Excursion Guidebook 1A*. Geological Society of South Africa, Johannesburg.

Munha J., Barriga F.J.A.S. & Kerrich R. (1986) High $^{18}O/^{16}O$ ore-forming fluids in volcanic-hosted base metal massive sulphide deposits: geologic, $^{18}O/^{16}O$, and D/H evidence from the Iberian Pyrite Belt; Crandon, Wisconsin; and Blue Hill, Maine. *Econ. Geol.*, **81**, 530–552.

NAMHO (1985) *NAMHO Guidelines*. National Association of Mining History Organizations, Matlock.

Nash C.R., Boshier P.R., Coupard M.M., Theron A.C. & Wilson T.G. (1980) Photogeology and satellite image interpretation in mineral exploration. *Miner. Sci. Eng.* **12**, 216–244.

Nebert K. (1978) Das Braunkohlenfuhrende Neogengebeit von Soma, West Anatolien. *Bull. Min. Res. Explor. Inst. Turkey*, **90**, 20–72.

Niblak W. (1986) *An Introduction to Digital Image Processing*. Prentice Hall, New York.

Nichol I., Closs L.G. & Lavin O.P. (1989) Sample representativity with reference to gold exploration. In Garland G.D. (ed.), *Proceedings of Exploration '87*, 609–624, spec. vol. **3**. Ontario Geological Survey, Toronto.

Nixon P.H. (1995) The morphology and nature of primary diamondiferous occurrences. *J. Geochem. Explor.*, **53**, 41–71.

Noetstaller R. (1988) *Industrial Minerals: a technical review*. The World Bank, Washington.

Northcote A.E.A. (1998) Scoping of feasibility studies In: *The Mining Cycle. Proceedings of the AusIMM 1998 annual conference held in Mount Isa, Queensland, 19–23 April 1998*. Australasian IMM publication series, no. 2/98, 105–110. Australasian Institute of Mining and Metallurgy, Carlton, Victoria.

Northern Miner (weekly) *Northern Miner Website*. http://www.northernminer.com.

Northern Miner (1990) Buried treasures. *Northern Miner Magazine*, **March**, 32–37.

Nowicki T.E., Carlson J.A., Crawford B.B., Lockhart G.D., Oshurst P.A. & Dyck D.R. (2003) Field Guide to the Ekati Mine. In Kjarsgaard B.A. (ed.) *VIIIth International Kimberlite Conference Field Trip Guidebook*. Geological Survey of Canada, CD-ROM.

Nunes P.D. & Pyke D. (1981) Time–stratigraphic correlation of the Kidd Creek Orebody with volcanic rocks south of Timmins, Ontario, as inferred from zircon U-Pb ages. *Economic Geology*, **76**, 944–951.

Obert L. & Duval W.I. (1967) *Rock Mechanics and the Design of Structures in Rock*. Wiley, New York.

O'Hara T.A. (1980) Quick guide to the evaluation of orebodies. *Can. Inst. Min. Metall. Bull.*, **Feb.**, 87–99.

Onions R.I. & Tweedie J.R. (1992) Development of a field computer data logger and its integration with the DATAMINE mining software. In Annels A.E. (ed.) *Case Histories and Methods in Mineral Resource Evaluation*, 125–133. Geol. Soc. spec. publ. **63**. Bath, UK.

Papenfus J.A. (1964) The Black Reef Series within the Witwatersrand Basin with special reference to the occurrence at Government Gold Mining Areas. In Haughton S.H. (ed.), *The Geology of Some Ore Deposits in Southern Africa*, 191–218. Geological Society of South Africa, Johannesburg.

Parasnis D.S. (1996) *Principles of Applied Geophysics*, 5th edn. Chapman & Hall, London.

Paterson N.R. & Halof P.G. (1991) Geophysical exploration for gold. In Foster R.P. (ed.), *Gold Metallogeny and Exploration*, 360–398. Blackie, Glasgow.

PDAC (Prospectors & Developers Association of Canada) (1998) *Mineral Property Valuation and Investor Concern*. Short Course Notes, March 1998.

Pearson G.N., Kelley S.P., Pokhilenko N.P. & Boyd F.R. (1997) Laser ^{40}Ar/^{39}Ar dating of phlogopites from southern African and Siberian kimberlites and their xenoliths: constraints on eruption ages, melt degassing and mantle volatile compositions. *Russian Geol. Geophys.*, **38**, 106–118 (English edition).

Pemberton R.H. (1989) Geophysical response of some Canadian massive sulphide deposits. In Garland G.D. (ed.), *Proceedings of Exploration '87*, 517–531, spec. vol. **3**. Ontario Geological Survey, Toronto.

Percival J.A. & Williams H.R. (1989) Late Archaean Quetico metasedimentary belt, Superior Province, Canada. *Can. J. Earth Sci.*, **26**, 677–693.

Perry G. (1989) *Trinity Mine Operations Report*. Unpublished Report for US Borax.

Peters E.R. (1983) The use of multispectral satellite imagery in the exploration for petroleum and minerals. *Phil. Trans. R. Soc. Lond.*, **A309**, 243–255.

Peters W.C. (1987) *Exploration and Mining Geology*, 2nd edn. Wiley, New York.

Phillips G.N. & Law J.D.M. (2000) Witwatersrand Gold Fields: geology, genesis and exploration. In Hagemann S.G. & Brown P.E. *Gold in 2000. Reviews in Economic Geology*, **13**, 439–500. Economic Geology Publishing Co., Littleton, CO.

Phillips G.N., Myers R.E. & Palmer J.A. (1987) Problems with the placer model for Witwatersrand gold. *Geology*, **15**, 1027–1030.

Pintz W. (1984) *Ok Tedi – Evaluation of a Third World Mining Project*. Mining Journal Books, Edenbridge, Kent.

Piper D.P. and Rogers P.J. (1980) *Procedure for the Assessment of the Conglomerate Resources of the Sherwood Sandstone Group*. Mineral Assessment Report **56**. Institute of Geological Sciences, Keyworth.

Pirajno F. (1992) *Hydrothermal Mineral Deposits: principles and fundamental concepts for the exploration geologist*. Springer-Verlag, Berlin.

Pirow H. (1920) Distribution of the pebbles in the Rand banket and other features of the rock. *Trans. Geol. Soc. S. Afr.*, **23**.

Pitard F.F. (1993) *Pierre Gy's Sampling Theory and Practice*. CRC Press, Boca Raton.

Popoff C. (1966) *Computing Reserves of Mineral Deposits; principles and conventional methods*. US Bureau of Mines Information Circular **8283**.

Potts D. (1985) Guide to the Financing of Mining Projects. *Trans. Instn Min. Met., Sect. A*, **94**, A127–A133.

Prain K.A.R. (1989) Application of MWD in the oil field drilling industry. *Miner. Ind. Int.*, **987** (March), 10–12.

Preston K.B. & Sanders R.H. (1993) Estimating the in-situ relative density of coal. *Aust. Coal Geol.*, **9**, 22–26.

Pretorius C.C., Jamison A.A. & Irons C. (1989) Seismic exploration in the Witwatersrand Basin, Republic of South Africa. In Garland G.D. (ed.) *Proceedings of Exploration '87*, 241–253, spec. vol. **3**. Ontario Geological Survey, Toronto.

Pretorius C.C., Trewick W.A. & Irons C. (1997) Application of seismics to mine planning at Vaal Reefs Gold Mine, Number 10 Shaft, Republic of South Africa. In Gubins A.G. (ed.) *Geophysics and Geochemistry at the Millenium: Proceedings of Exploration '97*, 399–408. Prospectors & Developers Association of Canada, Toronto.

Pretorius D.A. (1991) The sources of Witwatersrand gold and uranium: a continued difference of opinion. In Huchinson R.W. & Grauch R.I. (eds) *Historical Perspectives of Genetic Concepts and Case Histories of Famous Discoveries*, 139–163. Monograph **8**. Econ. Geol.

Price M. (1985) *Introducing Ground Water*. Chapman & Hall, London.

Reed S.J.B. (1993) *Electron Microprobe Analysis*. Cambridge University Press, Cambridge.

Reedman J.H. (1979) *Techniques in Mineral Exploration*. Applied Science Publishers, London.

Reeves C.V. (1989) Geophysical mapping of Precambrian granite-greenstone terranes as an aid to exploration. In Garland G.D. (ed.), *Proceedings of Exploration '87*, 254–266, spec. vol. **3**. Ontario Geological Survey, Toronto.

Reeves P.L. & Beck L.S. (1982) Famous mining camps: a history of uranium exploration in the Athabasca Basin. In Hanula M.R. (ed.), *The Discovers*, 153–161. Pitt Publishing, Toronto.

Reflections (2004) *Reflections Mining and Exploration Website*. http://www.reflections.com.au/MiningandExploration/.

Regan M.D. (1971) *Management of Exploration in the Metals Mining Industry*. MS thesis, Massachusetts Institute of Technology, Cambridge.

Reim K. (1988) *Trinity Ore Reserve, Task Force Report*. Unpublished Report for US Borax.

Reimann C. (1989) Reliability of geochemical analysis: recent experiences. *Trans. Instn Min. Metall., Sect. B*, **98**, B123–B129.

Reimer T.O. & Mossman D.J. (1990) Sulfidization of Witwatersrand black sands: from enigma to myth. *Geology*, **18**, 426–429.

Reinecke L. (1927) The location of payable ore-bodies in the gold-bearing reefs of the Witwatersrand. *Trans. Geol. Soc. S. Afr.*, **30**, 89–119.

Richardson S.H., Erlank A.J., Harris J.W. & Hart S.R. (1990) Eclogitic diamonds of Proterozoic age from Cretaceous kimberlites. *Nature*, **346**, 54–56.

Richardson S.H., Gurney J.J., Erlank A.J. & Harris J.W. (1984) Origin of diamonds in old enriched mantle. *Nature*, **310**, 198–202.

Rickard D. (1987) Proterozoic volcanogenic mineralization styles. In Pharaoh T.C., Beckinsale R.D. & Rickard D. (eds), *Geochemistry and Mineralization of Proterozoic Volcanic Suites*, 23–35, spec. publ. **33**. Geological Society, London.

Riddler G.P. (1989) The impact of corporate amd host country strategy on mineral exploration programmes. In Chadwick J.R. (ed.), *Paper 15, Mineral Exploration Programmes '89*. International Mining, Madrid.

Ritchie W., Wood M., Wright R. & Tait D. (1977) *Surveying and Mapping for Field Scientists*. Longman, New York.

Robb L. (2004) *Introduction to Ore-forming Processes*. Blackwell Publishing, Oxford.

Robb L.J., Davis D.W., Kamo N.L., et al. (1990) U-Pb ages on single detrital zircon grains from the Witwatersrand Basin, South Africa: constraints on the age of sedimentation and the evolution of granites adjacent to the basin. *J. Geol.*, **98**, 311–328.

Robb L.J. & Meyer F.M. (1990) The nature of the Witwatersrand hinterland: conjectures on the source area problem. *Econ. Geol.*, **85**, 511–536.

Robb L.J. & Robb V.A. (1995) Gold in the Witwatersrand basin. In: Wilson M.G.C. & Anhaeusser C.R. (eds) *The Mineral Resources of South Africa*. Council for Geoscience Handbook **16**, 294–349. Pretoria, South Africa.

Robinson D.N., Gumey J.J. & Shee S.R. (1984) Diamond eclogite, graphite eclogite xenoliths from Orapa, Botswana. In Kornprobst J. (ed.), *Kimberlites II: the mantle, crust-mantle relationships*, 11–24. Elsevier, Amsterdam.

Roberts R.G. & Sheahan P.A. (1988) *Ore Deposit Models*. Geological Association of Canada Reprint Series **3**, St Johns, Newfoundland.

Robinson S.C. (1952) Autoradiographs as a means of studying distribution of radioactive minerals in thin section. *Am. Miner.*, **37**, 544–547.

Rogers P.J., Chaterjee A.K. & Aucott J.W. (1990) Metallogenic domains and their reflection in regional lake sediment surveys from the Meguma Zone, southern Nova Scotia. *J. Geochem. Explor.*, **39**, 153–174.

Rombouts L. (2003) Assessing the diamond potential of kimberlites from discovery to evaluation bulk sampling. *Mineral. Deposita*, **38**, 496–504.

Rona P.A. (1988) Hydrothermal mineralization at oceanic ridges. *Can. Mineral.*, **26**, 431–465.

Rose A.W., Hawkes H.E. & Webb J.S. (1979) *Geochemistry in Mineral Exploration*. Academic Press, London.

Ross G.M., Parrish R.R., Villeneuve M.E. & Bowring S.A. (1991) Geophysics and geochronology of the crystalline basement of the Alberta Basin, western Canada. *Can. J. Earth Sci.*, **28**, 512–522.

Rothschild L. (1978) Risk. *Listener*, 30 November, 715.

Roux A.T. (1969) The application of geophysics to gold exploration in South Africa. In *Mining and Groundwater Geophysics*, 425–438. Econ. Geol. Rept **26**. Geological Survey, Canada.

Royle A.G. (1995) *The Sampling and Evaluation of Gold Deposits*. Unpublished PhD thesis, University of Leeds.

Rush P.M. & Seegers H.J. (1998) Ok Tedi Copper-Gold Deposits. In Hughes F.E. (ed.), *Geology of the Mineral Depoits of Australia and Papua New Guinea*, Monograph **14**, 1747–1754. Australasian Institute of Mining and Metallurgy, Melbourne, Victoria.

Saager R. (1981) Geochemical studies of the origin of detrital pyrites in the conglomerates of the Witwatersrand Goldfields, South Africa. In Armstrong F.C. (ed.), *Genesis of Uranium- and Gold-Bearing Precambrian Quartz-pebble Conglomerates*, L1–L17, prof. paper **1161**. US Geological Survey.

Sabins F.F. (1997) *Remote Sensing. Principles and interpretation*, 5th edn. Freeman, New York.

Salminen R. and the FOREGS Geochemistry Group (2004) *Geochemical Atlas of Europe*. http://www.eurogeosurveys.org/foregs/.

Sangster D.F. (1980) Quantitative characteristics of volcanogenic massive sulphide deposits: 1. metal content and size distribution of massive sulphide deposits in volcanic centres. *Can. Inst. Min. Metall.*, **73**, 74–81.

Sangster D.F. & Scott S.D. (1976) Precambrian, strata-bound, massive Cu-Zn-Pb sulphide ores in North America. In Wolf K.H. (ed.), *Handbook of Strata-bound and Stratiform Deposits*, vol. 6, 129–222. Elsevier, Amsterdam.

Sato T. (1977) Kuroko deposits: their geology, geochemistry and origin. In *Volcanic Processes in Ore Genesis*, spec. publ. **7**. Geological Society, London.

Sawkins F.J. (1984, 1990) *Metal Deposits in Relation to Plate Tectonics*. Springer-Verlag, Berlin.

Schandl E.S. & Wicks F.J. (1993) Carbonate and associated alteration of ultramafic and rhyolitic rocks at the Hemingway property, Kidd Creek volcanic complex, Timmins, Ontario. *Econ. Geol.*, **88**, 1615–1635.

Schowengerdt R.A. (1983) *Techniques for Image Processing and Classification in Remote Sensing*. Academic Press, New York.

Schunnesson H. & Holme K. (1997) Drill monitoring for geological mine planning in the Viscaria Copper Mine, Sweden. *CIM Bull.* **Sept.**, 83–89.

Scott A.C. (ed.) (1987) *Coal and Coal-bearing Strata: recent advances*, spec. publ. **32**. Geological Society, London.

Scott F. (1981) Midwest Lake uranium discovery, Saskatchewan, Canada. In IAEA (eds), *Uranium Exploration Case Histories*, 221–239. International Atomic Energy Agency, Vienna.

Scott-Russell H., Wanblad G.P. & De Villiers J.S. (1990) The impact of modified oilfield technology on deep level mineral exploration and ultimate mining systems. In Ross-Watt R.A.J. & Robinson P.D.K. (eds), *Technical Challenges in Deep Level Mining*, 429–439. South African Institution of Mining and Metallurgy, Johannesburg.

Select Committee (1982) Memorandum by the Inst. Geol. Sci., Strategic Minerals. In *Strategic Minerals Select Committee*, HL Rep; 1981–82 (217) xii.

Selley R.C. (1989) Deltaic reservoir prediction from rotational dipmeter patterns. In Whateley M.K.G. & Pickering K.T. (eds), *Deltas: sites and traps for fossil fuels*, 89–95, spec. publ. **41**. Geological Society, London.

Sengor A.M.C., Gorur N. & Saroglu F. (1985) Strike-slip faulting and related basin formation in zones of tectonic escape: Turkey as a case study. In Biddle K.T. & Christie-Blick N. (eds), *Strike-slip Deformation Basin Formation and Sedimentation*, 227–264, spec. publ. **37**. Soc. Econ. Paleo. Mineral, Houston.

Settle M., Abrams M.J., Conel J.E., Goetz A.F.H. & Lang H.R. (1984) Sensor assessment report. In Abrams M.J. (ed.), *Joint NASA/Geosat Test Case Report*, 2–1 to 2–24. American Association of Petroleum Geologists.

Severin P.W.A., Knuckey M.J. & Balint F. (1989) The Winston Lake, Ontario, massive sulphide discovery – a successful result of an integrated exploration program. In Garland G.D. (ed.), *Proceedings of Exploration '87*, 60–69, spec. vol. **3**. Ontario Geological Survey, Toronto.

Sichel H.S. (1966) The estimation of means and associated confidence limits for small samples from log normal populations. In *Symposium on Mathematical Statistics and Computer Applications in Ore Valuation*, 106–122. South African Institution of Mining and Metallurgy, Johannesburg.

Siegal B.S. & Gillespie A.R. (1980) *Remote Sensing in Geology*. Wiley, New York.

Siegel S. (1956) *Nonparametric Statistics for the Behavioural Sciences*. McGraw-Hill, New York.

Sillitoe R.H. (1991) Intrusion-related gold deposits. In Foster R.P. (ed.) *Gold Metallogeny and Exploration*, 165–209. Blackie, Glasgow.

Sillitoe R.H. (1995) *Exploration and Discovery of Base- and Precious-metal Deposits in the Circum-Pacific Region During the Last 25 Years*. Resource Geology spec. issue **19**. Tokyo.

Sillitoe R.H. (2000) *Exploration and Discovery of Base- and Precious-Metal Deposits in the Circum-Pacific Region: late 1990s update*. Resource Geology spec. issue **21**. Tokyo.

Sinclair A.J. (1976) *Probability Graphs*, spec. publ. **4**. Association of Exploration Geochemists, Toronto.

Sinclair A.J. (1991) A fundamental approach to threshold estimation in exploration geochemistry: probability plots revisited. *J. Geochem. Explor.*, **41**, 1–22.

Slade M.E. (1989) Prices of metals: history. In Carr D.D. & Herz N. (eds), *Concise Encyclopedia of Mineral Resources*, 343–356. Pergamon Press, Oxford.

Smith A. (1988a) Cyanide degradation and detoxification in a heap leach. In Van Zyl D., Huchison I. & Kiel J. (eds) *Introduction to Evaluation, Design and Operation of Precious Metal Heap Leaching Projects*, 293–305. Society of Mining Engineers, Denver.

Smith B.H. (1977) Some aspects of the use of geochemistry in the search for nickel sulphides in lateritic terrain in Western Australia. *J. Geochem. Explor.*, **8**, 259–281.

Smith C.B., Atkinson W.J. & Tyler E.W.J. (1990) Diamond exploration in Western Australia, Northern Territory and South Australia. In Glasson K.R. & Rattigan J.H. (eds), *Geological Aspects of the Discovery of Some Important Mineral Deposits in Australia*, 429–454. Australasian Institution of Mining and Metallurgy, Melbourne.

Smith D.M. (1988b) Geology of silver deposits along the western cordilleras. In *Silver Exploration, Mining and Treatment*, 11–24. Institution of Mining and Metallurgy, London.

Smith J. (1991) How companies value properties. *Can. Inst. Min. Metall. Bull.*, **84**, 953, 50–52.

Smith M.R. & Collis L. (2001) *Aggregates: sand, gravel and crushed rock aggregates for construction purposes*, 3rd edn, spec. publ. **17**. Geological Society of London Engineering Geology.

Smith R.J. & Pridmore D.F. (1989) Electromagnetic exploration for sulphides in Australia. In Garland G.D. (ed.), *Proceedings of Exploration '87*, 504–516, spec. vol. **3**. Ontario Geological Survey, Toronto.

Smith R.S. & Fountain D.K. (2003) Geophysics and diamond exploration – a review. *FUGRO Website*. http://www.dighem.com.

Smith R.S., Annan A.P., Lemieux J. & Pedersen R.N. (1996) Application of a modified Geotem system to reconnaissance exploration at Point Lake Area, N.W.T., Canada. *Geophysics*, **61**, 82–92.

Snow G.G. & Mackenzie B.W. (1981) The environment of exploration: economic, organizational and social constraints. In Skinner B.J. (ed.), *Economic Geology: Seventy-Fifth Anniversary Volume*, 871–896. Society of Economic Geologists, Littleton, Colorado.

Snyder J.P. (1987) *Map Projections: a working manual.* U.S.G.S. Professional Paper **1395**. Washington, DC.

Solomon M. (1976) 'Volcanic' massive sulphide deposits and their host rocks – a review and an explanation. In Wolf K.H. (ed.) *Handbook of Strata-Bound and Stratiform Deposits*, vol. 6, 21–54. Elsevier, Amsterdam.

Solovov A.P. (1987) *Geochemical Prospecting for Mineral Deposits.* Mir, Moscow.

Souch B.E., Podolsky T. & Geological Staff (1969) The sulphide ores of Sudbury: their particular relationship to a distinctive inclusion-bearing facies of the nickel irruptive. In Wilson H.D.B. (ed.), *Magmatic Ore Deposits Symposium*, 252–261, Econ. Geol. Monogr. **4**. Society of Economic Geologists, Littleton, Colorado.

South Deep (1990) *Prospectus for Flotation of South Deep Exploration Company Ltd.* South Deep Exploration Co., Johannesburg.

South Deep (1992) *Annual Report for 1991*. South Deep Mining and Exploration Co., Johannesburg.

South Deep (2001) *Annual Report for 2000.* South Deep Mining and Exploration Co., Johannesburg.

South Deep (2004) *South Deep Web Site.* http://www.placerdome.com/southdeep.

Spectrum Mapping (2003) *Remote Sensing: Hyperspectral Imaging: AISA Sensor*, Denver, Colorado. http://www.spectrummapping.com/rem-hyper-aisa.html.

Speight J.G. (1983) *The Chemistry and Technology of Coal.* Marcel Dekker, New York.

Spock L.E. (1953) *Guide to the Study of Rocks.* Harper, New York.

Spooner E.T.C. & Barrie C.T. (1993) A special issue devoted to Abitibi ore deposits in a modern context. *Econ. Geol.* **88**, 1307–1322.

Sprigg G. (1987) Advances in image analysers. *Microscopy and Analysis*, **Sept.**, 11–13.

Springett M. (1983a) Sampling and ore reserve estimation for the Ortiz Gold Deposit, New Mexico, USA. In *AIME Precious Metal Symposium, Sparks, Nevada, USA.*

Springett M. (1983b) *Sampling Practices and Problems*, preprint **83–395**. Society of Mining Engineers of AIME, New York.

Stach E. (1982) *Stach's Textbook of Coal Petrology.* Borntraeger, Berlin.

Stanistreet I.G. & McCarthy T.S. (1991) Changing tectono-sedimentary scenarios relevant to the development of the late Archean Witwatersrand Basin. *J. Afr. Earth Sci.*, **13**, 65–81.

Stanton R.L. (1978) Mineralization in island arcs with particular reference to the south-west Pacific region. *Proc. Australas. Inst. Min. Metall.*, **268**, 9–19.

Stanton R.L. (1991) Understanding volcanic massive sulfides – past, present and future. In Huchinson R.W. & Grauch R.I. (eds), *Historical Perspectives of Genetic Concepts and Case Histories of Famous Discoveries*, 82–95, monograph **8**. Econ. Geol., El Paso.

Steiger R. & Bowden A. (1982) Tungsten mineralization in southeast Leinster, Ireland. In Brown A.G. (ed.), *Mineral Exploration in Ireland, Progress and Developments 1971–1981*, 108–114. Irish Association of Economic Geology, Dublin.

Stephens J.D. (1972) Microprobe applications in mineral exploration and development programmes. *Miner. Sci. Eng.*, **3**, 26–37.

Stephenson P.R. (2003) *The JORC Code – maintaining the standard.* http://www.jorc.org/pdf/stephenson3.pdf

Stewart B.D. (1981) Exploration of the uranium reefs of Cooke Section, Randfontein Estates Gold Mining Company (Witwatersrand) South Africa. In International Atomic Energy Agency (eds), *Uranium Exploration Case Histories*, 141–170. International Atomic Energy Agency, Vienna.

Stewart J.H. (1980) *Geology of Nevada*, spec. publ. **4**. Nevada Bureau of Mines and Geology, Reno, Nevada.

St Pierre M. (1999) Geophysical characteristics of BHP/Dia Met Kimberlites, NWT, Canada. In Lowe C., Thomas M.D. & Morris W.A. (eds) *Geophysics in Mineral Exploration: fundamentals and case histories*. Geological Association of Canada Short Course Notes **14**. St John's, Newfoundland.

Strauss H. (1989) Carbon and sulfur isotope data for carbonaceous metasediments from the Kidd Creek massive sulfide deposit and vicinity, Timmins, Ontario. *Econ. Geol.*, **84**, 959–962.

Sutherland D.G., & Dale M.L. (1984) Methods of establishing the minimum size for sampling alluvial diamond deposits. *Trans. Instn Min. Metall., Sect. B*, **93**, B55–58.

Swan A.R.H. & Sandilands M. (1995) *Introduction To Geological Data Analysis*. Blackwell Science, Oxford.

Taufen P.M. (1997) Ground waters and surface waters in exploration geochemical surveys. In Gubins A.G. (ed.) *Geophysics and Geochemistry at the Millenium: Proceedings Of Exploration 97*, 271–284. Prospectors & Developers Association of Canada, Toronto.

Tankard A.J., Jackson M.P.A., Eriksson K.A., Hobday D.K., Hunter D.R. & Minter W.E.L. (1982) *Crustal Evolution of Southern Africa*. Springer-Verlag, Berlin.

Taylor B.E. & Huston D.L. (1998) Hydrothermal alteration of oxygen isotope ratios in quartz phenocrysts, Kidd Creek mine, Ontario: magmatic values preserved in zircons. *Geology*, **26**, 763–764.

Taylor C.D., Zierenberg R.A., Goldfarb R.J., Kilburn J.E., Seal R.R. II & Kleinkopf M.D. (1995) Volcanic-associated massive sulfide deposits In du Bray E. (ed.) *Preliminary Compilation of Descriptive Geoenvironmental Mineral Deposit Models*, 137–144. USGS Open File 95-831 http://pubs.usgs.gov/of/1995/ofr-95-0831

Taylor H.K. (1989) Ore reserves – a general overview. *Miner. Ind. Int.*, **990**, 5–12.

Telford W.M., Geldart L.P., Sheriff R.E. & Keys D.A. (1990) *Applied Geophysics*, 2nd edn. Cambridge University Press, Cambridge.

Thatcher J., Struhsacker D.W. & Kiel J. (1988) Regulatory aspects and permitting requirements for precious metal heap leach operations. In Van Zyl D., Hutchison I. & Kiel J. (eds), *Introduction to Evaluation, Design and Operation of Precious Metal Heap Leaching Projects*, 40–58. Society of Mining Engineers. Society of Mining Engineers, Littleton, Colorado.

Theron J.C. (1973) Sedimentological evidence for the extension of the African continent during the late Permian–early Triassic times. In Campbell K.S.W. (ed.), *Gondwana Geology*, 61–71. A.N.U. Press, Canberra.

Thiann R. Yu (1983) Rock mechanics to keep a mine productive. *Can. Min. J.*, **April**, 61–66.

Thomas E. (2003) *Exploration for Diamonds in Canada*. Powerpoint presentation. Resource Expo 2002, Calgary.

Thomas L. (1992) *Handbook of Practical Coal Geology*. Wiley, Chichester.

Thompson A.J.B. & Thompson J.F.H. (eds) (1996) *Atlas of Alteration: a field and petrographic guide to hydrothermal alteration*. Geological Association of Canada, Mineral Deposits Division, St John's, Newfoundland.

Thompson M. (1982) Control procedures in exploration geochemistry. In Howarth R.J. (ed.), *Statistics and Data Analysis in Exploration Geochemistry. Handbook of Exploration Geochemistry*, vol. 2, 39–58. Elsevier, Amsterdam.

Thompson M. & Walsh N. (1989) *Handbook of Inductively Coupled Plasma Spectrometry*. Blackie, London.

Thomson I. (1987) Getting it right. In Fletcher W.K., Hoffman S.J., Mehrtens M.B., Sinclair A.J. & Thomson I. (eds), *Exploration Geochemistry: design and interpretation of soil surveys. Reviews in Economic Geology*, vol. 3, 1–17. Society of Economic Geologists, Littleton, Colorado.

Thorpe R. & Brown G.C. (1993) *The Field Description of Igneous Rocks*. Wiley, New York.

Tilton J.E., Eggert R.G. & Landsberg H.H. (eds) (1988) *World Mineral Exploration*. Resources for the Future, Washington.

Tona F., Alonso D. & Svab M. (1985) Geology and mineralization in the Carswell Structure – a general approach. In Laine R., Alonso D. & Svab M. (eds), *The Carswell Structure Uranium Deposits*, 1–18, spec. paper **29**. Geological Association of Canada, St Johns, Newfoundland.

Travis G.A., Keays R.R. & Davison R.M. (1976) Palladium and iridium in the evaluation of nickel gossans in Western Australia. *Econ. Geol.*, **71**, 1229–1242.

Tregoning G.W. & Barton V.A. (1990) Design of a wide orebody mining system for a deep-level mine. In Ross-Watt R.A.J. & Robinson P.D.K. (eds), *Technical Challenges in Deep Level Mining*, 601–623. South African Institution of Mining and Metallurgy, Johannesburg.

Tucker M.E. (1988) *Techniques in Sedimentology*. Blackwell Scientific Publications, Oxford.

Tucker M.E. (2003) *Sedimentary Rocks in the Field*, 3rd edn. Wiley, New York.

Tucker R.F. & Viljoen R.P. (1986) The geology of the West Rand Goldfield. In Annhaeusser C.R. & Maske S. (eds), *Mineral Deposits of Southern Africa*, 649–698. Geological Society of South Africa, Johannesburg.

Turnbull N. (1999) *Internal Control: guidance for directors on the combined code*. Consultative Committee of Accounting Bodies, London.

Turner B.R. (1985) Uranium mineralization in the Karoo Basin, South Africa. *Econ. Geol.*, **80**, 256–269.

USBM (1976) *Coal Resource Classification System of the US Bureau of Mines and US Geological Survey*. US Geol. Surv. Bull. **1450-B**.

USGS (1976) *Principles of a Resource/Reserve Classification for Minerals*. US Geol Surv Bull. **1450-A**.

USGS (2003) *Gemstones 2003*. minerals.usgs.gov/minerals/pubs/commodity/gemstones/.

USGS (2004) *US Geological Survey Mineral Deposit Models Web Page*. minerals.cr.usgs.gov/team/depmod.html.

Ussher W.A.E. (1912) *The Geology of the Country Around Ivybridge and Modbury*. Memoirs of the Geological Survey of England and Wales, Sheet **349**. HMSO, London.

Van Blaricom R. (1993) *Practical Geophysics II*. Northwest Mining Association, Spokane.

Vearncombe S. & Vearncombe J.R. (2002) Tectonic controls on kimberlite location, southern Africa. *J. Struct. Geol.*, **24**, 1619–1625.

Vikre P.G. (1989) Fluid mineral relations in the Comstock Lode. *Econ. Geol.* **84**, 1574–1613.

Viljoen R.P. (1990) Deep level mining – a geological perspective. In Ross-Watt R.A.J. & Robinson P.D.K. (eds), *Technical Challenges in Deep Level Mining*, 411–427. South African Institution of Mining and Metallurgy, Johannesburg.

Viljoen R.P., Viljoen M.J., Grootenboer J. & Longshaw T.G. (1975) ERTS-1 imagery – an appraisal of applications on geology and mineral exploration. *Min. Sci. Eng.*, **7**, 132–168.

Visidata (2004) *Visidata Website*. http://www.visidata.com.au/.

Walker J. & Howard S. (2002) *Finding the Way Forward – how could voluntary action move mining towards sustainable development?* World Business Council for Sustainable Development, Conches, Switzerland. (http://www.wbcsd.ch/).

Walker R.R., Matulich A., Amos A.C., Watkins J.J. & Mannard G.W. (1975) The geology of the Kidd Creek Mine. *Econ. Geol.*, **70**, 80–89.

Walsham B.T. (1967) Exploration by diamond drilling for tin in west Cornwall. *Trans. Instn Min. Metall., Sect. A*, **76**, A49–56.

Walters W.H. (1999) Building and maintaining a numeric data collection. *J. Documentation*, **55**, 271–287.

Wang C. (1995) *Application of Geographical Information Systems to the Interpretation of Exploration Geochemical Data and Modelling of Gold Prospects, South Devon, England*. PhD Thesis, University of Leicester.

Wanless R.M. (1982) *Finance for Mine Management*. Chapman & Hall, London.

Ward C.R. (ed.) (1984) *Coal Geology and Coal Technology*. Blackwell Scientific Publications, Oxford.

Ward M-C. & Lawrence R.D. (1998) Comparable transaction analysis: the market place is always right. In PDAC (Prospectors & Developers Association of Canada) *Mineral Property Valuation and Investor Concern*. Short Course Notes, March 1998.

Weaver T.A., Freeman S.H., Broxton D.E. & Bolivar S.L. (1983) *The Geochemical Atlas of Alaska*. Los Alamos National Laboratory, Los Alamos.

Webb J.S., Howarth R.J., Thompson M., et al. (1978) *Wolfson Geochemical Atlas of England and Wales*. Clarendon Press, Oxford.

Wellmer F.-W. (1998) Problems related to cut-off levels. In *Statistical Evaluations in Exploration for Mineral Deposits*, 151–159. Springer-Verlag, Berlin.

Wellmer F.W. & Becker-Platen J.D. (2002) Sustainable development and the exploitation of mineral and energy resources: a review. *Int. J. Earth Sci.*, **91**, 723–745.

Wells J. (1999) *Bre-X: the inside story of the world's biggest mining scam*. Texere Publishing, New York.

Werdmuller V.W. (1986) The Central Rand. In Antrobus E.S.A. (ed.), *Witwatersrand Gold – 100 Years*, 7–48. Geological Society of South Africa, Johannesburg.

West G. (1991) *The Field Description of Engineering Soils and Rocks*. Open University Press, Milton Keynes.

Western Areas (2003) *Western Areas Annual Report 2002*. http://www.westernareas.co.za.

Western Mine Engineeering (2004) Website. www.westernmine.com.

Western Mining (1993) *News Release to Shareholders*, Western Mining Corp., Adelaide.

Whateley M.K.G. (1991) Geostatistical determination of contour accuracy in evaluating coal seam parameters: an example from the Leicestershire

Coalfield, England. *Bull. Soc. Géol. France*, **162**, 209–218.

Whateley M.K.G. (1992) The evaluation of coal borehole data for reserve estimation and mine design. In Annels A.E. (ed.), *Mineral Resource Evaluation*, 95–106, spec. publ. **63**. Geological Society, London.

Whateley M.K.G. (2002) Measuring, understanding and visualising coal characteristics – innovations in coal geology for the 21st century. *Int. J. Coal Geol.*, **50**, 303–315.

Whateley M.K.G. & Harvey P.K. (eds) (1994) *Mineral Resource Evaluation II: methods and case histories*, spec. publ. **79**. Geological Society, London.

Whateley M.K.G. & Spears D.A. (eds) (1995) *European Coal Geology*, spec. publ. **82**. Geological Society, London.

Wheat T.A. (1987) Advanced ceramics in Canada. *Can. Inst. Min. Metall. Bull.*, **80**(April), 43–48.

Whitchurch K.D., Gillies Saunders A.D. & Just G.D. (1987) A geostatistical approach to coal reserve classification. *Pacific Rim Congress*, **87**, 475–482.

White A.H. (1997) *Management of Mineral Exploration*. Andrew White and Associates, Moggill, Queensland & Australian Mineral Foundation, Glenside, South Australia.

White C.J. (1989) *A Summary of the Mineral Sands Industry of Western Australia Including an Exploration and Feasibility Project*. Unpublished BSc Dissertation, Leicester University.

White M.E. (2001) Feasibility studies: scope and accuracy. In: Edwards A.C. (ed.), *Mineral Resource and Ore Reserve Estimation: the AusIMM guide to good practice*. Australasian IMM monograph series, **23**, 421–434. Australasian Institute of Mining and Metallurgy, Carlton, Victoria.

Whitely R.J. (ed.) (1981) *Geophysical Case Study of the Woodlawn Orebody, Australia*. Pergamon Press, Oxford.

Williams H.R. & Williams R.A. (1977) Kimberlites and plate tectonics in West Africa. *Nature*, **270**, 507–508.

Willis R.P.H. (1992) The integration of new technology in South African gold mines as a survival strategy. In Jones M.J. (ed.), *Mines, Materials and Industry*, 509–524. Institution of Mining and Metallurgy, London.

Wills B.A. (1985, 1988) *Mineral Processing Technology*. Pergamon Press, Oxford.

Wills B.A. (1997) *Mineral Processing Technology*. 3rd edn. Pergamon Press, Oxford.

Woodall R. (1984) Success in mineral exploration. *Geoscience Canada*, **11**, 41–46, 83–90, 127–133.

Woodall R. (1992) Challenge of minerals exploration in the 1990s. *Min. Eng.*, **July**, 679–684.

Zarraq G. (1987) *Comparison of Quality and Reserve Estimation Methods on Soma Lignite Deposit, Turkey*. Unpublished MSc Dissertation, University of Leicester.

Zeegers H. & Leduc C. (1991) Geochemical exporation for gold in temperate, arid, semi-arid and rain forest terrains. In Foster R.P. (ed.), *Gold Metallogeny and Exploration*. Blackie, Glasgow, 309–335.

Zussman J. (1977) *Physical Methods in Determinative Mineralogy*. Academic Press, London.

INDEX

Page numbers in *italics* refer to figures; page numbers in **bold** refer to tables. The letter B after a page number refers to boxes.